Spatial Complexity, Informatics, and Wildlife Conservation

Samuel A. Cushman • Falk Huettmann
Editors

Spatial Complexity, Informatics, and Wildlife Conservation

Springer

Editors
Dr. Samuel A. Cushman
US Forest Service
Rocky Mountain Research Station
800 E Beckwith, Missoula
MT 59801, USA

Dr. Falk Huettmann
EWHALE lab- Biology and Wildlife Department
Institute of Arctic Biology
University of Alaska-Fairbanks
419 IRVING I, Fairbanks
AK 99775-7000, USA

ISBN 978-4-431-54755-6 ISBN 978-4-431-87771-4 (eBook)
DOI 10.1007/978-4-431-87771-4
Springer Tokyo Berlin Heidelberg New York

Softcover re-print of the Hardcover 1st edition 2010

Front cover:
Center: Prediction of Arctic Plankton Distribution; courtesy of Imme Rutzen, M.Sc.

Printed on acid-free paper

Springer is part of Springer Science+Business Media (www.springer.com)

I dedicate this book to my parents.

Samuel A. Cushman

I devote this work to my daughter Sophia and to all future generations.

Falk Huettmann

Preface

Earth is facing the greatest mass extinction in 65 million years, driven by the rapid growth in human consumption and population and related environmental effects, such as global warming, habitat loss, and fragmentation. The scientific community has a critical role to play in the urgent task of redirecting human civilization toward a sustainable trajectory. We believe that scientists must take a stronger role in informing the public and political decision makers about the science of ecology and its contributions to global sustainability. In addition to being a global biodiversity crisis, the present is a moment of tremendous foment and emergence in ecological science. The integration of powerful computing, massive spatial databases, rapid advances in modeling, online delivery, and computation are leading to revolutionary changes in how ecological research is done, the kinds of questions it addresses, and its power to predict and explain phenomena in nature. This book is an expression of our current understanding of these developments. We focus on the feedback between theory, method, and data in science, and posit that rapid technological advances are enabling rapid expansion in each of these areas, with synergistic feedback leading to the emergence of a really new science of ecology. This emerging philosophy of ecology we argue is based on linking ecological mechanisms directly to responses at their operative scales across broad extents in complex landscapes. We describe the technical advances that are allowing this emergence, and discuss the main conceptual, methodological, and philosophical perspectives that are contributing. We hope that the vision we present may resonate with practicing scientists and funding agencies who are embedded in the tumult of the field, and that it provides some interesting ideas for the ongoing discussion that is ecological science.

Given the urgency of the global sustainability crisis, we also hope that rapid advances in technical ecological knowledge will lead to simultaneous leaps in the role science plays in policy and management. Over the past 15 years there have been tremendous technical advances in ecological science worldwide, across a wide array of taxa, habitats, and applications. Given this record of improvement it seems puzzling that there has been so little progress toward ecological sustainability achieved on a global, national, regional, and local level; institutions still have not adjusted in their infrastructure and underlying philosophies. Baby steps have been seen on many fronts. Various problems across all levels are identified and studied;

but results are often not publicly available, not synthesized, not implemented in the management community, buried in the fine print, removed from the public agenda, or remain largely unknown by the public, and thus do not get addressed appropriately. This can be seen by many environmental, social, and economic metrics.

It is in this conceptual framework that this book and its research was compiled, with the goal of provoking discussion within the ecological community about the state of our knowledge and to propel ecologists into the public and intense debate about ecological sustainability and the future. If we want to fulfill the mandate of global sustainability we believe that many things have to change. Some of this change has to be the responsibility of scientists; for example, to look beyond myopic perspectives on scientific disciplines and traditional methodology to embrace the possibility of new generations of tools, approaches, and perspectives; to look beyond the narrow confines of producing knowledge and reach out to inform the public and the political debate. Efficiency and time are critical.

We would like to acknowledge that this book is based on our symposium "S53: Advanced GeoScience Applications" at the 9th International Mammalogical Conference (IMC) in Hokkaido 2005. The support from Takashi Saitoh and his team is highly appreciated.

Samuel A. Cushman
Falk Huettmann

Contents

Part I Theory and Practice

1 **Introduction: Ecological Knowledge, Theory and Information in Space and Time** 3
Samuel A. Cushman and Falk Huettmann

2 **Space and Time in Ecology: Noise or Fundamental Driver?** 19
Samuel A. Cushman

3 **The Problem of Ecological Scaling in Spatially Complex, Nonequilibrium Ecological Systems** 43
Samuel A. Cushman, Jeremy Littell, and Kevin McGarigal

4 **Landscape Ecology: Past, Present, and Future** 65
Samuel A. Cushman, Jeffrey S. Evans, and Kevin McGarigal

5 **The Gradient Paradigm: A Conceptual and Analytical Framework for Landscape Ecology** 83
Samuel A. Cushman, Kevin Gutzweiler, Jeffrey S. Evans, and Kevin McGarigal

Part II Method and Data

6 **Data on Distribution and Abundance: Monitoring for Research and Management** 111
Samuel A. Cushman and Kevin S. McKelvey

7 **Animal Movement Data: GPS Telemetry, Autocorrelation and the Need for Path-Level Analysis** 131
Samuel A. Cushman

8 **Using a Random Forest Model and Public Data to Predict the Distribution of Prey for Marine Wildlife Management** 151
Steffen Oppel and Falk Huettmann

9 **Landscape Genomics: A Brief Perspective** 165
Michael K. Schwartz, Gordon Luikart, Kevin S. McKelvey, and Samuel A. Cushman

10 **Spatial Information Management in Wildlife Ecology: Adding Spatially Explicit Behaviour Data to the Equation?** 175
Kim Jochum and Falk Huettmann

11 **Critical Remote Sensing Contributions to Spatial Wildlife Ecological Knowledge and Management** 193
Gregory J. McDermid, Nicholas C. Coops, Michael A. Wulder, Steven E. Franklin, and Nicole E. Seitz

12 **Spatial Data Management Through Metadata: Global Concepts, Formats, Tools and Requirements** 223
Vivian Hutchison

13 **Free Database Availability, Metadata and the Internet: An Example of Two High Latitude Components of the Census of Marine Life** 233
Bodil Bluhm, David Watts, and Falk Huettmann

14 **Components of Spatial Information Management in Wildlife Ecology: Software for Statistical and Modeling Analysis** 245
Hawthorne L. Beyer, Jeff Jenness, and Samuel A. Cushman

15 **Spatial Analysis of Wildlife Distribution and Disease Spread** 255
Marie-Josée Fortin, Mark R.T. Dale, and Stefania Bertazzon

16 **Current State of the Art for Statistical Modelling of Species Distributions** 273
Troy M. Hegel, Samuel A. Cushman, Jeffrey Evans, and Falk Huettmann

17 **Landscape Genetics** 313
Kevin S. McKelvey, Samuel A. Cushman, and Michael K. Schwartz

Part III Examples

18 Using Geographical Mapping and Occupancy Modeling to Study the Distribution of the Critically Endangered Leopard (*Panthera pardus*) Population in Armenia 331
Igor G. Khorozyan, Alexander G. Malkhasyan, Shushanik G. Asmaryan, and Alexei V. Abramov

19 Mapping Landscape Resistance to Identify Corridors and Barriers for Elephant Movement in Southern Africa 349
Samuel A. Cushman, Michael Chase, and Curtice Griffin

20 Habitat Fragmentation Effects Depend on Complex Interactions Between Population Size and Dispersal Ability: Modeling Influences of Roads, Agriculture and Residential Development Across a Range of Life-History Characteristics 369
Samuel A. Cushman, Bradley W. Compton, and Kevin McGarigal

21 Linking Cetaceans to Their Environment: Spatial Data Acquisition, Digital Processing and Predictive Modeling for Marine Spatial Planning in the Northwest Atlantic 387
Simon J. Pittman and Bryan Costa

22 Multi-spectral Satellite-Airborne Management of Ice Form Marine Mammals and Their Habitat in the Presence of Climate Change Using a "Hot Spots" Approach 409
Vladimir V. Melentyev and Vladimir I. Chernook

23 How Spatial Information Contributes to the Conservation and Management of Biodiversity 429
Dawn Robin Magness, John M. Morton, and Falk Huettmann

24 Future and Outlook: Where Are We, and Where Will the Spatial Information Management in Wildlife Ecology Be in 50 Years from Now? 445
Samuel A. Cushman and Falk Huettmann

Index 451

Part I
Theory and Practice

Chapter 1
Introduction: Ecological Knowledge, Theory and Information in Space and Time

Samuel A. Cushman and Falk Huettmann

A central theme of this book is that there is a strong mutual dependence between explanatory theory, available data and analytical method in determining the lurching progress of ecological knowledge (Fig. 1.1). The two central arguments are first that limits in each of theory, data and method have continuously constrained advances in understanding ecological systems and second that recent revolutionary advances in data and method are enabling unprecedented expansion of ecological investigation into areas of inquiry previously unapproachable due to lack of fine-detail, broad scale data on environmental conditions, the distribution and performance of organisms, the lack of sufficient computational power to process and analyze such voluminous data sets, and inadequate analytical tools to investigate pattern–process relationships among many interacting entities over large, spatially complex landscapes.

1.1 Mutual Dependence of Theory, Method, Data

There is a strict interdependence in science between theory, method and data. It is not possible to decouple these in the practice of science. In some sense it would be desirable if one could. When each corner of this triangle (Fig. 1.1) is dependent and limited by the others there is a feedback where the limitations of each further limit progress in the others. If these could be decoupled conceptually it would perhaps improve the rate of scientific advance. Classic conceptions of the scientific method

S.A. Cushman (✉)
US Forest Service, Rocky Mountain Research Station, 800 E Beckwith, Missoula, MT 59801, USA
e-mail: scushman@fs.fed.us

F. Huettmann
EWHALE lab- Biology and Wildlife Department, Institute of Arctic Biology, University of Alaska-Fairbanks, 419 IRVING I, Fairbanks, AK 99775-7000, USA
e-mail: fffh@uaf.edu

S.A. Cushman and F. Huettmann (eds.), *Spatial Complexity, Informatics, and Wildlife Conservation*
DOI 10.1007/978-4-431-87771-4_1, © Springer 2010

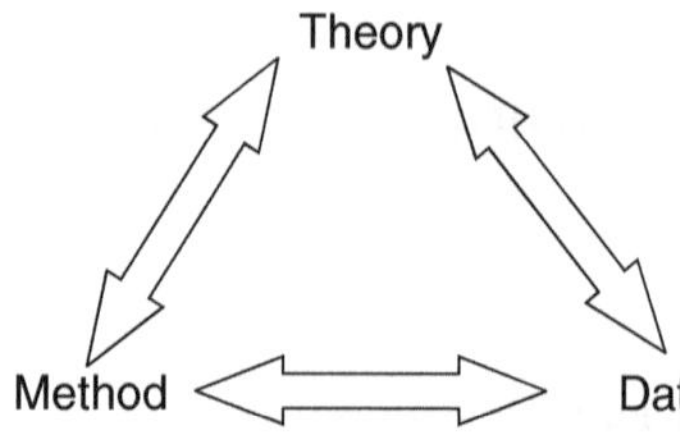

Fig. 1.1 There is a mutual interdependence between methods of observation and analysis, kind and character of data collected, and theories used to explain phenomena. Importantly, there is no possible decoupling by which they will be independent, lending a kind of circularity to the logic of scientific justification. Sometimes, this promotes a self-confirmatory process, with a theory proposing a method designed to produce data that will confirm the theory

Fig. 1.2 Baconian process of induction by which data are collected "objectively", which then suggest appropriate methods for analysis and interpretation, which then suggest the correct theory for explanation

typically assume one of two decouplings. First, the "Baconian" inductive approach proposes a decoupling in which observations (data) are a reflection of reality uncontaminated by implicit theories and unaffected by methods of data collection, such as sampling and measuring (Fig. 1.2). In this conception, the mind, its pre-conceptions and biases is seen as an obstacle to true understanding and instead the scientist collects data dispassionately and then theory regarding causation emerges inductively from the observations (Bacon 1620). In contrast, the "Cartesian" approach proposes a converse decoupling in which a sentient observer imagines processes governing ideal systems (Descartes 1637). In this conception, it is observation that is unreliable and ideal and eternal conceptions of theory are truly reliable (Fig. 1.3).

Each of these decouplings between theory, method and data are easily refuted. Despite the fact that they are over 400 years old, neither is a realistic view of any actual process used by a practicing scientist to link method, data and theory to build understanding. In the former case, it is easily argued that observations are always "infected" by implicit theory and affected by methods of sampling and measuring. Therefore, it is virtually impossible to obtain purely objective data from which to induce generalizable theory. In addition, due to the logical fallacy of affirming the consequent, patterns observed through induction do not provide proof for a theory with which they may be consistent (Fig. 1.4). The latter case assumes theories are created by the mind, independently from the historical context of current and past explanation. They would be unaffected by the scope and limits of available empirical observations related to the entities and processes related to the theory or by the methods of measurement and analysis that these data are customarily subjected to. These seem severe and unjustifiable assumptions. In addition, strict Cartesian distrust for observation makes empirical evaluation of theory difficult.

Fig. 1.3 Cartesian process of deduction through which a theory is proposed a priori. In strict Cartesian argument, the process stops there, as an ideal theory is seen as superior to the noisy and imperfect methods and data of actual fact. In practice, methods are selected to evaluate the theory, data is collected with these methods, and then used to verify the theory

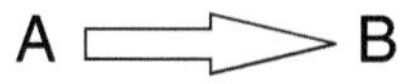

AFFIRMING THE CONSEQUENT	DENYING THE ANTECEDANT
If A is true then B is true	If A is true then B is true
B is true	A is not true
Therefore A is Indeterminate	Therefore B is Indeterminate
MODUS PONENS	MODUS TOLLENS
If A is true then B is true	If A is true then B is true
A is true	B is not true
Therefore B is certainly true	Therefore A is certainly not true

Fig. 1.4 Four logical syllogisms central to scientific reasoning. All four have the same major premise, if A is true then B is true. The bottom two, modus ponens and modus tollens, are logically correct determinate judgments, while the top two, affirming the consequent and denying the antecedent, are indeterminate. Affirming the consequent has special prominence in scientific reasoning and is an abiding challenge to obtaining reliable knowledge

A major focus of this book will be on reasoning within the practice of science, particularly in regard to common logical errors that lead to incorrect conclusions. Figure 1.4 lists four major forms of logical argument from a major and a conditional premise. Two of these (Modus ponens and Modus tolens) are logically correct; the other two (Affirming the Consequent and Denying the Antecedent) lead to logical indeterminacy. A graphical depiction of these four conditional syllogisms may be helpful in gaining an intuitive understanding of the subtle logical traps of affirming the consequent and denying the antecedent (Fig. 1.5). The area outside oval X is –X and the area outside oval Y is –Y. Elements l, m, n are particular members of X, Y, and –Y. In Fig. 1.5, the statement "If X then Y" is represented by the Venn diagram in that all elements of X are also included in Y, so if X is true then Y is also true by overlap, and is a form of the Modus ponens argument. Likewise, the statement "If Y is not true then X is also not true" is represented in that no elements of X exist outside of Y, and is a form of the Modus tolens argument. In contrast, the statement "If Y then X" is not correct in this case, as there are elements of Y (for example m)

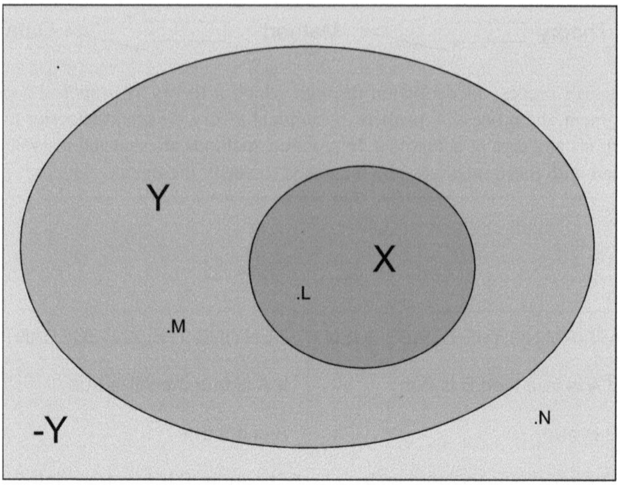

Fig. 1.5 Venn diagram schematic showing the logic behind the four syllogisms listed in Fig. 1.4

that are not also elements of X. This is a case of affirming the consequent, because if X is true then Y would also be true. But it is possible for Y to be true without X being true, as there are elements of Y which do not overlap X. In addition, the conclusion that if X isn't true then Y also isn't true is an example of denying the antecedent, as there are elements of Y which exist outside of X.

One of the inferential challenges of scientific research is that the researcher does not have knowledge of the true relationships between premises a priori. That is, the Venn diagram for a given ecological research question corresponding to Fig. 1.5 is hidden to the observer. That is one reason why data mining and related methods become so popular (Breiman 2001), and model selection so intensely debated (Burnham and Anderson 1998). In deductive reasoning the researcher proposes conceptual Venn diagrams corresponding to hypothetical relationships between causal and response factors. Then observations may be compared to the conceptual model and if they do not match the model may be rejected. In inductive reasoning, the researcher collects data, building evidence, in an effort to confirm universal conditions from a collection of consistent particular observations.

The overall goal of this introductory chapter is to try to link the data–method–theory interdependency (Fig. 1.1) to these four forms of argument within the context of ecological reasoning. The remainder of the book is focused on the details of ecological methods, data and theory, especially in regard to where errors of affirming the consequent and denying the antecedent commonly are made. No definite conclusion can be drawn when the antecedent is denied or the consequent is affirmed. In inferential science both of these errors are extremely common, both when proceeding in an inductive path from data to infer theory, or from a deductive path from theory to confirming data.

The celebrated, so-called "hypothetico-deductive" approach is a modern attempt to partly reconcile deductive and inductive approaches and through a partial fusion

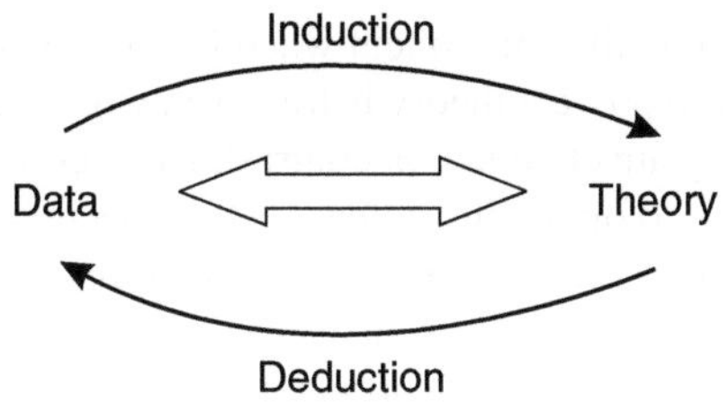

Fig. 1.6 The relationship between data and theory and the pathways of inference of induction and deduction. In induction, theories are induced from a collection of observations, while in deduction a theory is proposed and then data is compared to the theory to see if the theory is consistent or can be falsified

remedy their respective weaknesses (Fig. 1.6). The basic idea is that data without theory are uninformative, while theory without data is not compelling. A scientist proposes a hypothesis to explain some phenomenon, develops a crucial experiment to test it, with the critical attribute that the outcome of the experiment can produce results inconsistent with the hypothesis, and thereby lead to its rejection. The scientist then collects data required by the experiment, conducts the experiment, then either rejects or fails to reject the hypothesis. We believe that the dominant, if often implicit and subconscious, mode of investigation and explanation in modern western science is based on this idea, as elaborated by Karl Popper (1959, 1963) and others, and discussed in a famous paper by Platt published in Science in 1964. While the approach seems to reconcile Baconian induction with Cartesian deduction by linking data to theory through the crucial experiment and refutation, several issues remain. An obvious one is the validity of crucial experiments and falsification.

The criterion of falsification relates to how well defined the research question is and how meaningful is the proposed theory. For example, suppose there are four theories for temperature, T1, T2, T3, T4. The application of these four theories to a particular location and time imply four implications: (I1) Someday the temperature will be above freezing, (I2) tomorrow the temperature will be above freezing, if the conditions are right, (I3) tomorrow the temperature will be above freezing, (I4) tomorrow the temperature will reach at least 20°C. If the temperature tomorrow reaches 18° we will reject theory T4, but not T1–T3. Both T3 and T4 are well-defined statements, in that they stipulate unambiguous criteria for rejecting the theory. However, this is not the case for T1 and T2. Without an independent specification of the necessary and sufficient "if the conditions are right", I2 is true by definition and hence is circular. Also, I1 cannot be falsified because it implies an indefinite period for the observation to occur. Therefore, T1 and T2 are not testable theories, resulting in dubious statements and actions if they should be used as a base for management actions (Hillborn 1997).

An ecologist proposes a theory T for some phenomenon, and identifies and implication I. This then produces a major premise of the form: "if theory T, then implication I." The outcome of the research is data (D), which supplies the minor premise of the conditional syllogism proposed by the theory. Scientists face the indeterminacy of affirming the consequent when D is like I. This is the reason why scientists cannot claim to prove a theory is true even when data consistent with the implication of the theory are obtained. In contrast, when D is inconsistent with I the research conclusion, based on modus tollens, is the rejection of theory T.

The ecologist traditionally has taken a refutationist approach to advancing knowledge. As discussed above there is no means to logically prove a theory through refutation. Thus, empirical science is always in the realm of affirming the consequent. Instead of proving theory, the traditional refutationist paradigm is based on "theory corroboration", through which a researcher is expected to make a concerted and repeated effort to refute the theory, with confidence in the theory supposedly increasing with additional failed efforts to falsify the theory. In this view, science is a process of eternal skepticism, built around the endless quest to disprove theories. This approach in ecology is represented by inferential statistics based on the null hypothesis. The null hypothesis is the implication expressed in data if the theory is not true. It usually is expressed as no difference between treatments in an ANOVA framework or no slope in a regression framework (Hillborn 1997; Zar 1996).

There are at least six major potential criticisms of this "theory corroboration" paradigm using inferential statistics and null hypotheses. First, one may question whether the null hypothesis is a meaningful statement, in terms of the criterion of falsification, described above. We argued that for a theory to be meaningful it must make specific predictions that can be disproved. The traditional null hypothesis has a very weak criterion for falsification, as it is usually expressed in the broadest terms of difference from randomness. Observing statistically significant difference from randomness is a very weak platform to build theory corroboration. This is particularly true for the life sciences, in which true random processes are rare, and there exists a deep foundation of accumulated knowledge of structure–function relationships. Informative hypothesis should at least be based on the latter (O'Connor 2000).

Second, most inferential statistical analysis using null hypotheses has focused on Type I error, or rejecting the null hypothesis when it is true, and not considered Type II errors nearly as carefully. Type II errors are failure to reject the null hypothesis when it is false, and is expressed as rate beta. One minus beta is power, or the probability that one will correctly reject the null hypothesis when it is false. In practice, given the tradeoff between the two error types, the traditional focus on Type I error, and the expense of collecting representative sample sizes large enough to obtain low rates of both error types, actual statistical power is often quite low. With low statistical power the researcher will be unlikely to reject a null hypothesis even if it is false. This clearly cripples one's ability to refute hypotheses, and seriously undercuts the confidence one should have in "corroboration" through repeatedly failing to reject a null hypothesis.

Third, one must question the dedication of practicing scientists to negative inference and skepticism. It's well known that negative results are undervalued and difficult to publish; publication decisions are clearly biased towards positive findings. However, logically, the refutationist approach only advances through rejection of null hypotheses, and does not confirm explanation. However, in practice researchers often treat significant statistical tests as proof of the veracity of the theory. Even when they tacitly admit falsification does not prove a hypothesis, researchers rarely report comprehensive results from many different tests of the same hypothesis to build a measure of theory corroboration. This in large part is due to the nature of scientific publishing in peer-reviewed journals, where a premium is given to space and speed – there is not the space for comprehensive reporting of multiple efforts to refute hypotheses with complementary analyses, nor is there time given the competitive nature of science for researchers to conduct such large, and complex analyses. An extension of this argument would be that even if a researcher were to diligently test a theory against a null hypothesis a number of times with complementary methods and extensive independent data, to what degree does this increase our confidence in the result? One of the insidious attributes of affirming the consequent is that data measured in many ways in many systems may appear consistent with an incorrect theory. There are potentially infinite theories to explain any phenomenon, and a large portion of these might be expected to produce implications measured in data that are equivocal. Thus there is an under-determination of theories by facts (Quine 1953). The data might consistently be consistent with multiple mutually exclusive alternative theories. If this is the case then no amount of "theory corroboration" through failure to refute a null hypothesis really increases our confidence in the theory. We present an example of this in Chapter 17 on landscape genetics.

Fourth, is the challenge of obtaining large, representative samples of ecological data from spatially and temporally complex systems. McGarigal and Cushman (2002) conducted an extensive review of studies to quantify the ecological effects of habitat fragmentation. They noted a dramatic paucity of replicated and controlled studies in which the units of observation were appropriate for landscape-level inferences about pattern–process relationships. The vast majority of published studies used mensurative rather than manipulative study designs focused on relationships between habitat patch size and shape and ecological process. Very few studies combined manipulative design, replication, statistical control, with inference to landscape-level pattern–process relationships. One of the major themes of this book is that spatial and temporal context are fundamentally important to pattern process relationships. The severe shortage of field studies that provide statistically powerful replicated and controlled data sets on ecological responses to spatial and temporal variability at appropriate scales is a major limitation to current knowledge.

Fifth, one of the main themes of this book, which we begin to develop in the following chapter, is that spatial and temporal variability fundamentally alter pattern process relationships such that these are not stationary across space and time. We argue in the following chapters that this implies that research should focus directly on mechanisim–response relationships in context and in particulate. This has major implications for traditional inferential refutationist analysis. For example, if spatial and temporal variability are fundamental factors it will be difficult or impossible to obtain representative and independent samples for inferential analysis. Inferential statistics is based on being able to draw independent and representative observations from a population. This implicitly assumes an urn-model kind of population, where a random draw will produce a representative and independent observation. However if spatial and temporal variability are fundamental we will not be able to obtain either statistical independence or representativeness through random sampling in space. More importantly, if ecological processes are nonstationary in space and time then it will be difficult to define a meaningful population from which to sample. The entire principle of random sampling from a population presupposes that there is a discretely bounded and internally stationary population to sample. If pattern process relationships vary continuously across space and through time, as we will argue throughout this book, then there is no objectively defined population from which to draw independent and representative samples. Instead, as we argue in Chapters 2, 3 and 4, analysis may have to abandon the notion of representative sampling from populations and explicitly adopt a model in which the time-space location of each sample in a large network is explicitly included in the analysis. This poses a tremendous challenge to inferential, population-based, null-hypothesis refutation, a challenge that is not substantially lessened by including time in a repeated measures design or space in a mixed-model form. In addition, linear analysis approaches often cannot cope with these data situations (Craig and Huettmann 2008).

Sixth, naïve falsification has been questioned on several other philosophical grounds. Popper (1963) noted that nearly any statement can be made to fit the data, with the addition of ad hoc adjustments. In addition, the so-called Quine–Duhem thesis holds that for any given set of observations there are an innumerably large number of explanations. Specifically, Quine's (1951) doctrine of confirmation holism emphasizes the under-determination of theories by data, noting that because of the principle of affirming the consequent there may be many equally justifiable alternative explanations for a given observation set. Thus empirical evidence cannot force the revision of a theory. As such, the Quine–Duhem thesis is seen as a refutation of Popper's criterion of falsification as a reliable means of obtaining knowledge.

In recognition of the logical and practical insufficiency of refutationist methods, an alternative paradigm of confirmation has developed. In this approach, instead of proposing a single theory and then seeking disproof, the scientist proposes multiple alternative hypotheses and then seeks to evaluate which explanation is most likely. The exchange is of the questionable logic of critical experiments and decisive refutation for weighing the relative support for multiple alternatives. Likelihood theory, Bayesian belief and the Information Theoretic Approach are all examples of this alternative paradigm. While this appears to circumvent the dilemma of refutation, it embraces an equally challenging dilemma of confirmation. Specifically, there is no logical means to evaluate to what degree the most likely explanation evaluated in a confirmationist approach conforms with any true underlying process. There is just as much inherent risk of affirming the consequent or denying the antecedent as in the refutationist case. Observing that one theory is more consistent with a set of data than another does not prove that theory is correct. As discussed above, there may be any number of alternative theories that could explain the data as well or better (e.g. Quine 1951).

In our view, neither the refutationist nor the confirmationist models of scientific inference address the fundamental interdependence between theory, method and data. We feel that much progress has been made in advancing knowledge using both approaches; yet we also feel both approaches are contingent and particular, rather than providing universal, necessary and certain knowledge of underlying cause and effect. A scientist stands embedded in history, on the shoulders of giants. From that vantage, the view of the scientific landscape, including what questions to ask, on what topics, what methods to use, and likely explanatory hypotheses, are all affected in profound ways by the context of accepted knowledge, prevailing theory, and predominant scientific methods and techniques. Within that context, the prospect for reliably identifying correct relationships between cause and effect is daunting. If spatial and temporal variability are fundamental, and if pattern-process relationships are dependent on them in highly scale-dependent ways, then there may be a strong case for an entirely new approach to ecological inference based on integrating contextual pattern–process relationships explicitly across space and through time, as we discuss in the following several chapters.

There is no possible decoupling of theory, method and data. There is no data that is not infected by theory. There is no theory that is not built on observation. There is no observation that is not dependent on method. There is no method that is not informed and chosen in the context of theory. In the absence of decoupling of method, data and theory, it is difficult to avoid a boring-in of tradition, where theory proposes methods, which produce data, which support the theory, in an endless, inward spiral of self-confirmation. This of course is antithetical to the logical arguments about the risk of affirming the consequent presented above. The most important tool a scientist can possess is skepticism, and the most essential skill is to dispassionately direct this skepticism at one's own theories and observations. This also is the most difficult psychological challenge a scientist faces. Scientists' work is evaluated by their peers within a community which shares common paradigms and assumptions. Thus, even should a scientist have the temerity to challenge the basic assumptions of the field it is very likely their work will receive harsh criticism, while less controversial and more conventional work will usually receive positive reviews as "confirmation" of established belief. This perhaps is the central message of Kuhn's Structure of Scientific Revolutions (1970), that while we stand on the shoulders of giants, we are also shackled by them to triads of self-confirmatory, interdependent traditions of theory, method and data. Thus, there truly is a central role of revolution and paradigm shift in science. Self-confirmatory, boring-in of theory–method–data–theory produces a detailed exegesis of the products of the underlying assumptions of that particular tradition of theory–method–data. But, as in a debate about how many angels can dance on the head of a pin, often further progress in expanding and advancing knowledge of nature depends on breaking out of that inward spiral through questioning assumptions, proposing radical new theories, or developing new methods to obtain novel data and produce new kinds of analyses.

This is the core focus of this book. We believe that the confluence of unprecedented increases in computational power and profound improvements in the ability to measure and sample attributes of the environment at fine scales, across vast spatial expanses and overtime are providing revolutionary changes to two corners of this conceptual triangle. In our conceptual model, this coming revolution in ecological science is being pushed by immense changes in method, specifically computational

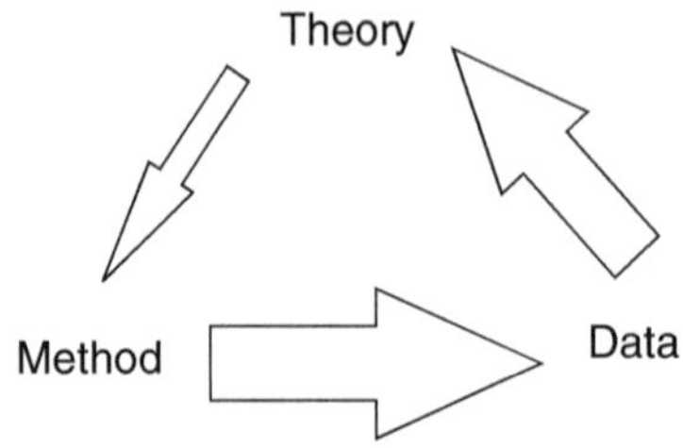

Fig. 1.7 Modern technological advances have enabled revolutionary changes in the kinds and extent of data that can be collected. This in turn has led to enormous advances in methodologies needed to handle, store and analyze these data. This in turn is providing a "bottom-up" forcing on reexamining ecological theory within the context of these new methodologies and data, with new methodlogical advances the dominant driver of change and progress. In contrast, classic ecology of the twentieth century was led by theory, with method and data trailing after. This is because of the combination of the tremendous success of simple and parsimonious mathematical models creating internally consistent explanations, with the lack of ability to collect and analyze large data sets across big spatial extents and over time

power of fast computers, advanced spatial and multivariate statistical and modeling approaches, geographic information systems, remote sensing, and landscape ecology (Fig. 1.7). These, in turn, are vastly altering the amount, kind and scope of data available for analysis, and the kinds of analyses that are possible. At this point in time we are in the birth-throes of new paradigms, with method and data far surpassing theory. The goal of this book is to step through each corner of this triad, to discuss the revolutions of measurement, computation, analysis methodologies, to describe the vast new kinds of data these methodologies produce, and finally, and most difficult, look forward and describe the directions into which we believe these revolutions of method and data will enable the expansion of theory.

The history of ecological science in the twentieth century is marked by the emergence of ecology as a quantitative science. This past revolution in ecological science was led by theory, rather than method or data. It was characterized by the development of ideal mathematical models to describe a broad range of ecosystem and population processes. Importantly, these models provided for the first time formal description of ecological processes that enable prediction and investigation of cause and effect relationships. However, like all ideal models, they typically presented a simplified caricature of processes in which complicating factors, such as spatial variability and temporal disequilibrium, were ignored. Implicit in this is the assumption that equilibrium processes acting non-spatially are sufficient to characterize the major attributes of ecological structure and function, and that interactions of space and time could be shunted into the error term as unexplained and unimportant noise. We believe that nearly all classic quantitative ecological theories of the past century share this characteristic, including Lotka–Volterra predatory prey models, theories of habitat compositional analysis based on proportional use, carrying capacity, predator–prey models, maximum sustainable yield (msy), approaches to analyzing sequential animal movement data based on assumptions of the random use of fixed home ranges, landscape ecology of discrete, categorical patch mosaics, genetics of ideal Hardy–Weinberg populations, population dynamics, dispersal, connectivity, and evolution.

In the final two decades of the twentieth century the sufficiency of these ideal models was broadly called into question. The essence of the concern was whether spatial and temporal variability were in fact noise clouding otherwise ideal non-spatial, equilibrium processes, or whether spatial complexity and temporal variability were central to the basic functioning of ecosystem and population processes in ways that require their explicit incorporation into theory and analysis. This upsurge in theoretical concern about spatial complexity and temporal disequilibrium was enabled by the rise in powerful computing, which allowed scientists to address complex processes in unprecedentedly sophisticated ways. One of the main motivations for adopting ideal, non-spatial, equilibrium models in ecology is that they are mathematically tidy and easy to understand and use to make predictions. On the other hand if the contingencies of history and complexities of spatial pattern across scale are fundamentally important, then these simple, tidy mathematical representations of the world do not suffice. Further, it would be very likely that there will not exist tractable, closed-form alternatives that correspond to a spatially complex and temporally disequilibrial world. We believe this requires a change from seeking closed form equations for processes in ideal systems, to understanding the processes in particulate and in context. By in particulate, we mean understanding the process at the level of the individual entities by representing their actions and interactions explicitly. By in context, we mean representing these particulate processes within the spatial and temporal domains. The advance of powerful computing enabled researchers to simulate processes that could not be represented in closed form equations, and investigate how spatial complexity and temporal dynamics influenced the actions and interactions of ecosystem and population processes. This change in method enabled a change in theory. Prior to the ability to simulate processes in spatially complex and temporally variable systems, few scientists ever asked if idealized models were sufficient, and if they did they likely quickly concluded that addressing spatial and temporal complexity in a hyperdimensional predictor space at the at the level of individual entities was intractable and thus not worth spending time and energy contemplating.

1.2 Informatics and Biology: A Very Powerful But Not Fully Used Melding of Disciplines

Informatics has become a well-established modern scientific discipline; it is taught at most major universities and has emerged as a primary framework in public and private sectors alike for efficiently storing, updating and analyzing vast datasets to address challenging research and economic issues. It provides a general digital and quantitative framework and powerful tools for collecting, compiling, processing and analyzing large and complex datasets and models. However, most of the past informatics applications have not dealt with detailed animal biology or ecology.

Traditionally, informatics has had a strictly technical and mathematical focus, with most past applications in the military and industrial contexts. Sustainability questions were virtually ignored and business applications dominated.

The use of GIS started in North America in the early 1960s: Early efforts were motivated primarily by military, but also by agricultural and landuse issues. GIS is a rather complex topic involving geographic projections, digital databases, spatial analysis and the display of data. Given the technical and informatic challenges posed by large, integrated spatial data storage and analysis, development was rather slow, taking more than 15 years until robust and established software platforms became widely available, allowing to spread GIS applications of relevance for animal ecology. The North American and Australian wildlife community played a major role in applying and implementing GIS into the discipline. However, countries in the Eastern Block, China, India, Africa and Central- and South America were not able to contribute much to GIS, enforcing the digital divide that exist to the very day. GIS applications benefited from the Clinton administration's suspension of Selective Availability for Global Positioning Systems (GPS), which allowed public and commercial use of GPS with high accuracies. This greatly facilitated the collection of highly accurate spatial ecological data in the field, which is a foundation for all efforts to integrate space and time into ecological analysis. Specifically, wildlife surveyors realized that spatially referenced digital databases can offer huge advantages for analysis and quick turn-around of results. Because their data is inherently spatial, GIS provides an optimal platform for storing and analyzing spatial ecological data (Huettmann 2004 for an example). Lastly, many museums and natural-history collections world-wide started to digitize their data holdings into databases, and also realized quickly the value when spatial information is available (Graham et al. 2004). These efforts not only benefit from developments of GIS, GPS, the internet and informatics as a whole, but also from geo-referencing tools such as BioGeoMancer (http://www.biogeomancer.org/), which enable plotting of many data points onto a map and with a quantitative measure of accuracy.

It was quickly realized that large and sophisticated spatial databases require extensive documentation to be reliable and interpretable for multiple users. This in turn prompted the development of global formats and standards for Metadata. Metadata started with the library community, then got added a spatial components (Content Standard for Digital Geospatial Metadata CSDGM by the Federal Geographic Data Committee FGDC http://www.fgdc.gov/) and biological components (Biological Data Profile BDP by the National Biological Information Infrastructure NBII www.nbii.com). They now form the absolute foundation for best available online delivery for a global audience to make best use of the offered huge raw data amounts (Huettmann 2005, 2007b), and for a better ecology and management of natural resources.

Another major development in ecological informatics can be traced to the 1992 Rio Convention, which proposed that digital biodiversity information is a global property and heritage, and thus should be administered as a common good by global agencies such as U.N. and its environmental branches (UNEP; Strong 2001). The Global Biodiversity Information Forum (GBIF, http://www.gbif.org/) was a direct outlet of this movement and it represents the hosting facility for any digital biodiversity information world-wide. Such globally networked ecological databases provide a revolutionary opportunity to collect, share and update information on the global environment as well as a means to compile hyper-variate, multi-scale databases essential for robust ecological analysis and modeling (e.g. Chapter 6). Despite major questions of copyright and information use (e.g. Stiglitz 2006), this represents a new and fresh look to overcome the digital divide, at how informatics can serve Earth and its people for global well-being and sustainability (Huettmann 2005). Secondly, it finally moves informatics back to a tool serving human needs foremost, instead of being used as an intimidating tool 'for the sake of a tool', or to contribute to maintaining unequally distributed wealth. Ecoinformatics will only be effective at contributing to ecology and global sustainability when it is understood and utilized by large portions of the global citizenry within a global democratic process.

It is in this philosophical framework that EcoInformatics and similar disciplines are positioned. Foremost, EcoInformatics deals with the entire ecosystem. As outlined earlier, an ecosystem is a very complex, interconnected entity which provides free ecological services, such as clean air, clean water, healthy climate, pollination and agriculture productivity. Entire human societies, and its wealth, rises and falls with ecological services (Taber and Payne 2003). EcoInformatics provides a framework for society to assure that the most optimal solution is reached for humans managing the Earth. An EcoInformatics and subsequent policy that ignores equal access to information for all citizens, and does not promote environmental justice cannot stand (Stiglitz 2006). It is here where BioInformatics, a simplistic combination of traditional informatics and biology, has not yet achieved the promise of global sustainability. Most university biology departments these days are split into a Biochemical section and a traditional Animal/Wildlife and Plant section. Biology as a discipline suffers from this situation tremendously, and so does BioInformatics, which widely defines itself as using Informatics for the use of Genetic analysis, culminating in applications with an industrial focus such as Genbank and data mining for specific genes occupying high-performance computing centers.

However, Biology is a much wider discipline and foremost, and its disciplines like wildlife management hold the answer to global sustainability, well-being, wealth, and subsequently, human happiness (Taber and Payne 2003). Deep frustration about the narrow one-sidedness in the traditional Biology Departments has already led to split from BioInformatics and its underlying philosophy, and create new disciplines such as EcoInformatics, Biodiversity Informatics, Wildlife Informatics, Ocean Informatics, Polar Informatics, Landscape Informatics, Conservation Informatics etc. This follows the still ongoing and rather problematic paradigm of narrow specialization, which has already set us up for major problems in the sciences, our universities, funding agencies and beyond (Taber and Payne 2003). However, it is hoped that it allows for new and more balanced developments and approaches. Hopefully, these 'new' disciplines can join forces, be fully accessible, compatible and interdisciplinary, and thus, form a new global movement, Sustainability Informatics, where Ecology and technological tools, linked with high-performance computing and online initiatives at it's core, allow to reach global sustainability for the sake of human well-being that fully takes nature into account. The use of decision-making tools would present one of such endavours (Huettmann 2004, 2007c). This new call for 'Sustainability Informatics' is extremely relevant, if our approaches are to make sense and assure long-term survival of animals, biodiversity as a whole, habitats and humans alike. Templates for a sound management exist already: Adaptive Management (Walters 1986). However, the current scheme lacks the inclusion of sound Ecology and Informatics. If these three would get combined (Huettmann 2007d) and also could get manifested in our day-to-day teaching efforts to reach the new generation (Huettmann 2007b), it should allow us to make the best possible decisions, using the available tools and sciences.

This rapid advance of method and data collection enabled a reconsideration of many of the basic assumptions of the preceding dominant paradigm, particularly the sufficiency of equilibrium, non-spatial idealized process models and null-hypothesis inferential statistics. In the chapters that follow we will discuss each corner of this triad. We begin in Chapter 2 where we discuss the history, context and change of a foundational concept in ecology, the species–environment relationship and the structure of biological communities. Chapter 3 addresses the fundamentally important and exceptionally challenging task of handling scale and scaling in ecological relationships. In Chapter 4 we outline emergence of the science of landscape ecology and discusses its potential to provide a coherent framework for the development of a new paradigm of spatial and temporal ecology, and the challenges to be overcome in integrating particulate and contextual processes into existing landscape ecological science. Chapter 5 focuses on the ongoing evolution of landscape ecology, particularly in its conceptual expansion to more effectively address the issues discussed in Chapters 2 and 3. In Chapter 6 we present a framework for data collection to provide flexible, multi-variate, multi-scale, spatially synoptic, temporally dynamic databases to feed modeling and analysis of scale dependent ecological processes in spatially complex and temporally variable environments. Thereafter come a number of chapters on particular kinds of data, with reviews of the attributes, scope and limitations of data on organism occurrence, movement and genetics. Next, we focus explicitly on the next corner of the triad, with four chapters discussing a selection of methodological tools to facilitate spatial and temporal analysis of ecological systems, including remote sensing, geographical information systems, database compilation, software and geostatistics. In the following three chapters we then review current spatial and temporal modeling approaches, focusing particularly on landscape genetics and multi-model resistance modeling. We finally present eight examples from current research on terrestrial and marine ecosystems, and close with chapters on the interaction between public policy and scientific progress and a view into the future and discussion of the outlook for rapid progress to expand theory–method–data to most effectively increase our understanding of the ecological world.

This may seem a daunting and overly ambitious project. We are not trying to invent new paradigms of science, but simply to uncover what has already come into being and give it our interpretation. Such interpretations are necessarily limited by the knowledge, perceptiveness and biases of the interpreters. We hope these thoughts and perspectives may be of use in advancing discussion and promoting debate, leading to improved sciences and decision-making. Despite massive losses of ecological services world-wide, this is a fascinating moment in the history of our science, with tremendous advances in data acquisition, computing, and statistics enabling entirely new approaches to research and opening new avenues to understanding. Simultaneously, we face the unprecedented challenges regarding the state of the world, with climate change, population growth, ballooning consumption all contributing to a global biodiversity crisis. This brave new century appears to be destined to be the century of extinction unless there are dramatic changes to the global socio-economic system. The science of ecology must take a leading role in informing the path of these changes, providing knowledge about ecological systems, human impacts on them, and ways to mitigate our impacts on the biosphere. We hope that this book contributes for these noble goals.

References

Bacon F (1620) The New Organon (Cambridge Texts in the History of Philosophy)/Francis Bacon (ISBN 0521564832)

Breiman L (2001) Statistical modelling: the two cultures. Statistical Sci 16: 199–231

Burnham KP, Anderson DR (1998) Model Selection and Inference: a Practical Information-Theoretic Approach. Springer, New York

Craig E, Huettmann F (2008) Using "blackbox" algorithms such as TreeNet and Random Forests for data-mining and for finding meaningful patterns, relationships and outliers in complex ecological data: an overview, an example using golden eagle satellite data and an outlook for a promising future. Chapter IV in: W. Hsiao-fan (Ed.). Intelligent Data Analysis: Developing New Methodologies Through Pattern Discovery and Recovery. IGI Global, Hershey, PA, USA

Czech B. (2008) Prospects for reconciling the conflict between economic growth and biodiversity conservation with technological progress. Cons Bio 22: 1389–1398

Descartes R (1637) A Discourse on method. Pomona Press

Diamond, J (1999) Guns, Germs and Steel: The Fates of Human Societies. Norton, New York

Graham CH, Ferrier S, Huettmann F, Moritz C, Peterson AT (2004) New developments in museum-based informatics and applications in biodiversity analysis. Trends Ecol Evol 19: 497–503

Hillborn R (1997) The Ecological Detective: Confronting Models with Data. Princeton University Press, Princeton, NJ

Huettmann F (2004) Towards a Marine Environmental Information System (MEnvIS) for the Northwest Atlantic: Experiences and suggestions from a multi-disciplinary GIS conservation research project using public large scale monitoring and research data bases from the internet. 4th WSEAS (World Scientific and Engineering Academy and Society) Proceedings, Tenerife/Spain, December 2003

Huettmann F (2005) Databases and science-based management in the context of wildlife and habitat: Towards a certified ISO standard for objective decision-making for the global community by using the internet. J Wildlife Manage 69: 466–472

Huettmann F (2007a) German Environmental Informatics vs. Nature Conservation, GBIF, Rio, Statistics and Sustainability: Are we riding a dead horse? (in German). Rundbrief des Fachausschusses der Umwelt Informatik, Deutsche Gesellschaft fuer Informatik (GI) Nr 42: 9–14

Huettmann F (2007b) The digital teaching legacy of the International Polar Year (IPY): Details of a present to the global village for achieving sustainability, in: M. Tjoa and R.R. Wagner (Eds). Proceedings of the 18th International Workshop on Database and Expert Systems Applications (DEXA). IEEE Computer Society, Los Alamitos, CA, pp. 673–677

Huettmann F (2007c) Constraints, suggested solutions and an outlook towards a new digital culture for the oceans and beyond: experiences from five predictive GIS models that contribute to global management, conservation and study of marine wildlife and habitat, in: E. Vanden Berghe et al. (Eds.). Proceedings of 'Ocean Biodiversity Informatics': An International Conference on Marine Biodiversity Data Management Hamburg, Germany, 29 November–1 December, 2004. IOC Workshop Report, 202, VLIZ Special Publication 37: pp. http://www.vliz.be/vmdcdata/imis2/imis.php?module=ref&refid=107201

Huettmann F (2007d) Modern adaptive management: Adding digital opportunities towards a sustainable world with new values. Forum Public Policy: Climate Change Sustain Dev 3: 337–342

Kuhn TS (1970) The Structure of Scientific Revolutions. University of Chicago Press, Chicago, IL

McGarigal K, Cushman SA (2002) Comparative evaluation of experimental approaches to the study of habitat fragmentation effects. Ecol Appl 12: 335–345

O'Connor R (2000) Why ecology lags behind biology. Scientist 14: 35

Platt JK (1964) Strong inference: Certain systematic methods of scientific thinking may produce much more rapid progress than others. Science 146: 347–353

Popper K (1959) The Logic of Scientific Discovery. Basic Books, New York

Popper K (1963) Conjectures and Refutations. Routledge, London

Quine WVO (1951) Two dogmas of empiricism. Philos Rev 60: 20–43

Quine WVO (1953) From a Logical point of view. Harvard University Press. ISBN 0-674-32351-3

Stiglitz JE (2006) Making Globalization Work. Norton, New York

Strong M (2001) Where on Earth Are We Going ? Texere, New York

Taber RD, Payne NF (2003) Wildlife, Conservation, and Human Welfare: A United States and Canadian Perspective. Krieger, Malabar, FL

Verner J, Morrison ML, Ralph CJ (1984). Wildlife 2000: Modeling Habitat Relationships of Terrestrial Vertebrates: Based on an International Symposium Held at Stanford Sierra Camp, Fallen Leaf Lake, CA, 7–11

Walters C (1986) Adaptive Management of Renewable Resources. Blackburn, Caldwell, NJ

Weizenbaum J (1984) Computer Power and Human Reason. Pelican, London

Zar JH (1996) Biostatistical Analysis. 4 ed., Prentice-Hall, Upper Sadle River, NJ

Chapter 2
Space and Time in Ecology: Noise or Fundamental Driver?

Samuel A. Cushman

In this chapter I frame the central issue of the book, namely is spatial and temporal complexity in ecological systems merely noise around the predictions of non-spatial, equilibrium processes? Or, alternatively, do spatial and temporal variability in the environment and autogenic space–time processes in populations fundamentally alter system behavior such that ideal models of nonspatial and equilibrium processes do not represent the fundamental dynamics of ecological systems?

If the former is correct, then the task for ecologists is seems relatively simple, if practically daunting. If variation across space and through time are noisy but not transformational, then the task of ecologists is simply one of increasing the scope of inference to maximize precision. That is, with the additional noise inhering to ecological processes due to spatial and temporal variability, larger and more extensive empirical samples will be needed to obtain precise estimates of underlying parameters through either inferential or Bayesian approaches. This would emphasize the critical need for broad scale, consistent, large sample data collection efforts. It would also put a fundamental limit on the precision of predictions that would be possible for a specific fine-scale location at a specific time. Similarly, such a relationship between spatial and temporal variability and system behavior would require only modest changes to nonspatial and equilibrium theoretical and predictive models. Under such a scenario, the expected value of ideal, nonspatial and equilibrium models would be unbiased, assuming correct identification of important driving variables and proper parameterization. However, the precision of the estimates of such models would be questionable, as the amount of variability in observed phenomena due to spatial variability and temporal fluctuations would likely often be substantial.

Scale would play a particularly critical role in determining precision of estimates from both empirical inferences and predictions of ideal nonspatial, equilibrium models. Specifically, if spatial and temporal variability are simply noise, then there will be a consistent relationship between scale and the precision of estimates

S.A. Cushman
US Forest Service, Rocky Mountain Research Station, 800 E Beckwith, Missoula, MT 59801, USA
e-mail: scushman@fs.fed.us

S.A. Cushman and F. Huettmann (eds.), *Spatial Complexity, Informatics, and Wildlife Conservation*
DOI 10.1007/978-4-431-87771-4_2, © Springer 2010

from empirical data and of predictions of theoretical models. Simply put, if spatial and temporal variability are noise clouding ideal processes, then precision will increase monotonically with the extent of measurement across space and time. At fine spatial scales and over short time periods variability will be high and reliability of empirical data to infer underlying process, or of process models to predict empirical patterns, will both be low. However, given an unbiased expected value, this variability will average out and precision will greatly increase as extent increases in both temporal and spatial dimensions, and as sample sizes increase in density within space and time.

However, there is a catch to this solution to variability through expansion of the scope of analysis. In a word it is nonstationarity. In the previous paragraph we argued that increasing extent of analysis in space and time will average out variability due to spatial and temporal complexity. This is true, so long as one remains within a stable "scale domain" in which relationships between pattern and process are constant. However, one of the most basic concepts in ecology is that ecological patterns vary across space and time along complex gradients of spatially structured environmental drivers. Pattern and process relationships are likely, more often than not, to be spatially nonstationary. And this tendency to nonstationarity is directly and monotonically related to scale as well, and unfortunately in the inverse way as precision. The larger the spatial and temporal scope of analysis, the higher the chance of severe nonstationarity in the major pattern–process relationships governing the system. Therefore, we face a dilemma, with large scope of analysis needed to deal with spatial and temporal variability, but small scope of analysis needed to ensure a sufficiently coherent and stable set of pattern–process relationships. This is a major practical challenge to ecologists hoping to obtain high precision predictions across space and through time. Fortunately, there are approaches to incorporating space, time and scale into analysis to address this challenge, which is the topic of the next chapter.

If, on the other hand, spatial and temporal variability fundamentally alter pattern–process relationships, then the challenge is even more severe. Instead of being noise around an unbiased expected value predicted by nonspatial and equilibrium processes, spatial and temporal variability in this case would fundamentally alter pattern process relationships. It is important to clearly articulate what this means. In the previous paragraph we discussed nonsationarity of pattern process relationships, but implicitly assumed that there could in principle be a scale domain at which one could assume a stable and stationary relationship between pattern and process. If on the other hand, pattern and processes relationships are sensitive to spatial and temporal complexity across the full range of spatial and temporal scale then this collapses all the way down. In such a case one cannot assume nonspatial and equilibrium processes will be sufficient at any combination of scales. In such a case it will therefore be necessary to incorporate spatial and temporal factors directly into the theory proposed, the method used, and to ensure that the data collected are measured at scales at which pattern–process relationships are strongest.

In the introductory chapter we argued that ecology is at a major transition, marked by rapid advances in methodology which enable much broader scale collection of detailed spatial data, much more vast data storage and more sophisticated data organization, and vastly more powerful data handling, manipulation and modeling. These changes in method and data are now feeding back to theory, and providing some traction on the very difficult challenges posed by spatial and temporal variability in ecological systems. The remainder of this chapter is a kind of historical retrospective on the emergence of awareness of the critical role spatial and temporal variability play in ecological systems and how ecology has struggled to begin incorporating them

2.1 Space, Time and Ecological Communities

Ecological communities are not crisply defined, discrete and stable entities. Indeed, the concept of community is an abstraction made to facilitate investigation. It is useful; its definitions provide the scale and boundaries within which hypotheses can be developed and tested. However, it should always be kept in mind that the "community" is an aggregation in flux, varying through space and over time (Levin 1989). The spatial and temporal dynamics of ecological systems affect the interactions of organisms with each other and their environments in complex and synergistic ways.

The functions of temporal and spatial processes within communities are poorly understood. One key question is, how precisely does variability in space and time affect species interactions? Particularly, in what ways are competitive and predator–prey dynamics influenced by habitat geometry, stochastic events, and temporal trends or cycles? Are community structures largely a result of history, chance and the influences of spatial and temporal heterogeneity, or do they reflect underlying nonspatial processes and tend to an ultimate equilibrium? Resolving these questions is among the most challenging and important problems in modern ecology (Huston 1979; Kareiva 1989; Roughgarden and May 1989). In the remainder of this chapter we review the discoveries and theoretical advances that lead to the formation of these questions and consider current theory regarding the effects of heterogeneity in space and time on species interactions and community structure.

2.1.1 Lotka–Volterra, Gause, Hutchinson and Whittaker

Lotka (1932) and Volterra (1926) demonstrated mathematically that two species which are limited by a common resource cannot coexist in a finite system. The simplicity and elegance of their formulation was seized upon by ecologists who hoped that, as in physics, universal laws of ecology could be expressed in simple formulas. The Lotka–Volterra model gained support from the experimental work of Gause (1934, 1935). Gause's experiments indicated that when two species with common resource requirements are forced to coexist in an undiversified environment one will inevitably become extinct. This "law" became known as the "competitive exclusion" or competition principle. The competition principle provided a theoretical explanation for the observation by community ecologists that the species which coexist in any system are a small subset of what might co-occur (Roughgarden and Diamond 1986). This "limited membership" (Elton 1927) was believed to result primarily from exclusion of potential members by established species.

Competitive exclusion, coupled with the idea of limited membership gave studies in community ecology a focus, but it was not until the ecological niche concept was formalized by Hutchinson (1957) that a truly powerful theory of community organization emerged. Hutchinson defined a species' fundamental niche as an n-dimensional hypervolume, in which each dimension represents the range a necessary condition or resource could have and meet the requirements of the species for survival and reproduction (Fig. 2.1). The fundamental niche, then, describes the "space" in which an organism's requirements are met, and thus "where" the organism can potentially survive. Importantly, space in this conception is "ecological space" defined in terms of environmental variables and not in the familiar three dimensions of geographical space. The influences that merging these two models of "space" would have on ecological theory are critical, and will be elaborated upon later in this discussion.

To achieve or hold membership in an ecological community a species must have and maintain "niche space" in geographic space and through time (Fig. 2.2). To have niche space first requires that the resources and conditions needed by the species are available, and secondly that these resources are not co-opted and the conditions are not changed by the actions of other species. If interactions with one species reduce the resources available or change conditions needed by a second species then the two species compete. If this interaction reduces the niche space of one or both of the species then competition is a structuring factor in the community. If the realized niche of a species is sufficiently reduced then one or more of the resource dimensions will be insufficient and the species will be excluded. Thus the composition of any community is determined both by the availability of required resources through space and time (Gleason 1926; Whittaker 1956) and the ability of each species to maintain access to those resources and conditions in the face of interspecific competition (Gause 1935).

This is the central logic of the classical competition theory of community structure. The classical theory predicts that, given an initial set of potential members, the composition and structure of the community will trend toward equilibrium. Species that have similar resource needs compete and the inferior competitors are excluded from some or all of their fundamental niche space. A species excluded from all of its fundamental niche space will be excluded from the community. Over time the community is expected to stabilize at a sort of competitive climax which is conceptually similar to the "climatic climax" of Clements (1936).

The theory, in strict form, is an idealization. It is logically consistent, and therefore correct, within its assumptions. It is a useful mental model, but its assumptions are usually met only briefly and at fine spatial scales in the real world. Also, interspecific

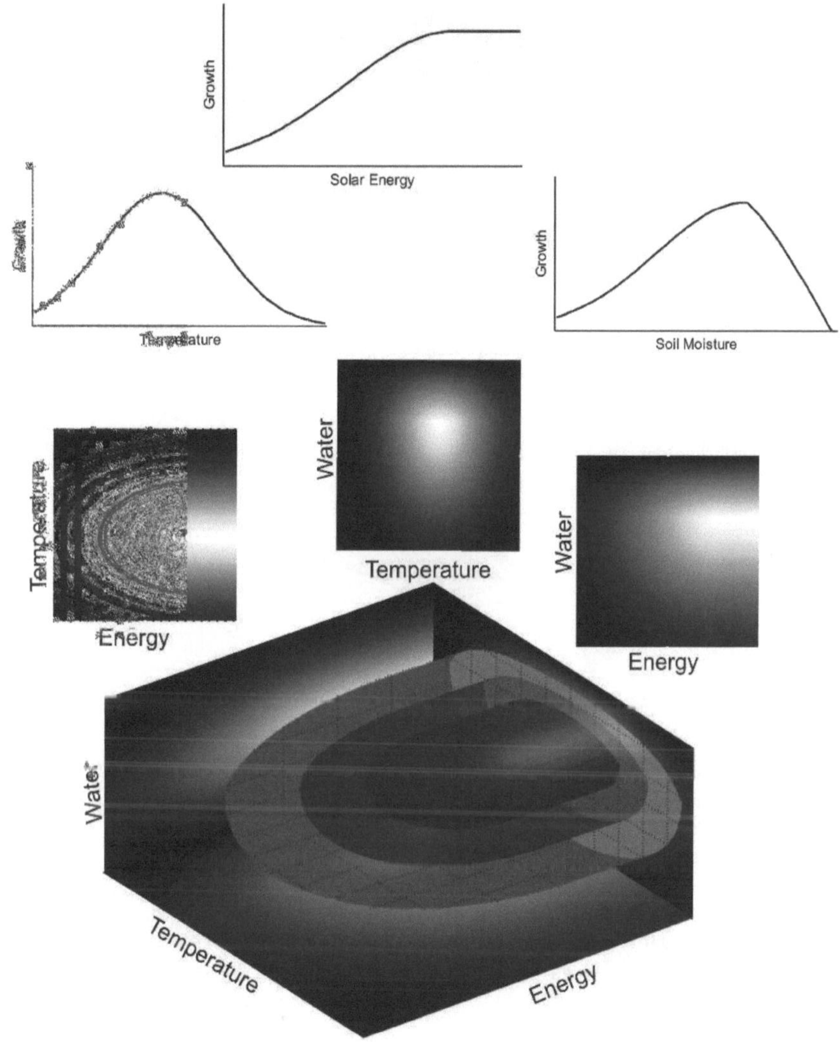

Fig. 2.1 Idealized fundamental niche as zone of tolerance and optimal fitness along three environmental dimensions

interactions are more complex than simple models of competition would suggest. Interactions between herbivory, predation, parasitism, direct, indirect and apparent competition are complex and have influences on determining niche space and ability to maintain membership in communities across space and time that are difficult to predict. This does not invalidate the logic of the model, nor does it necessarily indicate that new mechanisms beyond the niche concept and limited membership need be sought. It does suggest, however, that temporal and spatial dynamics of the environment, autogenic population processes driven by predators and parasites,

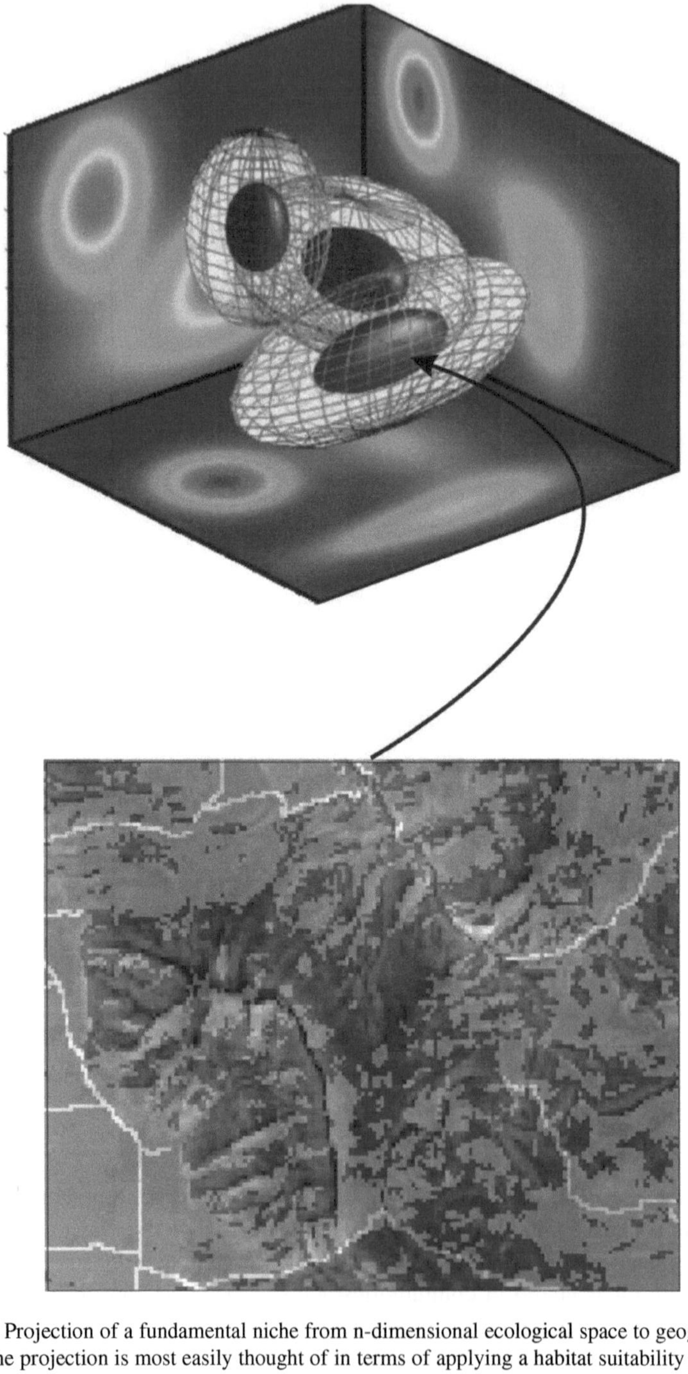

Fig. 2.2 Projection of a fundamental niche from n-dimensional ecological space to geographical space. The projection is most easily thought of in terms of applying a habitat suitability model to each location in a landscape in which each niche axis is a predictor variable. The outcome is a map of niche-suitability for the species of interest across geographical space

and their interactions, may create a spatially dynamic, non-equilibrium reality where the predictions of classical models are seldom if ever achieved.

2.1.2 Assumptions of the Competition–Equilibrium Theory

Hutchinson (1959) identified competition as the predominant process limiting species diversity and distribution, and hypothesized that it is what creates and maintains community structure. In the 1960s and 1970s many ecologists sought to apply this competition–equilibrium theory to explain community structure (e.g. Schoener 1974; Slatkin 1974; Armstrong and McGehee 1976). The theory predicted that if sympatric species were too similar then the less competitive species will go extinct according to the Lotka–Volterra model, and over time a stable equilibrium is achieved where the remaining species in the community each limit their own expansion through intraspecific competition more than they limit the population growth rates of other species (Hubbell and Foster 1986). It is hypothesized that coevolution of the member species over long periods leads to finer niche partitioning which allows for increased community diversity. Diversity is explained as the result of resource partitioning among species whose needs do not overlap completely. The concept of limiting similarity is evoked whenever two similar species coexist. As a net result, it was believed, coevolution, in conjunction with repeated invasions of "exotic" species that remixed the community yielded communities in which the theoretical limits of similarity were approximately achieved. This would give communities very predictable properties (Slatkin 1974; Paine 1984; Chesson and Case 1986).

The theory was formalized into a number of mathematical models which predicted the outcome of species interactions in model communities (MacArthur and Levins 1967; Cody 1974; Schoener 1974; Slatkin 1974). These models of community structure were tremendously influential in ecological theory in the 1970s; however, they make a number of limiting assumptions about species interactions which raise questions about their power to predict real world processes. As mentioned above (1) populations and communities are assumed to be at equilibrium set by resource scarcity; (2) competition for resources and selective pressure on resource-exploiting attributes is continuous and intense; (3) this competition is the major selective force determining the distribution of species (Wiens 1977).

2.2 Objections to the Strict Theory of Competition–Equilibrium

The validity of the theory's assumptions depends on several key conditions, including that physical and biotic interactions are of even frequency and intensity throughout the community and that historical effects, temporal trends, stochastic factors, spatial complexity and heterogeneity, migration and occasional perturbations are insignificant (Chesson and Case 1986). Space is assumed to be homogeneous and the system is assumed to be well mixed so that dispersal can be ignored (Wiens et al. 1986). However, in any real biological community the environment is a temporally varying patchwork or system of complex spatial gradients (Chap. 5, McGarigal and Cushman 2005), resulting from the space–time intersection of underlying environmental gradients with biotic and disturbance processes. In addition, organisms may often generate spatial pattern in community structure and composition autogenically through predator–prey and other interactions even in continuous and homogeneous environments (Roughgarden 1978; Kareiva 1987). Furthermore, the strength of interspecific competition is likely to vary among taxa (Schoener 1974). It is expected to be high in sessile organisms in crowded conditions, and low in phytophagous insects, for example. And for any taxonomic group the constancy of competition as a limiting factor over space and time is likely to vary markedly.

Not surprisingly, given these observations, a controversy ensued over whether violations of assumptions were "noise" obscuring the underlying competition–equilibrium processes, or whether the model itself failed generally to account for real-world community structure. A number of key papers investigated the effects on community processes when one or more of the assumptions are violated.

2.3 Key Papers: Challenges to Equilibrium Theory

Hutchinson (1957), in a paper which united the Lotka–Volterra competition model with the n-dimensional niche concept, wrote that the only conclusion that could be drawn was that although communities appear qualitatively to be constructed as if competition were regulating their structure, there are always difficulties in proving causality even in the most well studied systems. In the following 30 years a tremendous literature accumulated in clarifying this questions, with the same points being echoed and expanded prominently by Connell (1983) and Hairston (1985)

2.3.1 Fugitive Species and the Paradox of Plankton

Skellam (1951) coined the term "fugitive species" to describe situations where a poor competitor with a high reproductive rate coexists with a good competitor which reproduces slowly. Hutchinson (1951) identified stochastic disturbances as the key to the persistence of fugitive species. Persistence, in the fugitive species model, results from an equilibrium between the rate of exclusion and the rate of recolonization of newly created disturbance patches by the poorer competitor.

An important expansion of this idea is that disturbance and other changes in circumstances may interrupt or even reverse the direction of competition before it has run its course to an equilibrium state. Proctor (1957) in a study of two species of green algae found that in permanent pools one species inevitably excluded the other. In intermitted ponds, however, the other, more drought resistant species could persist because during dry periods it had the competitive advantage. This led Hutchinson (1957) to speculate that there could be many instances where the direction of competition is never constant enough to allow the elimination of competitors. He used this logic to resolve the so called "paradox of plankton" (Hutchinson 1941, 1944, 1961).

It had long been noted that many more species of plankton coexist than can plausibly be explained by the classic competition–equilibrium theory (Chesson and Case 1986). Hutchinson speculated that the lack of equilibrium could be an explanation. Physical conditions, such as temperature, light and nutrients, are highly variably temporally and spatially in the marine environment. This variability could promote species richness of phytoplankton by repeatedly changing the direction of competitive advantage of the component species across space and through time (Hutchinson 1961).

The resolution of the paradox of plankton was that space–time variability leads to interruptions or reversals of competitive interactions, and this slows or prevents competitive exclusion. Hutchinson (1961) argued that if environmental fluctuations occur with a period roughly equal to the time for competitive exclusion then the species would coexist and a community more diverse than expected would be maintained. But if the period between environmental changes is markedly different from the time for competitive exclusion then exclusion would sill occur (Horn and MacArthur 1972).

2.3.2 Intermediate Disturbance and Predator Mediated Coexistence

A number of researchers in the 1970s used simple mathematical models to explore the potential effects of temporal nonequilibrium in the environment on competitive interactions. For example, Chesson and Warner (1981) used simple mathematical models to show that environmental variability promoted coexistence in lottery competitive systems. Similarly, Koch (1974), Leigh (1975), Schoener (1976) and Levins (1979) showed that fluctuations in the environment could lead to coexistence by interrupting or reversing competition periodically.

Caswell (1978) modeled the influence of one species of predator on competition between two prey species in a "community" of 50 cells, each of which is equally accessible to individuals of each species. Without predation, one species inevitably became extinct. However, with predation the two competitors coexisted for over 1,000 generations. Caswell called this "exploiter mediated coexistence." Exploiter mediated coexistence was also observed earlier empirically in Paine's (1966) classic experiments. Paine (1966) demonstrated the importance of a keystone predator in maintaining the diversity of an intertidal community. When the sea-star predator was removed, a process of community succession occurred in which barnacles settled and excluded other species, and in turn were themselves excluded as mussels came to dominate the site. The community was reduced from 15 to 8 species following the removal of the sea-star. This classic work provided a dramatic illustration of the interaction between quasi-deterministic processes of competition for space and the effect of predation as a disturbance event (Paine 1984). Paine and Levin (1981) noted that stochastic physical disturbances had comparable effects to intermediate levels of predation in maintaining species within the community.

These results are explained by the "intermediate disturbance hypothesis" (Connell 1978). The highest levels of diversity are expected at intermediate levels of disturbance. Predation and stochastic physical disturbances can both lead to the coexistence of competitors. If the disturbance is too rare or not intense enough local competition tends toward equilibrium and the fugitive species are eliminated. If, on the other hand, disturbance is too frequent or too intense most species are eliminated. At intermediate frequencies competition is interrupted or reversed periodically and high diversity can be maintained (Kolasa and Pickett 1991). The important factors are the rate of disruption relative to the rate of exclusion (Connell 1978).

Cushman and McGarigal (2003) evaluated the effects of landscape-scale disturbances on avian diversity in the Oregon Coast Range. Their analysis found that the highest diversity of bird species were found in landscapes with a mixture of seral stages resulting from an intermediate level of disturbance. Interestingly, they found that in landscapes with low levels of disturbance the community structure changed in a highly predictable way, characterized by the selective loss of species associated with early seral habitats. Cushman et al. (2008) found that the early seral species in this system were much more highly related to fine-scale, plot-level habitat features, and less responsive to landscape-level patterns of heterogeneity and fragmentation. This perhaps suggests that they act as fugitive species in this system, such that they have lower competitive ability in mid and late seral forests that dominate the system, but that they may be superior colonizers given that their occurrence is not constrained by landscape composition or structure, but only by occurrence of required resources at the home-range scale.

2.3.3 Lottery Model

Sale (1977) studied the causes of community structure in coral reef fish communities. He emphasized that the reef community is both space and food limited. Nonetheless, the community is very diverse, with more species than predicted by the equilibrium hypothesis. This explanation follows from the fact that fish eggs and larva are pelagic. Coexistence results from the random process of larval settlement into the vacant space left when a predator or disturbance removes an established adult competitor (Roughgarden 1986). Thus population dynamics are governed by chance colonization which is similar to a lottery.

Hubbel and Foster (1986) note that in tropical forests the diversity of tree species is extremely high, and that many of the species seem to occupy similar niches. The stochastic process of gap formation by wind throw is thought to give all sub-canopy species, so long as they are evenly distributed, an equal probability of establishment in the canopy. Thus there is a lottery process, analogous to that on the coral reef, in which all species are given approximately equal chance of success and tend to coexist for long periods even though they may occupy similar niches.

Finding examples in communities as different as coral reefs and tropical forests suggests that such a lottery process may be in action in many communities in which there are limited and stochastically developing opportunities for establishment, and equally prolific and evenly distributed propagules or juveniles of the constituent species. A number of researchers have argued that in spatially extensive systems with many species chance events of mortality and colonization may result in systems in which exclusion of inferior competitors may take extremely long periods to achieve (e.g. Diamond 1979; Case and Sidell 1983)

2.3.4 *Huffaker and Others*

Huffaker (1958) performed a classic study on population dynamics of a predator–prey system in a heterogeneous environment. Working with a predatory mite and a prey mite that feeds on oranges, he manipulated spatial heterogeneity by varying the number of oranges, rubber balls, Vaseline barriers and wooden poles in a laboratory microcosm. When the system was homogeneous the predator invariably consumed all the herbivores and both species became extinct. However, when spatial complexity was added, sustained predator–prey oscillations resulted. Including sufficient spatial heterogeneity resulted in waves of predation which were out of phase with prey abundance (Huffaker 1958). This allowed the predatory–prey interaction to stabilize and the two species coexisted.

Huffaker's results inspired a number of studies on how spatial heterogeneity affects predator–prey and competitive interactions (e.g. Hassell and May 1974; Hilborn 1975; Caswell 1978; Hassell 1978; Hastings 1978; Levin 1978; McMurtie 1978; Crowley 1979; Atkinson and Shorrocks 1981; Ives and May 1985; Kareiva 1987). Patchiness was investigated with mathematical models which showed that fundamental properties of multi-species communities can be changed whenever spatial heterogeneity is important in interspecific interactions (Kareiva 1987). Levin (1974) demonstrated that subdivision of habitat into patches and interpatch dispersal may permit coexistence of species which would otherwise exclude one another. Other models repeatedly predicted that predator–prey or host–parasite interactions are stabilized if patterns of attack are sufficiently aggregated allowing some prey patches to escape predation (Kareiva 1986; Hassell and May 1974).

One example of Huffaker's work moved to the field is Kareiva's (1986, 1987) study of the effects of spatial heterogeneity on the interactions between predatory coccinellid beetles and herbivorous aphids. Patchiness resulted in aphid densities ten times those observed in a homogeneous control plot. Patchiness reduced ladybug predator dispersal rates by 50%. This led to more aphid colonies escaping predation for long periods. The net result was that with increasing patchiness of vegetation the ladybug–aphid interaction changed so that aphid populations increased, occasionally to autocatalytic outbreak densities.

Kareiva noted that while Huffaker's theory predicts that patchiness stabilizes predatory–prey interaction, in this case it led to destabilization, as the prey species oscillated in and out of large population explosions. This suggests that the actual effects of spatial heterogeneity depend on the details of demography and foraging behavior of the species involved in interaction with the scale and pattern of heterogeneity (Kareiva 1986). Kareiva also notes that insect populations rarely reach levels at which competition plays a major role. The effects of unpredictable fluctuations of environmental conditions predominate and there is little evidence of density dependent mortality (Kareiva 1986; Kolasa and Pickett 1991). Thus, this taxonomic group does not follow Lotka–Volterra community theory (Kareiva 1986).

2.4 A Non-equilibrium, Spatially Dependent World

It is clear that most of the assumptions of the "equilibrium–competition" theory do not hold in real biological communities. It is also clear that violations of the assumptions are not noise obscuring "true" underlying equilibrium processes. Spatial and temporal heterogeneity, historical effects and stochastic factors are key causal agents creating and maintaining community structure (Huston 1979; Kareiva 1986). Disturbance is so universal in natural communities that it is likely that truly equilibrium situations almost ever occur in nature (Begon et al. 1990). Even in the absence of clear disturbances and perturbations, spatial complexity in the environment and fluctuations in conditions overtime both alter population processes fundamentally in ways inconsistent with equilibrium–competition theory. Two questions thus arise: (1) does variability in space and time make the properties of communities unpredictable? (2) If no equilibrium point is achieved, will community structure be dominated by chance and historical effects? (Chesson and Case 1986). The answer is yes, on both points, and no.

2.4.1 Structure of Non-equilibrium Communities

Communities are not quasi-organisms. They are the collection of organisms of all species coexisting within some set of spatial bounds defined by the observer. Each one of these organisms has met the requirements for its survival. These are the dimensions of its fundamental niche. Assuming that an organism has geographic access to the region, its persistence at that location first depends on whether its fundamental requirements are available, and second whether other species control them or change them such that they no longer meet the needs of the organism. We believe the niche concept and interspecific competition are the foundation of any explanation of community structure, even though in the real world equilibrium are seldom if ever achieved.

It is futile to try to develop a general theory of communities, as communities are simply the collection of organisms living in one place at one time. To understand the “structure” of a “community” requires that you understand (1) the fundamental requirements of each species, (2) how these required conditions vary spatially, (3) how they vary temporally with directional environmental change, (4) how stochastic events are likely to affect them, (5) how their availability is influenced by the actions of other species and (6) how these influences vary spatially and temporally.

The individualistic concept of the community notwithstanding, we can make generalizations about some responses of species to common processes. When environmental conditions vary in space and time, many species may coexist on a single resource (Chesson 1985). This requires that fluctuations also occur in the competitive ranking of the species. Populations at one location may experience fluctuations which are out of phase with other adjacent locations (Chesson 1983, 1985; Comins and Noble 1987; Atkinson and Shorrocks 1981). Fluctuations in competitive ranking can occur in two ways. The first is differential variation of migration rates among species into particular patches, leading to flip-flops in numerical advantage within patches over time. The other is that changes in local environments may reverse the competitive advantage within a given patch (Chesson and Case 1986).

Coexistence in such a patchy environment depends on the geometry of he landscape and the relative dispersal abilities and habitat specific mortality rates of the interacting species. If there is too much movement among patches then the system behaves like one patch. If there is insufficient dispersal, populations will disappear from patches more rapidly than recolonization can occur (Kareiva 1987; Pulliam et al. 1992). Dispersal is always a key parameter, as it determines the coupling between patches, the degree of mixing in the environment and the ability of predators to follow prey populations (Kareiva 1989; Cushman 2006).

Nearly identical species will not coexist in an unvarying environment. In a spatially homogeneous or temporally unvarying environment, two species whose niches overlap will compete and the superior competitor will exclude the inferior one from the portion of its fundamental niche which overlaps. In a varying environment, the same process will always be in operation. However, the heterogeneity and temporal variability of the real world will often prevent equilibrium from occurring. At any moment, at any location, one species will have the advantage over the other in competition for their common resource. But, conditions may well change and shift the competitive balance before exclusion occurs. Alternatively, spatial heterogeneity in the pattern of competitive ability will enhance persistence. The combination of spatial heterogeneity of competitive advantage coupled with temporal fluctuations in this advantage across space are probably the major underlying factors maintaining biological diversity greater than predicted by classic equilibrium theory.

Also, abiotic or biotic factors, such as disturbance or predation, may reduce the numbers of the superior competitor, or of both species, below the point where interspecific competition is limiting. After that, both species will grow for a time without competing with each other (Hanski 1981). The metapopulation process of local "patch" extinction due to environmental factors, predation or chance may permit competing species to colonize and grow for a time without interspecific competition. So long as the spatial pattern and temporal frequency of these "extinctions" match the demographic and dispersal abilities of the fugitive species it will persist indefinitely even though it is always a poorer competitor.

Whereas it was once thought that environmental variability acted primarily to eliminate species, we now know that variability in may instances may act to promote diversity. Stochastic environments can convert competitive exclusion into competitive coexistence (Kareiva 1989). The key is the combination of a fluctuating, spatially variable environment and the possession of certain life-history traits (Chesson and Case 1986). This is an explanation of the coexistence of strongly competing species from different taxonomic groups (i.e. harvester ants, finches and rodents). Differences in life history, physiology and behavior in these groups lead to different responses to environmental pattern and change. Thus, a fluctuating spatially complex environment can promote coexistence.

A biological community is defined as a collection of organisms which all maintain niche space within a given area. Their effectiveness at holding this "space" at a given time is a measure of the Darwinian fitness of individuals of the species. Individuals which hold space less effectively relative to their con-specific conspecific competitors will produce fewer offspring, and a species which in general fails to hold space will be excluded by other species, and will become locally extinct. If conditions fluctuate around a stable mean value, then niche space of a species will expand and contract in phase with the fluctuation. While the space for one species expands due to a given fluctuation, that of a competitor may be contracting. Meanwhile, in an adjacent patch environmental conditions may be fluctuating in a different manner, leading to expansions and contractions that are spatially out of phase. This combination of spatial and temporal environmental variability promote coexistence by interrupting or redirecting the direction of competition.

However, when the mean value of conditions change over time and space the situation is different. Real world environments don't often fluctuate around a stable environmental mean. Slow, directional changes are constantly occurring in physical conditions due to climate change and geomorphic processes (Hubbel and Foster 1986). If the changes are sufficiently slow, then most organisms in the community can track them equally well through evolutionary change. However, if some organisms track better than others, community structure will change. Species which track poorly will experience reduced access to their fundamental niche, or their competitive ability to hold it relative to other community members will decrease. Since the relative abundances of the species present in a community are constantly adjusting to new conditions, but never complete the adjustment before conditions change, past abundances of species remain relevant. History plays an important role in the present abundances and composition of organisms in real communities (Chesson and Case 1986).

As conditions change continually, organisms are constantly decaying with respect to current conditions. The fitness of all organisms is expected to decrease with changes. But some species may be impacted more than others. Some species are more tolerant of change than others due to physiological or other ecological characteristics. For example some species plastic than others, and can behaviorally differentiate their niche space in response to changes. Other species can track environmental change more effectively through evolutionary change. The real world is a Red Queen world. Species in a community always "run" to keep up with environmental and competitive change. Species are affected individualistically, but not independently. Community structure reflects the moment to moment fluctuation of the niche space of individual species about a drifting mean value that is punctuated by stochastic disturbances.

Community structure at any moment is a snap-shot of this process. Organisms do not anticipate the future, and prepare for coming changes. Nor are they the best they can be given current conditions. They never have time to fully adjust. They were simply good enough individually in the past to survive, reproduce and maintain membership in the pool of species persisting at this place at this time. At present, under current conditions, a species may be at equilibrium, expanding relative to competitors, or declining toward extinction. We can say little about their fitness, or the composition of the community, in the future unless we know how conditions will change over time and space and how each organism will respond.

2.5 Irony of the Red Queen

We can now justify the answers to the two questions posed above. The questions were: (1) does variability in space and time make the properties of communties unpredictable, and (2) if no equilibrium point exists, will community structure be dominated by chance and historical effects? We are clearly dealing with a vastly more complicated world than classical ecological theory assumes or is equipped to deal with. The answers are clearly yes. Community structures will always be unpredictable to a degree because individual species responses will be determined in part by unpredictable, chance events such as disturbance, colonization and predation. These chance events undoubtedly play a much larger role in community structure than has been previously believed (Kareiva 1989). The answer is also yes because of spatial complexity. Even holding the environment constant, spatial complexity of the environment can lead to apparently chaotic and unpredictable population dynamics of multi-species communities (e.g. Huffaker 1958; Kareiva 1987). Real environments vary dramatically at multiple spatial scales, and the challenge of sufficiently representing the spatial pattern of key limiting resources for multiple interaction species across spatial extents relevant for population processes is enormously daunting. Even should we be able to measure and quantify such patterns, linking them through mechanisms and responses to population dynamics is exceptionally difficult. The challenge is even greater when one considers the interaction of variability in space with variability in time. We have argued that variability in either of these fundamentally alters outcomes in biological communities from the predictions of classical theory, and in many cases may make changes exceptionally difficult to predict and thus make current patterns exceptionally difficult to explain based on observed processes. In real biological communities, however, these two act in concert. Spatial patterns vary continuously at multiple scales, and temporal fluctuations occur across these complex landscapes continuously and at multiple scales. We believe the fundamental four-dimensional complexity of nature makes the future at any given location in space in large part unpredictable.

But in another sense the answer is no. Given any event, be it physical disturbance, directional environmental change, predation or colonization, the response of the community is the sum of the responses of the individuals of the individual species within it. These responses will be determined by the ways the events affect the resources and conditions within each species' fundamental niche, and how they affect their ability to maintain access to that niche space in the face of interactions with other species. Thus, if an event is known and its effects on each species across space are known, hen the response of the "community" can be determined. But, as conditions are changing continually through space and time, each moment and each location will yield different predictions about the future. With infinite knowledge of each location at all moments, and infinite knowledge about how these conditions will affect the interactions of individual organisms, it is in principle possible to predict future outcomes. This "in principle" possibility has been hopefully termed "Consilience" by Wilson (1998).

This digression has had a purpose. It takes us back to the thesis we laid out at the beginning of the chapter. The classical theories of ecology are based on equilibrium and non-spatial idealizations of systems and processes. The classical theory of community structure predicts that, given an initial set of potential members, the composition and structure of the community will tend toward equilibrium. A species excluded from all of its fundamental niche will be excluded from the community. Overtime, the community is expected to stabilize at a sort of "competitive climax". This is precisely what may be occurring everywhere, at every moment.

At any moment, within any unit of space, community structure will tend toward and equilibrium. But, it will rarely if ever get there. Chance disruptions of that trend will set it off on another course. Intermittent reversals of competitive advantage and transient most limiting factors will prevent progress toward equilibrium. Spatial variability in the environment will interact with these temporal fluctuations in complex ways. At any moment in any unit of space the observed trajectory of change is completely predictable, in principle. The classical theory is correct entirely in an ideal world, but almost always fails in real life. We believe we know the dominant mechanisms and processes. No new theory of cause and effect need be invoked. But we will always fail to understand the causes of current community structure and fail to predict future conditions because we can never know enough about how relationships in complex systems change over time. The mystery is dead; long live the mystery.

Is this a message of despair, or a message of inspiration? The reality of nature is marvelous and mysterious. We believe this abiding mystery is a source of inspiration.

Not in a mystical sense of worshiping an unknowable mystery, but in the sense of the challenge to understand, the mystery to be explored. There are several main messages that this chapter has been intended to evoke. First, idealized models often effectively describe the nature of important processes, but fail basically to predict how these processes act the context of a spatially and temporally complex world. Second, given the failure of such models, future progress in ecology we believe will focus on particulate and contextual analysis. As we discussed at the beginning of the chapter, we believe analysis will have to "step-into" spatial and temporal complexity, by applying process models to particulate reality, rather than globally. By particulate, we mean the interactions of particular individual organisms with the details of their particular environment. The second part of this "stepping-into spatial and temporal complexity is by conducting analysis contextually, by studying the interactions of particular organisms, in particular locations, at particular times. Third, in order for particulate and contextual analysis to be informative it must also be integrative, in a sense similar to the calculus. In calculus integration is the process of predicting the integrated response of continuous differential functions. In a similar way, in this view of ecology, integration will be a "stitching together" of a vast number of particulate and contextual pattern–process relationships across space and time. Many of these particulate relationships will act at multiple spatial scales, some of which will be continuous at fine spatial scales. Thus, conceptually this integration is much like calculus. The difference is that in calculus one is integrating an equation and seeking a closed form solution.

In ecology this will rarely be possible. In ecology, we will be integrating space–time processes continuously across space and time, where response and dependent variables both change both in value and in kind over space and time, and pattern–process relationships may change fundamentally as response and dependent variables change. One of the primary reasons for the popularity and influence of many classical non-dynamic theories in ecology is that they do provide tidy mathematical solutions. These tidy solutions may apply at fine spatial scales over short time periods in a spatially and temporally complex world. The challenge is integrating these across fundamentally nonstationary spatial and temporal domains. Thus, the fourth message we wish to evoke is the difficulty of achieving global prediction from closed-form equations. If ecological reality is particulate and contextual, and therefore nonstationary, it will generally be impossible to derive closed-form theory to make predictions across complex systems.

This therefore raises two critical themes which are twin backbones of this book. The first is the central and critical need for high quality *data* over *space* and through *time* on *environmental* conditions and *species* distribution, abundance, fitness, and behavior. If ideal theory will not provide reliable predictions, then we will have to rely on data to a much larger degree. In classical ecology, theory was supreme and data was secondary, in a sense, called upon mainly to parameterize and validate theory. Now, if integrating theory across particulate and contextual non-stationary space–time domains is found to be intractable, then data will be primary, with theory being applied locally and temporally, but relying explicitly on particulate and contextual data to be useful.

The second backbone is the ascending role of computational simulation. In the absence of closed-form solutions to theoretical models across particulate and contextual environments, simulation will be the primary tool to investigate complex ecological systems. Simulation does not require closed-form, global solutions or stationarity of space–time processes. The rise of powerful computing environments enables researchers to directly investigate the interaction of spatial complexity, temporal variability and population processes in particulate and in context. By coupling extensive spatial and temporal data collection to powerful spatial modeling environments it will be possible to integrate local and ephemeral dynamics across space and time (Cushman 2006).

Detailed, quality data gathered continuously through space and time and powerful, spatio-temporal simulation modeling are not alternative solutions to the problem of complexity in ecology. Rather they are complementary and mutually interdependent (Fig. 2.1). At the beginning of the chapter we asserted models without data are not compelling, while data without models are not informative. We can now further explain our meaning. Detailed, quality data on environmental conditions, and species distribution, abundance, and fitness can be used in statistical modeling to infer process from patterns. This is the meaning of the top left–right arrow in Fig. 2.1, namely that detailed spatial and temporal data on ecological systems is necessary to provide material to infer processes inductively from the data. The data alone, however, are not informative regarding process. Data alone is mere pattern. Statistical modeling seeks to link these patterns to potential explanatory processes (through the bottom left-right arrow). However, regardless of whether a refutationist, inferential analytical approach or a confirmationist, likelihood/Bayesian approach is used to infer this pattern–process linkage, there is an inevitable logical inconclusiveness. It is known as affirming the consequent through induction. The analysis evaluates the consistency of the data with one or many alternative explanations, or the likelihood of several explanations given the data. In either case, observing consistency of the pattern with an inferred process does not in any sense prove that causation. In the inferential case, the best we can do is fail to reject the hypothesis, which logically is entirely different than proving it. In the confirmationist case, the best we can do is rank the relative support of the alternative explanations, which inevitably are a vanishingly small fraction of possible alternatives. In either case asserting a proof of causality from an empirical correlation is an example of affirming the consequent through induction, in that any number of untested alternatives may exist, leading to a large probability of failing to evaluate the correct alternative in the confirmationist approach and the impossibility of rejecting all reasonable alternative hypotheses in the refutationist approach.

Conversely, through deduction we can go the other way, from process to pattern, from mechanism to response. This is the bottom right–left arrow in Fig. 2.1. In the context of this discussion, this is done through space–time simulation of ecological processes, given our expectation that ideal closed-form mathematical models will not apply to non-equilibrium and nonstationary ecological systems. In this effort, we design a space–time system consisting of entities located in space, space characterized by one or several driving variables, and time. The simulation proposes a space–time process, a set of rules which govern the interaction of entities with each other as functions of distribution and density of entities of different kinds, their behavior, the structure of the environment and how these change over time. The simulation then allows the "actualization" of the proposed process through the execution of the process in particulate and contextually within the "world" represented by the simulation. We asserted that models without data are not compelling. Without realistic data on patterns and processes to develop a simulation model, it is extremely unlikely that it its results will have much kinship with reality. Further, without reliable data of the appropriate variables at appropriate scales in space and time it is not possible to evaluate the degree to which model results conform to actual patterns in the real world. For a model to be compelling it must simulate a reasonably meaningful system and produce results reasonably consistent with observations in the real world. However, as in the case of inferring process from pattern, there is an unavoidable logical inconclusiveness in using process to recreate observed patterns. It is called affirming the consequent through deduction. Proposing a process governing a system, simulating that process, and observing the results are consistent with the observed pattern does not prove that causation. The model evaluated a single or perhaps several alternative processes. Observing that one of these recreated a response pattern that is consistent with an empirical observation does not materially lessen the likelihood that there could be any number of other processes that could also recreate the pattern, leading to a large probability of failing to model the correct process.

We have argued that the particulate, contextual and nonstationary nature of ecological systems strongly invokes the need for quality spatial and temporal ecological data collection, and sophisticated space–time simulation approaches. However, it would appear that in the previous two paragraphs we have torn down inference from data and deduction from process as viable routes to reliable knowledge. Theories are always underdetermined by facts; facts never fully confirm a theory. Here is our main point regarding the interdependent role of inference from data and deduction from process. While it is never possible to avoid the logical risk of affirming the consequent in science, the combination of inference from data directly with deduction from process is a potentially powerful framework to greatly lessen the risk. Through the collection of large spatial and temporal samples of appropriate ecological data it is possible through both refutationist and confirmationist methods to identify plausible explanatory processes. In and of themselves, these are relatively weak logically due to affirming the consequent through induction, as discussed above. However, this hypothesized explanatory process then can be used as a starting point for simulation, in which the process is assumed and the model then recreates the pattern the data would have if the process were correct. As discussed above, this in and of itself is logically weak due to affirming the consequent through deduction. However, if the simulation based on an inferred process is able to recreate the pattern seen in the original data, there is much less likelihood that the explanation is an error of affirming the consequent.

Specifically, affirming the consequent through induction and affirming the consequent through deduction are independent errors, and therefore the chance of both errors occurring together, simultaneously with regard to a single pattern–process hypothesis is geometrically less than either alone. Thus, regarding induction from data and deduction from theory, both are necessary for reliable knowledge.

Thus, to summarize, we believe that spatial and temporal variability in ecological systems are not noise clouding ideal pattern–process relationships, but rather that they are fundamentally important to the processes themselves. Therefore, understanding ecological systems will require a reconceptualization of pattern–process relationships to focus on particulate interactions of organisms with their environments in the context of spatial location and temporal moment. This simultaneously will require a paradigm shift in the three areas of method, data and theory. First, we believe ideal theories will seldom be able to provide reliable predictions for large systems, and that pattern–process relationships will collapse down to local and ephemeral dynamics in space and time. Second, to address this will require development of new methodology. The recent explosion of computational technology, remote sensing, GIS, spatial statistics and space–time modeling have enabled rapid development of key components of this new methodology. Third, these advances have enabled vast improvements in our ability to collect, store, integrate and analyze large, fine-scaled spatial and temporal data on environmental patterns and organisms distribution, abundance and performance (Fig. 2.3). Finally, this feedback between method, data, and theory enables us to be hopeful that this a message of inspiration rather than despair; that this positive feedback between explosive progress in computing, computational simulation, remote sensing and other methodological advances, the extent, frequency and quality of data being collected and our ability to represent explanatory processes will enable rapid progress. We believe this progress will be built on a conception of logical linkage between inferring process from data, and then recreating data patterns from process, and involve the application of ecological theory in particulate and contextual context, at the level of the interactions of individual organisms with each other and their physical environment, integrated across heterogeneous space and over temporally dynamic time.

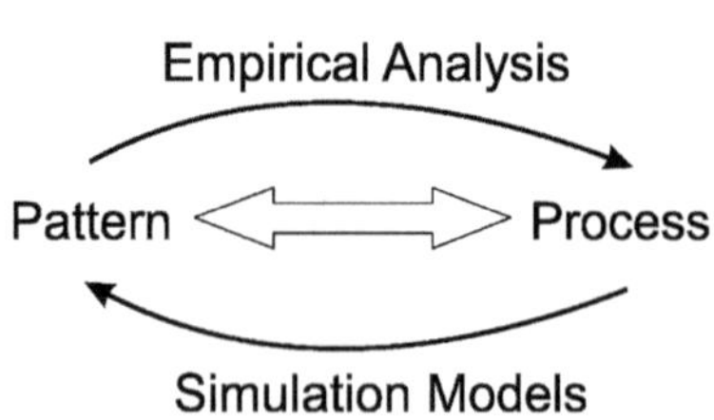

Fig. 2.3 Feedback between theoretical models and empirical data collection and analysis. Empirical analysis is used to infer process from patterns observed in data. Theoretical and process models are used to evaluate the ability of the inferred process to create the observed pattern. The feedback between these two enables much stronger inferences about pattern–process relationships than either alone and markedly reduces the severity of risk of affirming the consequent

References

Armstrong RA, McGehee R (1976) Coexistence of species competing for shared resources. Theor Popul Biol 9:317–328

Atkinson WD, Shorrocks B (1981) Competition on a divided and ephemeral resource: a simulation models. J Anim Ecol 50:461–471

Begon M, Harper JL, Townsend CR (1990) Ecology: individuals, populations and communities. Blackwell, Oxford

Case TJ, Sidell R (1983) Pattern and chance in the structure of model and natural communities. Evolution 37:832–849

Caswell H (1978) Predator mediated coexistend: a nonequilibrium model. Am Nat 112:127–154

Chesson P (1983) Coexistence of competitiors in a stochastic environment: the storage effect. In: Freedman, H, C. Strobeck (eds) Population biology. Springer, New York

Chesson P (1985) Coexistence of competitors in spatially and temporally varying environments: a look at the combined effects of different sorts of variability. Theor Popul Biol 28:263–287

Chesson P, Case T (1986) Environmental variation and the coexistence of species. In: Diamond J, Case T (eds) Community ecology. Harper & Row, New York

Chesson P, Warner R (1981) Environmental variability promotes coexistence in lottery competitive systems. Am Nat 117:923–943

Clements FE (1936) Nature and structure of the climax. J Ecol 24:252–284

Cody ML (1974) Competition and the structure of bird communities. Princeton University Press, Princeton, NJ

Comins HN, Noble IR (1987) Dispersal, variability and transient niches: species coexistence in a uniformly variable environment. Am Nat 126:706–723

Connell JH (1978) Diversity in tropical rain forests and coral reefs. Science 199:1302–1310

Connell JH (1983) On the prevalence and relative importance of interspecific competition: evidence from field experiments. Am Nat 122:661–696

Crowley P (1979) Predator-mediated coexistence: one equilibrium interpretation. J Theor Biol 80:29–144

Cushman SA (2006) Effects of habitat loss and fragmentation on amphibians: a review and prospectus. Biol Conservat 128:321–240

Cushman SA, McKelvey K, Flather C, McGarigal K (2008) Testing the use of forest communities to evaluate biological diversity. Frontiers in Ecology and the Environment 6:13–17.

Cushman SA, McGarigal K (2003) Landscape-level patterns of avian diversity in the Oregon Coast Range. Ecol Monogr 73:259–281.

Diamond JM (1979) Community structure: is it random, or is it shaped by species differences and competition. In: Anderson R, Turner B, Taylor L (eds) Population dynamics. Blackwell, Oxford

Elton C (1927) The ecology of animals. Reprinted 1966 by Science Paperbacks and Methuen, London

Gause GF (1934) The struggle for existence. Hafner, New York.

Gause GF (1935) La théorie mathématique de la lutte pour la vie. Harmann, Paris

Gleason H (1926) The individualistic concept of the plant association. Bull Torrey Bot Club 53:1–20

Hairston NG (1985) The interpretation of experiments on interspecific competition. Am Nat 125:321–325

Hanski I (1981) Coexistence of competitors in patchy environments. Ecology 64:493–500

Hassell MP (1978) The dynamics of arthropod predator-prey systems. Princeton University Press, Princeton, NJ

Hassell M, May R (1974) Aggregation in predators and insect parasites and its effects on stability. J Anim Ecol 43:567–594

Hastings A (1978) Spatial heterogeneity and the stability of predator-prey systems: predator mediated coexistence. Theor Popul Biol 14:380–395

Hilborn R (1975) The effect of spatial heterogeneity on the persistence of predator-prey interactions. Theor Popul Biol 8:346–355
Horn HS, MacArthur R (1972) Competition among fugitive species in a harlequin environment. Ecology 53:749–752
Hubbell SP, Foster RB (1986) Canopy gaps and the dynamics of neotropical forests. In: Crawley MJ (ed) Plant ecology. Blackwell, Oxford
Huffaker C (1958) Experimental studies on predation: dispersion factors of predator-prey oscillations. Hilgardia 27:343–355
Huston M (1979) A general hypothesis of species diversity. Am Nat 113:81–101
Hutchinson GE (1941) Ecological aspects of succession in natural populations. Am Nat 75:406–418
Hutchinson GE (1944) Limnological studies in Connecticut. Ecology 25:3–26
Hutchinson GE (1951) Copepodology for the ornithologist. Ecology 32:571–577
Hutchinson GE (1957) Concluding remarks. Cold Springs Harbor Symp Quantitative Biol 22:415–427
Hutchinson GE (1959) Homage to Santa Rosalia, or why are there so many kinds of animals. Am Nat 93:137–145
Hutchinson GE (1961) The paradox of the plankton. Am Nat 93:145–159
Ives AR, May RM (1985) Competition within and between species in a patchy environment: relations between microscopic and macroscopic models. J Theor Biol 115:65–92
Kareiva P (1987) Habitat fragmentation and the stability of predator-prey interactions. Nature 321:388–391
Kareiva P (1989) Renewing the dialog between theory and experiments in population ecology. In: Roughgarden J, May RM (eds) Perspectives in ecological theory. Princeton University Press, Princeton, NJ
Koch AL (1974) Coexistence resulting from an alternation of density independent and density dependent growth. J Theor Biol 44: 373–386
Kolasa J, Pickett STA (1991) Ecological heterogeneity. Springer, New York
Leigh EG (1975) Population fluctuations, community stability and environmental variability. In: Cody M, Diamond J (eds) Ecology and evolution of communities. Harvard University Press, Cambridge
Levins R (1979) Coexistence in a variable environment. Am Nat 114:765–783
Levin S (1974) Dispersion and population interactions. Am Nat 108:207–228
Levin S (1978) Population models and community structure in heterogeneous environments. In: Levin S (ed) Studies in mathematical biology, part 2, American Mathematical Association, Washington, DC
Levin S (1989) Challenges in the development of a theory of community and ecosystem structure and function. In: Roughgarden J, May RM (eds) Perspectives in ecological theory. Princeton University Press, Princeton, NJ
Lotka AJ (1932) Elements of physical biology. Wiliams & Wilkins, Baltimore, MD
MacArthur RH, Levins R (1967) The limiting similarity, convergence and divergence of coexisting species. Am Nat 101:377–385
McGarigal K, Cushman SA (2005) The gradient concept of landscape structure. In: Wiens J, Moss M (eds) Issues and perspectives in landscape ecology. Cambridge University Press, Cambridge
McGarigal K, Tagil S, Cushman SA (2009) Surface metrics: an alternative to patch metrics for the quantification of landscape structure. Landscape Ecol 24:433–450
McMurtie R (1978) Persistence and stability of single-species and prey-predator systems in spatially heterogeneous environments. Math Biosc 39:11–51
Paine RT (1966) Food web complexity and species diversity. Am Nat 100:65–75
Paine RT (1984) Ecological determinism and the competition for space. The first MacArthur lecture. Ecology 65: 339–1348
Paine RT, Levin SA (1981) Intertidal landscapes: disturbance and the dynamics of pattern. Ecol Monogr 51:145–178

Proctor VW (1957) Studies on algal antibiosis using haematococcus and chlamydomonas, Limnol Oceanogr 2:23–139
Pulliam HR, Dunning JB, Liu J (1992) Population dynamics in complex landscapes: a case study. Ecol Appl 2:165–77
Roughgarden J (1978) Influence of competition on patchiness in a random environment. Theor Popul Biol 14:185–203
Roughgarden J (1986) A comparison of food-limited and space-limited animal competition communities. In: Roughgarden J, May RM (eds) Perspectives in ecological theory. Princeton University Press, Princeton, NJ
Roughgarden J, Diamond J (1986) Overview: the role of species interactions in community ecology. In: Diamond J, Case TJ (eds) Community ecology. Harper & Row, New York
Roughgarden J, May RM (1989) Perspectives in ecological theory. Princeton University Press, Princeton, NJ
Sale P (1977) Maintenance of high diversity in coral reef fish communities. Am Nat 111:337:359
Schoener T (1974) Resource partitioning in ecological communities. Science 185:27–39
Schoener T (1976) Alternatives to Lotka-Volterra competition: models of intermediate complexity. Theor Popul Biol 10:309–333
Skellam JG (1951) Random dispersal in theoretical populations. Biometrika 38:196–218
Slatkin M (1974) Competition and regional coexistence. Ecology 55:128–134
Volterra V (1926) Variazioni e fluttuazioni del numero d'individui in specie animali conviventi. Mem R Accad Naz dei Lincei Ser VI 2
Whittaker RH (1956) Vegetation of the great smoky mountains. Can J Forest Res 26:1–80
Wiens JA (1977) On competition and variable environments. Am Sci 65:590–597
Wiens JA, Addicott JF, Case TJ, Diamond J (1986) Overview: the importance of spatial and temporal scale in ecological investigations. In: Diamond J, Case TJ (eds) Community ecology. Harper & Row, New York
Wilson EO (1998) Consilience. Vintage, New York

Chapter 3
The Problem of Ecological Scaling in Spatially Complex, Nonequilibrium Ecological Systems

Samuel A. Cushman, Jeremy Littell, and Kevin McGarigal

In the previous chapter we reviewed the challenges posed by spatial complexity and temporal disequilibrium to efforts to understand and predict the structure and dynamics of ecological systems. The central theme was that spatial variability in the environment and population processes fundamentally alters the interactions between species and their environments, largely invalidating the predictions of ideal models of community structure and population processes. In addition, we argued that temporal variability enormously amplifies the challenge of prediction, by altering and reversing species–species and species–environment relationships over time. Typically these fluctuations do not occur globally across space in synchrony; rather change in time is spatially dependent on location in the environment, and thus interacts in highly complex and nonlinear ways with spatial heterogeneity in influencing ecological processes. Given these challenges, we proposed focusing on the interactions between species and their immediate environments in the context of current and past conditions. However, given critical sensitivity of ecological processes to spatial and temporal factors, it is also necessary to consider their action within the context of a broader landscape of conditions, constraints and drivers. This therefore seems a catch-22, with fine-scale understanding of process required at the scale where ecological entities (e.g. organisms) directly interact with each other and their environments, and also integration of these fine-scale processes across complex and temporally varying broad-scale environments. This challenge fundamentally relates to scale and scaling ecological processes.

One of the challenges that oft plagues efforts to discuss and integrate concepts related to multiple-scale analysis is inconsistency and contradiction of terminology

S.A. Cushman (✉)
US Forest Service, Rocky Mountain Research Station, 800 E Beckwith, Missoula, MT 59801, USA
e-mail: scushman@fs.fed.us

J. Littell
Research Scientist, CSES Climate Impacts Group, Box 355672, University of Washington, Seattle, WA 98195-5672, USA

K. McGarigal
Department of Natural Resources Conservation, University of Massachusetts, Amherst, MA, USA

S.A. Cushman and F. Huettmann (eds.), *Spatial Complexity, Informatics, and Wildlife Conservation*
DOI 10.1007/978-4-431-87771-4_3,

and the resulting confusion of meanings (Allen and Hoekstra 1992; Schneider 1994; Peterson and Parker 1998). In this chapter we discuss concepts of scale and concepts of hierarchy, and how they are related in the context of ecosystems within landscapes. In our usage, scale is a metric spatial or temporal attribute. It is a continuous property of observation defined by units of length, area, volume or time. For our purposes we focus on two key attributes of scale, Grain and Extent. Grain refers to the resolution of the data in terms of the smallest increment of measurement. Temporal grain can be measured in seconds, minutes, hours, etc. Spatial grain can be measured in length (m), area (m^2), or volume (m^3). The grain defines the finest resolution of measurement and thus the finest resolution of patterns that can be described in the data. The extent refers to the temporal or spatial span of the data. Temporal extent refers to the duration of time at which sampling occurs. Spatial extent refers to the length, area, or volume over which sampling occurs. Extent defines the domain in space or time for which data are available and thus the scope of the inferences that can be drawn from them.

Scale refers to a continuous property measured in common units, and therefore is not a discrete property defined by levels in the ecological hierarchy. There is no constant plot scale, patch scale, or landscape scale. Plots, patches or landscapes may be represented in different scales, but they are not scales themselves. Often researchers are interested in relationships among things, such as organisms, populations, communities, stands, and landscapes. These things can be arranged in hierarchies to organize thinking and facilitate analyses. In our usage we discuss hierarchical levels of system organization based on forest ecosystem organization. The levels we include are (1) organism, (2) stand, (3) landscape, (4) eco-region, (5) biome, (6) planet. These categories were chosen because they reflect the major traditional way of conceptualizing forest ecosystems (Turner et al. 2001). This organizational hierarchy is one of convenience and provides a reference framework for discussing entities and relationships among them. One beneficial characteristic of this organizational hierarchy is its nestedness. Conceptually, stands are composed of organisms, landscapes are composed of stands, ecoregions are composed of landscapes, etc. A key attribute of organizational levels, as opposed to scale, is that as organizational level changes, the thing itself is changing – stands are not biomes. However, moving from organism to stand, the entities, processes and structures of interest change; the variables measured change; the required scale of measurement changes both in grain and extent. In this chapter, we recognize that the size of landsacpes varies with the organism(s) of interest, so we use the term "landscape scale" generally – and consistently – to refer to the scales of mechanism and response that are required to understand landscape pattern and evolution, particularly with regard to organisms and their environments.

The fact that measurements change across scale is a challenge to ecology. It is difficult to translate relationships from one scale to another. A much greater challenge, however, is translating relationships from one organizational level to another. When one only changes scale, the thing being measured is unchanged. All that is changed is the grain and/or extent over which it is measured. In contrast, when one changes levels, one simultaneously changes scale, but also changes the apparent entities being measured and the apparent processes governing their behavior (Fig. 3.1).

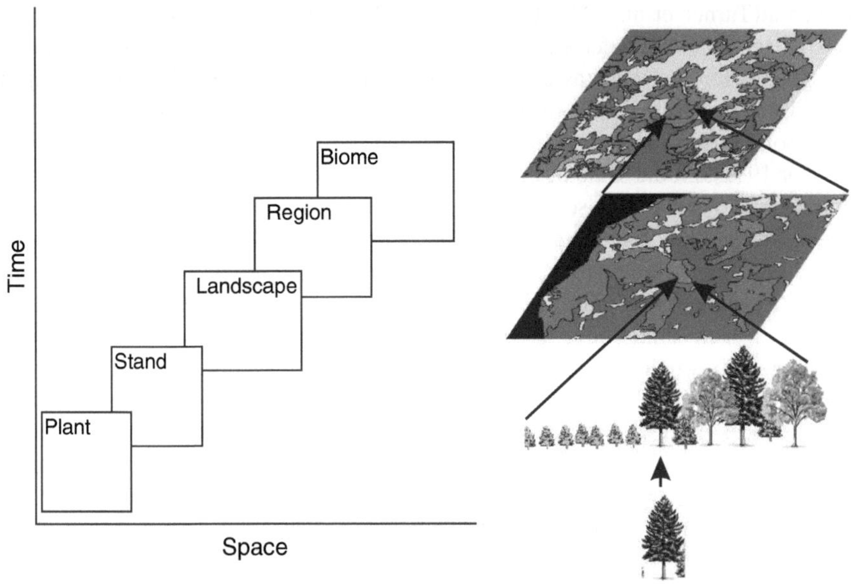

Fig. 3.1 Traditional hierarchical model of ecosystem organization. The system is characterized by a nested or partly nested series of levels. Each level occupies a distinct portion of the space–time scale space. Also each level consists of different entities, defined in part by the definition of that level of organization. There thus is simultaneous change in scale in both space and time and change in entities considered as one moves through a hierarchical model of ecological systems. When processes change across scale we face a challenge of translating the effects of a process on the entity at one scale to the effects of a different process on the entity at another scale. When entities change across scale we have the challenge of linking entities at lower levels of organization to aggregate properties of entities at higher levels. When both entities and processes change it is intractably difficult to rigorously predict relationships between entities and driving processes across scale. Combination of differing process and changing entity is a fundamental obstacle to understanding

Linking relationships across levels of organization therefore faces three simultaneous challenges: (1) change in scale, (2) change in entity, and (3) change in processes.

The purpose of this chapter is to explore the challenge of integrating the ecological pattern-process relationships across space and through time, and the roles that ecological scale and scaling serve in that effort. Despite much effort, there has been limited success in simultaneously traversing scale and hierarchy from fine scale to large scale and vice versa. Moving upward in ecological hierarchy from individual trees to stands to basins and landscapes, apparent ecological patterns change, as to the apparent relationships between patterns and processes (Wiens 1989, Levin 1992), and the scales of measurement appropriate in one may fail to capture the important processes in another. The cost of sampling very large geographical areas at a fine grain is prohibitive, and methods of integrating these fine grain data to produce broad scale predictions are challenging (King 1991, Rastetter et al. 1992, Schneider 1994). Moving downward in hierarchy from continents and ecoregions to landscapes and watersheds, broad scale relationships between regional climate, vegetation community structure, and disturbance processes become less coherent(Turner et al. 2001, Baker 2003). This is because variance due to local spatial and temporal heterogeneity becomes small relative to variance due to major exogenous forcings such as climatic drivers. However, downscaling to landscapes and watersheds, the details of the spatial structure of the landscape (e.g., topography, soils, historical events such as disturbance and succession) become critical (Baker 1995). At the scale of forested landscapes, it is logistically and analytically intractable to extend fine-scale mechanistic relationships between individual organisms and their environments across space and time, and, simultaneously, regional and continental scale relationships between climate and productivity and life-form lose explanatory power due to increasing spatial and temporal variance.

The reason that understanding ecological mechanisms and responses on landscapes is so challenging is rooted in scale: downscaling fails to capture the spatial and temporal contingencies that affect local responses, but integration of local processes does not neatly add up to landscape dynamics. Unfortunately, landscape evolution is fundamentally driven by time-variant, meso-scale interactions of endogenous and exogenous mechanisms, and so the variation in abiotic and biotic dynamics central to understanding current conditions and projecting future conditions is in this middle-number domain (Cushman et al 2007). This is fundamentally different than merely contrasting fine scale and coarse scale. Fine scale relationships are tractable because they isolate relationships between ecological mechanisms and organisms at scales in space and time where it is possible to control variation through sampling and experimental designs. For example, ecological experiments at the site or stand level obtain reliable inferences through replication and control (McGarigal and Cushman 2002). Replication and control ensure reliable site level inferences by statistically isolating patterns and processes at the site level from patterns and processes at other scales. This enables inferences at one scale, but disables inferences from that scale to be extended to other scales.

At broad scales, spatial and temporal variation in pattern and process are small relative to fine grain patterns of some ecological responses such as net primary production, carbon storage, life-form distribution. However, these relationships are unable to provide reliable inferences at landscape scales for several reasons. First, at landscape scales the spatial and temporal variance of ecological patterns is generally high due to interactions between biophysical gradients, vegetation gradients, and disturbance processes (Cushman et al. 2007). The increase in spatial and temporal variance at landscape scales makes inferences based on broad-scale patterns unreliable. Second, the parameters addressed in broad scale work are rarely those most pertinent at the landscape scale. Broad scale studies typically focus on questions such as global range limits for life-forms or species, net primary production, and carbon storage. Most research at landscape scales focuses on questions such as rates and patterns of disturbance processes and how they interact with landform and existing vegetation, distributions and abundance of species with respect to environmental gradients and disturbance history, and growth and regeneration of species across the landscape with respect to biotic and abiotic gradients. Such questions cannot be tackled using broad scale generalizations.

The structure of landscapes in terms of topography, soils, and existing biotic communities interacts with regional climate to create complex spatial patterns of biophysical gradients (Whittaker 1967, Fig. 3.2). These spatial patterns fundamentally influence the ecological processes occurring across the landscape (Turner 1989). The feedback between pattern and process across landscapes is not a statistical nuisance to be averaged away by aggregation to a higher organizational level, but is fundamentally important to understanding relationships between ecological mechanisms and responses (Wiens 1989; Levin 1992; Schneider 1994). The details of landscape structure influence relationships between forest ecosystems, climate and disturbance regimes in complex and interacting ways (Cushman et al. 2007). Fine scale environmental structure has strong influences on species distributions, dominance, and succession (Whittaker 1967; Tilman 1982; ter Braak and Prentice 1988). The biophysical context of a location within a landscape also strongly influences growth rates and regeneration (Bunn et al. 2003, Bunn et al. 2005). Furthermore, the probability of different kinds of disturbances (Risser et al. 1984; Runkle 1985; Baker 2003), and patterns of recovery from disturbances (Finegan 1984; Glenn and Collins 1992) are strongly dependent on the pattern of environmental variation across the landscape.

Temporal variation in ecological conditions across space is also fundamentally important to many ecological processes (Turner 1989). Seasonal to interannual

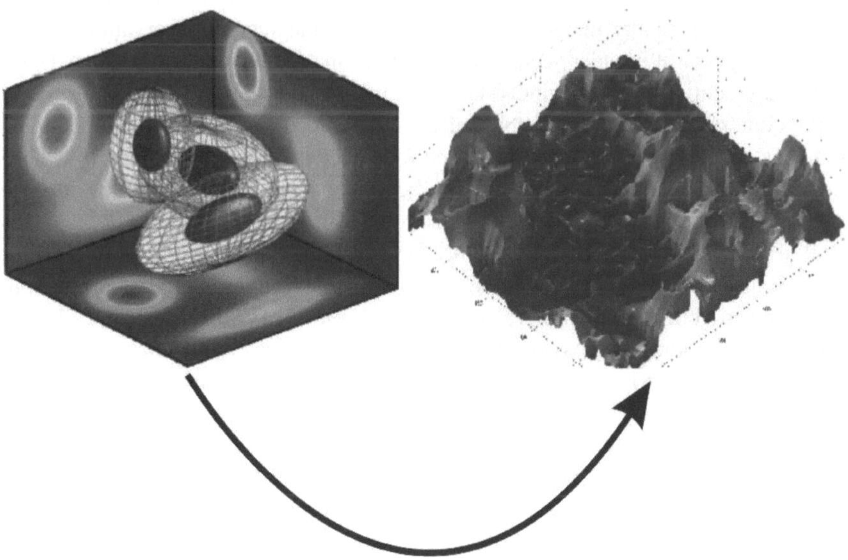

Fig. 3.2 One of the major themes of modern ecology is that each organism exists within a limited range of conditions which satisfy its ecological niche. This zone of "ecological space" can be projected onto the physical environment, producing spatial depiction of the quality of each location for each individual species. The interactions between multiple ecological attributes at a variety of scales in space, plus fluctuations in time, with the ecological tolerances and requirements of each member of the biotic community are fundamental drivers of ecological system composition and process

variation in populations, disturbances and climate cause the complex biophysical structure of landscapes to also change continuously through time. Thus, forest landscapes are fundamentally disequilibrial. Conditions continually change through both space and time, and biotic responses are always responding in kind, and usually lagging. Biophysical conditions change through time with drifting climate, species invasion and succession, and the type, extent, frequency and severity of past disturbances (Peet and Loucks 1976; Pickett 1980). Long-lived trees usually survive short-term climatic fluctuations, and species best adapted to current climate may colonize only following disturbance (Dunwiddie 1986). Thus, present biophysical conditions and disturbance regime and may not explain forest structure established earlier (Clark 1990). In contrast, disturbances serve to maintain community structure and ecosystem function in many ecological systems by preventing progress toward equilibrium (White 1979; Mooney and Godron 1983; Sousa 1984; Glenn and Collins 1992; Collins et al. 1998).

3.1 Scaling Ecological Knowledge

Most ecological research has focused at the level of organisms and the scale of their direct interactions with immediate ecological conditions (Kareiva and Anderson 1986). This is a fundamentally important scale (Wiens et al. 1993; Schneider 1994). However, many of the phenomena of interest to managers, conservationists and scientists appear at broader scales in space and time (Kareiva and Anderson 1986) and it is essential to extend understandings from fine scales in space and time to wider contexts (Shugart et al. 1988; Jeffers 1988). As a result, nearly all ecological analyses entail up-scaling from measured to expected values (Schneider 1994).

Scaling ecological processes up to predict phenomena at broader scales in space and time is one of the fundamental challenges in ecology (Levin 1992; Wiens et al. 1993; Schneider 1994). There are several issues that complicate the effort. First, translation across scales should simultaneously consider changes in both spatial heterogeneity and temporal scale (Risser 1986, 1987; Rosswall et al. 1988). Second, extrapolation across scale domains often involves transmutation (O'Neill 1979; Chesson 1981) where the relationships between patterns and processes change qualitatively. These nonlinearities preclude deriving patterns at a given scale by some simple function that amalgamates average values from observations at a finer scale (Gardner et al. 1982; Cale et al. 1983; Welsh et al. 1988; King 1991). Third, up-scaling usually involves changes in the organizational-level of observation and inference, for example from organisms to stands, stands to landscapes or landscapes to regions. Moving across organizational levels changes the grain and extent of observations in space and time, as well as the entities observed, variables measured and the processes governing phenomena (Cushman et al. 2007, Fig. 3.1). Thus, scaling involves up to six simultaneous problems: (1) changes in spatial extent, (2) changes in spatial grain, (3) changes in temporal duration, (4) changes in temporal grain, (5) changes in entities measured or predicted, (6) changes in variables measured or predicted, and (7) changes in the processes linking entities and variables.

A number of approaches have been suggested for translating ecological relationships across scales (Gates 1985; King 1991; Rastetter et al. 1992; Wiens et al. 1993; Schneider 1994; McGarigal and Cushman 2002). Methods exist to either downscale (Gates 1985) from coarser to finer scale, or upscale from fine-scale data to coarser scale predictions (King 1991; Rastetter et al. 1992; Wiens et al. 1993, Fig. 3.3). Downscaling (Gates 1985) from continents and regions to landscapes and basins, the broadscale relationships between regional climate, vegetation community structure and disturbance processes become unstable (Baker 2003). This is because at very large scales variance due to local heterogeneity in conditions in space and time becomes small relative to variance due to major climatic drivers. However, as you downscale to landscapes and basins, the details of the spatial structure of the landscape in terms of topography and soils becomes critical, as do historical events such as disturbance and succession, and the spatio-temporal interplay between them (Baker 1995). At landscape scales it is logistically and analytically challenging to extend fine-scale mechanistic relationships between individual organisms and their environments across space and time, and, simultaneously, regional and continental scale relationships between climate and productivity and life-form lose explanatory power due to increasing spatial and temporal variance. For example, temperature and precipitation patterns are predicted by global simulation models have spatial resolution of 1–2°. However, precipitation and temperature vary substantially at much finer scales, and this variability is important to local ecological processes (Lynn et al. 1995; Kennedy 1997; Russo and Zack 1997; Turner et al. 2001).

While downscaling is a critically important challenge, most attention has focused on methods to infer aggregate properties of entities at higher organizational levels from the characteristics and interrelationships of entities at a lower level (e.g. King 1991; Rastetter et al. 1992, Fig. 3.3). Because of the difficulty of addressing large numbers of fine-scale components individually they are usually aggregated and treated collectively (Rastetteter et al. 1992). This kind of upscaling estimates properties of aggregates by averaging, integrating or otherwise combining information about the entities at a lower level of organization (Rastetter et al. 1992; Turner et al. 2001). The challenge is that aggregates generally do not behave the same way as the fine-scale components comprising them (O'Neill 1979). Translation from fine to coarse scale usually involves both an increase in extent and aggregation to coarser grain (Allen et al. 1984; Meentenmeyer and Box 1987; King 1991). These efforts are explicitly hierarchical and involve changes in both scale and organizational level, and thus changes in grain and extent, as well as in the entities, variables and processes being addressed (O'Neill 1988; Rastetter et al. 1992).

Rastetter et al. (1992), King (1991), and Schneider (1994) review methods for upscaling to aggregate properties. These authors discuss the strengths and limitations of several methods to up scale to aggregates including: (1) lumping, (2) extrapolation by expected value, and (3) calibration. Lumping (King 1991) refers to the process of estimating the mean value of a parameter for an aggregate by averaging

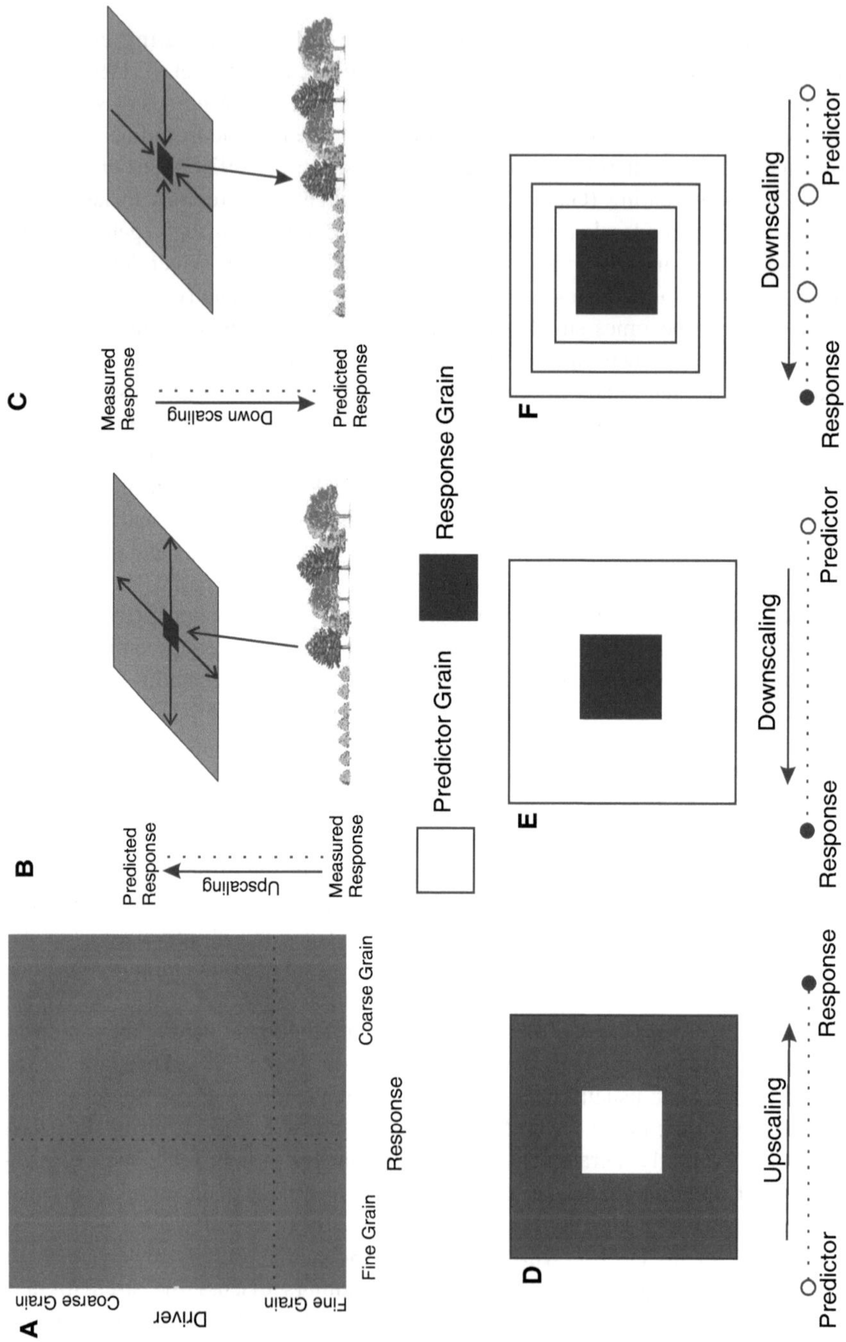
A
Driver
Coarse Grain
Fine Grain
Fine Grain
Coarse Grain
Response
B
Predicted Response
Upscaling
Measured Response
C
Measured Response
Down scaling
Predicted Response
Predictor Grain
Response Grain
D
Predictor
Upscaling
Response
E
Response
Downscaling
Predictor
F
Response
Downscaling
Predictor

Fig. 3.3 Up-scaling and down-scaling in ecological systems. The strength and nature of relationship observed between a response variable and a predictor variable will change as functions of the scale of each (A). Scaling response variables is problematic and requires either upscaling (e.g., plot-level vegetation response to predicted landscape-level aggregate properties) (B), or downscaling (e.g., 1-km scale measured response data to plot-level predicted response) (C). These extrapolations require information at multiple scales. Forest plots are rarely nested across a broad range of scales, for example. The errors associated with scaling response variables are avoided by making predictions at the same grain at which response variables were measured. Spatial scope can then be achieved by imputing model predictions across large spatial extents at the native grain of the response variable (see Imputation section). Scaling predictor variables, in contrast, is an essential task in modeling species–environment relationships. The grain at which environmental data are most strongly related to the response of the dependent variable may be finer (D) or coarser (E) than the grain of the response variable. Haphazard selection of scales of variables leads to equivocal conclusions. Ideally, the grain of the response variable should be fixed at an appropriate level and modeled across spatial and temporal extents (F). The predictor variables ideally would be measured at the finest possible grain over the largest possible extent. Partial bivariate scaling could then be used to identify the neighborhood extent around sampling units at which the predictor variable has the strongest relationship with the response variable over a sample of finer scale entities. The mean, or lumped value, is then used to represent large-scale aggregate expression of the finer-scale phenomena. Lumping assumes that the coarser scale aggregate is equivalent to the average finer-scale entity. This only holds if equations describing the system are linear (King 1991; Rastetter et al. 1992). This is rarely the case in ecological systems. Aggregates typically do not behave as simple multiples of their components (Schneider 1994), due to nonlinear interactions (Levin 1992; Rastetter et al. 1992; Bazzaz 1996; Reynolds et al. 1993). Lumping often produces biased predictions because it does not account for variability in the scaling process and ignores nonlinear changes in variable of interest with scale (Rastetter et al. 1992; Turner et al. 2001). Also, the interaction between sample density, spacing and spatial covariance among parameters influences the predictions. This may result in predictions inconsistent with those that would occur at another scale (Schneider 1994). However, statistical up scaling works reasonably well for quantities which do not interact spatially and can be related to attributes that can be measured remotely (Turner et al. 2001). Aggregation by lumping often involves inappropriate application of fine-scale relationships to predict aggregate properties, and is an example of the "fallacy of the averages" (Wagner 1969; Rastetter et al. 1992).

The statistical expectation operator is a theoretically attractive approach to transform fine-scale functions into coarse-scale functions (King 1991; Rastetter et al. 1992). The expectation operator works by quantifying variation among fine-scale components using a probability density function and then computing weighted sums to predict the expected value for the aggregate based on the frequency of different values of the parameter among its fine-scale constituents. In an ideal analysis, extrapolation by expected value is free of aggregation error and the aggregated system will exactly correspond to the original system of fine-scale relationships (Iwasa et al. 1987, 1989; Rastetter et al. 1992). However, all forms of variability among components being aggregated must be fully characterized with a joint, multivariate probability density function and incorporated into the aggregated equations (King 1991; Rastetter 1992). Implementing the statistical expectation operator is often impossible because (1) adequate statistical characterization of the fine-scale variability is problematic, and (2) each transformation makes the equations more complex and subsequent transformations become more and more difficult.

Calibration is a third method of upscaling to aggregate properties which avoids some of the problems facing lumping and the statistical expectation operator. In calibration, weighted sums are regressed against coarser scale measurements to produce empirically estimated scaling factors (Schneider 1994). Estimates of parameter values are produced through calibration of the coarse-scale relationship to the fine-scale data (Rastetter et al. 1992). While any parameter estimation procedure can be used to make these estimations, regression is the most common method. The main advantage of calibration is that it corrects for all sources of aggregation error simultaneously, including hidden and unknown sources (Rastetter et al. 1992). In addition, as calibration does not require the estimation of new parameters, the complexity of the model can be kept to a minimum. However, there are three major limitations: (1) Calibration requires coarse-scale data, which may be difficult to acquire; (2) parameter estimates are valid within the range of the calibration data, but the reliability of the calibration is unknown outside this range; (3) calibration is limited to variables measured at both coarse and fine scales, which often do not include the functionally most important factors. In practice, a combination of approaches are usually recommended for upscaling to aggregate properties (King 1991; Rastetter et al. 1992; Wiens et al. 1993; Schneider 1994). In particular, approaches using partial transformations under the statistical expectation operator followed by calibration may be particularly effective.

There are a number of difficulties in reliably upscaling to aggregate properties regardless of the methods employed (Fig. 3.3). First, in each of the aggregation procedures except calibration, variability among the fine-scale components must be characterized and incorporated into corrected equations. This increases the number of parameters in the corrected equations, and error associated with these estimates decreases precision. Also, aggregation methods increase complexity of the predicting equations resulting in a loss in heuristic value (Rastetter et al. 1992). Second, there are major limitations in the abilities of aggregation methods to address complex processes with multiple parameters. Many ecological systems and pattern–process relationships are strongly characterized by disequilibria and transient drivers. None of the aggregation methods discussed offer a means of dealing with disequilibrial dynamics. Third, aggregation assumes that fine-scale processes that vary spatially across the landscape can be treated as random variables, and their joint probability distribution defines the spatial heterogeneity of the landscape (Rastetter et al. 1992). This is limited to situations where fine-scale phenomena at one site is independent of the phenomena at other locations. Spatial correlations or complex spatial patterns may produce very complex joint probability distribution functions, making it difficult to solve the aggregation equations. Fourth, the goal of aggregation methods is to subsume variation among fine-scale entities to produce a single prediction for an emergent coarser-scale entity. Aggregation assumes that spatial pattern of the parameter value within the aggregate does not matter. However, the fine-scale spatial structure of ecological systems is often fundamentally important to mechanistic relationships between organisms and environment (Levin 1992; Schneider 1994). Aggregation to composite predictions results in the loss of information on internal structure and behavior of the system below the level of aggregation. Fifth, scaling up presupposes the existence of an aggregate landscape property that can be derived from the finer scale information (King 1991). This is consistent with a hierarchical model which views the landscape as an integrated entity with aggregate properties and dynamics linked to properties and dynamics of the finer-scale entities that comprise it (Webster 1979; Allen and Starr 1982; Patten 1982; Allen et al. 1984; O'Neill et al. 1986).

Hierarchy is often proposed as a framework for addressing scale translations (Urban et al. 1987, O'Neill 1988, Shugart and Urban 1988, King 1991, Figure 3.1). However, ecological systems may often be more accurately described as multi-scale gradient systems than as categorical hierarchies (Hutchinson 1957, Whittaker 1967, Cushman et al. in press, McGarigal et al. in press, Figure 3.4). Ecological parameters follow complex patterns of covariation across space and time. These gradients

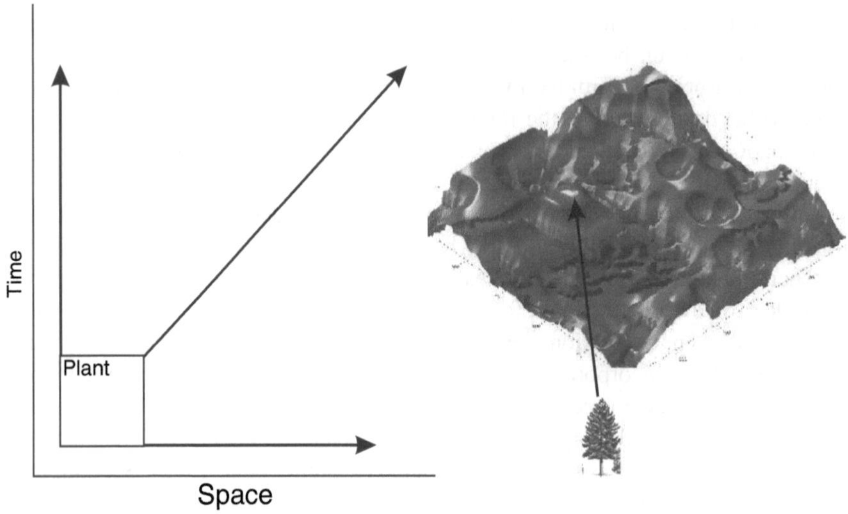

Fig. 3.4 As an alternative to a hierarchical model of ecological system organization, with its attendant challenges of meaningfully defining entities across levels of organizations, and handling simultaneous change in scale and entity in analysis, we believe a gradient model is often more powerful, tractable and consistent with fundamental ecological theory than hierarchical models of system organization. Instead of inventing entities and struggling with the problem of defining them, their boundaries and characteristics, focus instead is on organisms and their interactions with fundamental driving factors across spatial and temporal scale. This removes one of the two major problems facing the traditional ecological organization: translation across entities. The second problem of translation across processes with scale is greatly facilitated. The relationship between processes and organism responses can be modeled continuously across scale. This provides a picture of how different processes interact and the nature of their influence as a function of scale

in landscape structure range from strong to weak, and continually shift and change (Schneider 1994, Cushman et al. in press). Imposing hierarchical organization on a spatially complex and temporally dynamic system may often obscure more than it reveals. It is often difficult to objectively defend definitions used to define the aggregate entities defining levels in a hierarchy (McGarigal and Cushman 2005). Simultaneously, it is challenging to translate phenomena at one level to an entirely different set of entities at another level in a conceptual hierarchy (Cushman et al. in press, Figure 3.5). Transmutation across scale, where pattern process relationships qualitatively change with changes in scale, may largely be an artifact of the hierarchical framework adopted for analysis. Gradients lend themselves to measurement of change in conditions over change in space and time, and they can be multi-dimensionally related to each other – even nested. Most importantly, the biophysical mechanisms that affect organism responses and limit or facilitate many aspects of their biology and ecology can be relatively conveniently expressed at multiple scales.

Linking mechanisms and responses across scale may benefit from approaches that use scalable rather than logical units. Multiscale analysis is facilitated by analysis in ratio scale units that can be expanded in extent, decimated in grain, and combined

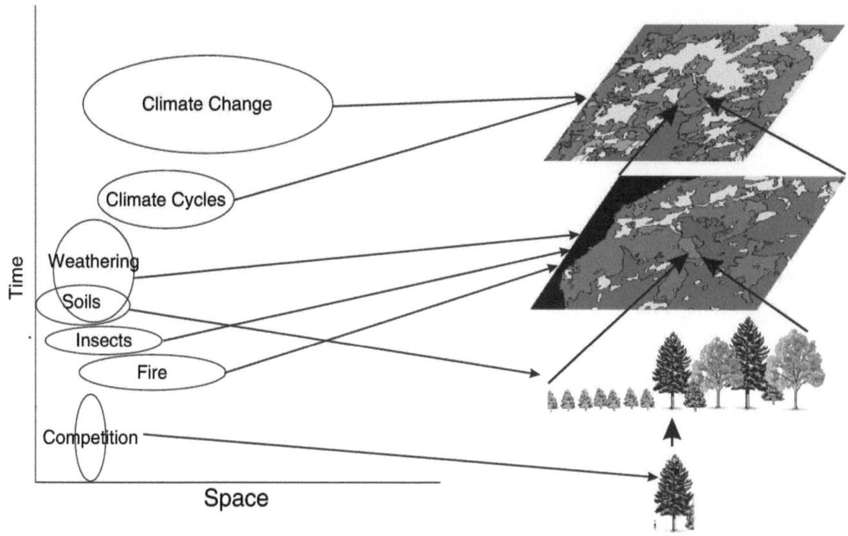

Fig. 3.5 Functional relationships across hierarchical levels are difficult to translate as there is a logical discontinuity as one transmutes across scale and level simultaneously. This results in basic disconnect in prediction of process relationships across the system as a function of level and scale. Linking the traditional hierarchical model of ecosystem structure to processes typically associates specific processes with specific levels of the hierarchy. For example, competition is expected to act at the level of the community, influencing organisms. Fire and insects are expected to act at the level of the landscape, affecting stands. Climate change is expected to act at the level of the region, affecting the landscape

to make new units via multiplication and division (Schneider 1994, Fig. 3.6). Instead of focusing on efforts to estimate emergent properties of aggregate entities, it may be more fruitful to focus attention directly at the level of most biological interest, such as the organisms and its immediate interactions with the environment. Understanding relationships between mechanisms and ecological responses at that level may be facilitated by measuring relationships between organisms and their environment in ratio scale units at a fine grain and over broad extent. Such measurements of multiple parameters across space and through time form a gradient cube in which pattern–process relationships can be investigated simultaneously for a range of organisms and processes, and across a range of scales without the need to recode or reclassify the data (Cushman et al. in press).

3.2 Gradient Theory and Extrapolation Across Scales

There are alternatives to up-scaling to aggregates which may produce better predictions across broad scale ranges. One alternative is to adopt a gradient perspective in lieu of hierarchical models of system organization (Cushman et al.

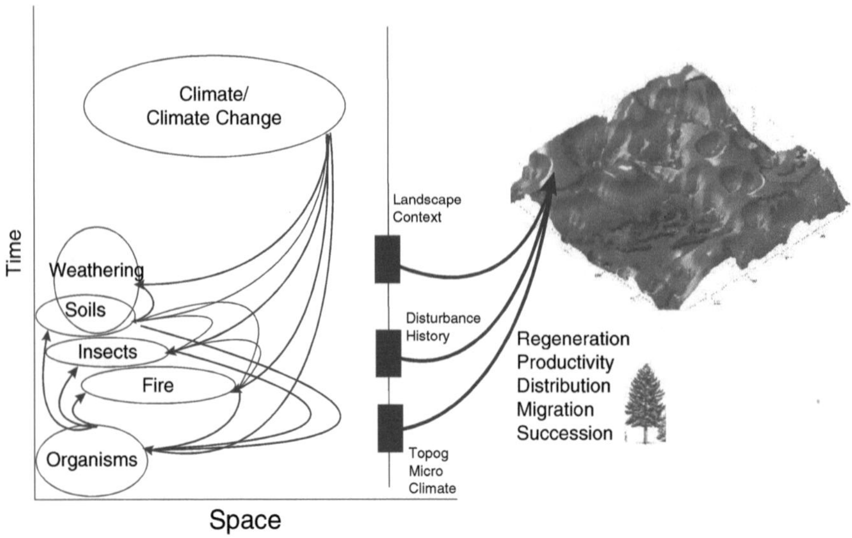

Fig. 3.6 Functional relationships across scales in a continuous gradient system are tractable given measurement of appropriate variables relating mechanisms to responses at appropriate extents and grains. This is because there is no discontinuity in the system's definition across scale, making extension of functional relationships much more direct and tractable. In contrast to the traditional hierarchical model, the gradient model of ecosystem structure does not suppose that processes act at particular levels of organization. Processes act across characteristic ranges of spatial and temporal scale, and interact with other processes. However, instead of supposing their effects to be directed at a particular emergent entity at a particular hierarchical level, the gradient approach emphasizes that the effects of ecological processes at any scale must be considered from the context of how they influence organisms. This returns focus fundamentally to the definition of ecology, which is the interaction between organisms and their environment

in press; McGarigal et al. in press, Figs. 3.2, 3.4, 3.6). In a gradient perspective, a system of hierarchically organized aggregate subsystems is not assumed. Rather, emphasis is on directly measuring response variables and the factors that drive their behavior, and modeling the relationships between them across space, time and scale (Fig. 3.2). In a gradient method one can fix grain of analysis and define entities and variables that do not qualitatively change across scale (Fig. 3.4). For example, one may fix grain at the scale of forest plots and landsat pixels, and use individual organisms as focal entities. This allows one to model interactions between mechanisms and responses across scale through a gradient hypercube incorporating space, time, disturbance and biophysical gradients within a single analytical framework (Cushman et al. 2007). There are several advantages to this (Fig. 3.6): (1) grain in space and time is fixed, (2) the challenge of defining meaningful aggregates at multiple organizational levels is obviated, (3) challenge of compounding error rates due to aggregation and classification errors is eliminated, (4) challenge of translating patterns and processes across organizational levels is reduced, (5) ability to address threshold, non-linear, and multivariate interactions across space and time is greatly improved. Importantly, in contrast to hierarchical approaches, there is no need to redefine entities, variables or units of observation in multi-scale analysis, greatly simplifying the task of robustly linking patterns with processes across scale.

In the next chapter we discuss a potentially unifying paradigm, focusing on mechanistic relationships between organisms and their environments at multiple spatial scales (McGarigal and Cushman 2005; Cushman et al. 2007; McGarigal et al. in press). The environment is characterized as a system of covarying and interacting gradients that vary across space and through time due to local biotic interactions, abiotic processes, conditions, and disturbance. The level of the organism and the scale of its direct interactions with the environment is the fundamental focus (Wiens et al. 1993; Schneider 1994). The goal of analysis is to link patterns of distribution and abundance with environmental drivers across space, through time, and across a range of spatial and temporal scales to account for the influences of spatial complexity, temporal dynamics, disequilibria, disturbance and transient drivers. The essence of the multi-scale gradient modeling approach is to sample driving and response variables on large networks of sample plots, develop multi-scale gradient models linking mechanisms to responses, and use the models to predict species distributions and ecological conditions across space through imputation (Ohmann and Gregory 2002; Cushman et al. 2007; Evans and Cushman 2009, Fig. 3.7). The gradient models can be linked to ecosystem dynamics modeling and landscape dynamics simulation to quantitatively address temporal dynamics and large-scale disturbance (Cushman et al. 2007, Fig. 3.8).

This paradigm is not a new invention, but rather a synthesis and reemphasis of several existing perspectives and approaches. In particular, it draws heavily from community ecology and ideas of niche relationships along biophysical gradients (Hutchinson 1957; Whittaker 1967). It extends community ecology to the landscape level by explicitly addressing spatial complexity and temporal disequilibria, and by adopting a multi-scale approach. It extends landscape ecology by linking analysis of spatial and temporal patterns directly to organism responses to spatially and temporally varying environmental gradients. The traditional methods and theories of community and landscape ecology are different and have contributed to the long-standing disjunction between the fields. However, it is clear that a quantitative and conceptual synthesis between landscape and community ecology is essential if we are to address the key issues of how variation through space and time and across scale influence the relationships between organisms and their environments (Cushman et al. in press; McGarigal et al. in press). Schneider (1994) notes that rapid progress was made in meteorology and physical oceanography when fluid dynamics was taken out of pipes and put into a geophysical grid (Batchelor 1967; Pedlosky 1979) with attention to time and space scales (Stommel 1963). Similarly, the key to addressing spatially and temporally complex ecological systems is adopting a multi-scale, mechanistic paradigm. Linkage with geophysical gradient theory has been suggested as a means to accomplish this (Risser et al. 1984; Shugart et al. 1988; McGarigal et al. in press).

The approach incorporates aspects of a number of strategies suggested by other researchers. First, multivariate gradient modeling predicts all parameters simultaneously, accounting for their covariation. As noted by Rastetter et al. (1992),

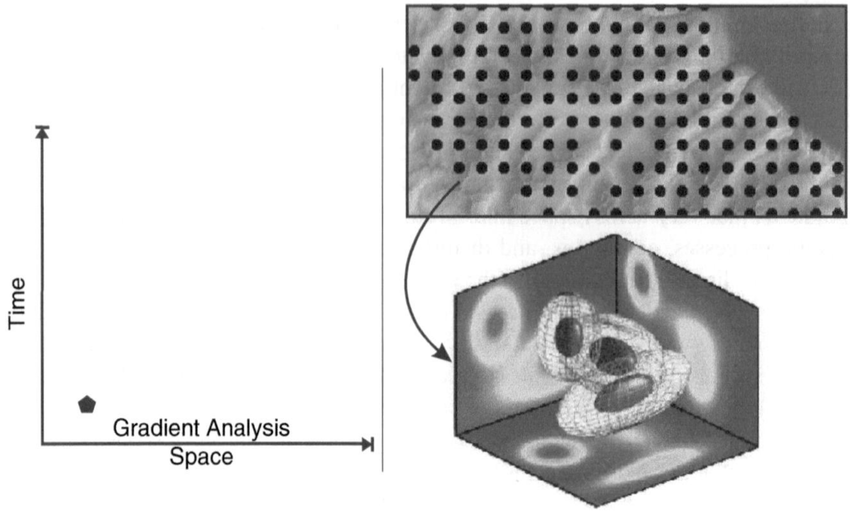

Fig. 3.7 There are three steps in the multi-scale gradient modeling approach. The first two are shown here. One is to sample driving and response variables on large networks of sample plots; two is to develop multi-scale gradient models linking mechanisms to responses

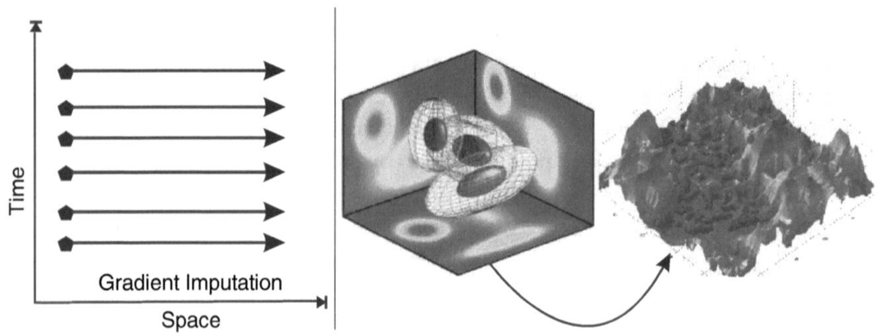

Fig. 3.8 The third step in the multi-scale gradient modeling approach is to use the models to predict species distributions and ecological conditions across space and over time through imputation

simultaneous estimation of all parameters can substantially reduce propagated error (Hornberger and Cosby 1985). In addition it combines the principle of similitude and scope extrapolation suggested by Schneider (1994) with the method of extrapolation by increasing model extent discussed by King (1991). Extrapolation by increasing model extent uses a model for a single site to simulate the same processes for a collection of sites across the landscape. Each individual simulation requires data on the independent variables required by the fine-scale model. Thus, it is a case of using variables with large spatial scope to calculate a variable with a more limited scope (Schneider 1994). As King (1991) notes, model structure is unaltered, grain is not changed and there is no averaging or aggregating of the data.

In the principle of similitude, spatial or temporal variation in a variable is expressed as a function of location or time (Schneider 1994). Dimensionless ratios are formed and these are used to scale measurements of limited scope to expected values at larger scope. Due to the spatial complexity of the physical environment and the typically low correlation between spatial and environmental gradients the principle of similitude would seem to have limited direct application to predicting ecological phenomena across complex landscapes. However, gradient imputation can be considered a method of similitude in environmental space rather than geographical space. Multi-scale gradient models predict species responses as functions of continuously varying environmental gradients. The gradient space depicted in the models is continuous in n-dimensions without break or deviation in rate of change. Gradient modeling involves predicting response variables across space for locations where they have not been sampled based on where those locations occur in environmental gradient space, and what the expected value of the response variable is at that location in gradient space. The connection to similitude is that variation as a function of location in 'gradient space' is used to develop scaling ratios for inferring values not sampled from values sampled.

3.3 Summary and Conclusions

Ecological processes are difficult to predict at the landscape level. Inconsistency of concepts, sampling, and analytical approaches across scales and organizational levels make it difficult to extend knowledge up the ecological hierarchy from plots and stands to landscapes, or down the hierarchy from biomes and regions to landscapes. Moving down the organizational hierarchy from continents and regions to landscapes, broad scale relationships between regional climate and regional productivity and other forest parameters become unstable. As one moves downward to landscapes, the details of the spatial structure of the landscape in terms of topography and soils becomes critical, as do historical events such as disturbance and temporal processes such as succession. Simultaneously, it is logistically and analytically difficult to extend fine-scale mechanistic relationships between individual organisms and their environments across space and time. Thus researchers face the challenging task of linking relationships across levels of organization, and translating between measurements and relationships of different entities at different scales. This simultaneous challenge of translating among scales and organizational levels is the fundamental challenge to reliable prediction of forest ecosystems across spatial and temporal scale. There are at least four important parts of this challenge. First is a problem of transfer and deals with scale mismatches between drivers and responses. Second is a problem of heterogeneity and deals with spatial patterns of vegetation and the environment. Third is stationarity and deals with the transience in the drivers. Fourth is a problem of extrapolation that results from nested gradients interacting across scales in space and time. Understanding and predicting the responses of forest resources to changing climate and disturbance regimes requires approaches that link mechanisms to responses across scales while accounting for spatial and temporal variability.

The dominant paradigm that has guided most research in this arena is the hierarchical model of ecological systems (Allen and Starr 1982; O'Neil et al. 1986). In this model, ecological systems are conceptualized as nested collections of aggregate subsystems. Each subsystem is contains qualitatively different entities than those existing above and below, and is subject to influences of processes acting at different spatial and temporal scales (O'Neil et al. 1986). Levels of organization in a nested hierarchy occur within monotonically increasing scales of time and space, with lower levels characterized by smaller finer scales and finer temporal scales (King 1991). Urban et al. (1987) defined one such nested hierarchy for forest systems consisting of gaps, stands, watersheds and landscapes.

However, the hierarchical model faces a number of challenges, particularly in its ability to address interacting processes across a range of scales. First is the challenge of defining meaningful aggregates at multiple organizational levels (Schneider 1994). In many ecological analyses, there is a high degree of subjectivity, imprecision and loss of information in defining aggregate entities, and prediction may often be improved by adopting a gradient concept of ecological organization (Cushman et al. in press). A gradient approach greatly reduces the challenge of compounding error rates due to aggregation and classification. Also, the challenge of translating patterns and processes across organizational levels is reduced by adopting a gradient approach. In a hierarchical system changes in scale are accompanied by changes in organizational level. When one changes organizational level, the apparent entities, variables and processes characterizing the system all change qualitatively. It is exceedingly difficult to translate quantitative relationships across qualitative hierarchical levels. Gradient approaches retain a common quantitative framework, with unchanging grain, variables, and entities. This greatly facilitates analysis across scale in space and time. Importantly, gradients of ratio scale variables provide a ready means to address threshold, non-linear, multivariate interactions across space and time (McGarigal and Cushman 2005). In contrast to hierarchical approaches, there is no need to redefine entities, variables or units of observation in multi-scale analysis, greatly simplifying the task of robustly linking patterns with processes across scale. Reliable understanding of relationships between communities, disturbance regimes and climate change will only be possible using approaches that integrate mechanisms and responses, scale and pattern, space and time. Gradient approaches to multi-scale modeling facilitates this; hierarchical methods may impede it.

References

Allen TFH, Hoekstra TW (1992) Toward a unified ecology. Columbia University Press, New York

Allen TFH, Starr TB (1982) Hierarchy: perspectives in ecological complexity. University of Chicago Press, Chicago, IL

Allen TFH, O'Neill RV, Hoekstra TW (1984) Inter-level relations in ecological research and management: some working principles from hierarchy theory. USDA Forest Service Gen Tech Rep RM-110 Rocky Mountain Forest & Range Exp Stn, Fort Collins, Colorado

Baker WL (1995) Longterm response of disturbance landscapes to human intervention and global change. Landsc Ecol 10:143–159

Baker WL (2003) Fires and climate in forested landscapes of the U.S. Rocky Mountains. In: Veblen T, Baker WL, Montenegro G, Swetnam TW (eds.) Fire and climate change in ecosystems of the western Americas. Ecological studies 160. Springer, New York

Batchelor GK (1967) An introduction to fluid mechanics. Cambridge University Press, Cambridge

Bazzaz FA (1996) Plants in changing environments: linking physiological, population and community ecology. Cambridge University Press, Boston, MA

Bunn AG, Lawrence RL, Bellante GJ, Waggoner LA, Graumlich LJ (2003) Spatial variation in distribution and growth patterns of old growth strip-bark pines. Arctic, Antarctic and Alpine Research 35:323–330

Bunn AG, Graumlich LJ, Urban DL (2005) Trends in twentieth-century tree growth at high elevations in the Sierra Nevada and White Mountains, USA. Holocene 15:481–488

Cale WG, O'Neill RV Jr, Gardner RH (1983) Aggregation error in nonlinear ecological models. J Theor Biol 100:539–550

Chesson PL (1981) Models for spatially distributed populations: the effect of within-patch variability. Theor Popul Biol 19:288–325

Clark JS (1990) Integration of ecological levels: individual plant growth, population mortality, and ecosystem dynamics. J Ecol 78:275–299

Collins SL, Knapp AK, Briggs JM, Blair JM, Steinhauer EM (1998) Modulation of diversity by grazing and mowing in native tallgrass prairie. Science 280:745–747

Cushman SA, McGarigal K (2003) Hierarchical, multi-scale decomposition of species-environment relationships. Landsc Ecol 17:637–646

Cushman SA, McKenzie D, Peterson DL, Littell J, McKelvey KS (2007) Research agenda for integrated landscape modelling. USDA For Serv Gen Tech Rep RMRS-GTR-194

Dunwiddie PW (1986) A 6000-year record of forest history on Mount Rainier, Washington. Ecology 67:58–68

Evans JS, Cushman SA (2009) Gradient modeling of conifer species using random forests. Landsc Ecol 24:673–683.

Finegan B (1984) Forest succession. Nature 312:109–114

Gardner RH, Cale WG, O'Neill RV (1982) Robust analysis of aggregation error. Ecology 63:1771–1779

Gates WL (1985) The use of general circulation models in the analysis of the ecosystem impacts of climatic change. Climatic Change 7:267–284

Glenn SM, Collins SL (1992) Disturbances in tallgrass prairie—local and regional effects on community heterogeneity. Landsc Ecol 7:243–251

Hornberger GM, Cosby BJ (1985) Evaluation of a model of long-term response of catchments to atmospheric deposition of sulfate. In: Proceedings of the 7th IFAC symposium on identification and system parameter estimation. Pergamon, New York

Hutchinson GE (1957) Concluding remarks. Cold Spring Harbor Symp Quant Biol 22:414–427

Iwasa Y, Andreasen V, Levin S (1987) Aggregation in model ecosystems: I. perfect aggregation. Ecoll Model 37:287–302

Iwasa Y, Levin S, Andreasen V (1989) Aggregation in model ecosystems: II. approximate aggregation. IMA J Math Appl Med Biol 6:1–23

Jeffers JNR (1988) Statistical and mathematical approaches to issues of scale in ecology. In: Rosswall TR, Woodmansee RG, Risser, PG (eds) Scale and global changes: spatial and temporal variability in biospheric and geospheric process. Wiley, Chichester

Kareiva P, Andersen M (1986) Spatial aspects of species interactions: the wedding of models and experiments. In: Hastings A (ed) Community ecology: lecture notes in biomathematics. Springer, Berlin

Kareiva PM (1994) Space: the final frontier for ecological theory. Ecology 75:1
Kennedy AD (1997) Bridging the gap between general circulation model (GCM) output and biological microenvironments. Int J Biometeorol 40:119–122
King AW (1991) Translating models across scales in the landscape. In: Turner MG, Gardner RH (eds) Quantitative methods in landscape ecology, ecological studies, vol 82. Springer, New York
Levin SA (1992) The problem of pattern and scale in ecology. Ecology 73:1943–1967
Lynn BH, Rind D, Avissar A (1995) The importance of mesoscale circulations generated by subgrid-scale landscape heterogeneities in general-circulation models. J Clim 8:191–205
McGarigal K, Cushman SA (2002) Comparative evaluation of experimental approaches to the study of habitat fragmentation effects. Ecol Appl 12:335–345
McGarigal K, Cushman SA (2005) The gradient concept of landscape structure. In: Wiens JA, Moss MR (eds) Issues and perspectives in landscape ecology. Cambridge University Press, Cambridge
McGarigal K, Tagil S, Cushman SA (2009) Surface metrics: An alternative to patch metrics for the quantification of landscape structure. Landsc Ecol 24:433–450
Meetenmeyer V, Box EO (1987) Scale effects in landscape studies. In: Turner, MG (ed) Landscape heterogeneity and disturbance. Springer, New York
Mooney HA, Godron M (1983) Disturbance and ecosystems. Springer, New York
Ohmann JL, Gregory MJ (2002) Predictive mapping of forest composition and structure with direct gradient analysis and nearest-neighbor imputation in coastal Oregon, USA. Can J Forest Resources 32:725–741
O'Neill RV (1979) Transmutations across hierarchical levels. In: Innis SG, O'Neill RV (eds) Systems analysis of ecosystems. International Cooperative Publishing House, Fairland
O'Neill RV (1988) Perspectives in hierarchy and scale. In: Roughgarden J, May RM, Levin, SA (eds) Perspectives in ecological theory. Princeton University Press, Princeton, NJ
O'Neill RV, DeAngelis DL,Waide JB, Allen, TFH (1986) A hierarchical concept of ecosystems. Princeton University Press, Princeton, NJ
Patten BC (1982) Environs: relativistic elementary particles for ecology. Am N 119:179–219.
Pedlosky J (1979) Geophysical fluid dynamics. Springer, New York
Peet SK, Loucks OL (1976) A gradient analysis of southern Wisconsin forests. Ecology 58:485–499
Peterson DL, Parker VT (1998) Ecological scale: theory and applications. Columbia University Press, New York
Pickett STA (1980) Non-equilibrium coexistence of plants. Bull Torrey Bot Club 107:238–248
Rastetter EB, King AW, Cosby BJ, Hornberger GM, O'Neill RV, Hobbie JE (1992) Aggregating fine-scale ecological knowledge to model coarser-resolution attributes of ecosystems. Ecol Applic 2:55–70
Reynolds JF, Hilbert DW, Kemp PW (1993) Scaling ecophysiology from the plant to the ecosystem: a conceptual framework. In: Ehleringer JR, Field CB (eds) Scaling physiological processes. Academic Press, New York
Risser PG (1986) Report of a workshop on the spatial and temporal variability of biospheric and geospheric processes: research needed to determine interactions with global environmental change, Oct. 28-Nov.1, 1985. St. Petersburg, Fla. Paris: ICSU Press.
Risser PG (1987) Landsacpe ecology: state of the art. In: Turner, MG (ed) Landscape heterogeneity and disturbance. Springer, New York
Risser PG, Karr JR, Forman RTT (1984) Landscape ecology: directions and approaches. Special Publication Number 2. Illinois Natural History Survey, Champaign
Roswall T, Woodmansee RG, Risser PG (eds) (1988) Scales in global change. Wiley, New York
Runkle JR (1985) Disturbance regimes in temperate forests. In S.T.A. Pickett and P.S. White, eds. The ecology of natural disturbance and patch dynamics, pp. 17–34, Academic, New York
Russo JM, Zack JW (1997) Downscaling GCM output with a mesoscale model. J Environ Manag 49:19–29.
Schneider DC (1994) Quantitative ecology: spatial and temporal scaling. Academic, San Diego, CA

Shugart HH, Urban DL (1988) Scale, synthesis, and ecosystem dynamics. In: Pomeroy LR, Alberts JJ (eds) Concepts of ecosystem ecology. Springer, New York
Shugart HH, Michaels PJ, Smith TM, Weinstein DA, Rastetter EB (1988) Simulation models of forest succession. In: Roswall T, Woodmansee RG, Risser PG (eds) Scales in global change. Wiley, New York
Sousa WP (1984) The role of disturbance in natural communities. Annul Rev Ecol Systemat 15:353–391
Stommel H (1963) The varieties of oceanographic experience. Science 139:572–576
ter Braak CJF, Prentice IC (1988) A theory of gradient analysis. Adv Ecol Res 18:271–313.
Tilman D (1982) Resource competition and community structure. Princeton University Press, Princeton, NJ
Turner MG (1989) Landscape ecology: the effect of pattern on process. Annu Rev Ecol Systemat 20:171–197
Turner MG, Gardner RH, O'Neill RV (2001) Landscape ecology in theory and practice. Springer, New York
Urban DL, O'Neill RV, Shugart HH Jr (1987) Landscape ecology. Bioscience 37:119–27.
Wagner HM (1969) Principles of operations research. Prentice-Hall, Englewood Cliffs, NJ.
Webster JR (1979) Hierarchical organization of ecosystems. In: Halfon E (ed) Theoretical systems ecology. Academic, New York
Welsh AH, Peterson TA, Altmann SA (1988) The fallacy of averages. Am Nat 132:277–288.
White PS (1979) Pattern, process, and natural disturbance in vegetation. Bot Rev 45:229–299
Whittaker RH (1967). Gradient analysis of vegetation. Biol Rev 49:207–264
Wiens JA (1989) Spatial scaling in ecology. Funct Ecol 3:385–397
Wiens JA, Stenseth NC, Van Horne B, Ims RA (1993) Ecological mechanisms and landscape ecology. Oikos 66:369–380

Chapter 4
Landscape Ecology: Past, Present, and Future

Samuel A. Cushman, Jeffrey S. Evans, and Kevin McGarigal

4.1 Historical Origins of Landscape Ecology

In the preceding chapters we discussed the central role that spatial and temporal variability play in ecological systems, the importance of addressing these explicitly within ecological analyses and the resulting need to carefully consider spatial and temporal scale and scaling. Landscape ecology is the science of linking patterns and processes across scale in both space and time. Thus landscape ecology is, in a real sense, the foundational science for addressing the central issues of sensitive dependence of ecological process on spatial and temporal variability. This chapter reviews the historical origins and evolution of landscape ecology, discusses its current scope and limitations, and then anticipates the following chapter by looking forward to identify how the field could best expand to address the central challenges of ecological prediction in spatially complex, temporally disequilibrial, multi-scale ecological systems.

Landscape ecology has emerged as an integrative, eclectic discipline, focusing explicitly on spatial structure and dynamics of landscape systems (Turner 1987). Traditionally, ecological science has largely been limited to the study of relationships between the structure and function of entities assumed for simplicity to be spatially homogeneous and temporally stable (Pickett and Cadenasso 1995). Landscape ecology, in contrast, views spatial heterogeneity as a prime causal factor in ecological interactions. Pickett and Cadenasso (1995) state that the primary insight to emerge from landscape ecology is the realization that spatial heterogeneity at various

S.A. Cushman (✉)
US Forest Service, Rocky Mountain Research Station, 800 E Beckwith, Missoula, MT 59801, USA
e-mail: scushman@fs.fed.us

J.S. Evans
Senior Landscape Ecologist, The Nature Conservancy, NACR – Science

K. McGarigal
Department of Natural Resources Conservation, University of Massachusetts, Amherst, MA, USA

S.A. Cushman and F. Huettmann (eds.), *Spatial Complexity, Informatics, and Wildlife Conservation*
DOI 10.1007/978-4-431-87771-4_4, © Springer 2010

scales exerts important influences on many ecological interactions. They define the goal of landscape ecology as "showing how processes at various scales interact to shape ecological phenomena and exposing regularities that have wide explanatory potential." Godron and Forman (1981) identify primary areas of enquiry in landscape ecology as the study of the distribution patterns of landscape elements, flows of material, biological or energenic units between the elements and the dynamics of landscape morphology, while Naveh and Liberman (1994) define landscape ecology more broadly as the scientific basis for landscape study, planning and conservation.

The development of landscape ecology has been cosmopolitan and eclectic, borrowing perspectives from a host of biological and geographical sub-disciplines and also being the heir to separate traditions originating in Europe, Russia and the United States (Forman and Godron 1986; Naveh and Lieberman 1994). The roots of landscape ecology may be said to begin in the middle nineteenth century with the introduction of "landscape" as a scientific term by the explorer-geographer Alexander Von Humboldt (Naveh and Lieberman 1993). Humboldt viewed landscapes as exhibiting coherence in spatial distribution and interconnectedness of phenomena, and was a pioneer in the study of spatial relationships between biological and physical phenomena (Dickinson 1970). Von Humboldt's work laid the foundations for much of modern geography and led directly to advances in the study of landscape characteristics.

Extending Von Humboldt's work, S. Passarge proposed "landscape science" as a new subfield of geography in 1919 (Troll 1971). Passarge's framework for landscape science was adopted and expanded by a series of Russian geographers. C.S. Berg described a landscape as a "community of a higher order, consisting of communities of organisms…together with the complex of inorganic phenomnena" (Troll 1971).

In 1935 A.G. Tansley first proposed the scientific concept of the ecosystem (Tansley 1935). This prompted the German geographer Carl Troll to advance the term "landscape ecology" in 1939 (Troll 1971). Troll described landscape ecology as "the study of the main complex causal relationships between the life communities and their environment in a given section of the landscape." Troll's original goal was to show ecological distributions within landscapes. Importantly, the initial focus of the Russian and German efforts in landscape science and landscape ecology focused on continuous patterns of environmental variability and continuous population processes, in a way presaging the gradient concepts of American community ecology later proposed by Gleason (1926) and Whittaker (1967). However, landscape ecology shortly thereafter departed from this gradient framework, and instead evolved into an effort to divide landscapes into small components and ascertain the logic through which they were grouped and interacted as a landscape mosaic (Troll 1971).

Following World War II, landscape ecology emerged as quasi-independent disciplines in the Soviet Union and several central European nations (Naveh and Lieberman 1994). In Germany, E. Neef, J. Schmithusen, and G. Haase made the first major contributions to quantification of landscape structure (Forman and Godron 1986). In the 1960s The Institute of Care and Nature Protection at the Technical

University of Hanover made important contributions to methods for using landscape ecology as a tool for landscape management and planning (Naveh and Lieberman 1993).

At about the same time, in the Netherlands C. Van Leeuwen provided original insights into the linkage of temporal variation and spatial heterogeneity in landscapes (Forman and Godron 1986). Zonneveld (1972) stressed the importance of studying landscapes as holistic amalgamations of separate components. He viewed the structure of the abiotic environment as central to landscape ecology, and held that the field was more naturally considered a subunit of geography and not ecology. Important theoretical contributions to the conceptualization of landscapes as mosaics of discrete elements were also made by workers in Australia and Switzerland (Naveh and Lieberman 1993).

While in Europe landscape ecology was developing powerful techniques for the description and analysis of the physical structure of landscape mosaics, American scientists were following a different route. Late in the nineteenth century American ecologists began to focus for the first time on the rigorous study of communities of living organisms (Krebs 1994). This trend was given direction by S.F. Forbes' classic paper, "The Lake as a Microcosm" (Forbes 1887). Forbes proposed the idea of an ecological community as an organic complex of mutually interdependent entities, and focused on how a "balance of nature" was maintained by competition between species for limited resources. Forbes' community approach was very insightful, but assumed that communities functioned as homogeneous units isolated from surrounding ecological systems.

At the beginning of the twentieth century American ecologists were beginning to make connections between landscape structure and community function, yet the development of truly landscape level science was still far behind that of European investigators (Troll 1971). In 1925 the American geographer C.O. Sauer wrote "The Morphology of Landscape" which provided, for the first time in the United States, a critique of European Landscape science (Sauer 1965). He sought to give new vigor to American geography by introducing the patch mosaic landscape perspective as a central focus of geography (Forman and Godron 1986).

Work on plant succession by H.C. Cowles (1899) and F. Clements (1916) provided some of the first truly landscape level investigations of ecological phenomena in America. In particular, Cowles' "physiographic ecology" emphasized constant interactions between plant communities and underlying geological formations, and viewed flora in a landscape as an ever-changing panorama (Real and Brown 1991). However it wasn't until Egler's work on plant ecology in the 1940s that a holistic view of plant associations and their interactions with human influences came to prominence in America (Egler 1942).

What is commonly thought of as modern landscape ecology may be considered to have begun in the United States with the work of a number of ecologists in the 1950s and 1960s (Forman and Godron 1986). Dansereau (1957) developed a system of classification of landscape elements based on tropic levels of energy transfer, and argued that the landscape unit was the highest level in the hierarchy of ecological structure. Christian (1958) advanced a model of discretely defined land units forming a spatial mosaic. Dickinson (1970) developed an approach termed "regional ecology" which focused on qualitative description of spatial associations among phenomena over the earth's surface. Isard (1975) further added to the theoretical base of regional science, emphasizing social problems with regional and spatial characteristics. The emergence of transportation theory also added new insights to landscape ecology (Forman and Godron 1987). Taaffe and Gauthier's "Geography of Transportation" (1973) and Lowe and Moryada's "Geography of Movement" (1975) were path breaking in the study of network structure and connectivity, which would become one of the central interests of landscape ecology in the coming two decades. Also, their work set the foundations for the quantitative study of flow of ecological entities, such as nutrients, organisms and energy across landscape networks (Forman and Godron 1986).

The modern view of landscape structure as being composed of a mosaic of patches in a matrix was first formally put forth by Forman (1981) and Forman and Godron (1981). Modern landscape ecology is based on the patch mosaic paradigm, in which landscapes are conceptualized and analyzed as mosaics of discrete patches (Forman 1995; Turner et al. 2001). Sometimes the "patch mosaic" model is referred to as the "patch-corridor-matrix" model after Forman and Godron (1986) and Forman (1995) in order to recognize the different major landscape elements that can be present in a patch mosaic. Any reading of the published landscape ecology literature shows near uniformity in the adoption of this approach. Consequently, our current state of knowledge regarding landscape pattern-process relationships is based almost entirely on a categorical representation of spatial heterogeneity. The patch mosaic model has led to major advances in our understanding of landscape pattern-process relationships (Turner 2005) and has been applied to landscapes across the globe. Its strength lies in its conceptual simplicity and appeal to human intuition. In addition, it is consistent with well-developed and widely understood quantitative techniques designed for discrete data (e.g., analysis of variance). Furthermore, there is ample evidence that it applies very well in landscapes dominated by severe natural or anthropogenic disturbances (e.g. fire dominated landscapes and built landscapes).

Modern landscape ecology is characterized by several variant conceptions of the patch mosaic paradigm of landscape structure and change (Wiens 1994, With and King 1999). These perspectives largely differ in how the focal habitat is perceived and represented relative to other landscape elements, and whether the landscape structure is viewed as relatively static (i.e., unchanging) or dynamic (i.e., constantly changing). Although there are many variations, two paradigms have emerged that provide alternative frameworks for conceptualizing the habitat loss and fragmentation process.

4.1.1 Static Island Biogeographic Model

The first paradigm we call the "static island biogeographic model." In this model, habitat fragments are viewed as analogues of oceanic islands in an inhospitable sea or ecologically neutral matrix. Under this perspective, discrete habitat patches are seen as embedded in a uniform matrix of non-habitat. Connectivity is assessed by the size and proximity of habitat patches and whether they are physically connected via habitat corridors. The key attributes of the model are its representation of the landscape as a binary system of habitat and inhospitable matrix, and that, once lost, habitat remains matrix in perpetuity. In extreme cases, the process of habitat loss and fragmentation continues until the target habitat is eliminated entirely from the landscape. This scenario is perhaps best exemplified by urban sprawl and agricultural development, where remnant habitat fragments are maintained in an otherwise relatively static matrix or are eventually eliminated entirely from the landscape.

The static island biogeography paradigm has been the dominant perspective since the inception of the fragmentation concept. The major advantage of the island model is its simplicity. Given a focal habitat, it is quite simple to represent the structure of the landscape in terms of habitat patches contrasted sharply against a uniform matrix. Moreover, by considering the matrix as ecologically neutral, it invites ecologists to focus on those habitat patch attributes, such as size and isolation, that have the strongest effect on species persistence at the patch level. The major disadvantage of the strict island model is that it assumes a uniform and neutral matrix, which in most real-world cases is a drastic over-simplification of how organisms interact with landscape patterns. Not all matrix is created equal. Moreover, the strict island model usually assumes a static landscape structure, at least with respect to the matrix. Once habitat is lost, it remains matrix. This, too, is not realistic in many landscapes, especially those driven by natural disturbances and/or forest management activities.

The landscape transformation process as conceptualized under the idealized static island biogeographic model can be divided into several broad stages or phases that are demarcated by significant changes in the pattern or function of the landscape (Forman 1995). In reality, these phases are not strictly separate from each other since they may take place simultaneously; however, a dominant phase can often be identified.

1. *Perforation* – Often, the first stage of habitat loss and fragmentation involves the perforation of natural habitat through direct loss, usually resulting from conversion to other land uses (e.g., agricultural clearing, housing development, timber harvesting). Perforation creates holes in otherwise contiguous habitat. Here, there is both a direct loss of habitat and a change in the spatial distribution of remaining habitat. The degree of impact on habitat configuration will depend on the pattern of perforation (see below). However, at this stage, the habitat is still physically well connected.
2. *Dissection* – The second stage of habitat loss and fragmentation involves the dissection of natural habitat. In most cases, a perforated pattern will become a dissected pattern at certain threshold levels of habitat loss. Dissection may precede or occur in conjunction with perforation. A common route to habitat dissection is through the construction of roads or other transportation corridors that span the landscape. In most cases, there is relatively little reduction in habitat area caused by dissection. However, the resulting linear landscape elements can be a significant source of disruption to the natural community because

they can provide avenues for the intrusion of edge predators, invasive species, exotics, diseases and pathogens that adversely affect the organisms of interest. In addition, these linear elements can affect landscape connectivity by altering movement patterns of organisms. Perhaps most importantly, some dissecting agents such as roads provide human access to the natural habitats and establish a network by which future habitat loss and alteration will occur. Of course, as in all cases, the effects of dissection will depend on the habitat and the organism(s) of interest.

3. *Subdivision* – The third stage of habitat loss and fragmentation involves the subdivision of habitat into disjunct patches. Forman (1995) referred to this phase as "fragmentation", but like Jaeger (2000), who referred to this phase as "dissipation", we prefer to use the term fragmentation to refer to the entire sequence rather than a single phase. During this phase, the landscape undergoes an important phase transition from a landscape characterized by physically connected habitat to a landscape in which the habitat is broken up into disjunct fragments. At this point, the areal extent of habitat may still be quite large and may not yet be significantly limiting landscape function for the organism(s) of interest. However, at this point, the habitat is physically disconnected and may disrupt movement patterns of the target organism(s) and cause the subdivision of populations into separate units. The consequences of this population subdivision will be discussed later. Note, this phase may be confused or confounded with the "dissection" phase. The dissection phase, as idealized, typically occurs as a result of road-building in which the habitat is subdivided or dissected by linear features that do not result in significant reduction in habitat area. In contrast, the subdivision phase is typically characterized by concurrent habitat loss and results when the remaining habitat becomes subdivided into disjunct patches embedded within a matrix of "non-habitat."
4. *Shrinkage and Attrition* – The final stage of habitat loss and fragmentation involves the shrinkage and, in some cases, complete disappearance of the focal habitat. Here, the landscape is in a critical state with respect to the viability of the target habitat. As habitat patches are reduced in size and become more isolated from each other, the function of the landscape is seriously jeopardized for organisms associated with the target habitat. Under the island-biogeographic model, the remaining habitat fragments represent true islands in an inhospitable sea. Of course, the hostility of the matrix will depend on the organisms and how their life-history and vagility characteristics interact with landscape patterns, as discussed below.

This four-stage conceptualization of the landscape transformation represents an idealized and oversimplified view of habitat loss and fragmentation processes under the static island biogeographic model; no real landscape follows this trajectory exactly. Nevertheless, it depicts the general sequence of events characteristic of habitats undergoing reduction and fragmentation caused by urban and/or agricultural development. Although this simple conceptual model provides a useful framework, it is important to understand that there are many alternative scenarios or patterns of habitat loss and fragmentation associated with the above landscape transformation. Forman (1995) refers to such variations as "mosaic sequences."

5. *Random Model* – Habitat is lost progressively in a random pattern. Although not representative of any real-world pattern of habitat loss and fragmentation, it provides a useful null model against which to compare other scenarios.
6. *Contagious Model* – Habitat is lost in a contagious (i.e., aggregated) pattern. In this case, the fragmentation of habitat is controlled by the degree of contagion in the residual habitat. Thus, under a maximum contagion scenario, the residual habitat would be aggregated into a single patch, and the habitat would not be fragmented per se.
7. *Dispersed Model* – Habitat is lost in a dispersed (i.e., disaggregated) pattern. Under a maximum dispersion scenario, habitat would be perforated by dispersed 'openings' and would eventually be broken into discrete fragments.
8. *Edge Model* – Habitat is lost progressively in a wave-like manner, beginning on one edge of the landscape and moving progressively across the landscape. In this scenario, there is no fragmentation of habitat per se, since the original habitat is not subdivided into disjunct patches, but simply reduced in size steadily over time. This process is typical of urban expansion outward from a city or some large-scale forestry operations.
9. *Corridor Model* – Habitat is first bisected by corridor development (e.g., roads) and then lost progressively outward from the corridors. In this scenario, the habitat is both reduced and fragmented. This process is typical of rural and suburban residential development in many areas.
10. *Nuclear Model* – Habitat is lost progressively from nuclei that may be dispersed in a random, uniform, or clumped pattern. Perforations in the habitat grow steadily in size in radial fashion until eventually the habitat is subdivided (i.e., becomes disconnected). The rate and pattern of fragmentation per se will depend on the dispersion of nuclei. This process is typical of rural development and timber harvesting.

These models represent alternative patterns by which habitat may be lost from a landscape, and although idealized and oversimplified, they illustrate the wide range of possible patterns of habitat loss. More importantly, they illustrate the quantitative differences in habitat loss and fragmentation that can result under various scenarios (Forman 1995: page 425). For example, given the same trajectory of habitat loss, the edge model maintains the largest patches of habitat in the landscape without causing fragmentation. Conversely, the dispersed model results in the quick elimination of large patches from the landscape and, at some point, causes the fragmentation of the habitat that remains.

4.1.2 Dynamic Landscape Mosaic Model

The second major conceptual paradigm is the dynamic landscape mosaic model. In this paradigm, landscapes are viewed as spatially complex, heterogeneous assemblages of cover types, which can't be simplified into a dichotomy of habitat and matrix (Wiens et al. 1993; With and King 1999). Rather, the landscape is viewed from the perspective of a particular ecological process or organism. Habitat patches are bounded by other patches that may be more or less similar (as opposed to highly contrasting and often hostile habitats, as in the case of the island model) and the mosaic of patches itself changes through time in response to disturbance and succession processes. Connectivity is assessed by the extent to which movement is facilitated or impeded through different land cover types across the landscape. Land cover types may differ in their "viscosity" or resistance to movement, facilitating movement through certain elements of the landscape and impeding it in others. This perspective represents a more holistic view of landscapes, in that connectivity is an emergent property of landscapes resulting from the interaction of organisms with landscape structure.

The dynamic landscape mosaic paradigm has only recently emerged as a viable alternative to the static island biogeographic model. The major advantage of the landscape mosaic model is its more realistic representation of how organisms perceive and interact with landscape patterns. Few organisms exhibit a binary (all or none) response to habitat types, but rather use habitats proportionate to the fitness they confer. Moreover, movement among suitable habitat patches usually is a function of the character of the intervening habitats. Two suitable habitat patches separated by a large river may be effectively isolated from each other for certain organisms, regardless of the distance between them. In addition, the landscape mosaic model accounts for the dynamics in landscape structure due to the constant interplay between disturbance and succession processes. This is especially important in forested landscapes where natural disturbances and timber harvesting are the major drivers of landscape change. The major disadvantage of the landscape mosaic model is that it requires detailed understanding of how organisms interact with landscape structure; in particular, how the landscape mosaic composition and configuration affect movement patterns. Unfortunately, it is exceedingly difficult in practice to collect the needed quantitative information, rendering this model less practical. However, even in the absence of detailed information about how target organisms interact with entire landscape mosaics, it is often beneficial to characterize the landscape more realistically than as a simple binary map of habitat and matrix.

4.2 Spatial Components of the Patch-Mosaic Landscape Model

Whatever landscape paradigm one ascribes to, it is essential to understand what a given change in a landscape means physically. This requires explicit attention to the spatial components of landscape structure. There are conceptually many different spatial attributes which characterize a landscape mosaic pattern. Thus, a multivariate perspective is required and it is unreasonable to expect a single metric, or even a few metrics, to be sufficient.

Habitat loss and landscape fragmentation involve changes in both landscape composition and configuration (McGarigal and Marks 1995; Cushman and McGarigal 2002). *Landscape composition* refers to the presence and amount of each habitat type within the landscape, but not the placement or location of habitat patches within the landscape mosaic. *Landscape configuration* refers to the spatial character and arrangement, position, orientation, and shape complexity of patches in the landscape. We recognize five major components of landscape composition and configuration affected by habitat loss and fragmentation, even though the distinctions among these components can be somewhat blurry at times.

1. *Habitat Extent* – As noted previously, habitat loss and fragmentation are almost always confounded in real-world landscapes. Therefore, it is essential that habitat extent be considered in conjunction with any assessment of habitat fragmentation. Indeed, as described later, it is difficult, and in some cases impossible, to interpret many fragmentation metrics without accounting for habitat extent. Habitat extent represents the total areal coverage of the target habitat in the landscape and is a simple measure of landscape composition.
2. *Habitat Subdivision* – Habitat fragmentation fundamentally involves the subdivision of contiguous habitat into disjunct patches, which affects the overall spatial distribution or configuration of habitat within the landscape. Subdivision explicitly refers to the degree to which the habitat has been broken up into separate patches (i.e., fragments), *not* the size, shape, relative location, or spatial arrangement of those patches. Because these latter attributes are usually affected by subdivision, it is difficult to isolate subdivision as an independent component.
3. *Patch Geometry* – Habitat fragmentation alters the geometry, or spatial character, of habitat patches. Specifically, as patches are subdivided via habitat loss (Figs. 3 and 4), they become smaller, contain proportionately less core area (i.e., patch area after removing the area within some specified edge-influence distance), typically extend over less area, and often have modified shapes, although the nature of the change may vary depending on the anthropogenic agent (e.g., Krummel et al. 1987).
4. *Habitat Isolation* – Habitat fragmentation increases habitat insularity, or isolation. That is, as habitat is lost and fragmented, residual habitat patches become more isolated from each other in space and time. Isolation deals explicitly with the spatial and temporal *context* of habitat patches, rather than the spatial character of the patches themselves. Unfortunately, isolation is a slippery concept because there are many ways to consider context. In the temporal domain, isolation can be considered as the time since the habitat was physically subdivided, but this is fraught with practical difficulties. For example, rarely do we have accurate historical data from which to determine when each patch was isolated. Moreover, given that fragmentation is an ongoing process, it can be difficult to objectively determine at what point the habitat becomes subdivided, since this is largely a function of scale. In the spatial domain, isolation can be considered in several ways, depending on how one measures the spatial context of a patch. The simplest measures of isolation are based on Euclidean distance between nearest

neighbors (McGarigal and Marks 1995) or the cumulative area of neighboring habitat patches (weighted by nearest neighbor distance) within some ecological neighborhood (Gustafson and Parker 1992). These measures adopt an island biogeographic perspective, as they treat the landscape as a binary mosaic consisting of habitat patches and uniform matrix. Thus, the context of a patch is defined by the proximity and area of neighboring habitat patches; the role of the matrix is ignored. However, these measures can be modified to take into account other habitat types in the so-called matrix and their affects on the insularity of the focal habitat. For example, simple Euclidean distance can be modified to account for functional differences among organisms. The functional distance between patches clearly depends on how each organism scales and interacts with landscape patterns (With and King 1999); in other words, the same gap between patches may not be perceived as a relevant disconnection for some organisms, but may be an impassable barrier for others. Similarly, the matrix can be treated as a mosaic of patch types which contribute differentially to the isolation of the focal habitat. For example, isolation can be measured by the degree of contrast (i.e., the magnitude of differences in one or more attributes between adjacent patch types) between the focal habitat and neighboring patches.

5. *Connectedness* – Habitat loss and fragmentation affect the connectedness of habitat across the landscape. Connectedness integrates all of the above components and involves both a structural component (i.e., the amount and spatial distribution of habitat on the landscape; also referred to as "continuity") and a functional component (i.e., the interaction of ecological flows with landscape pattern; also referred to as "connectivity"). Structural connectedness refers to the physical continuity of habitat across the landscape. Contiguous habitat is physically connected, but once subdivided, it becomes physically disconnected. Structural connectedness is affected by habitat extent and subdivision, but also by the spatial extensiveness of the habitat patches (Keitt et al. 1997). Specifically, as habitat patches become smaller and more compact, they extend over less space and thus provide for less physical continuity of habitat across the landscape. Structural connectedness as considered here adopts an island biogeographic perspective.

What constitutes "functional connectedness" between patches clearly depends on the organism of interest; patches that are connected for bird dispersal might not be connected for salamanders. As habitat is lost and subdivided, at what point does the landscape become functionally "disconnected?" As With and King (1999) notes, "what ultimately influences the connectivity of the landscape from the organism's perspective is the scale and pattern of movement (i.e., scale at which the organism perceives the landscape) relative to the scale and pattern of patchiness (i.e., structure of the landscape);…i.e., a species' gap-crossing or dispersal ability relative to the gap-size distribution on the landscape" (Dale et al. 1994; With and Crist 1995; Pearson et al. 1996; With et al. 1997). Hence, functional connections might be based on: (1) strict adjacency (touching) or some threshold distance, e.g., a maximum dispersal distance); (2) some decreasing function of distance that reflects the probability of connection at a given distance; or (3) a resistance-weighted distance function, e.g., where the distance between two patches is computed as the least cost distance on a resistance surface, where each intervening location between habitat patches is assigned a resistance value based on its permeability to movement by the focal organism. Then various indices of overall connectedness can be derived based on the pairwise connections between patches.

4.3 Looking Forward – The Gradient Concept of Landscape Structure

In the past 20 years the patch mosaic paradigm of landscape ecology has been expanded through integration with spatial population theory (e.g. MacArthur and Wilson 1963; Terborgh 1976; Kareiva and Wennergren 1995), and emerging technologies such as remote sensing (Hobbs and Mooney 1990, and GIS (Haslett 1990; Haines-Young 1993). The revolutionary new computational capabilities of modern computers, coupled with the flexible raster analysis capabilities of GIS, and inexpensive, broadly available remotely sensed imagery provided new inspiration for the evolving field of landscape ecology. It is a dramatic case of the theory of landscape ecology lagging well behind the state-of-the-art in computation, spatial analysis and available spatial data. The landscape mosaic framework is well suited to systems which area dominated by clearly defined, internally homogeneous units, with the advantage of simplicity of representing and analyzing them. However, there are many situations when the patch mosaic model fails or is at best sub-optimal. The patch mosaic model does not accurately represent continuous spatial heterogeneity (McGarigal and Cushman 2005). Once categorized, patches subsume all internal heterogeneity, which may result in the loss of important ecological information. When applying the patch mosaic model in practice, it is prudent to ask whether the magnitude of information loss is acceptable. Most fundamentally, there appears to be a major disjunction between modern ecological theory in the fields of community and population ecology and the patch-mosaic conception of landscape structure. In the first chapter we discussed science as a historical process which is based on underlying conceptual caricatures of natural systems. Given the base assumptions of these underlying models, each field then proceeds to develop theory, collect data, propose relationships. However, rarely are these underlying conceptual models themselves the focus of scrutiny. This results in what we called a "boring-in" whereby these underlying paradigms become entrenched as quasi-dogmatic beliefs. We believe that the patch-mosaic model in landscape ecology is a classic example of this. In the previous two chapters we argued that spatial and temporal variation in ecological systems fundamentally alter pattern-process relationships in a highly scale-dependent way, and that reliable inferences are frequently only obtained by linking mechanisms with responses directly at operative scales and integrating these over space and time. Given this framework, in the following sections of this chapter, we evaluate the traditional patch-mosaic model of landscape structure relative to community ecology theory and propose an alternative framework for landscape ecology.

The modern scientific study of ecological communities is often traced to Clements (1907, 1916), who posited that the composition of species within a community is a deterministic product of regional climate and time since disturbance. Species composition within a community was thought to be highly predictable as deterministic as functions of regional climate and seral condition. However, the Clementsian view of communities as analogous to super-organisms was fundamentally challenged by Gleason (1917, 1926), who argued that identification of categorical vegetation types was inconsistent with the large amounts of heterogeneity in plant communities, arguing that areas of vegetation are actually similar to one another only by degrees and not in kind. He questioned delineating patch-mosaic maps of community types and opposed grouping of species in nameable associations. As an alternative, Gleason offered the individualistic concept of the plant association in which "the phenomena of vegetation depend completely upon the phenomena of the individual" species (Gleason 1917).

This individualistic concept of vegetation ecology is the foundation of modern community ecology. The fusion of individualistic community ecology (Gleason 1926; Curtis and McIntosh 1951; Whittaker 1967) with the Hutchinsonian niche concept (Hutchinson 1957) enabled a broad integration of ecological theory, spanning all the way from Darwinian evolution, to the niche characteristics of individual species, to the composition, structure and dynamics of ecological communities. Each species is seen responding to local environmental and biotic conditions. The biotic community in this context is conceived as a collection of species that are occurring together at a particular place and a particular time due to overlapping tolerances of environmental conditions and vagarities of history, rather than an integrated and deterministic mixture. Research in this paradigm focuses on extending the individual concept to quantitative analysis of species distribution along environmental gradients and the effort to quantify the fundamental niche of each species in terms of the range of resources and conditions needed for that species to survive. The natural level of focus of such analyses is the species, not community type, assemblage or patch type; the natural focal scale for such analyses is the location or pixel, rather than the stand or patch (McGarigal and Cushman 2005; Cushman et al. 2007).

4.3.1 Clementsian Landscape Ecology

Landscape ecology has been variously described as the study of the structure, function and management of large heterogeneous land areas (Forman 1995) or, more generally, the study of spatial pattern and process (Turner 1989, 2005). Likewise, landscapes are typically described in terms of patches, corridors, and matrix (Forman 1995). These definitions explicitly frame the scope of landscape ecology within a categorical patch-mosaic paradigm, and any reading of the published landscape ecology literature shows near uniformity in the adoption of this approach. In the patch-mosaic paradigm, each patch is implicitly treated like an individual of the super organism of each "patch type", and it is assumed that by measuring the area and configuration of patch types we can represent the most important attributes of the landscape, including the distribution and abundance of plant and animal species.

However, if biological communities are multivariate gradients of species composition, with each species responding individualistically to particular combinations of limiting factors, is a categorical patch-based representation appropriate? Put another way, isn't representing biological communities as categorical patches in a mosaic a de facto ratification of a Clementsian model of community composition at the landscape level? Most ecological attributes are inherently continuous in their spatial variation (at least at some scales; Wiens 1989), even in human-dominated landscapes. Consider soil properties such as depth, texture and chemistry, and terrain properties such as elevation, slope and aspect. These physical environmental properties typically vary continuously over space despite discontinuities in above-ground land cover that might exist due to natural or anthropogenic disturbances. Even above-ground land cover defined on the basis of vegetation more often than not varies continuously along underlying environmental gradients, except where humans have substantially modified it (Austin and Smith 1989; Austin 1999).

These observations have led several authors to propose alternatives to the patch mosaic model of landscape structure for situations where spatial heterogeneity is continuous rather than discrete. McIntyre and Barret (1992) introduced the "variegation" model as an alternative to the "island biogeographic" model, in which habitat is viewed as a continuous gradient instead of discrete patches within a homogeneous matrix. Later, Manning et al. (2004) defined the "continua-umwelt" model as a refinement of the variegation model in which habitat gradients are species-specific and governed by ecological processes in a spatially continuous and potentially complex way. Fischer and Lindenmayer (2006) offered an additional refinement of the continua-umwelt model by suggesting that the landscape be defined on the basis of four specific habitat gradients (food, shelter, space and climate) that are closely related to ecological processes that affect the distribution of animals. Importantly, these alternative conceptual models are all habitat-centric; that is, they propose a gradient model of "habitat"; they do not provide a general purpose model of landscape structure.

McGarigal and Cushman (2005) introduced a general conceptual model of landscape structure based on continuous rather than discrete spatial heterogeneity; they referred to this as the "landscape gradient" model. In this model, the underlying heterogeneity is viewed as a three-dimensional surface and can represent any ecological attribute(s) of interest. The most common example of a landscape gradient model is a digital elevation surface, but there are many other possibilities. Of course McGarigal and Cushman (2005) were not the first to recognize the need to characterize three-dimensional surfaces for ecological purposes. Geomorphologists, for example, have long sought ways to characterize land surfaces for the purpose of understanding the relationships between landforms and geomorphological processes (e.g., Strahler 1952; Schumm 1956; and Melton 1957), and biologists as early as 1983 have sought ways to assess topographic roughness for the purpose of characterizing fish and wildlife habitat (e.g., Beasom 1983; Sanson et al. 1995). To this end, many methods have been developed to quantify surface complexity (e.g., Pike 2000; Wilson and Gallant 2000; Jenness 2004). However, until recently these methods have focused almost exclusively on characterizing topographic surfaces at the scale of the individual pixel or cell (e.g., Moore et al. 1991; Jenness 2005), or as the basis for mitigating the source of error associated with the planimetric projection of slopes in the calculation of patch metrics (e.g., Dorner et al. 2002; Hoechstetter et al. 2008). Only recently has attention been given to the application of surface metrics for the purpose of quantifying surface heterogeneity at the scale of entire landscapes (McGarigal and Cushman 2005; Hoechstetter et al. 2008; McGarigal et al. in press).

Largely unknown to landscape ecologists, researchers involved microscopy and molecular physics have made large advances in the area of three-dimensional surface analysis, creating the field of surface metrology (Stout et al. 1994, Barbato et al. 1995; Villarrubia 1997; Ramasawmy et al. 2000). Over the past two decades structural and molecular physicists have been developing surface metrics which we believe will be highly applicable to landscape gradients (e.g., Gadelmawla et al. 2002). Until recently, however, there have been no landscape ecological applications of these surface metrics. Recently, McGarigal et al. (in press) described the use of surface metrics for quantifying landscape patterns. Specifically, they (1) clarify the relationship between the patch mosaic and gradient models of landscape structure and the metrics used to characterize landscapes under each model; (2) describe a variety of surface metrics with the potential for quantifying the structure of landscape gradients; (3) evaluate the behavior of a large suite of surface; and (4) discuss the challenges to the application of surface pattern metrics in landscape ecological investigations.

4.4 Conclusion

The analysis of landscape pattern to infer process is the underlying tenant in the field of landscape ecology (Forman and Godron 1986; Forman 1995; Turner et al. 2001). One's ability to effectively explain ecological processes therefore depends on correctly representing ecological patterns. Landscape ecology traditionally adopts a patch mosaic model of ecological patterns, implicitly assuming discretely bounded and categorically defined patches are sufficient to explain pattern-process relationships (McGarigal and Cushman 2005). However, most ecological attributes are inherently continuous and classification of species composition into vegetation communities and discrete patches provides an overly simplistic view of the landscape and limits our ability to explore the continuous nature of plant distributions (Cushman et al. in press, McGarigal et al. in press; Evans and Cushman in press).

The fusion of individualistic community ecology with the Hutchinsonian niche concept enabled a broad integration of ecological theory, spanning all the way from the niche characteristics of individual species, to the composition, structure and dynamics of ecological communities. Landscape ecology has been variously described as the study of the structure, function and management of large heterogeneous land areas. Any reading of the published landscape ecology literature shows near uniformity in the adoption of a categorical patch-mosaic paradigm. However, if biological communities are multivariate gradients of species composition, with each species responding individualistically to particular combinations of limiting factors, is a categorical patch-based representation appropriate?

If one adopts a niche-based (Hutchinson 1957), individualistic concept (Gleason 1926; Whittaker 1967) of biotic communities then it would be more appropriate to represent ecological patterns as continuous measures rather than the traditional abstraction into categorical community types represented in a mosaic of discrete patches (McGarigal and Cushman 2005; Cushman et al. 2007, Cushman et al. in press). Although the problem of categorizations of the landscape failing to represent continuous ecological patterns has been identified (McIntyre and Barret 1992; Manning et al. 2004; McGarigal and Cushman 2005; Cushman et al. 2007), few approaches have been proposed to predict gradients in a modeling environment (McGarigal et al. in press; Evans and Cushman in press). The next chapter presents a detailed evaluation of this gradient concept of landscape structure and how it ties into the concepts related to spatial and temporal complexity and scaling presented in the previous three chapters. Specifically, we discuss how a range of ecological questions, including mapping vegetation composition and structure, modeling wildlife habitat relationships, predicting habitat connectivity and measuring the genetic structure of populations all may best be addressed using flexible, multivariate, multi-scale gradient approaches. By moving from a landscape ecological paradigm based on categorical patches to one based on quantitative species and environmental responses across continuous space it may be possible to both produce much more effective predictions of species distributions and ecological processes and remove much of the disjunction between landscape ecology and mainstream community ecology theory.

References

Austin MP (1999) The potential contribution of vegetation ecology to biodiversity research. Ecography 22:465–484

Austin MP, Smith TM (1989) A new model for the continuum concept. Vegetatio 83:35–47

Barbato G, Carneiro K, Cuppini D, Garnaes J, Gori G, Hughes G, Jensen CP, Jorgensen JF, Jusko O, Livi S, McQuoid H, Nielsen L, Picotto GB, Wilening G (1995) Scanning tunnelling microscopy methods for the characterization of roughness and micro hardness measurements. Synthesis report for research contract with the European Union under its programme for applied metrology. European Commission Catalogue number: CD-NA-16145 EN-C Brussels Luxemburg

Beasom SL (1983). A technique for assessing land surface ruggedness. J Wildl Manag 47:1163–1166
Christian CS (1958) The concept of land units and land systems. Proc Ninth Pacific Sci Cong 20:74–81
Clements FE (1907) Plant physiology and ecology. Hentry Hold, New York
Clements FE (1916) The nature and structure of the climax. J Ecol 24:252–284
Cowles HC (1899) The ecological relations of the vegetation on the sand dunes of lake Michigan. Botanical Gazette 27:95–117
Curtis JT, McIntosh RP (1951) An upland forest continuum in the prairie-forest border region of Wisconsin. Ecol Monogr 32:476–496
Cushman SA, McGarigal K (2002) Hierarchical, multiscale decomposition of species-environment relationships. Landsc Ecol 17:637–646
Cushman SA, McKenzie D, Peterson DL, Littell J, McKelvey KS (2007) Research agenda for integrated landscape modelling. USDA For Serv Gen Tech Rep RMRS-GTR-194
Dale VH, O'Neill RV, Southworth F, Pedlowski M (1994) Modeling effects of land management in the Brazilian Amazonia settlement of Rondonia. Conserv Biol **8**:196–206
Dansereau P (1957) Biogeography: an ecological perspective. Roland, New York
Dansereau P (1975) Inscape and landscape: the human perception of environment. Columbia University Press, New York
Dickinson RE (1970) Regoinal ecology: the study of man's environment. Wiley, New York
Dorner B, Lertzman K, Fall J (2002) Landscape pattern in topographically complex landscapes: issues and techniques for analysis. Landsc Ecol 17:729–743.
Egler FE (1942) Vegetation as an object of study. Philos Sci 9:245–260
Evans JS, Cushman SA (2009) Gradient modeling of conifer species using random forests. Landscape Ecology 24: 673–683.
Fischer J, Lindenmayer DB (2006) Beyond fragmentation: the continuum model for fauna research and conservation in human-modified landscapes. Oikos 112:473–480
Forbes SA (1887) The lake as a microcosm. Bull. of the Scientific Association (Peoria, IL), 1887: 77–87
Formann RTT (1981) Patches and structural components for a landscape ecology. Bioscience 31:733–744
Forman RTT (1995) Land mosaics: the ecology of landscapes and regions. Cambridge University Press, Cambridge
Forman RTT, Godron M (1986) Landscape ecology. Wiley, New York
Gadelmawla ES, Koura MM, Maksoud TMA, Elewa IM, Soliman HH (2002) Roughness parameters. J Mat Process Tech 123:133–145
Gleason HA (1917) The structure and development of the plant association. Bull Torrey Bot Club 43:463–481
Gleason HA (1926) The individualistic concept of the plant association. Bull Torrey Bot Club 53:7–26
Gustafson E J, Parker GR (1992) Relationships between landcover proportion and indices of landscape spatial pattern. Landsc Ecol **7**:101–110
Haines-Young (1993) Landcape ecology and geographic information systems. Taylor & Francis, London
Haslett JR (1990) Geographic information systems: a new approach to habitat definition and the study of distributions. Trends Ecol Evol 5:214–218
Hobbs RI, Mooney HA (1990) Ecological studies 79: remote sensing of biosphere functioning. Springer, New York
Hoechstetter S, Walz U, Dang LH, Thinh NX (2008) Effects of topography and surface roughness in analyses of landscape structure – a proposal to modify the existing set of landscape metrics. Landsc Online 1:1–14
Hutchinson GE (1957) Concluding remarks. Cold Spring Harbor Symp Quantitative Biol 22:415–427
Isard W (1975) Introduction to regional science. Prentice-Hall, Englewood Cliffs, NJ

Jaeger JAG (2000) Landscape division, splitting index, and effective mesh size: new measures of landscape fragmentation. Landsc Ecol **15**:115–130
Jenness J (2004) Calculating landscape surface area from digital elevation models. Wildl Soc Bull 32:829–839
Jenness J (2005) Topographic Position Index (tip_jen.avx) extension for ArcView 3.x., Jenness Enterprises. http://www.jennessent.com/arview/tpi.htm.
Kareiva P, Wennergren U (1995) Connecting landscape patterns to ecosystem and population processes. Nature 373:299–302
Keitt TH, Urban DL, Milne BT (1997) Detecting critical scales in fragmented landscapes. Conservat Ecol 1:4
Krebs CJ (1994). Ecology: the experimental analysis of distribution and abundance, 4th ed. Harper Collins, New York
Krummel J R, Gardner RH, Sugihara G, O'Neill RV, Coleman PR (1987) Landscape configurations in a disturbed environment. Oikos **48**:321–324
Lowe JC, Moryadas S (1975) The geography of movement. Houghton Mifflin, Boston, MA
MacArthur RH, Wilson EO (1963) A equilibrium theory of insular zoogeography. Evolution 17:373–387
Manning AD, Lindenmayer DB, Nix HA (2004) Continua and umwelt: novel perspectives on viewing landscapes. Oikos 104:621–628
McGarigal K, Cushman SA (2005) The gradient concept of landscape structure. In: Wiens J, Moss M (eds) Issues and Perspectives in Landscape Ecology. Cambridge University Press, Cambridge
McGarigal K, Marks BJ (1995) FRAGSTATS: spatial pattern analysis program for quantifying landscape structure. US For Serv Gen Tech Rep PNW-GTR-351
McGarigal K, Cushman SA, Neel MC, Ene E (2002) FRAGSTATS: spatial pattern analysis program for categorical maps. Computer software program produced by the authors at the University of Massachusetts, Amherst www.umass.edu/landeco/research/fragstats/fragstats.html
McGarigal K, Tagil S, Cushman SA (2009) Surface metrics: An alternative to patch metrics for the quantification of landscape structure. Landsc Ecol 24:433–450
McIntyre S, Barrett GW (1992) Habitat variegation, an alternative to ragmentation. Conservat Biol 6:146–147
Melton MA (1957) An analysis of the relations among elements of climate, surface properties, and geomorphology. Columbia Univ, Dep Geol, Proj NR 389–042 Tech Rep 11, New York
Moore ID, Gryson RB, Ladson AR (1991) Digital terrain modeling: a review of hydrological, geomorphological, and biological applications. Hydrolog Process 5:3–30
Moore ID, Gessler PE, Nielsen GA, Petersen GA (1993) Terrain attributes: estimation methods and scale effects. In: Jakeman MB, McAleer (eds): Modeling change in environmental systems. Wiley, London
Naveh Z, Lieberman A (1994) Landscape ecology: theory and applications, 2nd ed. Springer, New York
Pearson SM, Turner MG, Gardner RH, O'Neill RV (1996) An organism-based perspective of habitat fragmentation. *In:* Szaro RC, Johnston DW (eds) Biodiversity in managed landscapes: theory and practice. Oxford University Press, New York
Pickett STA, Cadenasso ML (1995) Landsape ecology: spatial heterogeneity in ecological systems. Science 269:331–334
Pike RJ (2000) Geomorphometry – diversity in quantitative surface analysis. Progr Phys Geogr 24:1–20
Ramasawmy H, Stout KJ, Blunt L (2000) Effect of secondary processing on EDM surfaces. Surf Eng J 16:501–505
Real LA, Brown JH (1991) Foundations of ecology: classic papers with commentaries. University of Chicago Press, Chicago, IL
Sanson GD, Stolk R, Downes BJ (1995) A new method for characterizing surface roughness and available space in biological systems. Funct Ecol 9:127–135.
Sauer CO (1965) Land and life. University of California Press, Berkeley.

Schumm SA (1956) Evolution of drainage basins and slopes in badlands at Perth Amboy, New Jersey. Bull Geol Soc Am 67:597–646
Stout KJ, Sullivan PJ, Dong WP, Mainsah E, Lou N, Mathia T, Zahouani H (1994) The development of methods for the characterization of roughness on three dimensions. Publication no EUR 15178 EN of the Commission of the European Communities, Luxembourg
Strahler AN (1952) Hypsometric (area-altitude) analysis of erosional topography. Bull Geol Soc Am 63:1117–1142
Taaffe EJ, Gauthier HJ (1973) Geography of transportation. Prentice-Hall, Englewood Cliffs, NJ
Tansley AG (1935) The use and misuse of vegetatoinal terms and concepts. Ecology 16
Terborgh J (1976) Island biogeography and conservation: strategy and limitations. Science 193:1029–1030
Troll C (1971) Landscape ecology (geoecology) and biogeocenology: a terminological study. GeoForum 8:43–46
Turner MG (1987) Landscape heterogeneity and disturbance. Ecol Studies 64. Springer, New York
Turner MG (1989) Landscape ecology: the effect of pattern on process. Annl Rev Ecol Systemat 20:171–197
Turner MG (2005) Landscape ecology: what is the state of the science? Annu Rev Ecol Evol Syst 36:319–44
Turner MG, Gardner RH, O'Neill RV (2001) Landscape ecology in theory and practice. Springer, New York
Villarrubia JS (1997) Algorithms for scanned probe microscope, image simulation, surface reconstruction and tip estimation. J Nat Inst Stand Technol 102:435–454
Whittaker RH (1967) Gradient analysis of vegetation. Biol Rev 42:207–264
Wiens JA (1989) Spatial scaling in ecology. Funct Ecol 3:385–397
Wiens JA (1994) Habitat fragmentation: island v landscape perspectives on bird conservation. Ibis 137:S97–S104
Wiens JA, Stenseth NC, Van Horne B, Ims RA (1993) Ecological mechanisms and landscape ecology. Oikos 66:369–380
Wilson JP, Gallant JC (2000) Terrain analysis: principles and applications. Wiley, New York
With KA, Crist TO (1995) Critical thresholds in species' responses to landscape structure. Ecology 76:2446–2459
With, KA, and King, AW (1999) Extinction thresholds for species in fractal landscapes. Conserv Biol 13:314–326
With KA, Gardner RH, Turner MG (1997) Landscape connectivity and population distributions in heterogeneous environments. Oikos 78:151–169
Zonneveld IS (1972) Textbook of photo-interpretatoin, Vol. 7. ITC, Enschede

Chapter 5
The Gradient Paradigm: A Conceptual and Analytical Framework for Landscape Ecology

Samuel A. Cushman, Kevin Gutzweiler, Jeffrey S. Evans, and Kevin McGarigal

5.1 Introduction

Landscape ecology deals fundamentally with how, when, and why patterns of environmental factors influence the distribution of organisms and ecological processes, and reciprocally, how the actions of organisms and ecological processes influence ecological patterns (Urban et al. 1991; Turner 1989). The landscape ecologist's goal is to determine where and when spatial and temporal heterogeneity matter, and how they influence processes. A fundamental issue in this effort revolves around the choices a researcher makes about how to depict and measure heterogeneity (Turner 1989; Wiens 1989). Indeed, observed patterns and their apparent relationships with response variables often depend on the scale that is chosen for observation and the rules that are adopted for defining and measuring variables (Wiens 1989; Wu and Hobbs 2000; Wu and Hobbs 2004). Success in understanding pattern–process relationships hinges on accurately characterizing heterogeneity in a manner that is relevant to the organism or process under consideration.

To characterize heterogeneity, landscape ecologists have generally adopted a single approach – the patch-mosaic model of landscape structure (Forman and Godron 1986; Turner 1989; Forman 1995; Turner et al. 1989). In this model a

S.A. Cushman (✉)
US Forest Service, Rocky Mountain Research Station, 800 E Beckwith, Missoula, MT 59801, USA
e-mail: scushman@fs.fed.us

K. Gutzweiler
Baylor University, Waco, TX, USA

J.S. Evans
Senior Landscape Ecologist, The Nature Conservancy, NACR - Science

K. McGarigal
Department of Natural Resources Conservation, University of Massachusetts, Amherst, MA, USA

S.A. Cushman and F. Huettmann (eds.), *Spatial Complexity, Informatics, and Wildlife Conservation*
DOI 10.1007/978-4-431-87771-4_5, © Springer 2010

landscape is represented as a collection of discrete patches; major discontinuities in underlying environmental variation are depicted as discrete boundaries between patches, and all other variation is subsumed by the patches and implicitly assumed to be irrelevant. The patch-mosaic model provides a simplifying framework that facilitates experimental design, analysis, and management consistent with established tools (e.g., FRAGSTATS) and methods (e.g., ANOVA). The patch-mosaic model also is the foundation for the major axioms of contemporary landscape ecology (e.g., patch structure matters, patch context matters, pattern varies with scale). Yet, even the most ardent supporters of this model recognize that categorical representation of environmental variables often poorly represents the true heterogeneity of the system, which often consists of continuous multi-dimensional gradients of environmental attributes. We believe that advances in landscape ecology are constrained by the lack of methods and analytical tools for effectively depicting and analyzing continuously varying ecological phenomena at the landscape level.

In the sections that follow, we explain the limitations of categorical map analyses for landscape ecology and then discuss the gradient paradigm, and explain how it can be used to overcome many limitations of the patch-mosaic model. We finish by illustrating specific benefits of gradient approaches using real data. The patch-mosaic model has great heuristic value, and it is the appropriate model to use under many circumstances, such as when natural or anthropogenic forces have created sharp environmental discontinuities. But we argue below that a patch-mosaic model of landscape structure is prone to large errors and distortion of underlying environmental patterns that can obscure true pattern–process relationships and inhibit flexible analysis across scales. We also argue that a gradient based representation of landscape structure is much more consistent with fundamental ecological theory, and that to achieve the full potential of integrating spatial analysis with quantitative ecology the categorical patch-mosaic model should take its rightful place as a special case within a generalized gradient framework.

5.1.1 Limitations of Categorical Mapping

Many of landscape ecology's perspectives and techniques have their origins in classical cartographic analysis (Forman and Godron 1986). The first step in most landscape ecology analsyses is to map the system. It has become traditional in geography to abstract the world into non-overlapping regions, or polygons. In terms of observational scale, this kind of mapping truncates the intensity of measured variables into categories. Quantitative information about how variables vary through space and time is lost, leaving rigid, internally homogeneous patches. Though this perspective has been useful for many applications, it is important to recognize how it influences measurements and analyses.

In categorical mapping, discontinuities are presupposed; the world is assumed to be inherently discrete. When quantitative landscape variation is reduced to categories, four important representation and interpretation problems are generated. First, subjective decisions of what to characterize and how to define boundaries will constrain what patterns can be seen and what relationships can be inferred. Second, patch boundaries based on criteria defined by the observer may not be meaningful or even perceived by the organism in question. Third, once patches are created, all internal variability within and among patches of the same class is eliminated, and all interclass differences are reduced to categorical differences. Fourth, categorical patches define the regions of assumed homogeneity in a single or composite attribute. Once defined, all variability in that attribute not used to define the patch is discarded. The cumulative effect of these issues can result in any number of statistical problems associated with data aggregation including the Modifiable Ariel Unit Problem (Openshaw 1984; Jelinski and Wu 1996; Wu 2007), Ecological Fallacy (Robinson 1950; Wood and Skole 1998; Wu 2007), and misspecification (Guthrie and Sheppard 2001).

Two or more layers of patches can be overlaid and analyzed using map algebra. This is the standard approach to analyzing multi-level categorical map patterns. However, the boundaries of patches in different layers are often poorly related, as they reflect slices through the distributions of independently varying environmental attributes and are based on different classification rules. In the traditional patch-based model, analyzing many layers of patches results in intractably vast numbers of unique map categories. There often are many "sliver" patches resulting from poor matching of edges that are not indicative of any ecological process. This magnitude of this latter problem increases multiplicatively with additional choropleth layers. When a researcher attempts to predict a response variable, such as the habitat suitability for a particular species, as a function of a number of landscape-level attributes across several categorical data layers, prediction can only be based on combinations of categories.

With each combination of categorical data the information loss multiplies, as do the errors of misclassification. No such penalty is incurred for combining quantitatively scaled variables. If the same response variable is predicted on the basis of several layers of quantitative predictor variables, the prediction can be based on how the quantitative landscape-level variables covary along dimensions that are related to the species or process in question (McGarigal and Cushman 2005). In addition, preserving quantitative ecological factors reduces subjectivity. The subjectivity of boundary definition is replaced by the subjectivity of measurement resolution, which often involves less-restrictive assumptions than do decisions about category width and boundary definitions. Retaining the quantitative scale of ecological variables also enables one to analyze many response variables simultaneously, with each responding individually to multiple landscape gradients.

When categorical patch mosaics are derived specifically to correspond to the scale and sensitivities of a particular organism or ecological process, they may represent landscape heterogeneity in an ecologically meaningful way. In most cases, however, little is known about the scale and resolution of landscape variability that are pertinent, and patterns at several scales may simultaneously influence an organism or process. Reducing a continuous ecological surface to a patch mosaic causes representation and interpretation problems because of inaccuracies in boundary placement and class divisions (Openshaw 1984), or because ecological variation is important across several scale ranges (Wu 2007). Even if a constructed patch mosaic ideally represents an organism's ecological landscape, this mosaic is not likely to do so in an optimal way for a second or third organism, making comparisons between organisms based on a single landscape map questionable (Cushman et al. 2007).

5.1.2 Gradient Attributes of Categorical Patterns

Even when categorical data is appropriate, conventional analytical methods often fail to produce unbiased assessments of organism responses. Organisms often experience categorical environments as pattern gradients. For example, consider a species that responds to landscape structure as measured by the density of edges in the landscape weighted by their structural contrast. Traditional landscape pattern analysis would measure the total contrast-weighted edge density for the entire landscape. However, landscape patterns are rarely stationary, and there may be no place in the landscape with a contrast-weighted edge density equivalent to that calculated for the landscape as a whole. If the landscape is large relative to the organism's home range, the organism is unlikely to even experience the global average structure of the landscape. The organism responds to the local structure within its immediate perception, within its daily foraging area, and within its home range. Thus, a more useful description of landscape pattern would be a location-specific measure at a scale relevant to the organism or process of interest (Wiens 2001; Wu 2007). We propose that organisms experience landscape structure as pattern gradients that vary through space according to the distance at which a particular organism perceives or is influenced by landscape patterns. Therefore, instead of analyzing global landscape patterns, it is usually more appropriate to quantify the local landscape pattern across the space delimited by an organism's perceptual abilities.

Tools exist to calculate traditional patch based landscape metrics within a moving window (e.g. McGarigal et al. 2002). The window size should be selected such that it reflects the scale at which the organism perceives or responds to pattern. If this is unknown, one can vary the size of the window over several runs and empirically determine to which scale of a landscape variable an organism is most responsive. The window moves over the landscape one cell at a time, calculating the selected metric within the window and returning that value to the center cell. The result is a continuous surface which reflects how an organism of that perceptual ability would perceive the structure of the landscape as measured by that metric (Fig. 5.1). The surface then would be available for combination with other such surfaces in multivariate models to predict, for example, the distribution and abundance of an organism continuously across the landscape.

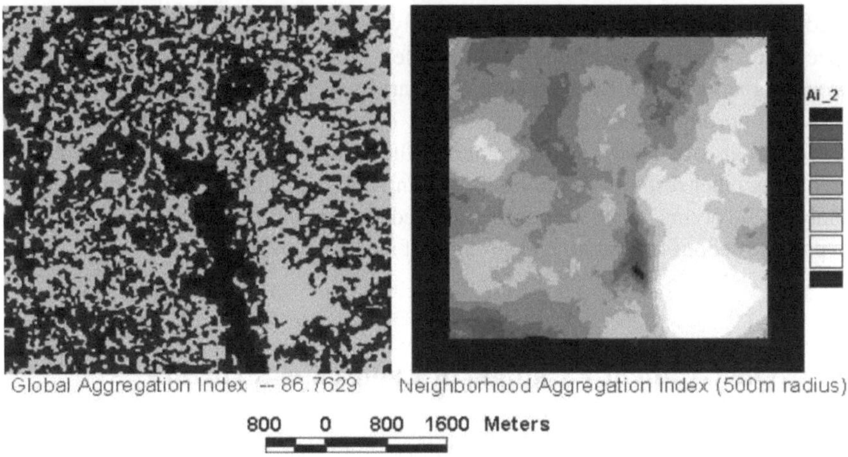

Fig. 5.1 Comparison of global and neighborhood-based calculation of a landscape metric for a categorical map. The Aggregation Index (AI) was calculated for the "forest" class (grey) in the binary map on the left for the landscape overall, and within 500-m-radius circular windows centered on each pixel. The moving window calculation, shown on the right, produces a surface whose height is equal to the neighborhood AI value. There is a border classified as "no data" around the edge of the landscape to a depth of the selected neighborhood radius. Higher AI values are light, lower values are dark

5.1.3 Gradient Analysis of Continuous Field Variables

When patch mosaics are not clearly appropriate as models of the variability of particular environmental factors, there are a number of advantages to modeling environmental variation as individually varying gradients. First, it preserves the underlying heterogeneity in the values of variables through space and across scales. The subjectivity of delimiting boundaries is eliminated. This enables the researcher to preserve in the analysis many variables that vary independently, avoiding the disadvantages of reducing the set to a categorical description of boundaries defined on the basis of one or a few attributes. In addition, the subjectivity of defining cut points for categories is eliminated. With gradient data, scale inaccuracy and boundary sensitivity are not issues because the quantitative representation of environmental variables preserves the entire scale range and the complete gradient. The only real subjectivity is the resolution at which to measure variability.

By tailoring the grain, extent, and intensity of the measurements to the hypotheses and system under investigation, researchers can develop a less equivocal picture of how the system is organized and what mechanisms may be at work. An important benefit is that one can directly assess relations between a continuous response variable for an organism with spatial and temporal patterns in the environment that are continuously scaled. By not truncating patterns of variation in landscape variables to a particular scale and set of categories, one can use a single set of predictor variables to simultaneously analyze many response variables, be they species responding individually to complex landscape gradients, or ecological processes acting at different scales. Comparison between organisms or processes is not compromised, because each can be optimally predicted by the surface or combinations of surfaces without altering the data in ways that limit its utility for predicting other response variables. Importantly, this facilitates efficient multivariate analyses involving many response and predictor variables simultaneously to test hypotheses about the nature and strength of system control.

5.2 The Gradient Paradigm of Landscape Structure

We propose a conceptual shift in spatial ecology that integrates categorical and continuous perspectives. We believe it will be useful for landscape ecologists to adopt a gradient perspective, along with a new suite of tools for analyzing landscape structure and the linkages of patterns and processes under a gradient framework. This framework includes, where appropriate, categorically mapped variables as a special case. In the sections that follow we outline how a gradient perspective can be valuable in several areas of landscape ecological research.

5.2.1 Evaluating A Categorical Mapping of Canopy Density

In this example we explore the differences between gradient and categorical representations of an important ecological variable, canopy density. Canopy density is a measure of the amount of canopy photosynthetic material per unit of ground surface area, and is correlated with a number of ecological processes of interest, including net primary productivity, carbon sequestration rate, and is an important habitat attribute for many wildlife species.

5.2.1.1 Mapping Approaches

Lidar

Lidar data was acquired in August of 2006 by Watershed Sciences, Corvallis, Oregon using a Leica-ALS50 sensor with a pulse repetition frequency of 80 kHz, a nominal point-density of ~48 points per/m^2, and a maximum scan angle of 14°. Ground measurements were identified using Multiscale Curvature Classification (Evans and Hudak 2007). Canopy density was calculated using the ratio of non-ground to ground measurements (Fig. 5.2a) within a 15 m cell size to make it directly comparison with VMAP. The ratio of non-ground/ground Lidar measurements

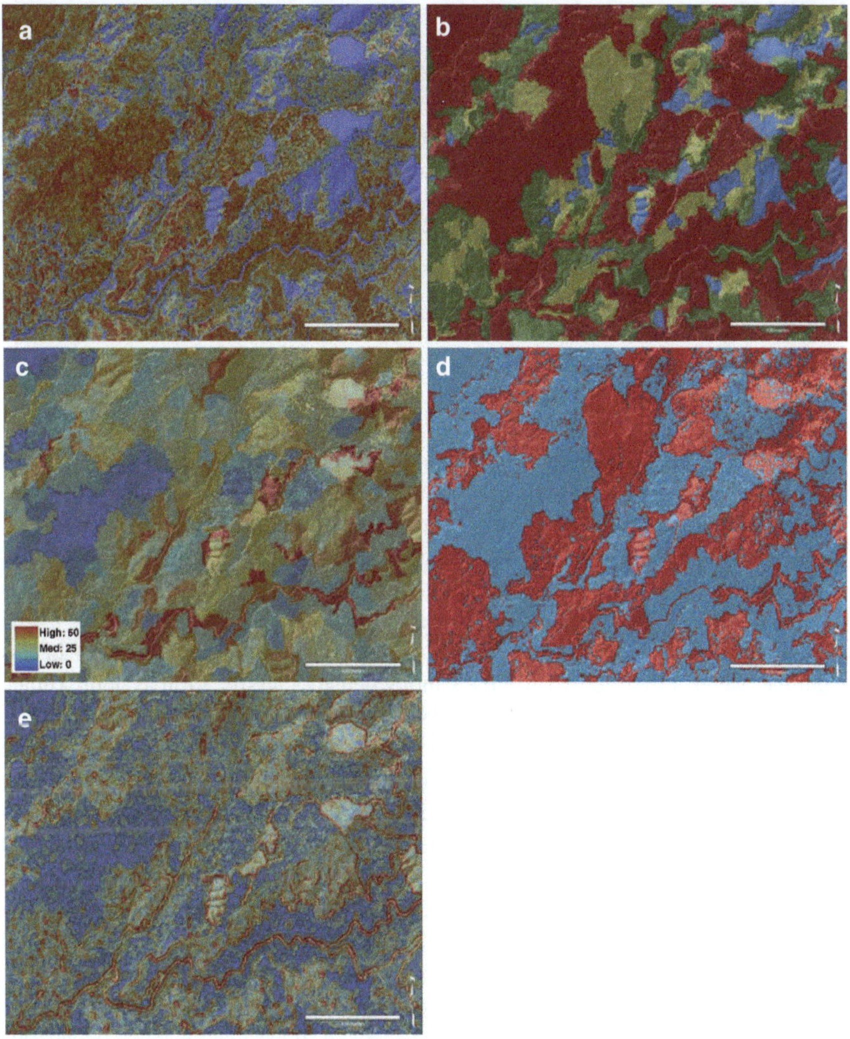

Fig. 5.2 (**a**) Lidar derived canopy density 15 m. Colors are ramped, blue (0%) to red (100%) using a standard deviation stretch. (**b**) VMAP derived canopy density. Colors are representative of 4 classes, blue (0–10%), green (25–59%), yellow (10–24%), and red (60–100%). (**c**) Standard deviation of lidar canopy density by VMAP polygon (**d**) Non agreement between VMAP and Lidar classification (45% error). Red is an error, blue is correct. (**e**) Maximum Rate of Change, Canopy Density 15 m. Colors are ramped, blue (low change) to red (high change) using a standard deviation stretch

accurately represent the amount of light reaching the ground and are directly comparable to traditional measures of canopy density. Correlation between Lidar derived and field measured canopy cover have been strongly supported (r = 0.97) in several studies.

R1-VMAP

The USDA Forest Service Region 1 VMAP project is a vegetation classification based on hierarchal image segmentation. Multi-temporal Landsat ETM + 7 spectral bands were fused with the panchromatic band (band 8) to create 15 m multispectral images. These images were used to create image object polygons using eCognition. Canopy density was classified into four classes; 1–9%, 10–24%, 25–59%, and 60–100% (Fig. 5.1b). Validation was conducted using photo-interpretation reporting a producer's accuracy based on omission error (65.4%) and a user's accuracy based on commission error (78.7%).

5.2.1.2 Analysis and Interpretation

Canopy density is an inherently continuous attribute which varies at scales at least as fine as the canopy width of individual trees. Continuous representation intuitively seems much more appropriate. Lidar effectively represents this continuous variability, given its fine sample resolution (~48 points per/m^2) and sensitivity to fine differences in measurement scale (continuous values from 0 to 100% in this case). It has demonstrated very high accuracy in predicting actual canopy density at a fine spatial scale (Leafsky et al. 1999; Means et al. 2000; Hudak et al. 2006). In this example, we will treat the 15 m^2 lidar canopy density classification as approximate truth and evaluate the deviation of the classified map from it.

This example provides a means to test three important questions. First, do the patch boundaries delineated in Fig. 5.2b correspond to discontinuities in the actual patterns of canopy density. In other words, do the patch boundaries correspond to hard boundaries or "breaks" in canopy density. Second, do the patches correspond to areas of homogenous canopy closure, such that categorical representation does not result in severe loss of information about internal variability. Third, is there a strong relationship between the value of canopy density predicted in the VMAP and lidar canopy classifications at the pixel level.

The patches in Fig. 5.2b do not strongly correspond to discontinuities in actual canopy closure. Visual comparison of Fig. 5.2a and b shows that the patch boundaries in 2b are largely artificial and arbitrary truncations of a continuously varying phenomenon and do not correspond to natural breaks in the pattern. Figure 5.2e further shows that these patches are artificial. The patterns of maximum rate of change of canopy closure in this landscape do not in general suggest the existence of natural boundaries that could meaningfully describe patches, and certainly do not correspond to the patch boundaries shown in Fig. 5.2b

Second, visual inspection of Fig. 5.2a and b also shows that the patches in 2b do not correspond to areas of homogeneous canopy density. This was more formally evaluated by computing the standard deviation of lidar canopy density by VMAP polygon (Fig. 5.2c). The majority of the landscape is covered by patches that have standard deviation of internal canopy density over 25%. Given the range of this value from 0 to 100%, a 25% standard deviation is very large. Over 15% of the landscape is occupied by patches with standard deviation over 50%. This analysis shows that the patches delineated by VMAP for canopy closure do not correspond to areas of internal homogeneity, and the internal heterogeneity is so high that the patches are largely meaningless.

The third question is the accuracy of the VMAP classification in representing the degree of canopy closure. In this comparison we evaluate how well the truncated ranges in the classified map correspond to the same artificial ranges imposed on the lidar map. We have already shown that these truncated ranges are artificial and do not represent natural breaks or areas of internal homogeneity. But a remaining question is; do they at least match pixel by pixel to the same range of values in the lidar map with some accuracy? The Lidar canopy density was classified into the same four classes as VMAP. A Boolean equality operation was performed in Workstation ArcInfo between the classified Lidar and VMAP. We conducted two evaluations of accuracy of VMAP in terms of matching lidar. First, a Persons correlation was calculated in R (R Development Core Team 2007) between the classified Lidar and VMAP adjusting for autocorrelation and degrees of freedom (Dutilleul 1993). The value of this correlation was r = 0.329, which indicates that only approximately 10% of the variation in truncated canopy closure values at the pixel level is explainable VMAP. This indicates that VMAP is a very poor predictor of even artificially truncated ranges of canopy closure. Second, we computed the error between the classified map and the lidar prediction (Fig. 5.2d), calculated as the proportion of cells incorrectly classified into the wrong truncated ranges of canopy closure. This analysis indicated that over 45% of the cells in the classified map were incorrectly assigned to one of the four ranges of canopy density.

The classified map fails each of these three critical questions. The patches do not represent discretely bounded discontinuities. Rather, the pattern of canopy density is continuous at a fine scale in a way that does not lend itself to the identification of discrete patch boundaries. Second, the patches do not represent areas of internal homogeneity, but instead subsume a level of heterogeneity that is nearly the same as that between putative patches. The bins used to truncate this continuous variable are artificial and arbitrary. But even if we assumed them to be meaningful, the classified map fails to accurately predict even these artificial truncations, based on the cell correlation and classification accuracy.

This analysis is a comparison of one environmental variable between one classified map product and one continuous representation. However, the classified map product was produced in a multi-million dollar landscape mapping effort using the best available classification techniques and imagery. The published validation reported a producer's accuracy based on omission error (65.4%) and a user's accuracy based on commission error (78.7%). This classified map thus can be considered representative of the upper end of expected quality among the population of such maps available to ecologists and managers. Its failure to represent the important attributes of this variables spatial pattern and cell-level value suggests that efforts to classify inherently variable ecological attributes into categorical maps are questionable. Even if the cell-level classification into truncated bins was highly accurate, it still would not satisfy questions one and two above, resulting in a distortion of pattern by artificially defining boundaries in a continuous landscape and obliterating the very high degree of internal variability. However, in this case the cell-level accuracy was so low that even if the classification levels were ecologically meaningful, the result is so inaccurate as to be of questionable value.

5.3 Multi-scale Gradient Concept of Habitat

In the previous example we considered how well categorical maps represent continuous attributes of vegetation structure. As we noted in that discussion, there are a great many ecological attributes which vary continuously across multiple spatial scales and for which a gradient approach to representing patterns and analyzing pattern-process relationships might be appropriate. One of the most important of these is habitat.

Ecological theory suggests that species exhibit a unimodal response to limiting resources in n-dimensional ecological space (Whittaker 1967; ter Braak 1988; Cushman et al. 2007b). A species not only requires a certain minimum amount of each resource but also cannot tolerate more than a certain maximum amount. Therefore, each species performs best near an optimum value of a necessary environmental variable and cannot survive when the value diverges beyond its tolerance (Shelford 1931; Schwerdtfeger 1977). The relationships between species' performance and gradients of critical resources and conditions describe its fundamental niche (Hutchinson 1957). The composition of biotic communities changes along biophysical gradients because of how the niche relationships of the constituent species interact with the spatial structure of the environment and competing species (Hutchinson 1957; Whittaker 1967; Austin 1985; Rehfeldt et al. 2006).

Most basically habitat is the resources and conditions necessary to allow survival and reproduction of a given organism (Hutchinson 1957). It is organism specific, and characterized as an n-dimensional function of multiple resources and conditions, each operative at particular spatial scales. Habitat relationships often change along a continuum of spatial scale reflecting the hierarchical nature by which animals select resources (Johnson 1980; Cushman and McGarigal 2004). Because relationships at finer scales may reveal mechanisms that are not apparent at broader spatial scales, a multiscaled, hierarchical approach is valuable (Cushman and McGarigal 2002). The volume of ecological space in which the organism can survive and reproduce defines its "environmental niche" (Hutchinson 1957; Rehfeldt et al. 2006).

The environmental gradients comprising the niche are clines in n-dimensional ecological space. In geographical space these gradients often form complex patterns across a range of scales (Wiens 2001; Wu 2007). The fundamental challenge to integrating landscape and community ecology is linking non-spatial niche relationships with the complex patterns of how environmental gradients overlay heterogeneous landscapes (Austin 1985; ter Braak and Prentice 1988; Cushman et al. 2007). Traditionally there has been a severe disjunction between n-dimensional gradient theory of niche structure and spatial analysis of habitat patterns (McIntyre and Barrett 1992; McGarigal and Cushman 2005). The majority of spatial analyses in habitat ecology have fallen into one of two camps, each of which is conceptually divorced in important ways from gradient theory of niche and habitat structure.

The first paradigm we call the "island biogeographic model" (See Chapter 4). In this model, habitat fragments are viewed as analogues of oceanic islands in an inhospitable sea or ecologically neutral matrix. Under this perspective, discrete habitat patches (fragments) are seen as embedded in a uniform matrix of non-habitat. The key attributes of the model are its representation of the landscape as a binary system of habitat and inhospitable matrix, and that, once lost, habitat remains matrix in perpetuity.

The static island biogeography paradigm has been the dominant perspective since its inception. Its major advantage is simplicity. Given a focal habitat, it is quite simple to represent the structure of the landscape in terms of habitat patches contrasted sharply against a uniform matrix. Moreover, by considering the matrix as ecologically neutral, it invites ecologists to focus on those habitat patch attributes, such as size and isolation, that have the strongest effect on species persistence at the patch level. A major disadvantage of the strict island model is that it assumes a uniform and neutral matrix, which in most real-world cases is a drastic over-simplification of how organisms interact with landscape patterns.

The second major conceptual paradigm is the landscape mosaic model. In this paradigm, landscapes are viewed as spatially complex, heterogeneous assemblages of cover types, which can't be simplified into a dichotomy of habitat and matrix (Wiens et al. 1993; With 2000). Connectivity is assessed by the extent to which movement is facilitated or impeded through different land cover types across the landscape. In this model, connectivity is an emergent property of landscapes resulting from the interaction of organisms with landscape structure.

Niether the island biogeographic nor the landscape mosaic model of habitat is consistent with the basic theory that habitat is organism specific, multiple scaled and characterized by a zone in n-dimensional environmental space that consists of resources and conditions necessary and sufficient for the survival and reproduction of the species. Conceptually, patch based models of habitat are Clementsian, in that patches are proposed as discrete entities, analogous to super-organism habitat types (Clements 1916). It seems ironic that modern landscape ecology has adopted this categorical, super-organismal patch based model, when gradient perspectives on species–environment relationships have been dominant in plant and community ecology for nearly 100 years (Gleason 1926; Whittaker 1967). This example will explore several attributes of this inconsistency. We will begin presenting an evaluation of the sufficiency of island biogeographic and patch mosaic representations of habitat for breeding birds in a forest environment. From this analysis we will argue that habitat quality is a continuous attribute that varies as a function of multiple resources and conditions, across a range of spatial scales, and representations that cast habitat as a categorical attribute in a patch mosaic risk serious error. We then will contrast a categorical representation of habitat with an alternative model in which habitat is represented as a continuous function of multiple variables at several spatial scales, and conclude that a gradient approach is consistent with basic ecological theory and less likely to result in spurious and misleading inferences about habitat amounts and patterns and their relationships with population processes.

The island biogeographic and landscape mosaic approaches implicitly assume that the environmental variation that is important to a species can be accurately represented as a mosaic of categorical patches. The analysis proceeds by proposing a series of landscape element types that are believed to comprise "habitat" for a species. In practice, these are usually represented as vegetation types. These are then classified into "habitat" vs. "nonhabitat" in the island biogeographic perspective, or are left as a mosaic of multiple cover types in the landscape mosaic perspective. We believe this categorical representation of habitat is fundamentally inconsistent with basic ecological theory in that it does not reflect species specific responses to multiple gradients of critical resources or conditions. All variability in environmental attributes is subsumed into a mosaic of patches that may or may not reflect attributes of importance to a species. Importantly, casting habitat as a categorical mosaic makes it very difficult to use the multi-variate and multi-scale methods that have been developed to construct niche-habitat models. This conceptual disjunction provides a major obstacle to linking the methods and theories of niche-relationships to spatial analysis of habitat pattern and its implications to population processes (Urban et al. 2002).

Our example evaluates the sufficiency of categorical representations of habitat patches in comparison to a multi-scale, multivariate approach. It is based on a multi-scale analysis of the habitat relationships of forest birds in the Oregon Coast Range (Cushman and McGarigal 2002, 2004; Cushman et al. 2007). The major question is whether categorical representation of habitat attributes as a patch mosaic is appropriate. For a patch mosaic of vegetation types to serve as an effective proxy for species abundance, Cushman et al. (2007) propose that several conditions must be met simultaneously. The most crucial are: (1) habitat is a proxy for population abundance, and (2) mapped vegetation types provide a proxy for the habitat of multiple species. The first assumption requires that species population sizes are strongly associated with environmental conditions, such that environmental conditions alone are a sufficient proxy for population status and trend. The second assumption states that broadly defined vegetation types provide an effective surrogate for the habitat requirements of each species.

Cushman et al. (2007) assessed habitat relationships across a range of organizational levels, including plot-level measurement of vegetation composition and structure, and the composition and configuration of a classified landscape mosaic of patches representing vegetation cover types and seral stages. They found that the sufficiency of vegetation community types as proxies for habitat was highly dependent on the classification attributes and spatial scales at which communities were defined, and varied greatly among species. Their multi-scale analysis revealed that a large proportion of the variance in species abundance could not be explained by mapped community types, no matter how they were defined, and that fine-scale measurements of abiotic conditions and vegetation composition and structure were essential predictors of species abundance (Cushman and McGarigal 2004; Cushman et al. 2007). This suggests that the patch mosaic model, in addition to being conceptually distant from fundamental theories of the factors that drive species–environment relationships, also fails in practice to provide a strong predictor of habitat quality. This is primarily because of two factors. First, habitat is a multi-dimensional attribute, uniquely defined for each species, based on the resources it requires and conditions it can tolerate. Second, each of these critical resources or conditions may affect a species at a particular characteristic set of spatial scales. A categorical mosaic is inappropriate for both of these considerations. It is difficult to represent an n-dimensional function of environmental variation as a categorical mosaic. It is likewise difficult to define a unique patch mosaic from the habitat perspective of each individual organism. In addition, it is challenging to integrate environmental variation at several spatial scales into a single categorical representation of habitat quality (McGarigal and Cushman 2005).

The fundamental challenge to integrating the niche theory of habitat with spatial ecology lies in linking non-spatial niche relationships with the complex patterns of how environmental gradients overlay heterogeneous landscapes (Austin 1985; McIntyre and Barrett 1992; Urban et al. 2002; Manning et al 2004; Cushman et al. 2007b). By establishing species optima and tolerances along environmental gradients, researchers can quantify the characteristics of each species' environmental niche. The resulting statistical model can be used to predict the biophysical suitability of each location on a landscape for each species (Rehfeldt et al. 2006; Evans and Cushman 2009) (Fig. 5.3). This mapping of niche suitability onto complex landscapes is the fundamental task required to predict individualistic species responses to complexes of environmental conditions across landscapes. Importantly, it is fundamentally a gradient modelling exercise and the results are predictions of expected probability of occurrence, relative density or some other measure of habitat quality as continuous functions of multiple resources measured at one to many spatial scales. The insufficiency of patch-based representations of environmental structure as surrogates for species habitat relationships and the essential information provided by fine-scale vegetation and abiotic factors (Cushman et al. 2007a), implies that spatial representations of habitat should represent the environmental factors that most strongly predict organism abundance or performance. These factors will likely act across a range of scales, from within stand vegetation structure and composition, to local and landscape biophysical gradients of temperature, water and energy (Cushman et al. 2007b).

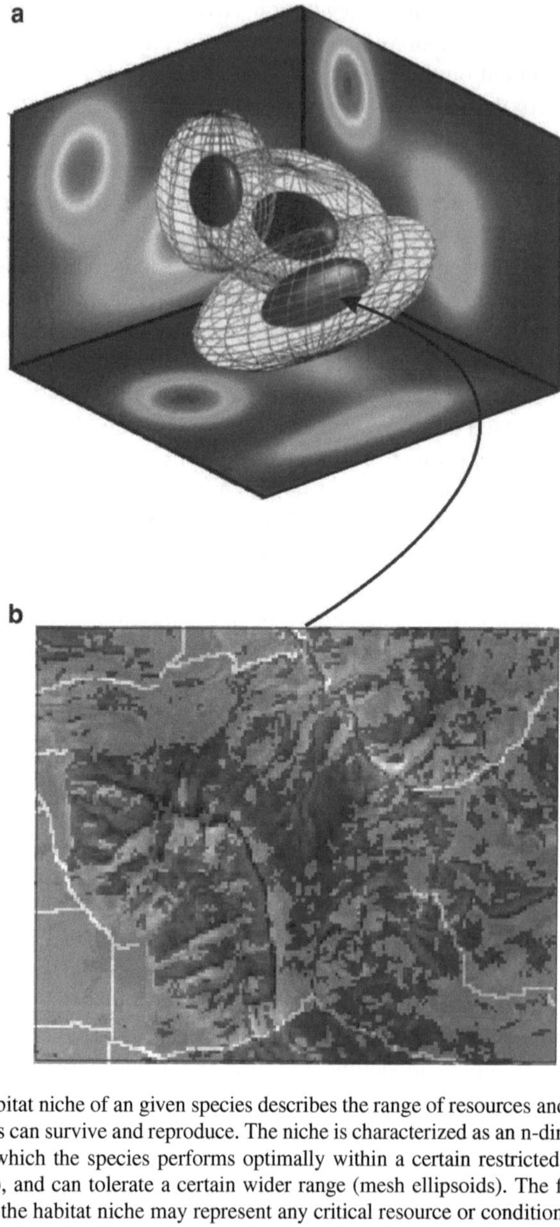

Fig. 5.3 The habitat niche of an given species describes the range of resources and conditions over which the species can survive and reproduce. The niche is characterized as an n-dimensional hyper-ellipsoid (**a**) in which the species performs optimally within a certain restricted zone (blue core ellipsoids above), and can tolerate a certain wider range (mesh ellipsoids). The factors that comprise the axes of the habitat niche may represent any critical resource or condition, many of which will likely best be described by continous environmental gradients, and may reflect environmental factors from a number of different spatial scales. Given the habitat relationship described by the niche model, it is possible in principle to evaluate the habitat quality each location in a complex landscape, by assessing where the complex of environmental conditions at that location reside within the habitat niche space of the organism (**b**). The map at bottom shows a hypothetical example where habitat quality is a continuous function of multiple environmental attributes. The map shows a grey scale gradient of habitat quality from very low (white) to very high (black)

5.4 Binary Compared to Multi-scale Gradient Representation of Habitat Quality

In this example we compare a typical categorical representation of habitat based on vegetation seral stage with a multi-variate and multi-scale gradient representation. The example is based on habitat suitability for a hypothetical organism that is associated with mature forests and high elevations, and avoids areas of fragmented forest with high edge density. A typical way to represent habitat for this species in the island biogeographic perspective is as a binary map of habitat vs. nonhabitat (Fig. 5.4a). In this map white areas are mapped late seral forest patches and black areas are covered by various conditions of non-forest and younger seral stages. In this map all locations in late seral forest are given equal quality (1) regardless of their context with respect to edges, elevation or other environmental conditions. Habitat is categorical. Likewise, all locations in non-habitat are given equal value (0) regardless

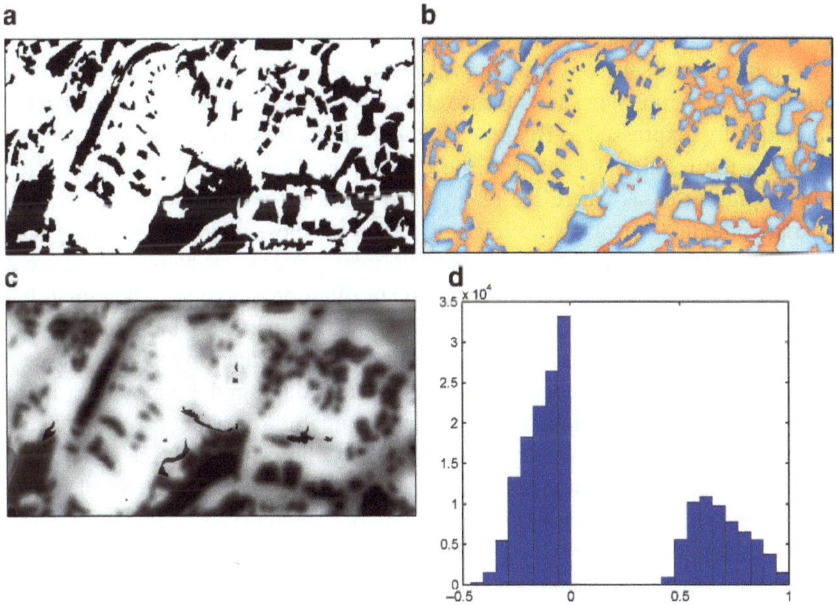

Fig. 5.4 (**a**) Binary representation of habitat (white) and non-habitat (black) for a late-seral dependent organism. (**b**) Gradient representation of habitat quality for the same organism, including multiple environmental attributes at a range of scales. (**c**) Pixel differences between the two maps, calculated as (**b**)–(**a**). Assuming that the gradient representation more faithfully represents the patterns of habitat suitability, negative values in (**c**) correspond to areas where the binary map overpredicts habitat quality. In the map these are shown as a color ramp from yellow to red. Positive values in (**c**) are areas where the binary map underpredicts habitat quality, and are shown in a scale from dark blue to light blue, with dark blue representing areas with the least difference between the two maps. (**d**) Histogram showing the frequency distribution of differences between the two maps. The histogram shows both extensive areas where habitat quality is over predicted by the binary map (positive values) and extensive areas where habitat quality is underpredicted (negative values)

of whether they are bare rock, young forest or mature forest, or whether they are surrounded by non-habitat or a small island surrounded by quality habitat.

An alternative representation of habitat quality using a multi-scale gradient representation is shown in Fig. 5.4b. In this representation, the organism is also primarily related to late successional forest, but habitat quality is also affected by several other environmental attributes. For example, habitat quality is affected by elevation, with quality decreasing in a Gaussian manner away from a peak at 1,800 m in elevation. Additionally, not all "matrix" patch types are equivalent. Some, such as bare rock and snow, have a 0 quality value, but others, such as young and mature forest, have some habitat value. Further, this organism is sensitive to the density of high contrast edges at the scale of its home range (630 m radius). This example combines categorical, gradient, and neighborhood attributes of habitat quality. In combination these factors produce a surface of hypothetical habitat quality that is continuously varying, without many hard edges (except those around the few patch types with 0 quality), which includes factors from several spatial scales. While this is a hypothetical example, it shows how the gradient perspective allows multi-variate combination of several environmental attributes measured at correct spatial scales with respect to the organism of interest.

A comparison of these two maps will illustrate several points which may be of general value. Figure 5.4c shows the pixel-by-pixel difference between the expected habitat quality (expressed as Binary – Gradient) of the two maps. The color scheme represents the relative deviation of the binary map from the gradient map. Blue colors represent areas where the binary map predicted lower habitat quality than the gradient map. Conversely, yellow to red areas are those in which the binary map over predicted habitat quality. There are two main patterns of interest. First, areas predicted as non-habitat in the binary map are often predicted as suboptimal, but not, nonhabitat in the gradient representation, and the degree of suboptimality varies as function of the vegetation type, elevation and landscape context (with respect to high contrast edges). Second, areas predicted as habitat in the binary map are of varying quality in the gradient map, such that the quality of habitat is overpredicted by the binary map for most locations, particularly those in which there are many high contrast edges and those at relatively lower elevations. This pattern of binary maps systematically over predicting the quality of habitat pixels and under predicting the quality of non-habitat is a general property of categorical patch mosaic representations of habitat and has important implications for assessing effects of landscape patterns on population processes (Fig. 5.4d).

Perhaps the most important implication of the difference between the two maps is how closely they predict habitat quality. If in either case the same general conclusion is reached about habitat amount and pattern then there would be little cost incurred for using a simple binary representation versus a more sophisticated multi-variate, and multi-scale approach. How similar are these two maps in their prediction of habitat quality? A basic measure of this is the pixel-by-pixel correlation between the two maps. The Pearson correlation between the maps is 0.358, which means that only about 13% of variance is shared between them. In other words, 87% of the information in the gradient map cannot be accounted for by the binary map, even though they are both based on the major influence of late succesional forest on habitat quality. Using one versus the other therefore in evaluating amounts and patterns of quality habitat would yield drastically different results.

5.5 Gradient Concept of Population Connectivity

Our final illustration of the gradient concept of landscape analysis centers on the question of population connectivity. One of the more immediate consequences of habitat fragmentation is the disruption of movement patterns and the resulting isolation of individuals and local populations. In the patch mosaic model of landscape structure, as habitat is fragmented, it is broken up into remnants that are isolated to varying degrees. If movement among habitat patches is significantly impeded, then individuals in remnant habitat patches may become reproductively isolated. In the patch mosaic model of categorical landscape structure, connectivity is assessed by the size and proximity of habitat patches and whether they are physically connected via habitat corridors. Patch edges may act as a filter or barrier that impedes or prevents movement, thereby disrupting emigration and dispersal from the patch (Wiens et al. 1985). In addition, the distance from remnant habitat patches to other neighboring habitat patches may influence the likelihood of successful movement of individuals among habitat patches.

However, in the previous example we argued that habitat often should not be represented as categorical patches due to the manner in which multiple environmental attributes combine across scale to influence site quality. Likewise, the factors that impede or facilitate movement may not best be represented as patch edges and inter patch distances. The influences of environmental structure on organism movement and population connectivity are species specific, and reliable inferences about population connectivity in complex landscapes requires assessing relationships between organism movement patterns and multiple environmental features across a range of spatial scales, rather than simplistic representation of habitat patch interiors, edges and inter-patch distances (Cushman 2006).

In practice it has been problematic to develop reliable inferences regarding how multiple environmental features influence movement of organisms across several spatial scales. The two traditional approaches to study animal movement have been mark-recapture and radio-telemetry (Cushman 2006). By quantifying movement rates, distances and routes of dispersing juveniles through complex environments researchers can describe species specific responses to environmental conditions. These methods are suited for incorporation in manipulative field experiments which provide the most reliable inferences about relationships between survival rates, movement and ecological conditions (McGarigal and Cushman 2002). Both of these methods are limited by logistical challenges that reduce their ability to test interactive effects of multiple landscape attributes on organism movements. The challenge in these studies is one of cost and sample sizes. It is very difficult to obtain a large sample size of individuals and then track their movements across many combinations of environmental conditions to provide data to infer patterns of movement in relation to landscape features.

Recent advances in landscape genetics have greatly facilitated developing rigorous, species-specific, and multi-variate characterizations of habitat connectivity for animal species (Manel et al. 2003; Holderegger and Wagner 2006; Storfer et al. 2007). Landscape genetic approaches largely mitigate the logistical and financial costs of extensive mark-recapture studies. To data, many population and landscape genetic studies have used *F*-statistics (Wright 1943) or assignment tests (Pritchard et al. 2000; Corander et al. 2003; François et al. 2006) to relate genetic differences among well-defined subpopulations to; distance relationships (Michels et al. 2001), putative movement barriers (Manni et al. 2004; Funk et al. 2005) or correlations with landscape features (Spear et al. 2005). This is an explicitly island-biogeographic perspective in which populations are assumed to be discretely bounded and relatively isolated, with no internal structure. Genetic differences are assumed to be a function of group membership entirely, with no effect of internal population structure, or the effects of distance or movement cost between populations.

Once discrete subpopulations have been identified, post hoc analyses are performed, correlating observed genetic patterns with interpopulation distance or putative movement barriers (e.g., Proctor et al. 2005). Populations, however, often have substantial internal structure (Wright 1943), and it is often difficult to rigorously define discrete boundaries between populations. In terrestrial landscapes it is more common to have species that are either continuously distributed or patchily distributed with low densities between populations (Cushman et al. 2006). Thus, in many situations, population structure is better defined as a gradient phenomenon than as a categorical, patch-based entity.

By sampling genetic material from a large number of organisms distributed across large and complex landscapes researchers can quantify neutral genetic variability among individuals (Storfer et al. 2007). Spatial patterns in this neutral variability are indicators of relative connectivity of the population across space (Holderegger and Wagner 2006; Cushman et al. 2006; Storfer et al. 2007). Individual-based analyses that associate genetic distances with alternative models of landscape resistance to gene flow offer a direct and powerful means to assess the affects of multiple landscape features across spatial scales on population connectivity. By comparing the least cost distances among individuals across alternative resistance hypotheses (Fig. 5.5) to genetic distances it is possible to evaluate alternative hypotheses, such as isolation by distance, barriers or landscape resistance gradients (Cushman et al. 2006; Storfer et al. 2007).

For example, Cushman et al. (2006) used least cost path analysis and causal modelling on resemblance matrices to test 110 alternative models of landscape resistance for American black bear (*Ursus americanus*). The approach framed landscape resistance as a gradient phenomenon whose total effect is a weighted combination of multiple landscape factors across a range of spatial scales. Importantly, the analysis framework provided an explicit test of isolation by population patches,

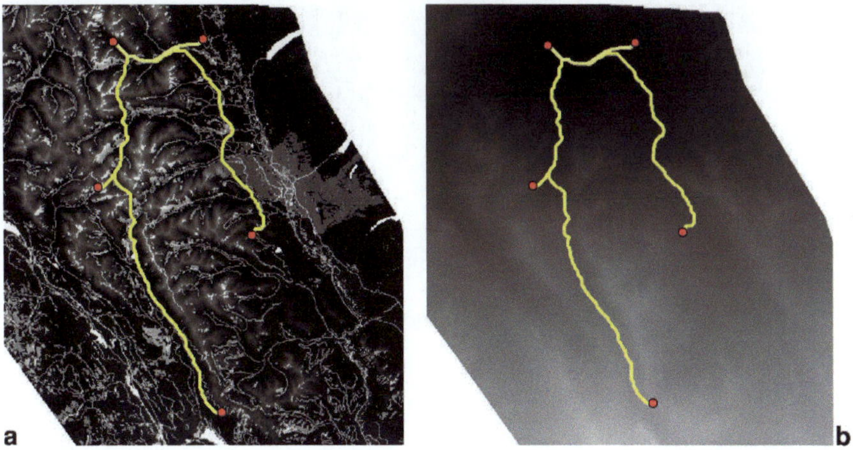

Fig. 5.5 Example of computing least cost paths to derive cost distances between individuals across a resistance hypothesis. (**a**) landscape resistance is a continuous spatial variable ranging from 1 (black) to approximately 65 (white). The locations of five individual animals are indicated by red dots. (**b**) least cost distance from the upper left individual across the costsurface (**a**). The least cost paths between the upper left individual and the other four animals are shown as yellow lines. Computing cost distances between all pairs of animals on this resistance models will create an independent variable matrix that can be associated with the genetic distances between all pairs of animals. By testing the degree of support for multiple alternative resistance models it is possible to identify the factors that facilitate or inhibit gene flow across complex landscapes

isolation by geographical distance and isolation by landscape resistance gradients. The results indicated that isolation-by-barrier and isolation-by-distance models are poorly supported in comparison to isolation by landscape-resistance gradients. Evaluating multiple competing hypotheses identified land cover and elevation as the dominant factors associated with genetic structure. Gene flow in this black bear population appears to be facilitated by forest cover at middle elevations, inhibited by nonforest land cover, and not influenced by topographical slope. The most supported model produced a map of resistance to gene flow when applied to the landscape (Fig. 5.6). This map shows that landscape connectivity is not a binary function of habitat and matrix, but is best characterized as a gradient of cell-level resistance as a function of several environmental variables.

Most population genetic studies have considered populations to be mutually isolated and internally panmictic. This is often an unrealistic model that imposes an artificial structure on analysis and can distort results. Actual populations usually exhibit continuous gradients of divergence across space and in relation to the resistance of landscape features (e.g. Cushman et al. 2006). Thus, it is often preferable to represent population structure as a gradient phenomenon rather than a categorical, patch-based entity. Representing the population structure in this way preserves internal information about how genetic characteristics vary across space, which would be lost in traditional closed-panmictic population analysis.

Fig. 5.6 Continuous landscape resistance map for black bears from Cushman et al. (2006). The most supported model of landscape structure indicated that resistance to gene flow was a continuous function of elevation and landcover. This map represents resistance to gene flow as a color ramp from white (high resistance) to black (low resistance)

Also, by representing population structure as a gradient phenomenon it is possible to compare population gradients with landscape resistance gradients. By representing both the genetic dependent variables and the landscape resistance variables as continuous gradients it is possible to test competing hypotheses of the effects of landscape structure on gene flow, in comparison to isolation by distance and putative barriers in one synthetic analysis. This would not be possible is populations were represented as categorical entities.

5.6 Gradient-Based Measures of Landscape Structure

Landscape ecologists often compare the structure of different landscapes, or the structure of the same landscape over time, and relate observed differences to some process of interest. When categorical maps are appropriate, conventional landscape metrics based on the patch-mosaic model are effective, and many metrics for this purpose exist (e.g., Baker and Cai 1992; McGarigal and Cushman 2002). However, when environmental variation is better represented as continuous gradients, it is not as simple to summarize the structure of each landscape in a metric because each landscape is represented as a continuous surface, or several surfaces corresponding to different environmental attributes.

The two fundamental attributes of a surface are its height and slope. The patterns in a landscape surface that are of interest to landscape ecologists are emergent properties of particular combinations of surface heights and slopes across the study area. The challenge is to develop metrics that characterize these aspects of surface patterns and that are effective predictors of organismic and ecological processes.

Geostatistical techniques can be used to summarize the spatial autocorrelation of such a surface (Webster and Oliver 2001). Measures such as Moran's I and semi-variance, for example, indicate the degree of spatial correlation in the quantitative variable (i.e., the height of the surface) at a specific lag distance (i.e., distance between points). These statistics are plotted against a range of lag distances to summarize the spatial autocorrelation structure of the landscape. The correlogram and semi-variogram can provide useful indices to quantitatively compare the intensity and extent of autocorrelation in quantitative variables among landscapes. Though these statistics can provide information on the distance at which the measured variable becomes statistically independent, and reveal the scales of repeated patterns in the variable, they do little to describe other important aspects of the surface. For example, the degree of relief, density of troughs or ridges, and steepness of slopes are not measured. Fortunately, a number of gradient-based metrics that summarize these and other important properties of continuous surfaces have been developed in the physical sciences for analyzing three-dimensional surface structures (Stout et al. 1994; Barbato et al. 1995, Villarrubia 1997). In the past 10 years, researchers in microscopy and molecular physics have made tremendous progress in this area, creating the field of surface metrology (Barbato et al. 1995).

Surface Metrology – In surface metrology, several families of surface-pattern metrics have become widely used. These have been implemented in the software package SPIP (2001). One such family of metrics quantifies measures of surface amplitude in terms of its overall roughness, skewness and kurtosis, and total and relative amplitude. Another family records attributes of surfaces that combine amplitude and spatial characteristics, such as the curvature of local peaks. Together, these families of metrics quantify important aspects of the texture and complexity of a surface. A third family measures certain spatial attributes of the surface associated with the orientation of the dominant texture. The final family of metrics are based on the surface bearing area ratio curve, also called the Abbott curve (SPIP 2001). The Abbott curve is computed by inversion of the cumulative height-distribution histogram. A number of indices that describe structural attributes of a surface have been developed from the proportions of this curve (SPIP 2001).

Many classic metrics for analyzing categorical landscape structure have ready analogs in surface metrology. For example, the major compositional metrics such as patch density, percent of landscape, and largest patch index correspond respectively to peak density, surface volume, and maximum peak height. Major configuration metrics such as edge density, nearest neighbor index, and fractal dimension index correspond respectively to mean slope, mean nearest maximum index, and surface fractal dimension. Many of the surface metrology metrics, however, measure attributes that are conceptually quite foreign to conventional landscape pattern analysis. Landscape ecologists have not yet explored the behavior and meaning of these new metrics; it remains for them to demonstrate the utility of these metrics, or develop new surface metrics better suited for landscape ecological questions.

Fractal Analysis – Fractal analysis provides a vast set of tools to quantify the shape complexity of surfaces. There are many algorithms in existence that can measure the fractal dimension of any surface profile, surface, or volume (Mandelbrot 1982; Pentland 1984; Barnsely et al. 1988). One such index that is implemented in SPIP calculates the fractal dimension along profiles of the surface from 0° to 180°. A number of other fractal algorithms are available for calculating the overall fractal dimension of the surface, rather than for particular profile directions. Variations on these approaches will yield metrics that quantify important attributes of surface structure for comparison between landscapes, between regions within a landscape, and for use as independent variables in modeling and prediction of ecological processes.

In addition, there are surface equivalents to lacunarity analysis of categorical fractal patterns. Lacunarity measures the gapiness of a fractal pattern (Plotnick et al. 1993). Several structures with the same fractal dimension can look very different because of differences in their lacunarities. The calculation of measures of surface lacunarity is a topic that deserves considerable attention. It seems to us that surface lacunarity, which would measure the 'gapiness' in the distribution of peaks and valleys in a surface rather than holes in the distribution of a categorical patch type, would be a useful index of surface structure.

Spectral and Wavelet Analysis – Spectral analysis and wavelet analysis are ideally suited for analyzing surface patterns. The spectral analysis technique of Fourier decomposition of surfaces could find a number of interesting applications in landscape surface analysis. Fourier spectral decomposition breaks up the overall surface patterns into sets of high, medium and low frequency patterns (Kahane and Lemarie 1995). The strength of patterns at different frequencies, and the overall success of such spectral decompositions can tell us a great deal about the nature of the surface patterns and what kinds of processes may be acting and interacting to create those patterns. They also provide potential indices for comparing among landscapes and for deriving variables that describe surface structure at different frequency scales that could be used for prediction and modeling (Kahane and Lemarie 1995; Cho and Chon 2006).

Similarly, wavelet analysis is a family of techniques that has many potential applications in landscape surface analysis (Bradshaw and Spies 1992; Chui 1992; Kaiser 1994; Cohen 1995). Traditional wavelet analysis is conducted on transect data, but the method is easily extended to two-dimensional surface data. Major advances in wavelet applications have occurred in the past several years, with many software packages now available for one- and two-dimensional wavelet analysis. For example, comprehensive wavelet toolboxes are available for R, S-Plus, MATLAB and MathCad. Wavelet analysis has the advantage that it preserves hierarchical information about the structure of a surface pattern while allowing for pattern decomposition (Bradshaw and Spies 1992). It is ideally suited for decomposing and modeling signals and images, and it is useful in capturing, identifying, and analyzing local, multiscale, and nonstationary processes. Because wavelet analyses score a range of kernels they area a robust tool for building multi-scale information directly into an analysis. It can be used to identify trends, break points, discontinuities, and self-similarity (Chui 1992; Kaiser 1994). In addition, the calculation of the wavelet variance enables comparison of the dominant scales of pattern among landscape surfaces or between different parts of a single surface (Bradshaw and Spies 1992).

Thus, wavelet decomposition and wavelet variance have great potential as sources of new surface-pattern landscape metrics and novel approaches to analyzing landscape surfaces.

5.7 Conclusions

The patch-mosaic model of landscape structure has provided a valuable operating framework for spatial ecologists, and it has facilitated rapid advances in quantitative landscape ecology, but further advances in spatial ecology are constrained by its limitations. We advocate a gradient-based paradigm of landscape structure that reflects continuously varying heterogeneity and that subsumes the patch-mosaic model as a special case. The gradient paradigm does not presuppose discrete structures, but it will identify them if they exist; it facilitates multi-scale and multivariate analyses of ecological relationships, and provides a flexible framework for conducting organism- or process-centered analyses. Through these advantages, the gradient paradigm of landscape structure will enable ecologists to represent landscape heterogeneity more flexibly in analyses of pattern–process relationships.

References

Allen TFH, Starr TB (1982) Hierarchy: perspectives for ecological complexity. University of Chicago Press, Chicago, IL

Baker WL, Cai Y (1992) The r.le programs for multiscale analysis of landscape structure using the GRASS geographical information system. Landsc Ecol 7:291–302

Barbato G, Carneiro K, Cuppini D, Garnaes J, Gori G, Hughes G, Jensen CP, Jorgensen JF, Jusko O, Livi S, McQuoid H, Nielsen L, Picotto GB, Wilening G (1995) Scanning tunnelling microscopy methods for the characterization of roughness and micro hardness measurements. Synthesis report for research contract with the European Union under its programme for applied metrology. European Commission Catalogue number: CDNA-16145 EN-C Brussels Luxemburg

Barnsely MF, Devaney RL, Mandelbrot BB, Petigen H, Saupe D, Voss RF (1988) The science of fractal images. Springer, New York

Bradshaw GA, Spies TA (1992) Characterizing canopy gap structure in forests using wavelet analysis. J Ecol 80:205–215

Cho E, Chon T (2006) Application of wavelet analysis to ecological data. Ecol Informat 1:229–233

Chui CK (1992) An introduction to wavelets. Academic, San Diego, CA

Clements FE (1916) The nature and structure of the climax. J Ecol 24:252–284

Cohen A (1995) Wavelets and multiscale signal processing. Chapman & Hall, London

Corander J, Waldmann P, Sillanpää MJ (2003) Bayesian analysis of genetic differentiation between populations. Genetics 163:367–374

Cushman SA, McGarigal K (2004) Hierarchical analysis of forest bird species-environment relationships in the Oregon Coast Range. Ecological Applications. 14(4):1090–1105.

Cushman SA, McKenzie D, Peterson DL, Littell J, McKelvey KS (2007) Research agenda for integrated landscape modelling. USDA For. Serv. General Technical Report RMRS-GTR-194.

Dutilleul P (1993) Modifying the t-test for assessing the correlation between two spatial processes. Biometrics 49:305–314
Evans JS, Hudak AT (2007) A multiscale curvature algorithm for classifying discrete return lidar in forested environments. IEEE Transactions on Geoscience and Remote Sensing, 45:1029–1038
Evans JS, Cushman SA (2009) Gradient modeling of conifer species using random forests. Landscape Ecology 24:673–683.
Ewe HT, Au WC, Shin RT, Kong JA (1993) Classification of SAR images using a fractal approach. Proceedings of Progress in Electromagnetic Research Symposium (PIERS) Los Angeles
Forman RTT (1995) Land mosaics: the ecology of landscapes and regions. Cambridge University Press, Cambridge
Forman RTT, Godron M (1986) Landscape ecology. Wiley, New York
François O, Ancelet S, Guillot G (2006) Bayesian clustering using hidden Markov random fields in spatial population genetics. Genetics 174:805–816
Funk WC, Blouin MS, Corn PS, Maxell BA, Pilliod DS, Amish S, Allendorf FW (2005) Population structure of Columbia spotted frogs (Rana luteiventris) is strongly affected by the landscape. Mol Ecol14:483–496
Gleason HA (1926) The individualistic concept of the plant association. Bull Torr Bot Club 53:7–26
Guthrie KA, Sheppard L (2001) Overcoming biases and misconceptions in ecological studies. J Roy Stat Soc Stat Soc 164:141–154
He HS, DeZonia BE, Mladenoff DJ (2000) An aggregation index (AI) to quantify spatial patterns of landscapes. Landsc Ecol 15:591–601
Holderegger R, Wagner HH (2006) A brief guide to landscape genetics. Landsc Ecol 21:793–796
Hutchinson GE (1957) Concluding remarks. Cold Spring Harbor Symp Quantitative Biol 22:415–427
Jelinski DE, Wu J (1996) The modifiable areal unit problem and implications for landscape ecology. Landsc Ecol 11:129–140
Johnson DH (1980) The comparison of usage and availability measurements for evaluating resource preference. Ecology 61:65–71.
Kahane JP, Lemarie PG (1995) Fourier series and wavelets.Studies in the Development of Modern Mathematics, vol 3. Gordon and Research Publishers
Kaiser G (1994) A friendly guide to wavelets. Birkhauser
Kotliar NB, Wiens JA (1990) Multiple scales of patchiness and patch structure: a hierarchical framework for the study of heterogeneity. Oikos 59:253–260
Levin SA (1992) The problem of pattern and scale in ecology. Ecology 73:1943–1967
Mandelbrot BB (1982) The fractal geometry of nature. W.H. Freeman, New York
Manel S, Schwartz MK, Luikart G, Taberlet P (2003) Landscape genetics: the combination of landscape ecology and population genetics. Trends in Ecol Evol 18:189–197
Manni F, Guerard E, Heyer E (2004) Geographic patterns of (genetic, morphologic, linguistic) variation: how barriers can be detected by using Monmonier's algorithm. Hum Biol 76:173–190
Manning AD, Lindenmayer B, Nix HA (2004) Continua and umwelt: novel perspectives on viewing landscapes. Oikos 104:621–628
McGarigal K, Cushman SA (2002) Comparative evaluation of experimental approaches to the study of habitat fragmentation. Ecol Appl 12:335–345
McGarigal K, Cushman SA (2005) The gradient concept of landscape structure. In: Wiens J, Moss M (eds) Issues and Perspectives in Landscape Ecology. Cambridge University Press, Cambridge
McGarigal K, Marks BJ (1995) FRAGSTATS: spatial analysis program for quantifying landscape structure. USDA For Serv Gen Tech Rep PNW-GTR-351
McGarigal K, Cushman SA, Stafford S (2000) Multivariate statistics for wildlife and ecology research. Springer, New York

McGarigal K, Cushman SA, Neel MC, Ene E (2002) FRAGSTATS: spatial pattern analysis program for categorical maps. Computer software program produced by the authors at the University of Massachusetts, Amherst www.umass.edu/landeco/research/fragstats/fragstats.html
McIntyre S, Barrett GW (1992) Habitat variegation, an alternative to fragmentation. Conservat Biol 4:197–202 McGarigal K, Cushman SA (2005) The gradient concept of landscape structure. In: Wiens J, Moss M (eds) Issues and Perspectives in Landscape Ecology. Cambridge University Press, Cambridge
Michels E, Cottenie K, Neys L, DeGalas K, Coppin P, DeMeester L (2001) Geographical and genetic distances among zooplankton populations in a set of interconnected ponds: a plea for using GIS modeling of the effective geographical distance. Mol Ecol 10:1929–1938
Moore ID, Gryson RB, Ladson AR (1991) Digital terrain modeling: a review of hydrological, geomorphological, and biological applications. Hydrolog Process 5:3–30
Murphy MA, Evans JS, Cushman SA, Storfer A (2008) Representing genetic variation as continuous surfaces: An approach for identifying spatial dependency in landscape genetic studies. Ecography 31:685–697
Ohmann JL, Gregory MJ (2002) Predictive mapping of forest composition and structure with direct gradient analysis and nearest-neighbor imputation in coastal Oregon, USA. Canadian Journal of Forest Resources 32:725–741.
O'Neill RV, DeAngelis DL, Waide JB, Allen TFH (1986) A hierarchical concept of ecosystems. Princeton University Press, Princeton, NJ
O'Neill RV, Johnson AR, King AW (1989) A hierarchical framework for the analysis of scale. Landsc Ecol 3:193–205
Openshaw S (1984) The modifiable areal unit problem. Norwich: Geo Books. ISBN 0–86094–134–5
Pentland AP (1984) Fractal-based description of natural scenes. IEEE Trans Patt Anal Mach Intel 6:661–674
Peterson DL, Parker VT (1998) Ecological scale: theory and applications. Columbia University Press, New York
Plotnick RE, Gardner RH, O'Neill RV (1993) Lacunarity indices as measures of landscape texture. Landsc Ecol 8:201–211
Pritchard JK, Stephens M, Donnelly P (2000) Inference of population structure using multilocus genotype data. Genetics 155:945–959
Proctor MF, McLellan BN, Strobeck C, Barclay RMR (2005) Genetic analysis reveals demographic fragmentation of grizzly bears yielding vulnerability by small populations. Proceedings of the Royal Society B: Biological Sciences 272:2409–2416
Rehfeldt GE, Crookston NL, Warwell MV, Evans JS (2006) Empirical analyses of plant-climate relationships for the western United States. Int J Plant Sci 167:1123–1150
Robinson WS (1950) Ecological correlations and the behavior of individuals. Am Socio Rev 15:351–357
Spear SF, Peterson CR, Matacq M, Storfer A (2005) Landscape genetics of the blotched tiger salamander (*Ambystoma tigrinum melanostictum*). Mol Ecol 14:2553–2564
SPIP (2001) The scanning probe image processor. Image Metrology APS, Lyngby, Denmark
Schneider DC (1994) Quantitative ecology: spatial and temporal scaling. Academic, San Diego, CA.
Stout KJ, Sullivan PJ, Dong WP, Mainsah E, Lou N, Mathia T, Zahouani H (1994) The development of methods for the characterization of roughness on three dimensions. Publication no EUR 15178 EN of the Commission of the European Communities, Luxembourg
Storfer A, Murphy MA, Evans JS, Goldberg CS, Robinson S, Spear SF, Dezzani R, Delmelle E, Vierling L, Waits LP (2006) Putting the 'landscape' in landscape genetics. Heredity 2006:1–15
ter Braak CJF (1988) Partial canonical correspondence analysis. In: Bock HH (ed.) Classification and related methods of data analysis. North-Holland, Amsterdam
Turner MG (1989) Landscape ecology: the effect of pattern on process. Annl Rev Ecol System 20:171–197

Turner MG, O'Neill RV, Gardner RH, Milne BT (1989) Effects of changing spatial scale on the analysis of landscape pattern. Landsc Ecol 3:153–162
Urban DL, O'Neill RV, Shugart HH (1991) Landscape ecology. BioScience 37:119–127
Urban DL, Goslee S, Pierce K, Lookingbill T (2002) Extending community ecology to landscapes. Ecoscience 9:200–212
Villarrubia JS (1997) Algorithms for scanned probe microscope, image simulation, surface reconstruction and tip estimation. J Nat Inst Stand Tech 102:435–454
Webster R, Oliver M (2001) Geostatistics for environmental scientists. Wiley, Chichester
Whittaker RH (1967) Gradient analysis of vegetation. Biol Rev 42:207–264
Wiens JA (1989) Spatial scaling in ecology. Funct Ecol 3:385–397
Wiens JA (2001) Understanding the problem of scale in experimental ecology. In: Gardner RH, Kemp WM, Kennedy VS, Petersen JE (eds) Scaling relations in experimental ecology. Columbia University Press, New York
Wiens JA, Stenseth NC, Van Horne B, Ims RA (1993) Ecological mechanisms and landscape ecology. Oikos 66:369–380
Wood CH, Skole D (1998) Linking satellite, census, and survey data to study deforestation in the Brazilian Amazon. In: Liverman D, Moran EF, Rindfuss RR, Stern PC (eds) People and pixels: linking remote sensing and social science. National Academy Press, Washington, DC
Wright S (1921) Correlation and causation. J Agric Res 20:557–585
Wright S (1943) Isolation by distance. Genetics 28:114–138
Wright S (1960) Path coefficients and path regressions: alternative or complementary concepts? Biometrics 16:189–202
Wu J (2007) Scale and scaling: a cross-disciplinary perspective. In: Wu J, Hobbs RJ (eds) Key topics in landscape ecology. Cambridge University Press, Cambridge
Wu J, Hobbs R (2002) Key issues and research priorities in landscape ecology: an idiosyncratic synthesis. Landsc Ecol 17:355–365

Part II
Method and Data

Chapter 6
Data on Distribution and Abundance: Monitoring for Research and Management

Samuel A. Cushman and Kevin S. McKelvey

In the first chapter of this book we identified the interdependence of method, data and theory as an important influence on the progress of science. The first several chapters focused mostly on progress in theory, in the areas of integrating spatial and temporal complexity into ecological analysis, the emergence of landscape ecology and its transformation into a multi-scale gradient-based science. These chapters weaved in some discussion about the interrelationships between method and these theoretical approaches. In particular, we discussed how powerful computing, large spatial databases and GIS cross-fertilized ecological theory by enabling new kinds of analyses and new scopes of investigation. However, up to this point we have given relatively little attention to the third leg of this triad, data. This and following chapters focus explicitly on data. The next several chapters discuss the advances in broad-scale data collection and analysis enabled by remote sensing, molecular genomics and satellite GPS telemetry, and how these data have made fundamental contributions to virtually all branches of ecology, especially spatial ecology, landscape ecology, and global scale research.

The goal of this chapter is to establish a framework for how data collection and management can best be designed to interact with modeling and analysis across both space and time. The chapter is divided into five sections. First, we discuss the fundamental importance of quality, large sample, spatially referenced, broadly distributed data for reliable inferences to advance research and to guide adaptive management. Second, we explore the challenge posed by limited quality, quantity and extent of data on species and environmental conditions over space and time and discuss the limitations this poses to effective monitoring to guide research and adaptive management. This then provides motivation for a discussion of the importance of monitoring resources themselves, in context and in particulate. We then introduce the concept of the four-dimensional monitoring data-cube, and argue that by collecting accurate data at a fine spatial scale and across large geo-

S.A. Cushman(✉) and K.S. McKelvey
US Forest Service, Rocky Mountain Research Station, 800 E Beckwith, Missoula, MT 59801, USA
e-mail: scushman@fs.fed.us

S.A. Cushman and F. Huettmann (eds.), *Spatial Complexity, Informatics, and Wildlife Conservation*
DOI 10.1007/978-4-431-87771-4_6, © Springer 2010

graphical extents for multiple resources we will be able to produce a flexible, multivariate, multi-scale data structure that will optimally support ecological analysis and meet monitoring needs for adaptive management. Finally, we discuss linking gradient modeling and integrated, multiple-scale monitoring. Because ecological systems are highly complex and vary dramatically across space and time, we need to think differently about data collection, monitoring and statistical analysis. As we discussed in Chapter 1, monitoring and analysis should not attempt to obtain replicated samples of representative individuals from unstructured populations, because such unstructured populations do not exist and independent sampling in space and time is usually impossible. Rather, the goal is to directly integrate space and time into a sampling frame, as described in the four-dimensional data cube idea described below, and link this to flexible, gradient modeling to infer condition and trend of ecological attributes over space and time.

6.1 Monitoring for Research and Adaptive Management

Adaptive management works by specifying resource goals, conducting management whose purpose is to create or maintain these desired conditions, and monitoring results to confirm that the system is behaving as expected and that resources are moving toward the desired conditions. This approach presupposes that the state of the system is well known across time. A good example of adaptivemanagement occurs whenever you drive a car. The general direction of travel is relatively constant, but, based on visual data, you constantly make small adjustments to keep the car on the road. Because you have precise data concerning where the road lies relative to your current direction of travel, this is easy and effective. However, if you were driving in dense fog or were driving blindfolded, you would quickly crash. For tracking the trajectories of ecological systems, monitoring data serve the same purpose as your eyes while driving. As a result, monitoring resource condition and trend has greatly elevated importance under the adaptive management paradigm. Cost effective, timely, representative, and broad-scale monitoring of multiple resources is the foundation on which adaptive management depends. Adaptive management literally cannot be "adaptive" without these data.

The adaptive management paradigm sets high priority on developing ongoing analyses, based on monitoring, to continually adjust or change land management planning decisions and thereby efficiently move toward desired conditions. The adaptive management cycle involves: (1) a comprehensive evaluation of current resource conditions, (2) frequent monitoring and evaluation of condition and trend relative to desired conditions, and (3) adaptation of management to improve performance in approaching or maintaining desired conditions. Multiple resource monitoring is critical for establishing ecologically meaningful and appropriate desired conditions, evaluating current conditions relative to these objectives, and evaluating effects of management over time to guide adaptive changes to the management regime. For monitoring to provide meaningful information to the adaptive management cycle it must provide statistically rigorous measurements of the condition and trend of multiple resources across the analysis area with sufficient temporal frequency to provide the periodic evaluations of resource condition and trend to guide management adaptation.

Broad-scale, large-sample, georeferenced measurement of multiple biological and abiotic environmental attributes is also the foundation for addressing spatial complexity and temporal variability in ecological research, as described in Chapters 2 and 3. In Chapter 3 we discussed the importance of focusing at the scale of organisms and their direct interactions with their environment, and then integrating pattern–process relationships at that grain across large spatial extents. This then implies sampling of species themselves and the environment at multiple scales, including direct measurement of occurrence, abundance, and population dynamics (where possible). The important point here is that, as spatial and temporal complexity are not noise to average away, but fundamentally important attributes of ecological systems, data collection for research and monitoring for adaptive management must be fine-grain, large extent, large sample, georeferrenced, measurement of multiple ecological attributes carefully chosen to directly represent the species, process or attributes of interest. As ecological systems are spatially complex, temporally dynamic, and scale dependent, frequent, multi-scale, spatially referenced data collection is the foundation for understanding. Given the spatial, temporal and contextual nature of ecological systems, frequent remeasurement across large spatial sampling networks is fundamentally important.

There are several critical attributes that a data collection or monitoring program must possess for it to be successful in providing reliable inferences about condition and trend to support analysis of ecological systems in particulate, in context, and integrated over space and time. We suggest that all monitoring initatives and existing programs be evaluated with respect to these essential attributes and that prioritization be given preferentially to monitoring efforts that provide statistically powerful inferences about condition and trend based on representative empirical samples.

Key attributes:

1. Based on empirical samples. For monitoring to provide any reliable information about condition and trend of a resource it must monitor the resource itself, or a proxy that has been reliably shown through rigorous scientific research to be a surrogate for the resource. Given that very little rigorous science exists relating resources to proxies, we strongly favor monitoring of resources themselves.
2. Based on representative samples. Samples must be collected in a representative manner from the target population. Representative sampling is essential to avoid biases in estimates of resource condition and trend. Monitoring inferences based on nonrepresentative samples are of unknown accuracy and thus of limited utility as a guide for adaptive management.
3. Provide sample sizes that are sufficient to provide statistically powerful inferences of condition and trend for the evaluation area at least every 5 years. This is perhaps the most daunting requirement for an acceptable monitoring program. In the past few monitoring efforts have evaluated their statistical

power to describe conditions and detect changes. However, without formal evaluation of statistical power in relation to sample size, analysis area, and temporal sampling period, the information provided by monitoring is of unknown value. Statistical power must be measured a priori to determine if a monitoring effort has the ability of describing current conditions and detecting changes with an acceptable probability. Acceptable confidence intervals and statistical power for estimates of condition will vary among resources given inherent variability of the data, importance and risk.

4. Spatially representative and well distributed. One of the key concepts in this book is that spatial pattern and temporal variation matter fundamentally to ecological processes. Therefore, in monitoring our goal is often not to estimate some mean attribute of some large heterogeneous area, but to measure, map and model the variation in the conditions of multiple resources across broad spatial extents. This has major implications for monitoring. Specifically, it requires that monitoring efforts be spatially informed, with representative sampling stratified by ecological strata across the spatial domain, with large sample sizes obtained without major spatial gaps in distribution, and with sufficient density to reflect the variability of ecological patterns and processes. Often spatial autocorrelation and spatial dependence will be of direct interest (see Chapter 7). In such cases, monitoring and other data collection must be guided by the desired precision in the spatial analyses that will follow, including details about the distances between all pairs of observations, and ensuring that these distances are to some degree optimized to accommodate autocorrelation and semi-variance analyses.
5. Based on recent samples. The age of data is a major issue in monitoring. Ecological conditions, species populations, and management activities all change over time. Data that is many years old has unknown relationships to current resource conditions. Thus, monitoring programs should continually collect new data and should base all inferences on recently collected data (perhaps within 5 years).
6. Include frequent remeasurement. Given the critical role temporal dynamics play in ecological processes (as discussed in Chapter 2), and the foundational role frequent remeasurement plays in adaptive management of ecological systems, it is essential that data collection be frequently repeated, using comparable methods to collect a consistent collection of variables at consistent scales. The frequency must be sufficient to provide meaningful guidance to managers in the adaptive management process, and for ecological research must be frequent enough to provide sufficiently precise tracking of ecological dynamics of both driver and response variables to reliably link mechanisms with responses.
7. Continue for long periods. To guide management and support understanding of ecological systems, it is critically important that data collection and monitoring efforts be maintained and continued over long periods of time, with consistent sampling in space, with comparable methodologies at consistent scales. This implies large and long-term commitments to maintaining data collection programs, with particular prioritization to maintaining permanent networks of spatially referenced sampling plots.

In the driving example, above, it is critical not only to know the current position of the car relative to the road, but also to be able to look ahead. In ecological systems, looking ahead involves modeling. There are 3 broad approaches to modeling the future. The first is simply to project current trends. The second is a statistical approach which uses past situations to infer the likelihood of various futures. The third is process modeling, where the model represents an animated hypothesis concerning the current state of the system and its dynamic properties. Dynamic models are flexible and allow a wide variety of future scenarios to be simulated, but generally lack statistical understandings of their validity. The purpose of this chapter is not to discuss these models in detail, but to note that none of these approaches has any validity without appropriate input data. For example, to project a current trend one must have data of high enough quality to produce the trend in the first place. An additional positive attribute for monitoring data is that it is consistent with both initializing and validating critical models.

6.1.1 Challenge of Limited Amount, Extent, and Quality of Spatially Referenced Ecological Information

The seven criteria listed above pose a major challenge to implementation. There are very few data collection programs which possess most of these attributes. Most data collection in ecological research is narrow in scope spatially and limited to a particular moment in time, or if the goal is to look at change, limited to a few temporal snap-shots. This severely limits ability to integrate pattern-process relationships across large, heterogeneous spatial extents and over time. Similarly, most ecological monitoring to guide natural resources management is severely limited by failure to adequately consider spatial sampling design, often failure to even establish spatially referenced sampling networks, inconsistencies across space and through time on what variables are measured, at what scale, and with what methods.

The general failure of most past efforts is largely due to issues associated with the nature of ecological data. Ecological processes and the landscapes they create are highly variable across space and time. The community composition, for example, at any location is determined by a myriad of factors at many scales. The microclimate, soils, juxtaposition, specific disturbance history, geographic location, and deeper history all play a part. Areas even a few meters away can be characterized by radically different communities due to the interactions of these factors in different combinations. For example, a slight change in aspect (microclimate) may lead to a site being overrun by an invasive weed if that site has been recently disturbed and if it is proximal to source populations of the weeds, is in an area within the range of the exotic, and if it lies in a continent or island where the native vegetation cannot compete with the weed and if no other weeds have already colonized the area. Most of these factors are both spatially and temporally variable: new road construction provides a proximal source population for the weeds; the weed continues to spread into new regions; a fire, windstorm, or insect outbreak produces necessary levels of disturbance; the local climate changes.

The highly variable and intrinsically multivariate nature of these data has historically relegated broad-scale monitoring to the collection of coarse data at large spatial scales. For example, in the weed example above, a national grid of vegetation plots with one plot every 10 km would only be able to directly speak to the weed spread at a broad scale. For example, in the county in which we reside, there would only be about 67 plots. If the climatic zone in which the weed could invade represented 10% of the total land area in the county, spread statistics would be based on at most 6–7 plots. Implementing this coarse grid, however, would be neither easy nor cheap. Nationwide there would be over 91,000 plots. At US$1,000 per plot (which is less than current vegetation monitoring systems cost per plot), this operation would require US$90,000,000.00 and a large and diffuse bureaucracy to implement. This combination of coarse data resolution coupled with high cost in turn leads to low levels of support and therefore inconsistent implementation across time and space.

This is a fundamental problem with collecting environmental data, and technologies will not cause it to go away. To address these issues, monitoring has turned to a variety of approaches both to increase the spatial and temporal resolution of the data and to reduce costs. Among these, are approaches based on monitoring system "macro-characteristics" rather than particulate and contextual data on individual species and key abiotic patterns and processes. These efforts often take one of two forms. The first is the surrogate approach to monitoring, in which a small collection of species or attributes are measured in the hope that their dynamics will represent those of the system. The second is the coarse filter approach, in which a few macro-characteristics, such as some broadly defined ecological community types, will provide sufficient information to infer the dynamics of the species and processes that act within them. In the two sections that follow we argue that neither of these approaches typically is sufficient to reliably track or predict the ecological condition and dynamics of populations and ecological processes across space and through time.

6.2 Species Surrogate Approaches

Species surrogacy has a long history in the field of conservation biology (Landres et al. 1988; Lambeck 1997; Wiens et al. 2008), with a number of different variants and implementations of the concept. Variants include ideas associated with shared habitat or functional requirements (guild membership), trophic dependencies (keystone species), area requirements (umbrella species), ecological function (engineer species), and ecological associations (focal species as defined by Lambeck 1997; See Noon et al. 2008 for a more complete list of surrogate types). Recently, Wiens et al. (2008) proposed that by grouping species using multivariate clustering it is possible to identify surrogates from each of the resulting groups to represent the group for purposes of monitoring. From a historical standpoint this concept is most similar to the guild indicator concept (Block et al. 1987).

The primary driver behind using surrogacy to monitor ecosystems is expediency. As Wiens et al. (2008) state: "…surrogate species or groups of species can be used as proxies for broader sets of species when the number of species of concern is too great to allow each to be considered individually." Implicit in this, of course, is the assumption that the surrogacy approach is effective. There have been few real tests of this, but when done, the results are seldom encouraging (Verner 1984; Landres et al. 1988; Andelman and Fagan 2000; Lindenmayer et al. 2002; Roberge and Angelstam 2004). This is obviously less onerous than monitoring all species individually, but entails large risks of bias if the chosen indicator does not sufficiently represent the abundances of other species within the group.

Cushman et al. (in press) evaluated the surrogate species concept for forest birds at two spatial scales and under two species grouping approaches. The overarching question was whether the abundance of a species across a large sample of locations provides a surrogate for the abundance of other species. They evaluate this based on an a priori grouping of species into life-history categories (Hansen and Urban 1992), and on an empirical grouping based on observed similarity of abundance patterns. In the first case, they tested whether there are strong surrogate relationships among species within objectively defined groups formed on the basis of ecological characteristics. In the latter case, the test was whether species in empirically formed groups derived from cluster analysis provide substantial surrogacy for the abundances of other group members. Data consisted of 72,495 bird observations on 55 species across 1,046 plots distributed across 30 sub-basins. They analyzed abundance patterns at two spatial scales (plot and sub-basin) and for two grouping rules. There were few significant indicator relationships at either scale or under either grouping rule, and those few found were unable to explain a substantial portion of the abundance of other species. They concluded that, coupled with the lack of proven efficacy for species surrogacy in the literature, these results indicate that the utility of indicators and similar types of surrogate approaches must be demonstrated rather than assumed.

Surrogacy has served as the lynchpin of ecosystem management and the movement away from single species paradigms. However, effective species surrogates appear to be rare. Fundamental ecological theory offers a possible explanation. No two species can long occupy the same niche (Gause 1934; Hutchinson 1957; Pulliam 2000). Thus, all coexisting, sympatric species must differ along at least one critical niche dimension. There must be some limit to the similarity of coexisting species (MacArthur 1967), and it is expected that species that are similar in some aspects of their niche will displace on others so to minimize competition. This would tend to lead to weak or negative patterns of co-occurrence for species sharing functional ecological characteristics, as was found in Cushman et al. (in press). Niche displacement processes would appear destructive to the stable existence of strong species surrogacy based on functional ecological characteristics.

As discussed in Chapter 2, community patterns are usually nonequilibrial in space and time, with frequent reversals in competitive balance, changes in relative fitness, the severity of predation and parasitism through time and over complex landscapes. Given this inherent disequilibrium in time and variability through space, simplistic ideas of species surrogacy seem unlikely to hold. The few evaluations, such as Cushman et al. (in press), have typically found poor performance of species surrogates. We believe that given the individualistic nature of species responses to scale dependent environmental and biotic interactions, it is necessary to consider ecological systems in particulate and in context, rather than seek surrogates to explain large components of the unsampled system.

6.2.1 Community Type as Surrogate for Population and Process

Coarse filter approaches to conservation assessment are attractive because of their potential efficiency and low cost. They attempt to place many species under the umbrella of one habitat assessment effort, and efficiency is obtained by monitoring a mosaic of community types as a surrogate for species viability. However, it is widely recognized that the sufficiency of coarse filter approaches are largely untested.

For community types to succeed as a coarse filter proxy for population performance at least four conditions have to be met simultaneously.

1. Habitat must be a proxy for population performance.
2. Mapped community types must be a proxy for habitat. Suitable habitat for any given species will involve ranges of critical resources and conditions across several spatial scales (Cushman and McGarigal 2004). Coarse filter conservation approaches are based on managing landscapes to provide certain amounts or configurations of vegetative community types, a great simplification both of habitat and of scale. Thus, for the coarse filter to proxy for habitat, mapped community types must be strongly related to species distributions, abundance, and population performance. Specifically, they must explain a large proportion of the species–habitat relationship for all species involved.
3. Suitable habitat is a species-specific characteristic. Each species, whether aquatic or terrestrial, has unique tolerances for different ranges of environmental conditions and unique requirements for critical resources. For the coarse filter to be successful, the mosaic of discrete community types which define the coarse filter habitat must be a good proxy for the responses of all species of interest.
4. In addition, these relationships must not decouple with management or natural disturbances. Specifically, the proxy of mapped community type for habitat may break down when management or natural disturbance occurs at a scale that will change habitat at a scale that species are directly responding to but not at a scale that results in changes in mapped community types. If this decoupling occurs, assumed correlations between the coarse community types and habitat may become invalid, resulting in the failure of the coarse filter to predict species performance. For example, stand-level vegetation manipulation (e.g. fuels treatment) in which the structure alone is modified such as decreasing density of the stand,

can have large effects on many vegetative and abiotic attributes of a stand, but may not change the cover type or seral stage as mapped by the coarse filter. In this case, the coarse filter is insensitive to important changes in habitat, resulting in a decoupling which may impair the sufficiency of the population performance proxy. In this case, additional attributes of forest structure (density or the relative openness of the forest) become important.

The community type coarse filter, therefore, represents four layers of proxies between what is measured and actual population performance. Because all four of these conditions need to be met simultaneously, errors are multiplicative. Consider an example where 75% of the variance in a population's performance is directly attributable to habitat, and, of that, 75% is directly related to broad cover types representing the utilized coarse filter, and coarse/fine scale decoupling only results in a 10% degradation of correlations. These relationships are all optimistic, as few published studies have reported explained variances of this magnitude. Nevertheless, the overall efficacy of the coarse filter will be 75*75*90% = 51% in this example. Thus it is by no means certain that coarse filter approaches will provide the required quality and resolution to perform with sufficient power and resolution to reliably represent the populations and processes they are intended to proxy, even if each level of the proxy chain performs well.

Cushman et al. (2007) conducted a multivariate analysis of bird community relationships to multiple scale habitat data in Oregon and Wyoming to test these assumptions of ecological community type as a proxy for the composition and abundance of the avian community. Their results indicate that the effectiveness of vegetation communities as proxies for population status is tenuous – even with the best, multi-scale vegetation data less than 60% of the variation in species presence is explained – and the effectiveness of landscape composition of vegetation types is highly dependent both on thematic resolution and how "community types" are defined. Their results suggest that, at a minimum, coarse filter community types should be characterized by a combination of detailed cover types across multiple seral stages. In addition, their results show that there are large differences in habitat relationships among species and that a single representation of coarse filter elements is unlikely to be an effective proxy for multi-species habitat. Furthermore, their results indicate that the relationship between habitat coarse filters and species distribution and abundance may decouple following management activities; species distribution and abundance may not be predictable based on mapped habitat in areas that are actively managed if the mapped habitat does not account for diversity in structure in terms of density, size, and multistory vs single story conditions.

6.3 Importance of Monitoring Resources Themselves

The arguments and examples given above speak directly to issues associated with using small subsets of species as surrogates, or mosaics of broadly defined cover types as coarse filters for multi-species terrestrial animal population performance.

The issues raised, however, are universal. We have argued that representative data that allow precise and unbiased estimation of ecosystem conditions are essential for adaptive management. We have also noted acquiring these data is both difficult and expensive. Given this, the loss of power and injection of largely unknown biases associated with using surrogate measures will generally lead to unacceptable power degradation. As demonstrated in the example above, this is true even if correlations between the surrogate and the resource of interest are high. In our investigations, and in the vast majority of similar investigations in the literature, these correlations are often weak and not significant. We therefore conclude that effective monitoring of a resource both to estimate its state and track its dynamics will almost always involve direct measurements of that resource. Power loss also occurs when a resource is modeled based on measured values. In a simple example, tree volume cannot be measured directly in the woods. Instead, it is modeled by measuring diameter and height, and by assuming a "form factor" to estimate taper. This is a type of surrogacy in which one attribute, in this case volume, is inferred through the measurement of two other variables. Volume estimates will contain the multiplicative errors associated with both diameter and height measurements and will contain unknown bias due to an unmeasured, but assumed, form factor. When making decisions on what to monitor, the practical ability to produce estimates with the necessary power will be directly linked to the degree to which the desired attributes and conditions can be framed in terms direct, easy, and accurate measurements.

6.3.1 The Multi-variate, Multi-scale Monitoring Data-Cube

In Chapter 5 we discussed a gradient-based paradigm for multiple-scale ecological analysis. The basic idea was that the state of any ecological system will be defined by the interactions of many processes, the behavior and distribution of many organisms, and abiotic environmental patterns across a broad range of spatial scales. There is no practical way to specify the state of the environment as a single attribute (e.g. classified map) that will optimally relate to multiple processes and how they affect the biotic and aiotic components of the system. Instead of seeking a few surrogates or a macro-attribute coarse filter which simultaneously answers all questions, we believe it is more practical and effective to retain multiple variables in the analysis at the native scale of their dominant interaction with the major ecological processes in the system. This is the multi-variate, multi-scale monitoring data cube idea. In this data cube the x, y dimensions represent space, and the z dimension represents multiple biotic entities, process or abiotic variables measured in a large spatial network of georeferenced locations (Fig. 6.1).

Importantly, these biotic entities, process and abiotic variables forming the y-dimension include a broad range of spatially referenced ecological data, from a variety of sources, at a variety of scales. As discussed at length earlier in the chapter, these will usually include measurements of multiple ecological variables across a spatial network of permanent plots. Such data could include vegetation inventory

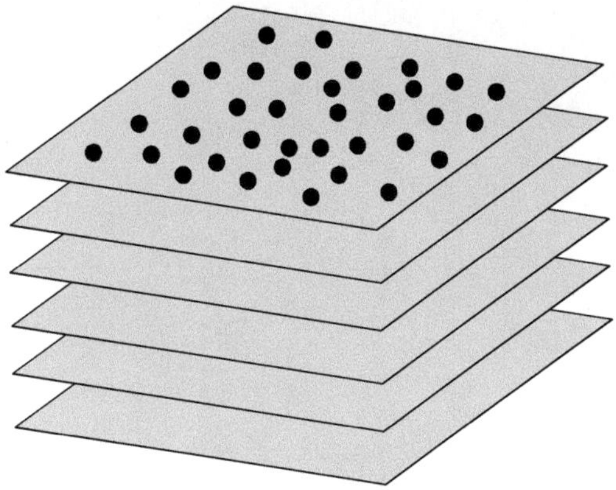

Fig. 6.1 Three-dimensional data cube for flexible multi-scale monitoring. Each layer represents a different ecological sampling dataset, such as point-level vegetation or wildlife occurrence data, topography, climate, soils, remotely sensed data, and disturbance and management history. Each layer is a spatial database, in that all records are represented at their geographical x–y coordinates

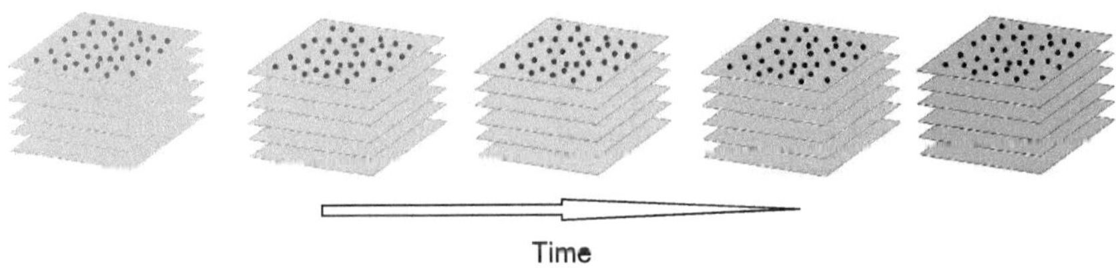

Fig. 6.2 Remeasurement of the three-dimensional data cube produces a four-dimensional data structure. Here we represent the fourth dimension as an arrow of time with remeasured data-cubes stepping out toward the right as time proceeds forward

on permanent vegetation plots, sampling wildlife occurrence or abundance, such as bird point counts, measuring soils depth, texture and chemistry, and recording microclimate variables such as temperature, soil moisture and relative humidity. These y-dimension variables of the data cube also include broader scale information derived from field sampling at stand or watershed scales, as well as spatial data stored as GIS layers on topography, geology, human management history, disturbance history, roads and other development, remotely sensed data such as satellite imagery, lidar, and aerial photography.

The fourth dimension would be repeat measurements of this 3-D data cube over time (Fig. 6.2). This cubic form provides an ability to directly integrate multivariate and spatial analyses, by having multiple measured variables georeferenced across a 2-D geographical space. The linkage to repeat measurements over time provides explicit ability to integrate spatial and temporal processes. The cube is multi-scale, in that the data are re-scaleable to provide optimal multivariate, spatial prediction for a range of dependent variables being inferred or predicted.

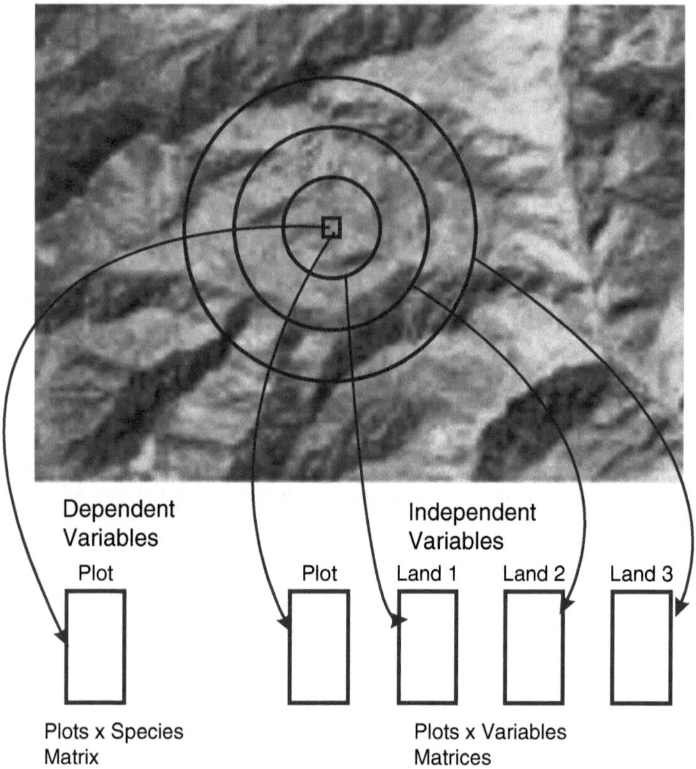

Fig. 6.3 Each layer in the multi-variate, spatial data-cube contains data measured at a particular native grain. Scaling relationships between that layer and a response variable of interest can be calculated by evaluating the strength of the statistical relationship between the response and predictor variable within differing neighborhood sizes and across variation in grain of the data through resampling

This last concept is particularly important, and deserves elaboration. Typically in landscape ecology and natural resources monitoring, what has been done in the past is to collect one or a few environmental attribute layers at a single scale. In landscape ecology it has typically been a single chloropleth classified map of a patch mosaic of landcover types. This data layer usually is the statistical product of a multivariate analysis of several spectral layers or other GIS data. This combination of multiple datasets into a single classified product involves many issues regarding the appropriateness of the classification for different research or monitoring questions, the appropriateness of the scale of the data in terms of grain and minimum mapping unit, the accuracy and meaningfulness of patch boundary definitions for management or research questions. As discussed in Chapter 3 and revisited above, no single-scale classified map can possibly represent the patterns and scales of environmental variability and the shifting importance of multiple organisms and simultaneous processes. This is the prime impetus for the multi-scale data cube (Fig. 6.3). Maintaining the data in a flexible, multi-variate, spatially explicit form facilitates the application of spatial and multivariate statistical models to produce the best predictions (given the

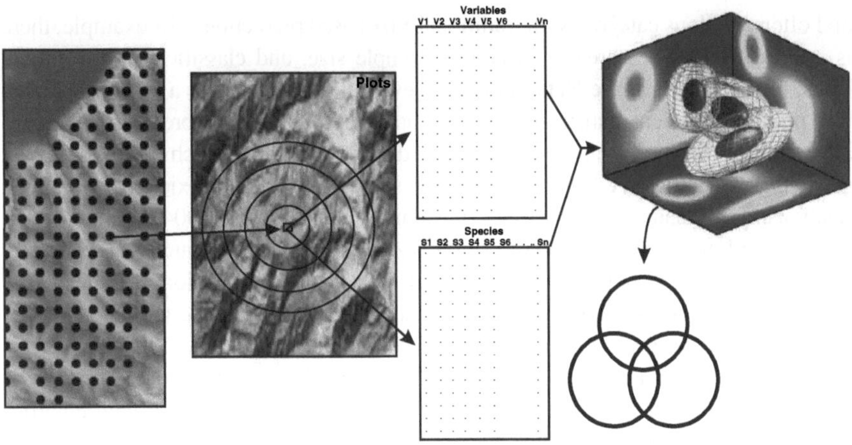

Fig. 6.4 The three-dimensional, spatial data-cube provides an ideal foundation for developing flexible, multi-scale gradient models linking multiple variables at different spatial sales to predictions of species occurrence, ecological process or other ecological attributes

data) for each species, process or attribute without losing information, introducing errors of classification, and collapsing scales (Fig. 6.4).

6.3.1.1 Limitations of Sample-Based Monitoring on Fixed Grids

One of the core concepts behind the multi-variate, spatial data-cube presented above is the central importance of large, spatially referenced, measurement plots for multiple variables. In the past most applications of large permanent re-measure plot systems have focused not on predicting spatial process-pattern relationships but on obtaining non-spatial estimates of mean and variance of some quantitative parameter within some (often large) region of space. It is important to consider the difference between this and the paradigm we are presenting. Instead of trying to estimate the mean value of some quantitative parameter within some spatial unit, our goal would be to adopt a multi-scale gradient perspective (Chapter 5) and represent variability of many ecological attributes simultaneously and continuously across space and through time.

Efforts to estimate mean values for monitored variables within relatively large spatial units are valuable, particularly in the historical sense that they were the impetus for the establishment of the large spatial sampling networks (e.g. FIA) that we presently have. However, they are not sufficient to meet data requirements for a flexible, multi-scale, multi-attribute approach to ecological analysis and adaptive management, as we described in Chapters 2, 3, and 4. First, sample-based statistical trend monitoring usually cannot provide spatial estimates of ecological variables across the analysis area; definitionally they collapse space to increase sample size and therefore power. As discussed in Chapter 3, this kind of upscaling is fraught with potential error and suffers from extreme loss of information, particularly because the spatial pattern of the data are completely ignored. Second, non-spatial sample based estimates of mean values of ecological variables are very often severely limited by sample size to broad, and often nebulous categories. In traditional FIA based predictions, for example, there is a trade-off between area of inference, sample size, and classification resolution. Thus, for inferences at the National Forest level, in order to obtain a sample size that will provide sufficient statistical power for an acceptable level of precision, it is usually necessary to limit analyses to very broad cover categories, such as cover type or seral stage. Such broad categories have been shown to be poor proxies for the habitat relationships of many wildlife species (Cushman and McGarigal 2004, Cushman et al. 2007). Third, the previous issue of sample size within an analysis area has an inverse dilemma. In order to obtain sufficient power for acceptable precision of a mean estimate for a variable of interest it is also usually necessary to make the analysis area very large. This results in the production of estimates of mean and variance for several ecological variables at very broad scales, scales much broader than usually can be linked to the interactions of ecological entities, such as organisms, and their environment through investigation of processes. Thus, estimating mean values from sample grids simultaneously suffers from two major handicaps, both of which seriously limit ability to analyze ecological pattern-process relationships flexibly across scale.

There are several other issues that limit traditional sampling for trend on fixed grids. First, these efforts are often exceptionally expensive due to the large sample size requirements, which represents a cost barrier to sampling many resources and species. In addition, if detection of population trend is the objective, repeated sampling for many years is usually required before any trend can be detected, even in a common species. Furthermore, sampling for trend provides no explanation for the causes of observed changes, which is essential if we are to use the information to understand ecological processes or guide management. Sampling for trend also provides no means to estimate resource condition across the landscape at areas not directly sampled. For monitoring to be useful for many research and management questions it should provide spatially explicit predictions of conditions across the analysis area. Finally sampling for trend provides no means to predict expected future changes in ecological conditions as a result of changing management or natural disturbance regimes. As discussed above, these issues have collectively served as major disincentives for planning and executing large scale representative monitoring.

6.3.2 Gradient Modeling and Integrated, Multiple-scale Monitoring

The above discussion of plot grids has historically presented a conundrum. Huge expense and long time frames are required to obtain data for few variables at scales too large to be useful. Overcoming these limitations, using traditional approaches, would require perhaps an order of magnitude more sampling effort, and would likely still be fraught with many of the same problems. Traditional attempts to overcome these problems through the use of coarse proxies have also proved ineffective. Thus the institutional response has been to do very little monitoring. We therefore find ourselves in the deplorable condition of entering into a period of rapid climate change with virtually no coherent, spatial baseline data concerning the condition and trends of our major natural ecosystems. This situation, however, could be corrected if there was a way around this historical conundrum. Here we propose a potential response which we believe has merit. The key is to keep all of the collected information, including spatial coordinates, in play (we have visualized this through the four dimensional data cube), and to achieve localization by linking these data through gradient models. Specifically, the combination of this kind of database and gradient modeling can identify the environmental and management factors related to each organism's distribution, and each processes controls, determine the causes of observed changes in ecological conditions, produce spatially explicit maps describing expected conditions across large spatial extents while maintaining grain at the scale of dominant pattern–process relationships, and provide spatially explicit predictions of expected future conditions resulting from altered management and natural disturbance regimes.

In this effort, models developed from current empirical data are used to produce current predictive maps of the condition of particular ecological variables continuously across the analysis area (Fig. 6.5).

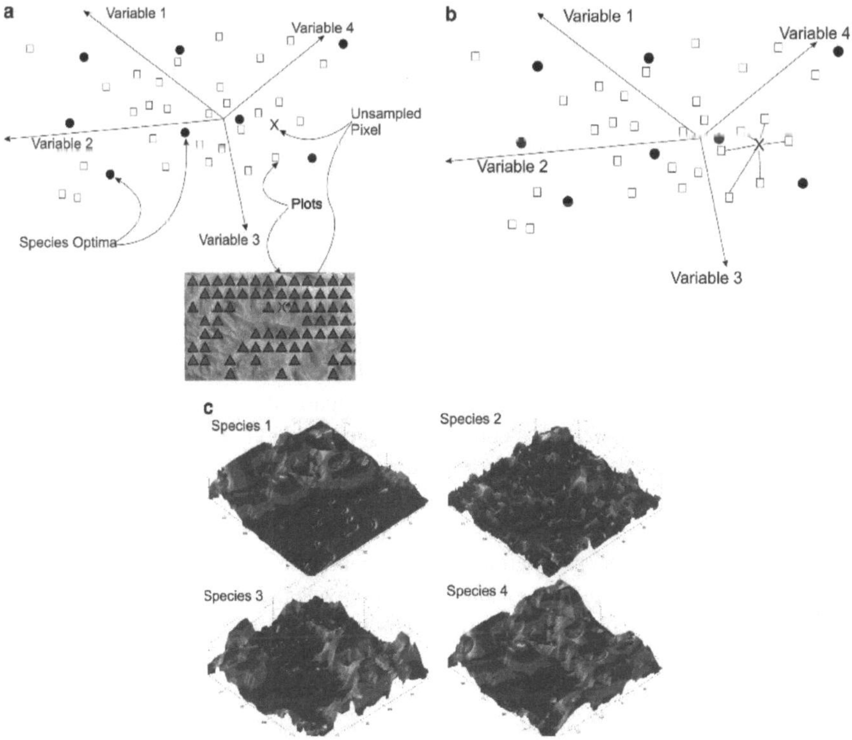

Fig. 6.5 The gradient models produced from the three-dimensional spatial data-cube can be used to produce predictive maps for each response variable through imputation mapping. (**a**) projecting unsampled locations into a gradient model, (**b**) imputing expected values for the unsampled location through imputation, (**c**) products are spatially synoptic predictions of each response variable

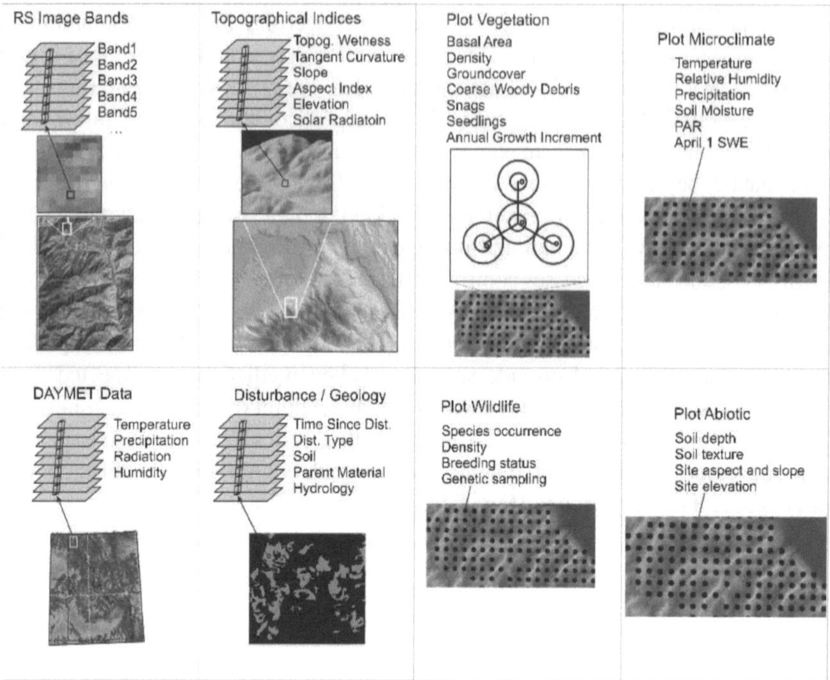

Fig. 6.6 The rich data which comprise the three-dimensional data-cube allow for integration of multiple ecological, management, abiotic and biotic predictiors within a coherent space–time framework for monitoring and prediction

In this approach, plot data, (FIA is an example) in conjunction with unclassified remotely sensed data, DEM, and other spatial data (Fig. 6.6) are used to develop statistical relationships to resources of interest (Fig. 6.4). These algorithms are then used to evaluate resource conditions at the plots (again like FIA) and can be used to produce maps depicting resource condition at fine spatial grain across large analysis areas (Fig. 6.5). In both cases, the reliability of products is directly computable. There are two keys to success (1) extensive and current samples of the condition of multiple resources at many locations across the analysis area, (2) a comprehensive and continually updated geospatial database containing spatial layers describing physiography, disturbance history, and other biophysical attributes, as well as radiometrically corrected bands from multiple remote sensing platforms.

This approach has several advantages over traditional approaches. First, it is fast. Maps are created in near real-time by transforming resource conditions into maps when they are needed. If, for instance, a resource is dependent on the proximity of roads, as soon as the road layer is updated the resource map also automatically updates. Importantly, unclassified satellite data are continually updated, so that changes in forest conditions due to harvest, fire, and other disturbances are immediately reflected in status of all modeled resources. Second, updating is relatively in expensive. Updating maps is in expensive – for many resources it is automatic and free. Developing the algorithms to produce the maps requires expenditure; applying these algorithms to a shifting landscape does not. Third, it uses fine grain, multivariate base data. The produced maps are therefore always custom designed for the resource of interest, and produced optimally from available multi-scale data. Fourth, it is maximally efficient. Because the approach produces optimized predictions from a large collection of multi-scaled, spatially referenced data layers, mapped descriptions of each modeled resource are the best that can be produced given the base data. Rather than there being a map that is optimal for one purpose and sub-optimal for all other purposes, each purpose has its own independent and optimal map.

A second advantage is localization. Conditions at very fine spatial scales can be inferred through imputation (Ohmann and Gregory 2002; Cushman et al. 2007) or other statistical interpolation approaches. In imputation, relationships between measured plot data, which are scattered spatially, and continuous spatial data, which therefore are available at fine grain sizes for all locations, are related through models to provide estimates of environmental conditions for virtually any desired spatial scale. The power of this approach lies in its ability to use data from the entire extent of measured plots to impute the resource conditions at any location. These predictive models provide rigorous assessment of relationships between the condition of each resource and environmental characteristics at multiple spatial scales and management actions. They further provide predictive maps of resource condition synoptically across space. This provides a means to identify the most likely areas to look for particular rare plant or animal species. It also provides a means to monitor changes in habitat over time, and to predict future changes in habitat amounts and qualities under alternative management and disturbance scenarios.

Gradient modeling can also be used to relate many other point-based data types to continuous data surfaces. For example, gene flow and population connectivity (Cushman et al. 2006; Holderegger and Wagner 2008; Balkhenol et al. 2008) can be modelled using similar methods providing the ability to identify corridors, if they are relevant, barriers, and the factors that influence gene flow for individual species. As described in more detail in Chapter 17, genetic gradient modeling is ideally suited to evaluating the factors that determine connectivity for a range of organisms. For example, it is an ideal approach to determine the factors that drive the spread of different invasive species and identifying the locations most likely to contain incipient populations of invasive species. Identifying these incipient invasions before they become entrenched is critical to effective control of the spread of invasive species. Genetic gradient modeling is also an extremely valuable approach to studying the connectivity of aquatic ecosystems for native and non-native fish and identifying the features that influence connectivity for each species. Furthermore, genetic gradient modeling provides a means of optimizing models of gene flow for a wide range of terrestrial wildlife, providing managers with detailed information to minimize the negative effects of management actions on population isolation and fragmentation if that is determined to be a problem, of species of concern and species of interest.

6.4 Summary and Conclusion

By linking empirical sampling of multiple resources, extensive geospatial databases and sophisticated spatial modeling, integrated resource monitoring has potential to provide estimates of current conditions, measure changes over time, and provide explanation of causes of observed changes, and predictions of expected future conditions. The approach avoids several major assumptions of the coarse filter. First, categorical patch mapping as proxy for habitat is avoided by representing ecosystem diversity as continuously varying gradients of vegetation composition, structure and biophysical variables rather than arbitrary patch mosaics (McGarigal and Cushman 2005; McGarigal et al. in press). Second, it does not assume that a single mapping of patches is an optimal surrogate of habitat for all species. Instead, each resource is predicted individualistically in response to key variables describing multiple biological and abiotic attributes across complex landscapes and over time. Third, it avoids the assumption that the scale of a particular mapped patch mosaic is ideal for all species. Instead, we relate individual species to individual driving variables at a range of spatial scales. Fourth, it provides timely updates of resource maps, using current data to predict resource distribution and condition across space. Fifth, combining gradient modeling with landscape simulation models (Cushman et al. 2007) provides a means to predict distribution and abundance in the future under altered climate, disturbance regime and management.

References

Andelman SJ, Fagan WF (2000) Umbrellas and flagships: efficient conservation surrogates or expensive mistakes? Proc Natl Acad Sci 97:5954–5959

Balkenhol N, Gugerli F, Cushman S, Waits L, Coulon A, Arntzen J, Holderegger R, Wagner H (2007) Identifying future research needs in landscape genetics: where to from here? Landscape Ecology 24:455–463

Block WM, Brennan LA, Gutierrez, RJ (1987) Evaluation of guild-indicator species for use in resource management. Environ Manag 11:265–269

Cushman SA, McGarigal K (2002) Hierarchical, multi-scale decomposition of species-environment relationships. Landsc Ecol 17:637–646

Cushman SA, McGarigal K (2004) Hierarchical analysis of forest bird species-environment relationships in the Oregon Coast Range. Ecological Applications. 14(4):1090–1105.

Cushman SA, Evans J, McGarigal K (in press) Do classified vegetation maps predict the composition of plant communities? The need for Gleasonian landscape ecology. Landsc Ecol

Cushman SA, O'Daugerty E, Ruggiero L (2003) Lanscape-level patterns of avian diversity in the Oregon Coast Range. Ecol Monogr 73:259–281

Cushman SA, O'Daugerty E, Ruggiero L (2004) Hierarchial analysis of forest bird species-environment relationships in the Oregon Coast Range. Ecol Appl 14:1090–1105

Cushman SA, McKelvey KS, Hayden J, Schwartz MK (2006) Gene-flow in complex landscapes: testing multiple hypotheses with causal modeling. Am Nat 168:486–499

Cushman SA, McKenzie D, Peterson DL, Littell J, McKelvey KS (2007) Research agenda for integrated landscape modelling. USDA For Serv Gen Tech Rep RMRS-GTR–194

Cushman SA, O'Daugerty E, Ruggiero L, McKelvey KS, Flather CH, McGarigal K (2008) Do forest community types provide a sufficient basis to evaluate biological diversity? Frontiers in ecology and the environment. 6:13–17
Cushman SA, O'Daugerty E, Ruggiero L (submitted) Sensitivity of habitat and species surrogacies to landscape disturbance. Ecology
Gause GF (1934) The struggle for existence. Williams & Wilkins, Baltimore, MD
Hansen AJ, Urban DL (1992) Avian response to landscape pattern: the role of species' life histories. Landsc Ecol 7:163–180
Holderegger R, Wagner HH (2008) Landscape genetics. Bioscience 58:199–208
Hutchinson GE (1957) Concluding remarks. Cold Spring Harbor Symp Quant Biol 22:414–427
Lambeck RJ (1997) Focal species: a multi-species umbrella for nature conservation. Conservat Biol 11:849–856
Landres PB, Verner J, Thomas JW (1988) Critique of vertebrate indicator species. Conservat Biol 2:316–328
Lindenmayer DB, Manning AD, Smith PL, Possingham HP, Fischer J, Oliver I, McCarthy MA (2002) The focal-species approach and landscape restoration: a critique. Conservat Biol 16:338–345
MacArthur RH (1967) Limiting similarity, convergence and divergence of coexisting species. Am Nat 101:338–345
McGarigal K, Cushman SA (2005) The gradient concept of landscape structure. In: Wiens JA, Moss MR (eds) Issues and perspectives in landscape ecology. Cambridge University Press, Cambridge
McGarigal K, Tagil S, Cushman SA (in press) Surface metrics: an alternative to patch metrics for the quantification of landscape structure. Landsc Ecol
Millspaugh JJ, Thompson III FR, editors (2009) Models for Planning Wildlife Conservation in Large Landscapes. Elsevier Science, San Diego, California, USA. 674 pages.
Noon BR, McKelvey KS, Dickson, BG (2008) Multispecies conservation planning on U S federal lands. In: Millspaugh JJ, Thompson FR (eds) Models for planning wildlife conservation in large landscapes.
Ohmann JL, Gregory MJ (2002) Predictive mapping of forest composition and structure with direct gradient analysis and nearest neighbor imputation in coastal Oregon, USA. Can J Forest Res 32:725–741
Pulliam HR (2000) On the relationship between niche and distribution. Ecol Lett 3:349–361
Roberge J-M, Angelstam P (2004) Usefulness of the umbrella species concept as a conservation tool. Conservat Biol 18:76–85
Schwartz MK, Luikart G, Waples RS, et al (2007) Genetic monitoring as a promising tool for conservation and management. Trends Ecol Evol 22: 25–33
Sherry TW (1979) Competitive interactions and adaptive strategies of American redstarts and least flycatchers in a northern hardwood forest. Auk 96:265–283
Verner J (1984). The guild concept applied to management of bird populations. Environ Manag 8:1–14
Wiens JA., Hayward GD, Holthausen RS, Wisdom MJ (2008) Using surrogate species and groups for conservation planning and management. Bioscience 58:241–252

Chapter 7
Animal Movement Data: GPS Telemetry, Autocorrelation and the Need for Path-Level Analysis

Samuel A. Cushman

In the previous chapter we presented the idea of a multi-layer, multi-scale, spatially referenced data-cube as the foundation for monitoring and for implementing flexible modeling of ecological pattern–process relationships in particulate, in context and to integrate these across large spatial extents at the grain of the strongest linkage between response and driving variables. This approach is powerful for developing information about the conditions of multiple ecological attributes continuously across the analysis area. However, there are a number of ecological questions that involve processes that are not functions of ecological conditions at point locations alone. Many of these involve spatial processes and mobile agents, such as the spread of disturbances, dispersal of propagules, and the movement of mobile animals. The focus of this chapter is on animal movement data.

In mobile animals, movement behavior is used to maximize fitness by maximizing access to critical resources and minimizing risk of predation. As organisms move through spatially complex landscapes, they integrate the distributions and conditions of multiple ecological attributes, adopting movement paths that optimize fitness benefits while minimizing fitness costs. There is exceptionally rich information in intersecting the space–time movement paths organisms follow with the kind of multi-variate, spatially explicit ecological data base described in the previous chapter. Until recently, however, it was not possible to obtain sufficiently precise records of movement paths of sufficient length to meaningfully associate movement behavior with ecological conditions across large and complex landscapes. However, that has changed dramatically in the past decade with the advent of GPS telemetry technology. It is a classic example of the phenomenon described in Chapter 1, of advances in methodology leading to entirely new kinds of data, which in turn drive the emergence of new analytical approaches and theoretical paradigm shifts. Also, as described in Chapter 1, the often acrimonious and frequently confused arguments relating to the statistical properties and proper analyses of such movement

S.A. Cushman(✉)
US Forest Service, Rocky Mountain Research Station, 800 E Beckwith, Missoula, MT 59801, USA
e-mail: scushman@fs.fed.us

S.A. Cushman and F. Huettmann (eds.), *Spatial Complexity, Informatics, and Wildlife Conservation*
DOI 10.1007/978-4-431-87771-4_7, © Springer 2010

data streams reflect to some degree a struggle between an old tradition and its procedures, assumptions and a new paradigm.

Location data obtained from radio and satellite telemetry have become very widely used in studies of animal space use, habitat selection and behavior (White and Garrott 1990; Aebisher et al. 1993; Cushman et al. 2005). These new data have both presented an unprecedented view into animal movement behavior and an unprecedented challenge to traditional thinking about statistical analysis of habitat use. Traditionally, analysis of habitat use has been based on comparing the habitat characteristics of a large number of statistically independent utilized locations with a large number of available or not utilized locations with compositional analysis or linear models, such as logistic regression. These approaches developed during the time when location data were obtained infrequently using VHF telemetry or networks of detection plots, such as remote cameras, hair snares or track plates. In the latter method, the investigator can establish a network of sampling stations which yield presence and pseudo-absence data. These can be effectively independent spatially if the design is carefully developed. In the former case, infrequent re-location of individuals through telemetry results in a sequence of utilized points which can then be compared to available points within some spatial extent, often the animals minimum convex polygon home range. These infrequent VHF points may appear to be statistically independent and not show significant autocorrelation. However, as Fortin and Dale (2005) note, sparse sampling of an autocorrelated data stream does not remove the effect of autocorrelation from the data, it only reduces power to detect it.

With the advent of GPS telemetry it is now possible to obtain relocation data as frequently as one desires. Sampling frequency of down to 5 min intervals is common. Animal movements are inherently highly autocorrelated because an animal's next step has to be to a location available from its current step, leading to a pathway in which locations are autocorrelated with previous locations for long time durations. This poses a major challenge to traditional point-based approaches which assume spatial independence of observations. In point based approaches, temporal autocorrelation of locations leads to underestimation of home range size and bias in predictions of habitat selection, core area, and intensity of resource use (Swihart and Slade 1985; Alldredge and Ratti 1986; Thomas and Tylor 1990; White and Garrott 1990; Cresswell and Smith 1992; Palomares and Delibes 1992; Litvaitis et al. 1994). The magnitude of these errors is proportional to strength of autocorrelation between observations across time-lags and will vary by species and by habitat (Swihart and Slade 1985; Harris et al. 1990). Accordingly, many scientists previously recommended that researchers calculate time to independence using time–distance curves (Litvaitis et al. 1994).

A priori, there is no way to determine what the time to apparent independence will be (Harris et al. 1990). Shoener's (1981) V statistic and Mantel correlograms (Cushman et al. 2005) have been recommended to produce correlograms to illustrate the structure and duration of autocorrelation. However, filtering data to achieve statistical independence often incurs heavy costs in terms of information loss. In addition, the apparent independence achieved may often merely indicate a loss of power to detect autocorrelation due to reduced sample size from a globally autocorrelated movement path (Fortin and Dale 2005; Cushman et al. 2005). In addition, many researchers have argued that such filtering may not be necessary (Swihart and Slade 1997; Rooney et al. 1998). For example, Rooney et al. (1998) argue that strict attempts to achieve statistical independence by subsampling result in substantial underestimation of range size and rates of movement. It also obliterates fine-scale patterns in habitat use that may be present in the data (Rooney et al. 1998). Also, even long sampling intervals do not guarantee independence (Cushman et al. 2005). The correct strategy for the best estimation of home range size, intensity of spatial use and quantification of fine-scale behavioral decisions may be to use the shortest possible sampling interval over the longest possible period (Rooney et al. 1998).

Swihart and Slade (1997) argue that regular sampling intervals resulting in autocorrelated data will not invalidate many estimates of home range size so long as the study time frame is adequate. Otis and White (1999) extend this argument and propose that the key requirement is to define a specific monitoring time frame appropriate for the study question and restrict inferences to the temporal and spatial scope justified by that time frame. However, if one uses a point-based analytical approach that assumes statistically independent spatial observations this is a conundrum in that the optimal estimate of utilization will come from the most frequent possible remeasurement, but the severity of violation of statistical assumptions will also increase with increasing frequency of observations. This suggests a necessary shift from point-based to path-based analyses, as discussed in Chapter 19.

Recently, there has been the realization that autocorrelation in many cases isn't a problem to be solved through carefully controlling sampling design, but a critical signal about the underlying structure of pattern–process relationships in spatially structured environments (Legendre 1993). Indeed, autocorrelation reflects the fact that ecological processes occur in a spatial context and their effects are expressed partly as functions of spatial location and distance. As we argued in Chapters 1 and 2, there is great advantage and enormous potential in addressing spatial complexity and temporal variability directly as critical attributes of ecological systems. Autocorrelation of animal movement data is a prime example of this opportunity.

The patterns of temporal and spatial autocorrelation of locations are an important part of the information that telemetry data provide (Cushman et al. 2005). Legendre (1993) and Legendre and Fortin (1989) argue that spatial autocorrelation is an informative ecological parameter that reveals scales and patterns of ecological processes which are often not resolvable through other methods. In terms of animal movements, addressing the details of the strength, scale and patterns of autocorrelation may illustrate details about animal use of space and its relationships to changes in the environment in response to perturbations, social interactions or seasonality (Cushman et al. 2005). In this chapter we review some of the issues relating to GPS telemetry data, investigate the structure and range of spatial autocorrelation of sequential relocation points under 18 different movement rules using simulation modeling, and then compare these to the structure and range of spatial autocorrelation in a mobile mammal, the African elephant. We use these analyses to argue against traditional point-based analyses of movement data, in favor of path-level analysis. This motivates the case study example on elephant movement presented in Chapter 19.

7.1 Methods

7.1.1 Movement Path Simulation

We simulated 18 different movement path varieties, consisting of a two-way factorial of path type and mean turning angle (Table 7.1). We replicated each of the 18 path varieties ten times. In all cases each path began by heading north, with subsequent steps all one unit in length. All paths were simulated to 1,000 steps. There were three path types, consisting of correlated random walk, bounded correlated random walk, and central place random walk. There were six turning angle distributions, each consisting of a normal random distribution with a specified mean. The mean turning angles for the six distributions were 1°, 6°, 12°, 24°, 45°, and 90°. In the correlated random walk the turning angle for each step from the previous step direction was taken from a normal random variable with the mean of the turning angle for that normal random distribution. This walk represents an unconstrained walk, with the only spatial parameter consisting of the normal random turning angle between steps. In the bounded correlated random walk the walk was calculated exactly the same, except with the additional constraint that it not exceed ten distance units from the origin. This reflects a correlated random walk within a fixed circular home range with radius of ten distance units. When a walk reached ten units distance from origin the next step was taken directly toward the origin, with subsequent steps correlated random walks from that new direction. The central place random walk consisted of correlated random walks of 100 steps, followed by an azmithal return to the origin in the next 100 steps, with this combination repeated five times over the 1,000 steps of the simulation. This simulates correlated random foraging paths out from a central place, such as a nest or natal den, followed by direct return, such as after capturing prey. Examples of each of these 18 path varieties are shown in Figs. 7.1–7.3, with Fig 7.1 showing correlated random walks, Fig. 7.2 showing bounded random walks, and Fig. 7.3 showing central place random walks.

Table 7.1 Factorial of the 18 path varieties simulated. They are a combination of three path types across six distributions of turning angle. The path types are unconstrained correlated random walk (UCW), random walks constrained to remain within ten step lengths of the origin by reflecting off the edge (BCW) and correlated random walks beginning at the origin and returning along the back azmuth at 100 steps (CCW). The six distributions of turning angle are obtained from normal random variables with mean of 1°, 6°, 12°, 23°, 45°, and 90°

	1°	6°	12°	23°	45°	90°
Unconstrained correlated random walk	UCW1	UCW6	UCW12	UCW23	UCW45	UCW90
Bounded correlated random walk	BCW1	BCW6	BCW12	BCW23	BCW45	BCW90
Central place correlated random walk	CCW1	CCW6	CCW12	CCW23	CCW45	CCW90

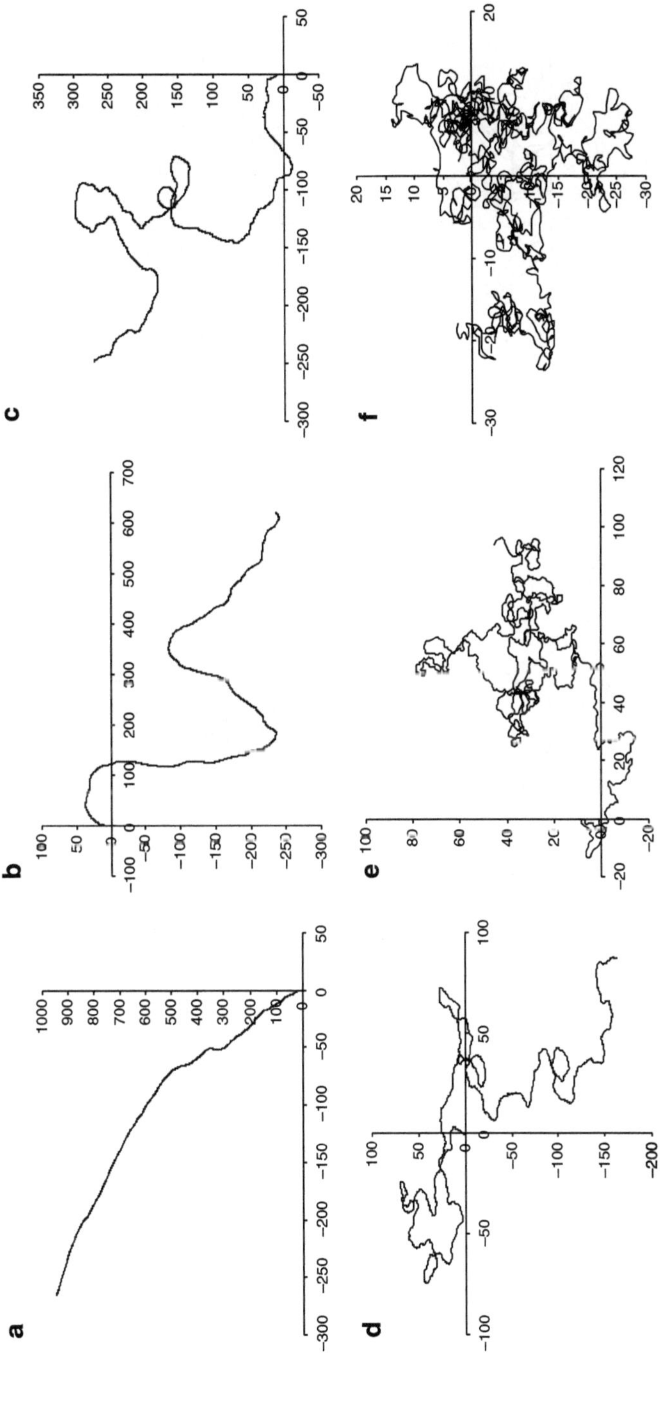

Fig. 7.1 Example paths for unconstrained correlated random walks for each of the six turning angle distributions

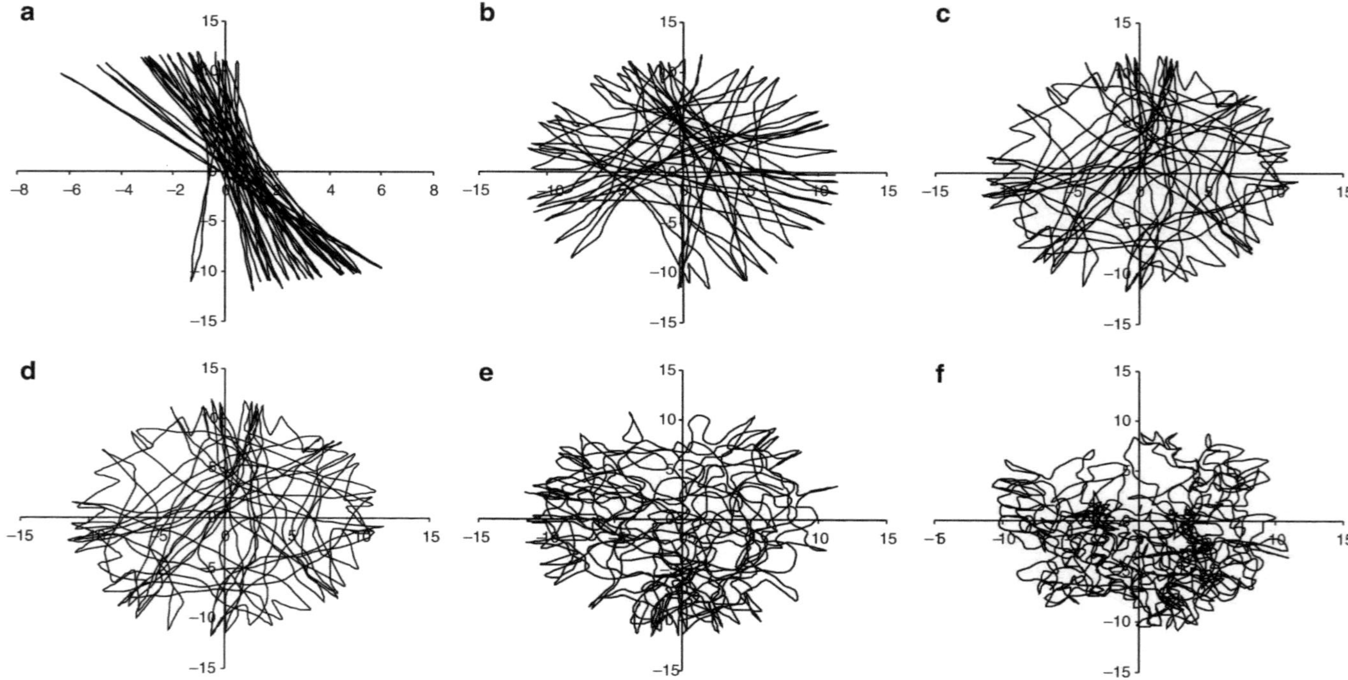

Fig. 7.2 Example paths for bounded correlated random walks for each of the six turning angle distributions

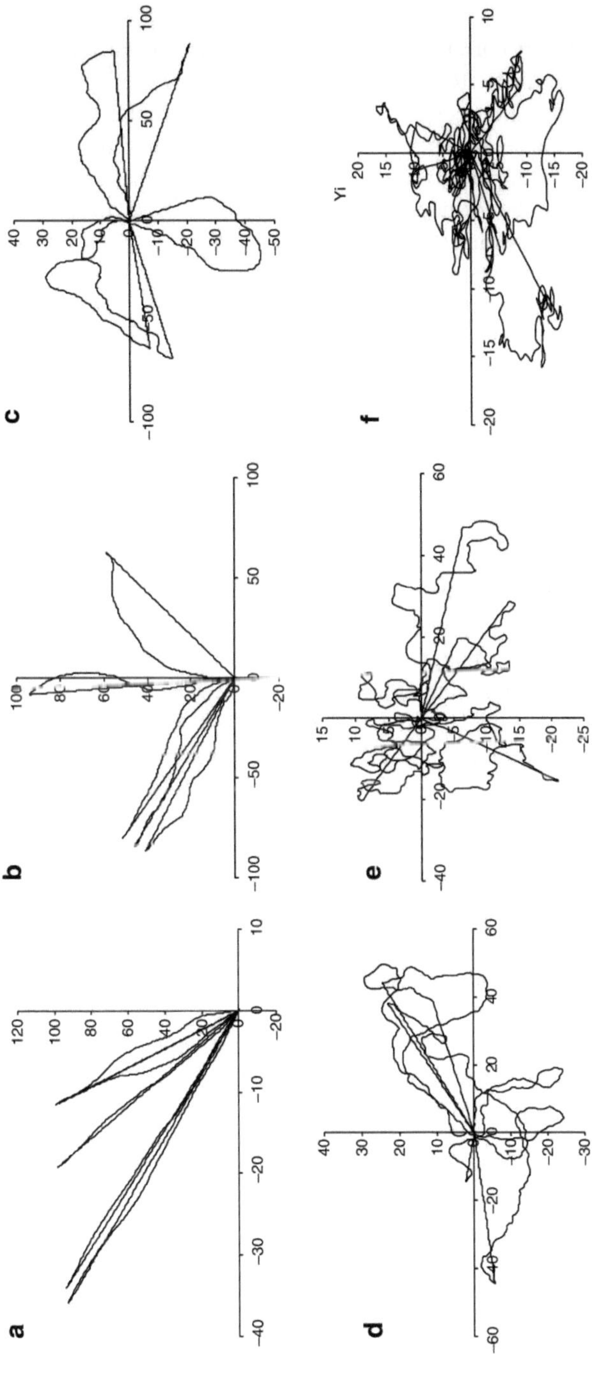

Fig. 7.3 Example paths for central place correlated random walks for each of the six turning angle distributions

7.1.2 Autocorrelation Analysis

We produced Mantel correlograms for each of the 180 movement paths using the Ecodist library in R (R-development team 2008). The Mantel test tests the degree of association between two distance matrices (Mantel 1967). In ecological research these distance matrices describe the pairwise dissimilarity or ecological distance between each pair of samples. Because any number of variables describing each sample can be included in the calculation of these distance matrices, the Mantel test is a multivariate test of the association between two data sets. When one of the distance matrices is coded as distance classes it is possible to construct a multivariate correlogram (Oden and Sokal 1986; Sokal 1986). The resulting correlogram shows the strength of correlation between the two multivariate distance mat rices across a range of lags between them. As discussed in Cushman et al. (2005), this is conceptually similar to univariate correlograms produced using Moran's I or Geary's C (Legendre and Legendre 1998), except that Mantel correlograms produce description of how multivariate correlations vary across several classes of lag-distance whereas univariate correlograms describe the relationships between one response variable and one lag variable across several classes of lagdistance.

In this study we create correlograms comparing the distance between organisms in geographical space with distance between them in time. Cushman et al. (2005) noted that the Mantel correlogram has a number of advantages over alternative ways of computing autocorrelograms for animal movement analysis. First, correlograms of the V statistic (Schoener 1981) do not have a significance test, and assume a bounded and elliptical home range. In contrast, the Mantel correlogram does not assume a fixed and elliptical home range and has both an asymptotically correct analytical significance test, and is readily tested non-parametrically with Monte-Carlo methods (Legendre and Legendre 1998).

These correlograms record the mantel correlation between distances between points across lag distances (e.g. Cushman et al 2005). The lag distances used in this analysis were 5, 15, 25,... 995, for a total of 99 lag distance classes. We calculated significance of mantel correlation at each lag distance using Monte Carlo permutation with 1,000 randomizations, for a point-wise significance level of 0.001. Our analysis included plotting the ten replicate correlograms for each of the 18 path varieties and overlaying the confidence intervals for statistically significant autocorrelation to determine the time to spatial independence, if any, for each path. This analysis provides a relatively comprehensive assessment of the structure, range and variability of autocorrelation in three common path types across a reasonably broad range of turning angle behavior. Such evaluation of simulated paths generated with known and controlled spatial processes is essential for evaluating autocorrelation in empirical movement paths.

7.1.3 *Empirical Analysis of Autocorrelation of Elephant Movement*

We also calculated Mantel correlograms for three elephants (*Loxodonta africana*) collared in Botswana (Cushman et al. 2005). Prior to analysis we preformed several data filtering tasks to retain the most accurate and least biased movement data. This reduction resulted in a final set of 709 locations for herd 55, 713 locations were retained for herd 56, 699 locations for herd 57 (Cushman et al. 2005). Next, we broke the three data sets into overlapping windows in time, to allow us to evaluate changes in autocorrelation patterns through seasons, and to reduce the effects of non-stationarity. We selected time windows of 60 days, and slid these across the 345 day sampling period, with 15 day time steps between the start of successive periods. This resulted in 20 overlapping time periods retained for the analysis; each 60 days long and beginning 15 days apart. For each of these windows we created two distance matrices for use in the Mantel correlogram analyses. First, we computed the geographical distances along the curvature of the earth between all pairs of points in each movement database. Next, we computed distance matrices for the same movement data, but for distance between points in time rather than in space. These time distance matrices were then recoded into distance class matrices, containing 120 distance classes each, corresponding to the number of 12 h periods over the 60 day sampling windows.

7.2 Results

7.2.1 *Structure, Range and Variability of Autocorrelation Within Simulated Movement Paths*

Each path variety had a characteristic autocorrelation structure (Figs. 7.4–7.6). Correlated random walks showed clines of increasing negative spatial autocorrelation between locations as time distance between them increased, with this pattern consistent across all six turn angle distributions. The six turn angle distributions were primarily related to variability among replicate correlograms, with variability increasing with larger mean turning angles. This is exactly as one would expect with correlated random walks. Importantly, for all turning angle distributions these correlograms indicate a global cline in autocorrelation such that there is never a lag distance at which points become spatially independent.

Bounded random walks all showed the same basic pattern of initial high positive autocorrelation of location, followed by a drop to negative autocorrelation and subsequent fluctuation between positive and negative autocorrelation. This fluctuation between positive and negative autocorrelation was highly sensitive to mean

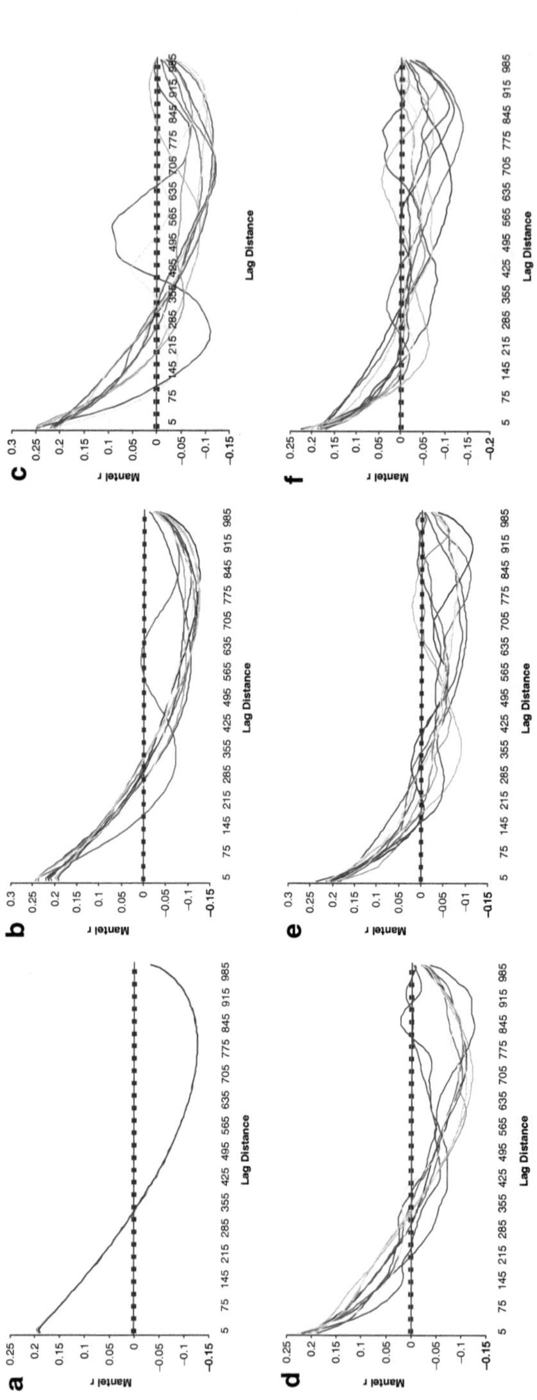

Fig. 7.4 Mantel correlograms for unconstrained paths across the six distributions of turning angle. Each plot shows ten overlain correlograms derived from the ten replicate realizations of each path type

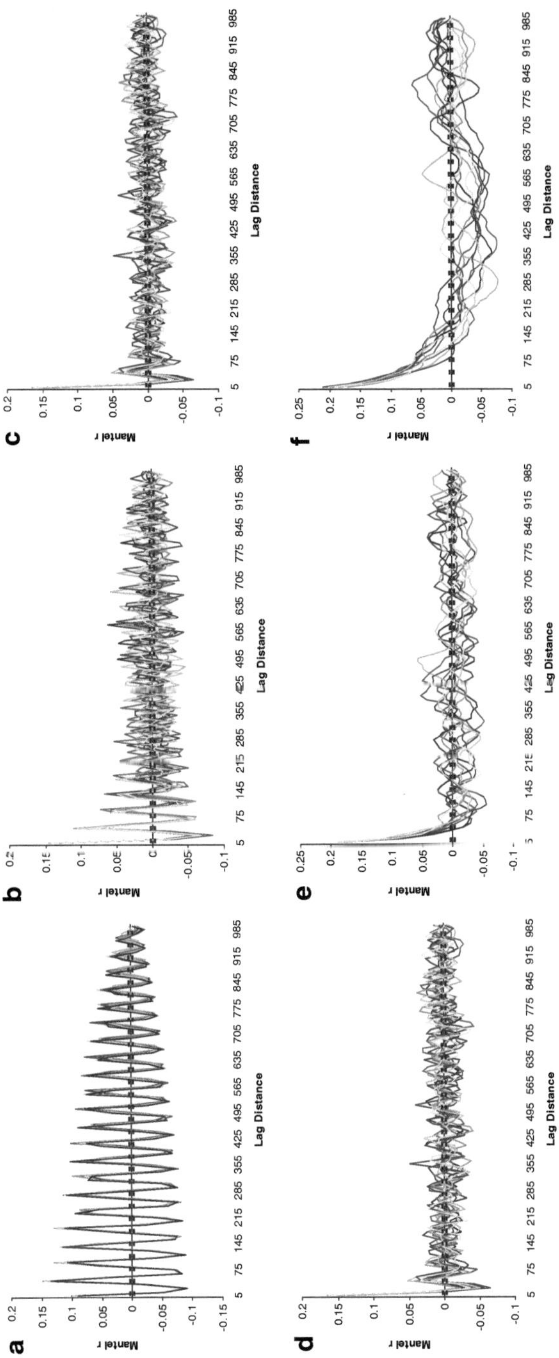

Fig. 7.5 Mantel correlograms for bounded paths across the six distributions of turning angle. Each plot shows ten overlain correlograms derived from the ten replicate realizations of each path type

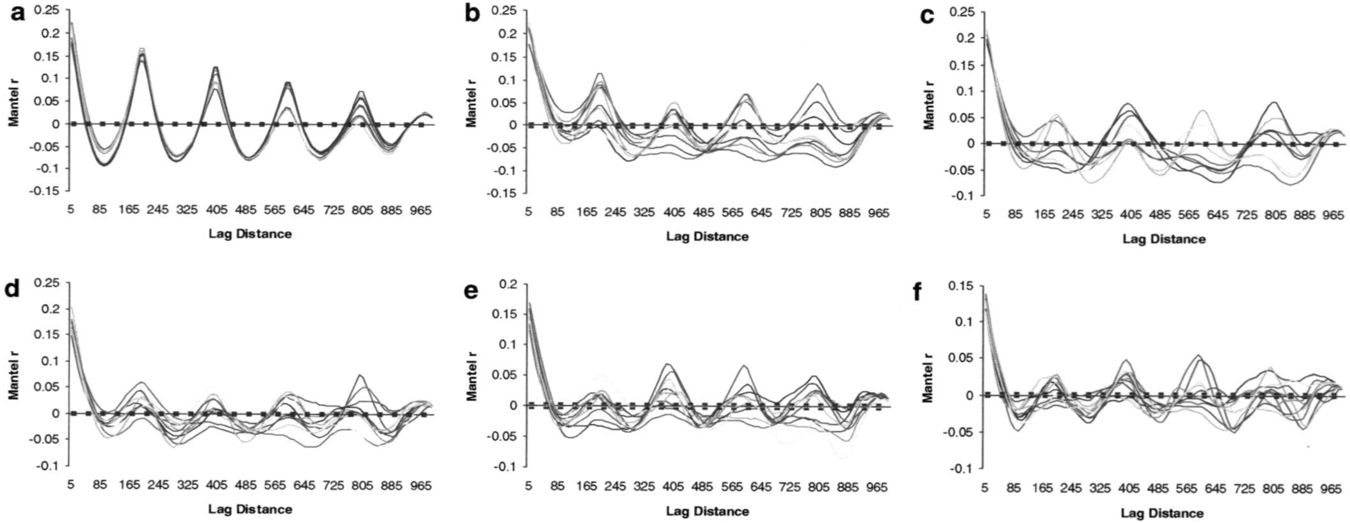

Fig. 7.6 Mantel correlograms for central place paths across the six distributions of turning angle. Each plot shows ten overlain correlograms derived from the ten replicate realizations of each path type

turning angle. For example, when mean turning angle was very small, such as 1°, this resulted in very strong periodic autocorrelation as the path zigzagged back and forth across the home range, bouncing off the frontier, with relatively little change in direction between. This periodic autocorrelation became more damped as the mean turning angle increased, but persisted in a detectable form up to at least a mean turning angle of 45°, and arguably all the way to a mean turning angle of 90°. As in the case of the unconstrained correlated random walks discussed above, variability among path correlograms increased with increasing mean turning angle. Also, as in the case of unconstrained correlated random walks, but perhaps surprisingly, there was no time to independence for bounded walks in any combination with turning angle distribution. This is surprising because random utilization of a fixed home range is held as the ideal and perhaps only case when there is an expectation of a time to spatial independence among sequential locations (Cushman et al. 2005). In this case, across all lag distances, there remained highly significant autocorrelation among locations. This is because a correlated random walk within a bounded home range does not equate to random utilization of a fixed home range. The utilization modeled here is a correlated random walk. This imparts randomness in turning angle, but not randomness in sequential location, as each location is contrained to be one step length from the previous and in a direction correlated to the previous step length. All real movement paths of actual organisms have this kind of constraint. This constraint results in very persistent and complex patterns of autocorrelation across very long time lags.

Central place random walks all shared a common basic form, characterized by repeated cycles between strong positive and strong negative autocorrelation with a period equal to 200 steps, or the time to complete a full cycle of foraging correlated random walk and azmithal return to the central place. As in the cases of unconstrained and bounded correlated random walks, increasing mean turn angle increased variability among replicate correlograms and damped the strength of the general pattern. Like the previous cases, there is no time lag sufficient to ensure spatial independence of locations. The periodic pattern of return to the central place burns in a cyclic pattern of positive to negative to positive autocorrelation that extends indefinitely.

7.2.2 *Autocorrelation of Elephant Movement*

Cushman et al. (2005) report long-term, spatially complex and seasonally variable patterns of autocorrelation among these elephants. Figures 7.7–7.9 show the correlogram surfaces for each collared elephant. The simulation analyses reported above provide a highly useful framework for understanding the structure of these empirical correlogram surfaces. This comparison indicates that across the year the pattern of autocorrelation of movement for each elephant show a full range of forms from smooth clines, strongly periodic use of central place, and occasional bounded correlated random use of temporarily fixed home ranges (Cushman

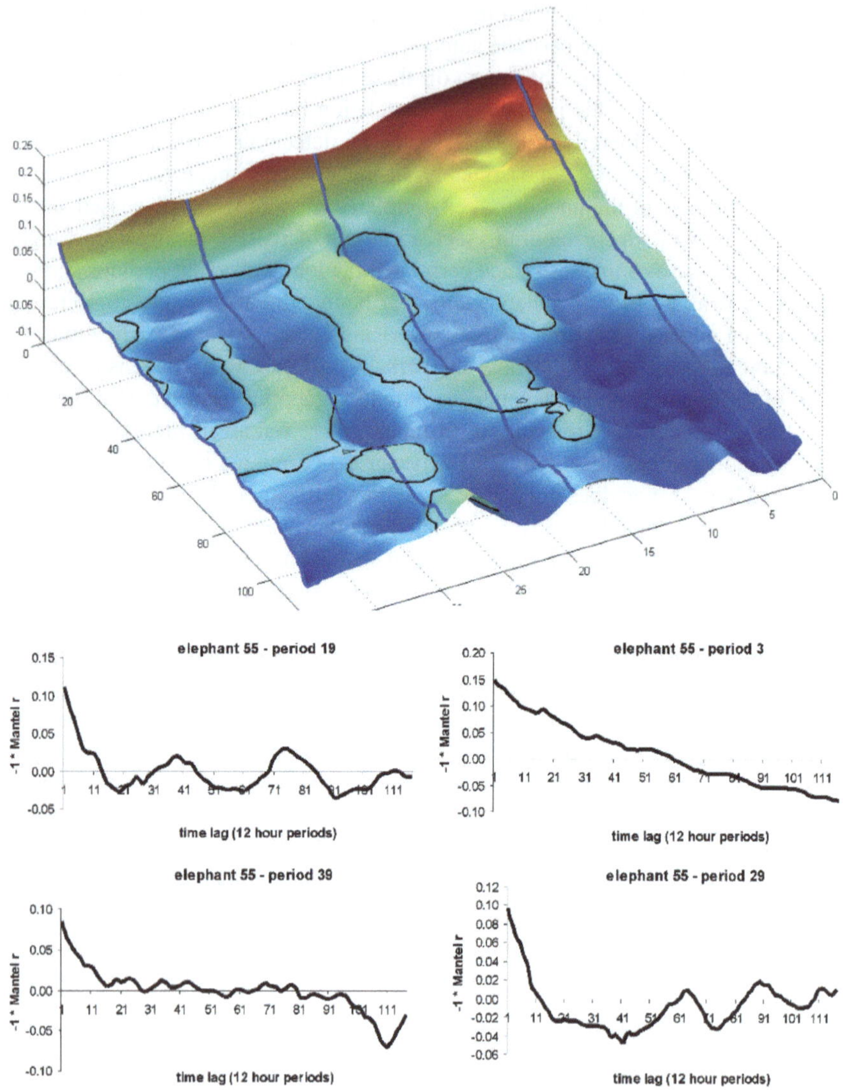

Fig. 7.7 Autocorrelation surface for elephant herd 55

et al. 2005). Interestingly, while the autocorrelogram surfaces for each elephant are highly complex, there is a strong degree of agreement between them, which in turn is related to the pattern of rainfall through the year (Cushman et al. 2005). This indicates that the structure of movement correlograms for these elephants provides highly interpretable information about movement behavior in relation to temporal changes in environmental conditions. During dry periods with little rain the elephants exhibit strongly periodic movement behavior corresponding

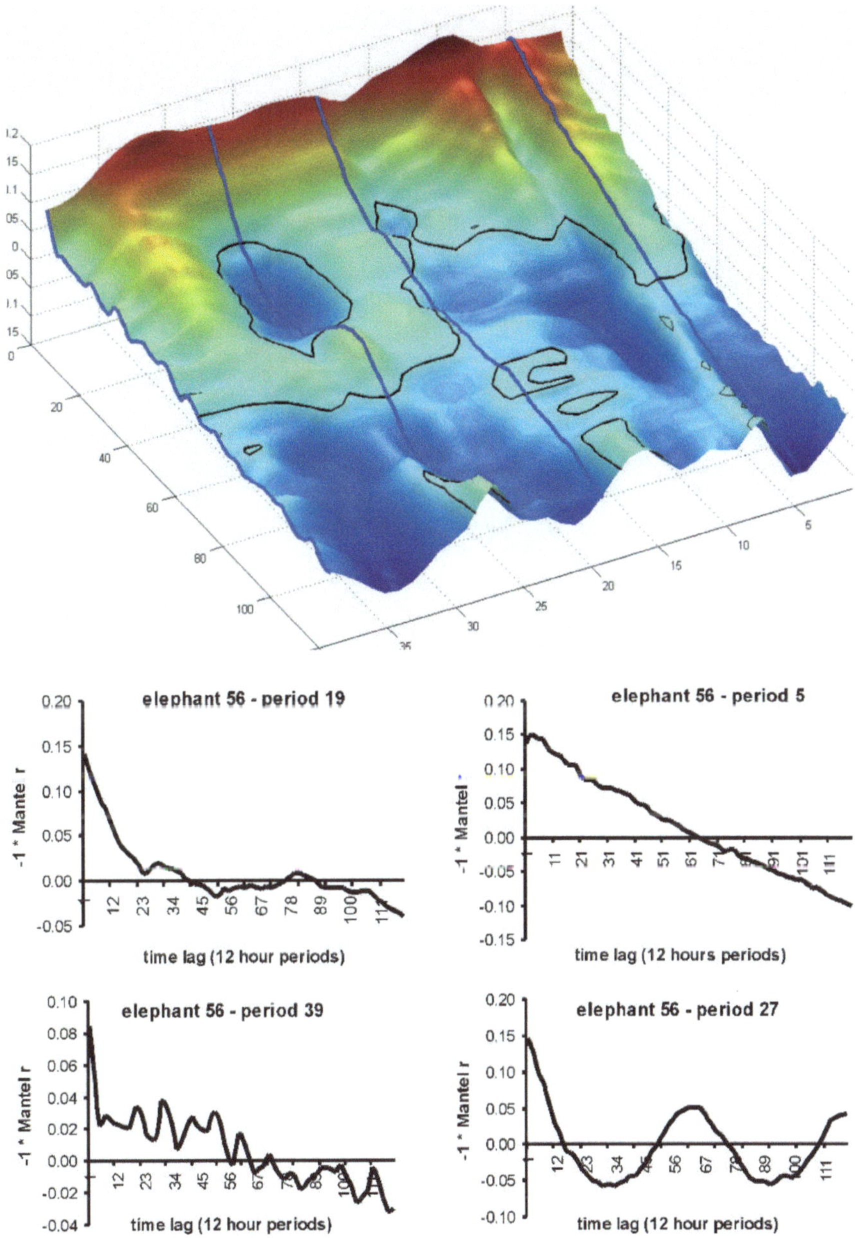

Fig. 7.8 Autocorrelation surface for elephant herd 56

to repeated visits to one or several permanent water holes interspersed with long foraging excursions away from the heavily over-grazed proximity of these water sources. During the rainy season when water is relatively continuously distributed,

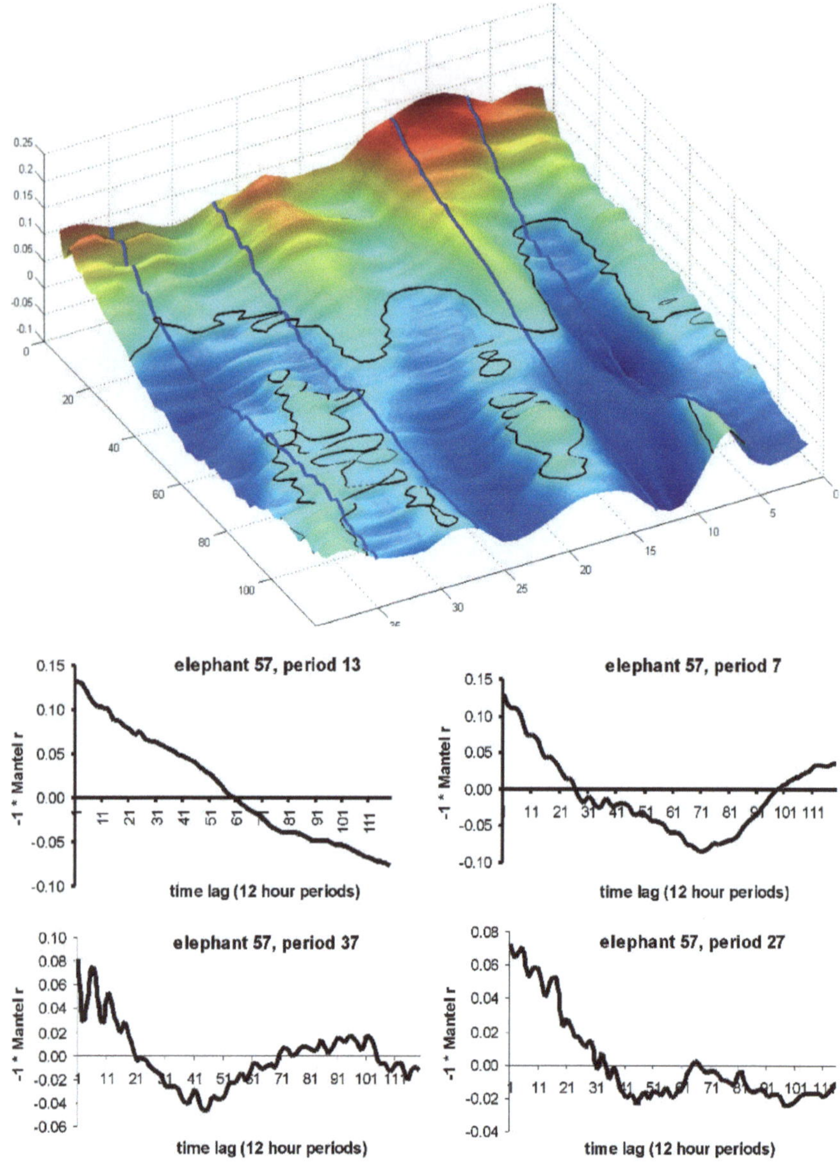

Fig. 7.9 Autocorrelation surface for elephant herd 57

the elephants fluctuation between unbounded correlated random walks, indicated by smooth clines in the correlogram surfaces, and random walks in temporarily bounded home ranges, as indicated by rapid drop to near zero followed by minor fluctuations around zero in the correlogram surfaces.

7.3 Discussion

7.3.1 Deep Autocorrelation in Animal Movement

Both simulation of movement paths with known process characteristics and analysis of empirical movement paths of a large and mobile mammal have clearly shown that autocorrelation is a fundamental attribute of movement data. In no simulated case did analyses show a meaningful time to independence. Even in the case of correlated random walks bounded within a fixed home range we found persistent periodic fluctuations between negative and positive autocorrelation as a result of the constraint that each step be within one step length of the previous one and in a direction to some degree correlated with the previous direction. No amount of subsampling of the data would remove the autocorrelation effects. In unconstrained correlated random walks and central place correlated random walks there are very strong patterns of autocorrelation that do not diminish appreciably over the full span of the walk. Subsampling data to a point where autocorrelation is no longer detected in such a case produces only an illusion of independence. It is conceptually similar to sampling a single point on the line. That point carries no information about the linear process that produced it, but it is entirely dependent on that process. Similarly, correlograms produced on subsampled or sparsely collected data on highly correlated pathway likely will show no significant long-range pattern of autocorrelation, even though there is underlying strong autocorrelation at every point and time lag.

7.3.2 Movement Paths Are Rich Source of Information

The consistent identification of persistent and often spatially complex patterns of autocorrelation in both controlled simulations and empirical analysis suggests that it will rarely be possible to satisfy assumptions of spatial independence among points for methods of analysis that require it. As mentioned in the introduction, this suggests the need for a change in focus from utilization points which are assumed to be a random sample of the full utilization distribution independent of sequential autocorrelation effects, to an explicit focus on the sequential, continuous nature of the movement paths themselves. The sequential spatial process of movement through complex landscapes contains rich information about the behavioral processes of foraging, resting, hiding, mating, defending territory, migrating and many other important behaviors in relation to underlying ecological gradients across a range of spatial scales. This linkage has largely been neglected, although some Markov chain approaches have begun to investigate it. In Chapter 20 we present an alternative approach, focusing on path-randomization and comparison of the ecological conditions encountered along utilized paths with those which would be encountered along potentially available paths of identical length and topology.

In addition, the strong and consistent structure of correlograms resulting from paths generated using different known spatial processes show that correlograms themselves can be highly useful tools to explore movement behavior and identify dominant movement processes, which then can be used to generate ecological and behavioral hypotheses. For example, the simulation results presented here clearly show three distinct correlogram shapes. The identification of one of these shapes in correlograms produced from empirical data will invalidate hypotheses about the existence of the process that would generate one of the other two correlogram forms. For example, if we hypothesize that an organism will exhibit random use of a fixed home range, and discover that its movement path produces a cline or periodic shaped correlogram we would be able to safely reject that hypothesis. Correlograms have some use, but less certainty, in confirming dominant underlying processes. For example, if we propose that an organism has an unconstrained correlated random walk, and discover its path produces a cline shaped correlogram, this is consistent with that explanation, but does not fully demonstrate its truth. This is a variety of the issue of affirming the consequent discussed in the first and second chapters. In this case, while a cline shaped correlogram is consistent with an unconstrained correlated random walk it may also be possible to obtain a similar cline shaped correlogram from other spatial processes, such as any walk in which locations tend to get farther away from each other in space as they become farther apart in time. Similarly, if one proposes a central place movement behavior and observes a strongly periodic correlogram, this observation is consistent with the hypothesis but does not prove it to be true. There are several ways one can obtain a similar periodic correlogram, including cyclic repeat of a similar route through a home range, or cyclic return to several foraging or resting locations. However, the form of correlograms are diagnostic for rejecting inconsistent movement hypotheses and are useful in determining the degree of support for alternative explanations of movement process.

References

Aebisher NJ, Robertson PA, Kenward RE (1993) Compositional analysis of habitat use from animal radiotracking data. Ecology 74:1313–1325

Allderedge JR, Ratti JT (1986) Comparison of some statistical techniques for analysis of resource selection. J Wildl Manag 50:157–165

Cresswell WJ, Smith GC (1992) The effects of temporally autocorrelated data on methods of home range analysis. In: Priede IG, Swift SM (eds) Wildlife telemetry: remote monitoring and tracking of animals. Ellis Horwood, London

Cushman SA, Chase M, Griffin C (2005) Elephants in space and time. Oikos 109:331–341

Fortin M-J, Dale M (2005) Spatial analysis: a guide for ecolgists. Cambridge University Press, Cambridge

Harris RB, Fancy SG, Douglas DC, et al. (1990) Tracking wildlife by satellite: current systems and performance. Tech Rep 30 US Fish and Wildlife Service

Legendre P (1993) Spatial autocorrelation: trouble or a new paradigm? Ecology 74:1659–1673

Legendre P, Fortin M-J (1989) Spatial pattern and ecological analysis. Vegetatio 80:107–138

Legendre P, Legendre L (1998) Numerical ecology. Developments in environmental modelling, 20. Elsevier, Amsterdam

Litvaitis JA, Sherburne JA, Bissonette JA (1986) Bobcat habitat use and home range size in relation to prey density. J Wildl Manag 50:110–117

Litvaitis JA, Titus K, Anderson EM (1994) Measuring vertebrate use of terrestrial habitats and foods. Invited contribution. In: T. Bookhout (Ed.) Research and management techniques for wildlife and habitats. The Wildlife Society, Washington, DC, pp. 254–274

Mantel N (1967) The detection of disease clustering and a generalized linear regression approach. Cancer Res 27:209–220

Neu CW, Byers CR, Peek JM (1974) A technique for analysis of utilization-availability data. J Wildl Manag 38:541–545

Oden NL, Sokal RR (1986) Directional autocorrelation: an extension of spatial correlograms to two dimensions. Syst Zool 35:526–530

Otis DL, White GC (1999) Autocorrelation of location estimates and the analysis of radio tracking data. J Wildl Manag 63:1039–1044

Palomares F, Delibes M (1992) Data analysis and potential bias in radio-tracking studies of animal habitat use. Acta Oecol 13:221–226

Rooney S, Wolfe MA, Kayden TJ (1998) Autocorrelated data in telemetry studies: time to independence and the problem of behavioral effects. Mammal Rev 28:89–98

Schoener TW (1981) An empirically based estimate of home range. Theor Popul Biol 20:281–325

Sokal RR (1986) Spatial data analysis and historical processes. In: Daly E, et al. (eds) Data analysis and informatics, IV. North-Holland, Amsterdam

Springer JT (1979) Some sources of bias and sampling error in radio triangulation. J Wildl Manag 43:926–935

Swihart RK, Slade NA, (1985) Influence of sampling interval on estimates of home-range size. J Wildl Manag 49:1019–1025

Swihart RK, Slade NA (1997) On testing for independence of animal movements. J Agric Biol Environ Statist 2:48–63

Thomas DL, Taylor EJ (1990) Study designs and tests for comparing resource use and availability. J Wildl Manag 54:322–330

White GC, Garrott RA (1990) Analysis of wildlife radio-tracking data. Academic Press, San Diego, CA

Chapter 8
Using a Random Forest Model and Public Data to Predict the Distribution of Prey for Marine Wildlife Management

Steffen Oppel and Falk Huettmann

8.1 Introduction

Modern wildlife management relies on studies investigating the distribution patterns and habitat selection of wildlife at appropriately large scales for decision-making. One important aspect to consider in the assessment of habitat suitability and the underlying mechanism of animal distribution is the spatial distribution of their food resources. In marine areas, where many mammal and bird species occur over large spatial scales, the analysis of habitat use and distribution patterns requires information on the distribution of food resources at appropriately large scales (Huettmann and Diamond 2006). An important food resource for several species of marine birds and mammals are invertebrate organisms that live on the bottom of the sea and are collectively described as the benthos, a community that is especially productive and diverse at high latitudes (Carey 1991; Piepenburg 2005; Starmans et al. 1999). The distribution and productivity of benthic foragers such as ice seals, walrus (*Odobenus rosmarus*), sea ducks (*Somateria* spp., *Melanitta* spp.), and gray whales (*Eschrichtius robustus*) is influenced by the distribution of accessible benthic prey resources (Kaiser et al. 2006; Lovvorn et al. 2003; Moore et al. 2003). Benthic invertebrates thus form a key component in the trophic structure of marine ecosystems, and the distribution of marine benthic invertebrates is of major interest to wildlife managers (Solan et al. 2004). Most of the species mentioned above are of management concern, and either are (Spectacled Eider, *Somateria fischeri*; Steller's

S. Oppel (✉)
Royal Society for the Protection of Birds, The Lodge, Sandy, Bedfordshire SG19 2DL, UK
email: steffen.oppel@rspb.org.uk

F. Huettmann
EWHALE lab- Biology and Wildlife Department, Institute of Arctic Biology, University of Alaska-Fairbanks, 419 IRVING I, Fairbanks, AK 99775-7000, USA
e-mail: fffh@uaf.edu

S.A. Cushman and F. Huettmann (eds.), *Spatial Complexity, Informatics, and Wildlife Conservation*
DOI 10.1007/978-4-431-87771-4_8, © Springer 2010

Eider, *Polysticta stelleri*) or have been proposed (walrus, ice seals) to be listed as 'threatened'. Therefore, the identification and delineation of critical habitat providing sufficient food resources for these species will become extremely important in the near future.

Mapping the distribution and abundance of benthic invertebrates is however logistically challenging; sampling is often confined to small spatial scales of $<100\,km^2$ and thus falls short of the spatial distribution of marine wildlife consuming benthic prey. Predicting the distribution of benthic fauna across large spatial scales ($>1{,}000\,km^2$) has recently been achieved by relating benthic faunal diversity and biomass to biological and physical properties of the environment using statistical modeling techniques (Ellis et al. 2006; Thrush et al. 2003; Ysebaert and Herman 2002). Traditionally, most approaches have been based on statistical data models like generalized linear models (GLM). But more recently, the use of algorithmic models has been advocated for ecological studies (Cutler et al. 2007; Elith et al. 2008; Hochachka et al. 2007). Algorithmic models derived from machine learning theory have been shown to be superior to conventional data models in modeling the distribution of terrestrial animals (Elith et al. 2006; Prasad et al. 2006), but their use in marine ecology and wildlife management has been limited so far (Denisenko et al. 2003; Leathwick et al. 2006). Here we demonstrate that an algorithmic modeling approach offers a simple and powerful alternative to predicting species and biomass distributions across large spatial scales in marine ecosystems, and thus provides a powerful tool to predict the distribution of food resources for larger wildlife.

The key difference between data models like GLMs and algorithmic models is the relationship between dependent variables such as the presence or abundance of benthic invertebrates and independent explanatory variables like physical or biological factors. When using data models, researchers need to specify an appropriate model a priori (e.g. a logistic curve, or a linear trend) without knowing the actual properties of the data, and then use data to estimate the parameters of this fixed model. In contrast, algorithmic models do not require the relationship between dependent and explanatory variables to be specified, but rather start virtually uninformed and use an algorithm to learn these relationships. Algorithmic models can thus elucidate extremely complex and unanticipated relationships in a dataset (Breiman 2001b). As the relationships between many oceanic properties and benthic invertebrate distribution are complex, and in many cases still poorly understood, we believe that algorithmic models provide a powerful tool for researchers to both model the spatial distribution of benthic invertebrates as well as to gain insight into functional relationships governing the distribution and biomass of benthic invertebrates if reliable data are available to train the model.

In this study we use a Random Forest algorithm (Breiman 2001a) and publicly available data to predict the spatial distribution of benthic invertebrate biomass across the Bering Sea. We compare our results from this model with available data from the region (Dunton et al. 2005; Grebmeier et al. 2006b) and outline how algorithmic models can be used in modeling relationships between marine fauna and environmental variables.

8.2 How Does a Random Forest Work?

Algorithmic models have several advantages over traditional data models (Breiman 2001b), such as high accuracy in complex datasets, ability to incorporate a large number of predictor variables, ability to automatically handle interactions, ability to handle missing data, and comparatively easy application and interpretation (Cutler et al. 2007; Elith et al. 2008; Hochachka et al. 2007). Algorithmic models are easily applied to many situations and thus have the potential to become a standard tool for wildlife managers.

The Random Forest algorithm is based on classification and regression tree analysis (Breiman et al. 1984; De'ath and Fabricius 2000). A classification or regression tree uses a series of rules to recursively split the data set into binary groups by identifying regions with the most homogenous set of a response to predictor variables. The model then fits a constant to each branch of the tree, with classification trees using the most probable category as the constant, and regression trees using the mean response observed in that subset of data. For each node the predictor variable and the split point are chosen to maximize the homogeneity of the data set along each of the two branches. Each branch can then be split again, either until a stopping criterion is reached, or until a user-specified number of terminal nodes is reached. The main advantage of trees is that predictor variables can be both categorical or continuous, and that irrelevant predictors are seldom selected for a split. Thus, there is no cost in including a large number of predictor variables. In contrast to GLMs, trees also incorporate, and benefit from, correlations and interactions due to the hierarchical structure within the tree. At each split the response depends not only on the value of the predictor at that split, but also on the predictors at all splits that occurred higher up in the tree. Further, trees are insensitive to outliers or missing values in a data set, which is a common occurrence in large spatial data sets (Craig and Huettmann 2008).

A Random Forest is an assemblage of a large number of classification or regression trees (Breiman 2001a). This technique is known to be robust against overfitting and yields highly accurate predictions (Breiman 2001b; Cutler et al. 2007; Elith et al. 2006; Prasad et al. 2006). The high predictive accuracy is achieved using two levels of randomisation in the construction of every tree in the Random Forest. First, each tree in the Random Forest is constructed from a random subset of the original data, either taken with a bootstrap sample with replacement or sampled randomly to a specified proportion of the entire data set. The data not chosen to construct the tree (termed 'out-of-bag' data, oob) are used to assess the predictive ability of that tree. Each tree thus provides both an algorithm to classify the data and an error estimate of predictive ability based on the oob data. Second, at each split within each tree a random subset m of the available predictor variables is used to partition the data set into two groups with minimal heterogeneity. This number m must be specified by the user, and should always be smaller than the total number of predictor variables available. Each tree recursively partitions the data using a random subset of predictor variables until homogeneity of the data in each terminal node cannot be increased by a further subdivision. After a user-specified number of trees has been constructed, each data point is run down every single tree in the Random Forest. Different trees may predict different outcomes for the same data point, and the most common classification across all trees is used to determine the predicted outcome of a data point.

To estimate the importance of predictor variables Random Forest uses a permutation procedure. In this procedure the values for a given variable are randomly permuted over the oob data set and the resulting reduction in model accuracy is assessed. Variable importance is inversely related to the reduction in model accuracy after permutation (Strobl et al. 2007). For easier interpretation, the variable importance can be standardized, with the most important variable being assigned a relative variable importance of 100%.

Important settings that need to be specified by a user are the number of trees to be constructed, the number of bootstrap samples taken for the training data of each tree, and the number m of variables randomly drawn at each split. The predictive ability of a Random Forest depends mainly on the correlation between any two trees in the forest, and the strength of each individual tree. Increasing the correlation increases the forest error rate, while increasing the strength of each tree decreases the forest error rate (Breiman 2001a). Reducing m reduces both the correlation and the strength, and vice versa. The optimal range of m is usually wide, and can be found using the oob error rate. Care needs to be taken when choosing m, as it is the only adjustable parameter to which Random Forests are somewhat sensitive.

8.3 A Benthic Biomass Model for the Bering Sea

The Bering Sea is an important habitat for many species of marine birds and mammals, including several that forage on benthic invertebrates. Proper management of these species demands a thorough understanding of the distribution of prey resources. To assess whether higher trophic level animals select locations with high benthic biomass as preferred foraging areas, a map describing the benthic biomass across the entire Bering Sea is needed. The benthic biomass distribution in the northern Bering and Chukchi Seas has so far only been mapped using coarse interpolation methods, where areas not sampled are assigned values from nearest sampling stations (Dunton et al. 2005; Grebmeier et al. 2006a). We used a Random Forest algorithm and globally available public data to predict benthic biomass across the entire Bering Sea based on local environmental conditions rather than interpolation.

8.3.1 Benthic Biomass Data

Benthic biomass data used to build the Random Forest algorithm were obtained from a variety of published and unpublished sources described in more detail in recent syntheses (Dunton et al. 2005; Grebmeier et al. 2006a). We used data from

Stoker (1981), from the Shelf-Basin Interaction program (available at http://www.eol.ucar.edu/projects/sbi/), and from a database provided by K. Coyle (Institute of Marine Science, University of Alaska Fairbanks). The data were collected over a time period of two decades, but detailed research at some of the sampling areas has shown that benthic communities in much of the Bering Sea are relatively stable at large spatial scales (Dunton et al. 2005; Grebmeier 1987). We standardized the data to a common unit (g wet weight m^{-2}), and divided the wet weight of benthic biomass at each sampling station into five categories to facilitate graphic presentation: <50, 50–100, 100–200, 200–500, and >500 g m^{-2}. We used the spatial coordinates of sampling stations to plot each sampling station on a map in ArcGIS v 9.2 (ESRI Inc., Redlands, CA).

8.3.2 Environmental Data

The distribution and abundance of benthic invertebrates is influenced by a number of environmental variables, like hydrography, chlorophyll *a* content, water temperature, sea ice scouring, and sediment grain size (Denisenko et al. 2003; Dunton et al. 2005; Feder et al. 1994; Grebmeier 1993; Grebmeier et al. 1988; Gutt 2001). We obtained environmental data for six important environmental predictor variables from publicly available sources. Further, we used computer-generated models of sea-bottom salinity and sea-bottom temperature (G. Panteleev, International Arctic Research Center, University of Alaska Fairbanks, unpublished data based on Panteleev et al. 2006). All these data were collated from a variety of remote sensing applications described in detail at the respective data source location (Table 8.1). We imported each environmental variable as a separate layer into ArcGIS, and used a point grid across the Bering Sea with a grid cell size of 10 × 10 km (Dunton et al. 2005). Thus, our data were ill suited to model local scale variations, but are appropriate for a model surface covering the entire Bering Sea.

We used sea ice extent data from late March 2000–2004 to calculate an average maximum sea ice extent across the study area. Each grid cell was assigned the averaged maximum sea ice extent (0–100%) at that location. We used coastline data to calculate the distance of each grid cell to the nearest coastline. Further, we used a multi-year average of August sea surface temperature, and a multi-year average of September chlorophyll *a* concentration. We were not able to obtain complete spatial coverage of sediment grain size data for the Bering Sea, and thus used sediment grain size in a separate model with only a subset of the data (presented separately).

8.3.3 Constructing the Random Forest Model

We overlaid each sampling station with the layers of all eight environmental predictor variables in ArcGIS, and thus obtained environmental data for each sampling station. Using the dependent variable (benthic biomass category 1–5) and eight predictor

Table 8.1 Environmental variables used to predict benthic biomass in the Bering Sea. Data were available at specified websites or were donated by recognized individuals

Abbreviation	Variable	Unit	Data provider/ organization	Website address
PHI	Sediment grain size	phi	National Geophysical Data Center	http://www.ngdc.noaa.gov/mgg/geology/size.html
BOTSAL	Sea bottom salinity	ppm	G. Panteleev, International Arctic Research Center	http://people.iarc.uaf.edu/~gleb/bering_sea/bering_sea.html
BOTTEMP	Sea bottom temperature	°C	G. Panteleev, International Arctic Research Center	http://people.iarc.uaf.edu/~gleb/bering_sea/bering_sea.html
DIST	Distance to coastline	km	National Geophysical Data Center	http://www.ngdc.noaa.gov/mgg/shorelines/shorelines.html
DEPTH	Water depth	m	National Geophysical Data Center	http://www.ngdc.noaa.gov/mgg/bathymetry/relief.html
CHLA	Chlorophyll *a* concentration	mg/l	National Oceanographic Data Center	http://www.nodc.noaa.gov/OC5/WOA05/pr_woa05.html
ICE	Maximum sea ice cover	%	National Ice Center	http://www.natice.noaa.gov/products/alaska/index.htm
SST	Sea surface temperature	°C	National Oceanographic Data Center	http://www.nodc.noaa.gov/OC5/WOA05/pr_woa05.html

variables for each of the 624 sampling stations we constructed 1,500 classification trees and used a random subset of 64% of the data without replacement to build single trees. We chose *m* to maximize classification accuracy, and report accuracy as the percentage of sampling stations for which the category of benthic biomass was predicted correctly. The importance of each environmental variable was calculated with a permutation procedure, and we report these results in relation to the most important (100%) variable identified by the model. We conducted our analyses in R 2.7.1 (http://www.r-project.org/index.html) with the add-on package randomForest, version 4.5–25.

8.3.4 Prediction of Benthic Biomass Across the Bering Sea

Unlike previous approaches to establish large scale maps of benthic biomass across the Bering and Chukchi Seas (Dunton et al. 2005; Grebmeier et al. 2006a), we did not use spatial interpolation approaches to project benthic biomass to areas where no sampling stations were located. Instead, we used the environmental variables available at a 10km grid cell resolution across the entire Bering Sea to predict benthic biomass in each grid cell based on the Random Forest algorithm developed from 624 sampling stations. We used R 2.7.1 to run the environmental data through the Random Forest model, and then plotted the resulting data in ArcGIS.

8.4 Results

At the 624 sampling stations used to construct the Random Forest model, benthic biomass ranged from 4.7–2,230.8 g m^{-2}. The model constructed with 1,500 trees and $m = 6$ classified 78.2% of sampling stations correctly. The model took 13 s to run on a 1700 MHz laptop with 512MB RAM. Different runs with varying m (3–7) resulted in marginally poorer classification accuracy (77.9% at $m = 7$, to 72.6% at $m = 3$). Obtaining such accurate predictions in very short time is of great value for wildlife managers.

The concentration of chlorophyll *a*, sea surface temperature, and depth were the most important variables for accurately predicting benthic biomass (Fig. 8.1). All the variables in our model had variable importance scores of >75% of the most important variable, indicating that we did not incorporate uninformative variables in the model. An algorithmic model also provides insight on functional relationships between environmental predictor and dependent variables. These relationships were nonlinear for the most important variables. Chlorophyll *a* concentration appeared to follow a threshold function with highest benthic biomass in areas with >5 mg/l chlorophyll *a*. Sea surface and sea bottom temperatures showed that the highest benthic biomass does not occur in the warmest waters, but in waters with cold-intermediate temperatures.

We then extended the model by including sediment grain size data, which was not available across the entire Bering Sea. We only used a subset of 564 stations with sediment grain size data, and built an equivalent model as above. This model classified 74.2% of sampling stations correctly (when using $m = 8$). The model took 10 s to run on a 1700 MHz laptop with 512MB RAM. Different runs with varying m (3–7) resulted again in marginally poorer classification accuracy (73.6% at $m = 7$, to 69.7% at $m = 3$). Including sediment grain size did not significantly change the importance of other variables, and sediment grain size was the least important variable in the set, with 64% of the importance score of chlorophyll *a* concentration (Fig. 8.1).

8.5 Discussion

Our model presents the first continuous map of predicted benthic biomass across the Bering Sea using environmental variables (Fig. 8.2). It offers a great tool for implementation in efficient wildlife management (Huettmann 2007).

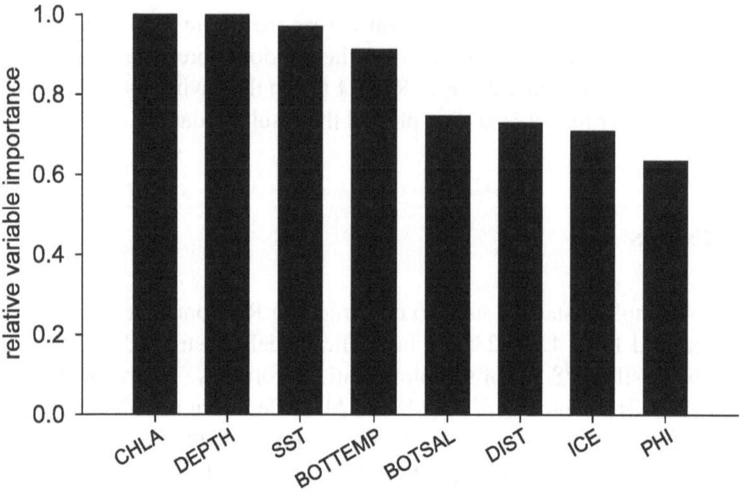

Fig. 8.1 Variable importance for Random Forest model predicting benthic biomass across the Bering Sea. See Table 8.1 for abbreviation of variables

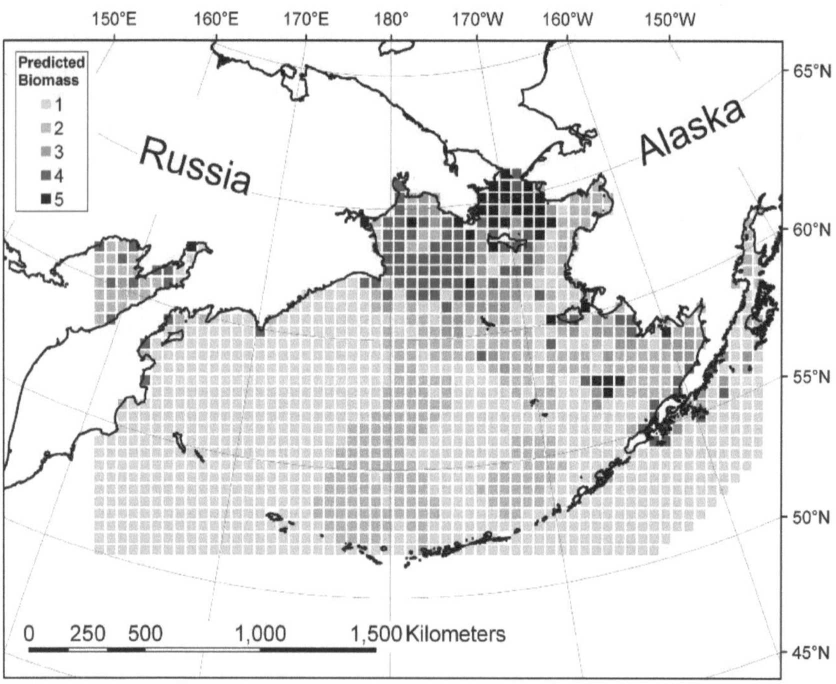

Fig. 8.2 Distribution of benthic biomass across the Bering Sea predicted from a Random Forest model trained with 624 sampling stations and projected using seven environmental predictor variables. Predicted biomass (wet weight in g m^{-2}): (1) <50, (2) 50–100, (3) 100–200, (4) 200–500, and (5) >500

The model was able to predict benthic biomass correctly at >75% of sampling stations. It agrees with previous maps (Dunton et al. 2005; Grebmeier et al. 2006a), but extends beyond areas sampled directly and shows wider variation to regional environmental conditions. In addition, this model and its map are available as GIS-layer to researchers and the public from our website (http://mercury.bio.uaf.edu/benthos), and we will continue to incorporate data and update the model available on the website. This model provides an approximation of the distribution of benthic biomass across the Bering Sea that can be used to relate the distribution or movements of higher trophic level consumers to the distribution of resources across large spatial scales (Oppel et al. 2009). However, our study was intended to demonstrate the approach of the modeling method for wildlife management, and we caution researchers that the results presented here may be biased at small spatial scales and require rigorous field validation. The continuous updating of existing models is a crucial component of our approach. Nonetheless, the model presented here will be helpful in directing future research, especially in those areas where currently no sampling stations exist and model predictions indicate unexpected patterns.

Our model of benthic invertebrate biomass distribution across the Bering Sea should be regarded as a first step towards a distribution map that could be considerably refined by the inclusion of additional variables that were not available to us. We achieved highest predictive ability when *m* was close to the total number of predictor variables, and none of the predictor variables had variable importance scores <50%. The importance of several variables indicates that interactions among variables play a major role. It further suggests that there was no redundancy among the variables included in the model, and further variables could significantly improve the predicted benthic biomass map for the Bering Sea. We encourage marine scientists to challenge and continue developing this model by using more benthic biomass training data from a wider geographic region, and by incorporating additional predictor variables as well as spatial autocorrelation (Dormann 2007). We used environmental variables at certain times of the year, but there is potential for chlorophyll *a* and temperatures to be important during other times of the year as well. We believe that expanding the model by including monthly averages of chlorophyll *a* and temperature, as well as mixed-layer depth and ocean layers of high resolution in time and space could substantially improve the model. Benthic communities are also influenced by biotic interactions and predation pressures, or disturbance effects of fisheries (Freese et al. 1999; Halpern et al. 2008; Nilsen et al. 2006; Sugden et al. 2008). Integrating all such factors would further strengthen the model, add realism, and would provide a useful opportunity to predict changes to benthic communities that are to be expected from an increase of commercial activities in the northern Bering Sea in the intermediate future (Hovelsrud et al. 2008; Kerr 2002).

Recent changes to benthic communities due to fisheries impacts and warming temperatures need to be considered when interpreting our results (Feder et al. 2003; Grebmeier et al. 2006b; Krajick 2007). A thorough evaluation of the input data used in the model may also lead to considerable improvement, especially if stations sampled decades ago were located in areas that may have undergone recent changes in benthic biomass (Gradinger 1995; Mueter and Litzow 2008; Ruhl 2007). It is important to note that the predictive accuracy of any model ultimately depends on the quality of the data used to construct it. A larger spatial and temporal coverage of benthic biomass data could therefore considerably improve the accuracy of the model presented here.

Besides providing a map of predicted biomass distribution, the algorithmic modeling approach presented here also enables the detection of nonlinear relationships between environmental predictor and response variables. We found that relationships between benthic biomass and environmental predictor variables in our model generally agree with known patterns of benthic biomass and diversity in relation to depth (Denisenko et al. 2003; Jones et al. 2007; Smale 2008), ice scouring (Gutt 2001; Smale 2007), food supply (Grebmeier et al. 1988; Lovvorn et al. 2005), and temperature (Tumbiolo and Downing 1994). However, the method used here also has the ability to reveal nonlinear threshold functions, which are extremely difficult to model with conventional GLMs (Hochachka et al. 2007). An interpretation of the mechanisms causing those relationships is beyond the scope of this contribution, but we encourage marine biologists to use algorithmic models to detect complicated relationships and thus improve our current understanding of the functional mechanisms governing the spatial distribution of benthic organisms.

Here, we used the broad and surrogate measure of benthic biomass to demonstrate the utility of the spatial modeling. However, benthic communities are composed of a large variety of different taxa which may respond differently to environmental conditions. Our model was not designed to make specific predictions about the composition of benthic communities or the occurrence of certain species, but such models could be constructed with the approach we presented here. For example, the presence, absence, or abundance of certain taxa (at the species or higher taxonomic level) could be modeled explicitly in relation to environmental variables and then predicted across large spatial scales from where those variables are available. By using this approach predictive models could be constructed for a large number of species or taxa. Aggregating those predictive distribution maps would then facilitate the development of large-scale maps of benthic community composition that could be verified by shipboard surveys. We believe that these maps would be extremely useful to project the changes in species composition resulting from climate changes, anthropogenic, and cumulative effects. Mapping such changes might be of special significance for economically important species such as mussels and crabs, as well as for the establishment of marine protected areas that are intended to serve as long-term refuges for threatened species (Lovvorn et al. 2009).

Acknowledgments We appreciate the support of data providers, most notably K. Coyle, G. Panteleev, and J. Grebmeier. I. Rutzen and D. Verbyla provided help with ArcGIS procedures, and we thank B. Bluhm, ArcOD, and J. Grebmeier for continued discussion about this work. We appreciate constructive criticism on an earlier version of this manuscript from J. Grebmeier and J. Lovvorn. This is EWHALE lab publication #63.

References

Breiman L (2001a) Random forests. Mach Learn 45:5–32

Breiman L (2001b) Statistical modeling: The two cultures. Stat Sci 16:199–215

Breiman L, Friedman JH, Olshen RA, Stone CG (1984) Classification and regression trees. Wadsworth International Group, Belmont, CA

Carey AG (1991) Ecology of North American Arctic continental shelf benthos. Cont Shelf Res 11:865–883

Craig E, Huettmann F (2008) Using 'blackbox' algorithms such as Treenet and Random Forests for data-mining and for finding meaningful patterns, relationships and outliers in complex ecological data: An overview, an example using Golden Eagle satellite data and an outlook for a promising future. In: Wang HF (ed) Intelligent data analysis: Developing new methodologies through pattern discovery and recovery. IGI Global, Hershey, PA, USA, pp 65–84

Cutler DR et al. (2007) Random Forests for classification in ecology. Ecology 88:2783–2792

De'ath G, Fabricius KE (2000) Classification and regression trees: A powerful yet simple technique for ecological data analysis. Ecology 81:3178–3192

Denisenko SG, Denisenko NV, Lehtonen KK, Andersin A-B, Laine AO (2003) Macrozoobenthos of the Pechora Sea (SE Barents Sea): community structure and spatial distribution in relation to environmental conditions. Mar Ecol Prog Ser 258:109–123

Dormann CF (2007) Effects of incorporating spatial autocorrelation into the analysis of species distribution data. Global Ecol Biogeog 16:129–138

Dunton KH, Goodall JL, Schonberg SV, Grebmeier JM, Maidment DR (2005) Multi-decadal synthesis of benthic–pelagic coupling in the western arctic: role of cross-shelf advective processes. Deep-Sea Res Pt II 52:3462–3477

Elith J et al. (2006) Novel methods improve prediction of species' distributions from occurrence data. Ecography 29:129–151

Elith J, Leathwick JR, Hastie T (2008) A working guide to boosted regression trees. J Anim Ecol 77:802–813

Ellis J et al. (2006) Predicting macrofaunal species distributions in estuarine gradients using logistic regression and classification systems. Mar Ecol Prog Ser 316:69–83

Feder HM, Naidu AS, Jewett SC, Hameedi JM, Johnson WR, Whitledge TE (1994) The northeastern Chukchi Sea - benthos-environmental interactions. Mar Ecol Prog Ser 111:171–190

Feder HM, Norton DW, Geller JB (2003) A review of apparent 20th century changes in the presence of mussels (*Mytilus trossulus*) and macroalgae in Arctic Alaska, and of historical and paleontological evidence used to relate mollusc distributions to climate change. Arctic 56:391–407

Freese L, Auster PJ, Heifetz J, Wing BL (1999) Effects of trawling on seafloor habitat and associated invertebrate taxa in the Gulf of Alaska. Mar Ecol Prog Ser 182:119–126

Gradinger R (1995) Climate change and biological oceanography of the Arctic Ocean. Philos Transact A Math Phys Eng Sci 352:277–286

Grebmeier JM (1987) The ecology of benthic carbon cycling in the northern Bering and Chukchi Seas. In: Institute of Marine Science. University of Alaska Fairbanks, Fairbanks, AK

Grebmeier JM (1993) Studies of pelagic-benthic coupling extended onto the Soviet continental shelf in the northern Bering and Chukchi seas. Cont Shelf Res 13:653–668

Grebmeier JM, McRoy CP, Feder HM (1988) Pelagic-benthic coupling on the shelf of the northern Bering and Chukchi seas I. Food supply source and benthic biomass. Mar Ecol Prog Ser 48:57–64

Grebmeier JM, Cooper LW, Feder HM, Sirenko BI (2006a) Ecosystem dynamics of the Pacific-influenced Northern Bering and Chukchi Seas in the Amerasian Arctic. Prog Oceanogr 71:331–361

Grebmeier JM et al. (2006b) A major ecosystem shift in the Northern Bering Sea. Science 311:1461–1464
Gutt J (2001) On the direct impact of ice on marine benthic communities, a review. Polar Biol 24:553–564
Halpern BS et al. (2008) A global map of human impact on marine ecosystems. Science 319:948–952
Hochachka WM et al. (2007) Data-mining discovery of pattern and process in ecological systems. J Wildl Manag 71:2427–2437
Hovelsrud GK, McKenna M, Huntington HP (2008) Marine mammal harvests and other interactions with humans. Ecol Appl 18:S135-S147
Huettmann F (2007) Modern adaptive management: Adding digital opportunities towards a sustainable world with new values. For Pub Pol Clim Change Sust Develop 3:337–342
Huettmann F, Diamond A (2006) Large-scale effects on the spatial distribution of seabirds in the Northwest Atlantic. Landsc Ecol 21:1089–1108
Jones D, Bett B, Tyler P (2007) Depth-related changes to density, diversity and structure of benthic megafaunal assemblages in the Fimbul ice shelf region, Weddell Sea, Antarctica. Polar Biol 30:1579–1592
Kaiser MJ et al. (2006) Distribution and behaviour of Common Scoter *Melanitta nigra* relative to prey resources and environmental parameters. Ibis 148:110–128
Kerr RA (2002) A warmer Arctic means change for all. Science 297:1490–1493
Krajick K (2007) Thriving Arctic bottom dwellers could get strangled by warming. Science 315:1527
Leathwick JR, Elith J, Francis MP, Hastie T, Taylor P (2006) Variation in demersal fish species richness in the oceans surrounding New Zealand: an analysis using boosted regression trees. Mar Ecol Prog Ser 321:267–281
Lovvorn JE, Richman SE, Grebmeier JM, Cooper LW (2003) Diet and body condition of spectacled eiders wintering in pack ice of the Bering Sea. Polar Biol 26:259–267
Lovvorn JR, Cooper LW, Brooks ML, De Ruyck CC, Bump JK, Grebmeier JM (2005) Organic matter pathways to zooplankton and benthos under pack ice in late winter and open water in late summer in the north-central Bering Sea. Mar Ecol Prog Ser 291:135–150
Lovvorn JR, Grebmeier JM, Cooper LW, Bump JK, Richman SE (2009) Modeling marine protected areas for threatened eiders in a climatically changing Bering Sea. Ecol Appl 19:1596–1613. doi:10.1890/08-1193.1
Moore SE, Grebmeier JM, Davies JR (2003) Gray whale distribution relative to forage habitat in the northern Bering Sea: current conditions and retrospective summary. Can J Zool 81:734–742
Mueter FJ, Litzow MA (2008) Sea ice retreat alters the biogeography of the Bering Sea continental shelf. Ecol Appl 18:309–320
Nilsen M, Pedersen T, Nilssen EM (2006) Macrobenthic biomass, productivity (P/B) and production in a high-latitude ecosystem, North Norway. Mar Ecol Prog Ser 321:67–77
Oppel S, Powell AN, Dickson DL (2009) Using an algorithmic model to reveal individually variable movement decisions in a wintering sea duck. J Anim Ecol 78:524–531
Panteleev GG, Stabeno P, Luchin VA, Nechaev DA, Ikeda M (2006) Summer transport estimates of the Kamchatka Current derived as a variational inverse of hydrophysical and surface drifter data. Geophys Res Lett 33:5
Piepenburg D (2005) Recent research on Arctic benthos: common notions need to be revised. Polar Biol 28:733–755
Prasad AM, Iverson LR, Liaw A (2006) Newer classification and regression tree techniques: bagging and random forests for ecological prediction. Ecosystems 9:181–199
Ruhl HA (2007) Abundance and size distribution dynamics of abyssal epibenthic megafauna in the northeast Pacific. Ecology 88:1250–1262
Smale D (2008) Continuous benthic community change along a depth gradient in Antarctic shallows: evidence of patchiness but not zonation. Polar Biol 31:189–198

Smale DA (2007) Ice disturbance intensity structures benthic communities in nearshore Antarctic waters. Mar Ecol Prog Ser 349:89–102

Solan M, Cardinale BJ, Downing AL, Engelhardt KAM, Ruesink JL, Srivastava DS (2004) Extinction and ecosystem function in the marine benthos. Science 306:1177–1180

Starmans A, Gutt J, Arntz WE (1999) Mega-epibenthic communities in Arctic and Antarctic shelf areas. Mar Biol 135:269–280

Stoker S (1981) Benthic invertebrate macrofauna of the eastern Bering/Chukchi continental shelf. In: DW H, AG C (eds) The Eastern Bering Sea Shelf: Oceanography and Resources, vol 2. Unversity of Washington Press, Seattle, WA, pp. 1069–1091

Strobl C, Boulesteix A-L, Zeileis A, Hothorn T (2007) Bias in random forest variable importance measures: Illustrations, sources and a solution. BMC Bioinformatics 8:25–45

Sugden H, Lenz M, Molis M, Wahl M, Thomason JC (2008) The interaction between nutrient availability and disturbance frequency on the diversity of benthic marine communities on the north-east coast of England. J Anim Ecol 77:24–31

Thrush SF, Hewitt JE, Norkko A, Nicholls PE, Funnell GA, Ellis JI (2003) Habitat change in estuaries: predicting broad-scale responses of intertidal macrofauna to sediment mud content. Mar Ecol Prog Ser 263:101–112

Tumbiolo ML, Downing JA (1994) An empirical model for the prediction of secondary production in marine benthic invertebrate populations. Mar Ecol Prog Ser 114:165–174

Ysebaert T, Herman PMJ (2002) Spatial and temporal variation in benthic macrofauna and relationships with environmental variables in an estuarine, intertidal soft-sediment environment. Mar Ecol Prog Ser 244:105–124

Chapter 9
Landscape Genomics: A Brief Perspective

Michael K. Schwartz, Gordon Luikart, Kevin S. McKelvey, and Samuel A. Cushman

9.1 Introduction

Landscape genetics is the amalgamation of population genetics and landscape ecology (see Manel et al. 2003; Storfer et al. 2007). In Chapter 17, we discuss landscape genetics and provide two examples of applications in the area of modeling population connectivity and inferring fragmentation. These examples, like virtually all extant landscape genetic analyses, were based on evaluating spatial genetic patterns using a relatively small number of selectively neutral (or nearly neutral) markers. Landscape genomics, on the other hand, is the simultaneous study of tens to hundreds of markers, ideally including markers in candidate adaptive genes (genes under selection), with georeferenced samples collected across a landscape. While landscape genomics is, in one sense, simply landscape genetics with lots of data (thus reduced variance and increased precision), the qualitatively different (adaptive, potentially non-independent) nature and analytical approaches associated with these data are different enough to produce a profoundly different field.

In the past year there has been a boom in molecular genetics technology and this has lead to an unprecedented amount of genomics data (Hauser and Seeb 2008;, Mardis 2008; Shendure and Ji 2008; Eid et al. 2009). Consider this: the Human Genome Project, whose goal it was to sequence one human genome cost US$3 billion and took 15 years (Collins et al. 2003), yet today a private company is offering to sequence a whole human genome for $350,000 in 2–3 months

M.K. Schwartz, K.S. McKelvey, and S.A. Cushman (✉)
US Forest Service, Rocky Mountain Research Station, 800 E Beckwith, Missoula, MT 59801, USA
e-mail: scushman@fs.fed.us

G. Luikart
Division of Biological Sciences, University of Montana, Missoula MT 59812, USA;
Centro de Investigação em Biodiversidade e Recursos Genéticos (CIBIO-UP), Universidade do Porto (CIBIO-UP), Portugal

S.A. Cushman and F. Huettmann (eds.), *Spatial Complexity, Informatics, and Wildlife Conservation*
DOI 10.1007/978-4-431-87771-4_9, © Springer 2010

(http://www.knome.com/home/). Other companies are promising a $100 human genome, produced within one hour, by the year 2012 (http://www.pacificbiosciences.com/index.php). This wealth of genetic information will lead to changes in the way we study animal populations across the landscape.

9.1.1 Population Genomics

To understand what genomics will do for the field of landscape genetics we first need to understand what genomics is, and the difference between selectively neutral markers that are currently being used in landscape genetics, versus adaptive makers that are under selection that will strongly impact the field of landscape genomics. Neutral markers, as we currently understand them, have no direct biological meaning (e.g., they do not influence survival or fecundity, the main components of individual fitness). However, we infer biological activity, such as animal movements, from characteristic genetic patterns derived from neutral markers (Manel et al. 2003).

Population genomics, in contrast, has been defined as the simultaneous study of numerous loci (markers on chromosomes), genes (coding for functional products) or genome regions to better understand the role of evolutionary processes such as genetic drift, selection and migration, that influence variation across genomes and populations (Luikart et al. 2003; Kohn et al. 2006). So, while to date most landscape genetic studies have been conducted with 5–20 neutral, microsatellite markers, in the very near future we will be able to examine hundreds, if not thousands, of regions of the genetic code in hundreds of individuals from across the landscape to make inferences as to the evolutionary forces in play – including natural selection. For example, in evolutionary studies we will be able to disentangle the influences of gene flow and genetic drift from natural selection in influencing the evolutionary trajectory of a population; in the field of conservation biology we will be able to better define Evolutionary Significant Units and Distinct Population Segments, which are the basis for legal protection of species in the United States (Waples 1995); and in the field of ecology we will better predict how climate change will influence continuously distributed populations subject to various selection regimes.

9.1.2 Neutral Versus Non-neutral Molecular Markers

One of the most important differences between the field of population genomics and population genetics lies in the active seeking and utilization of genetic markers under selection – that is areas of the genome that are associated with adaptive traits. Neutral molecular markers, which are not influenced by natural selection, are often used by population geneticists and landscape ecologists because they give unbiased estimates of genetic variation (e.g., heterozygosity), population structure, and gene flow – the core variables of interest in landscape genetics so far. Used in this context, markers under selection will bias estimates of variation, structure, gene flow, and population relationships. Even a few selected loci among tens of neutral loci can bias estimates of substructure and gene flow metrics such as F_{st} by 10–60% (Luikart et al. 2003; Storz and Nachman 2003) and change the relationships among gene and species trees.

Selection might also change the spatial pattern of relatedness between individuals, which in tern is the basis of most genetic distance measures often used in individual-based approaches in landscape genetics. For example, high levels of gene flow between two study areas should lead to similar allele frequencies in both areas (low F_{ST}), but if certain alleles were eliminated through natural selection (that is the individuals having these alleles died or have reduced fitness) in one of the two study areas, allele frequencies would remain divergent (high F_{ST}) regardless of rates of gene flow. For these reasons geneticists have largely viewed markers under selection as nuisances and thus things to be avoided.

In landscape genomics, however, many markers, both neutral and under selection, will be known for each individual across a landscape. By combining many neutral genetic marker results, allowing us to infer gene flow and drift, with results from markers under selection we could tease apart these evolutionary forces and understand how the landscape and environment are influencing our natural populations. Here the neutral model would serve as the null model (similar to using Euclidian distance as the null model in the wolverine example in Chapter 17) when testing for selection (or among alternative models of selection). Statistical methods to quantify these relationships are currently under development and pose significant challenges (potentially there are thousands of markers with varying degrees of selection, and different modes of selection, e.g. diversifying vs. directional selection). We believe, however, that variants of current methods used in landscape ecology to partial out factors may provide the statistical basis for these analyses. For example, there are well developed methods to associate species distributions with environmental gradients to infer species niche structure as zones of tolerance within an environmental hypervolume (see Chapters 2 and 16, this volume). Conceptually, by replacing spatial occurrence of species with occurrence of particular genetic variants at loci under selection within a particular species it will be possible to apply many of the same conceptual and analytical methods to modeling the patterns of variation in adaptive genes as functions of environmental selection gradients. Specifically, constrained ordination, such as redundancy analysis, is well suited to modeling simultaneous response of multiple genes to complex gradients of multiple environmental variables to identify main factors driving patterns of selection. In addition, the familiar logistic regression is a powerful tool for identifying non-random distributions of single genetic variants as functions of environmental gradients. These approaches will allow the identification of genetic markers potentially under selection by discriminating between those that vary randomly with respect to environmental gradients from those that have strong associations with different environmental conditions. This would suggest potential for differences in selection.

There will be three primary challenges in this effort. First is the challenge of identifying genes under selection from the vast background of genomic data (described further below). The second challenge will be identifying the interaction of multiple genes through epistasis, pleiotropy and gene expression on fitness (Foll and Gaggiotti 2008; Balkenhol et al. 2009). The third main challenge will be identifying the proper environmental variables at the proper spatial and temporal scales that drive the selection processes that result in spatial differentiation in these genes. Ultimately, the challenge is to associate patterns of adaptive genes within organisms with the environmental gradients primarily related to fitness differences of these alternative genetic states (Holderegger and Wagner 2008).

9.1.3 Finding Genes Under Selection

From the discussion of neutral and non-neutral markers above, it is clear that population genomics requires the identification of many neutral and non-neutral markers, and their clear separation. Furthermore, as already noted, confusing neutral markers with those under selection can lead to large errors in interpretation of results. While it is well established how we can obtain neutral molecular markers, the approaches to finding adaptive markers are just now being developed in natural populations. The ideal markers for studying adaptation will be directly involved in the genetic control of adaptive traits, will have a sequence of known function, and will have quantifiable variation (Gonzalez-Marinez et al. 2006). Markers that have these traits will be in or near important functional genes or in gene rich regions. Such markers can include microsatellites (Vasemägi and Primmer 2006, Luikart et al. 2008), but more likely will be Single Nucleotide Polymorphisms (SNPs), AFLPs, and DNA sequences (Box 9.1).

Box 9.1 Molecular Markers for Landscape Genomics

The ideal DNA analysis technology for landscape genomics should genotype hundreds of polymorphic markers (including neutral and adaptive gene markers) that cover the entire genome in a single, simple and reliable experiment. At present AFLPs, SNP multiplex genotyping, and massively parallel sequencing partially or completely fulfill these requirements.

AFLP (Amplified Fragment Length Polymorphism)

AFLP genotyping uses selective PCR to produce hundreds of polymorphic markers that cover the entire genome. However, AFLP markers sometimes cluster around chromosome centromeres. AFLPs have been used to identify

(continued)

Box 9.1 (continued)

markers associated with traits that are under selection in non-model plant and vertebrate species (Wilding et al. 2001; Bonin et al. 2006)). Variants of the classical AFLP protocol use one primer that contains a conserved sequence of a gene family (gene-targeted AFLP) or primers in widely-dispersed repeated sequences such as small inserted nuclear elements (SINEs; for example, Alu repeats; van Tienderen et al. 2002). Unlike the classical AFLP protocol, the SINE-based approach requires only a single PCR. Gene-targeted AFLP can facilitate the detection of selection signatures and adaptive genes. Gene targeting (or avoidance) can also be facilitated by using GC-rich (or GC-poor) restriction enzymes, which tend to cut genomic DNA in gene-rich (or gene-poor) regions. The main problems with AFLPs are they are dominant markers (making heterozygote identification difficult), repeatability can be questionable, and transferability between labs is problematic.

SNPs (Single Nucleotide Polymorphisms)

The rapid identification of numerous SNPs (including functional SNPs) is becoming feasible in non-model species (e.g., Vera et al. 2008), owing to the rapid growth of expressed sequence tag (EST) databases, data base mining software, large scale multiplex PCR strategies (Porreca et al., 2007), DNA "capture" techniques (Hodges et al. 2007) and, most importantly, massively parallel sequencing (Shendure and Ji 2008; see below). Recent improvements in SNP genotyping technology (Perkel 2008) make SNPs attractive for population genomics (Kim and Misra 2007). For example, 48 or 96 SNPs can be screened in 96 individuals in a few hours for a cost of only US$0.10–0.20 per SNP (Hauser and Seeb 2009; see also Illumina in online links box). A drawback of SNPs is that they are prone to severe ascertainment bias – bias in estimating genetic diversity and population parameters – which arises when choosing markers on the basis of their polymorphism level, when identifying SNPs using few individuals, or when transferring markers between populations (Morin et al. 2004).

Microsatellites

Microstallites often can be identified in or near genes thanks to genome sequences (Da Silva et al. 2008) and EST data bases (Vasemägi et al. 2005; see also Thurston and Field 2005). However the genotyping of hundreds of microsatellites would require too many DNA amplifications to be competitive with methods that allow a 'massively parallel' analysis (for example, AFLP, SNPs, and next generation sequencing).

(continued)

Box 9.1 (continued)

DNA Sequences

We can now generate sequence data for 100s of loci in dozens of individuals from non-model species relatively quickly (Shendure and Ji 2008; Meyer et al. 2008; Vera et al. 2008). Sequence data are desirable because ascertainment bias is reduced (compared to SNPs) or avoided, haplotypes can be identified (or inferred), and coalescent times and allele relatedness (genealogies) can be estimated. Difficulties with sequencing include the analysis of heterozygous sites, homopolymers (consecutive instances of the same base such as AAA), and insertion/deletion polymorphisms; DaSilva et al. 2008). Next-generation sequencing allows generation of hundreds of millions (or billions) of base pairs of sequence in days. Unfortunately, it is still difficult to sequence many individuals. Yet, techniques (such as gaskets), commercial kits and barcoding of primers are being developed to allow simultaneous sequencing of 10s of loci for approximately 10–200 individuals in an single run on a new generation sequencer (Meyer et al. 2008).

There are a growing number of statistical approaches to detect selection or molecular adaptation. Among the most widely used approach is based on F_{ST}-outlier tests (Beaumont 2005). However this approach assumes discrete populations and alternatives are needed. To detect selection, it will be extremely helpful to have markers in genes with functions related to environmental phenotypic gradients observed across the landscape meeting the criteria listed above. For most wild species we do not have the ability to conduct extensive experimental or captive studies over many generations to find these genes. Thus, we can turn to candidate gene approaches, where we use genes from model or semi-model species with known function and then use simulation modeling to test if these genes appear to be under selection in the wild (Antao et al. 2008). The second approach, often called model free methods, uses population genomic data and examine hundreds of markers with unknown function (Lawson Handley et al. 2007). Here there are several new techniques available to test for non-neutral loci (e.g. Joost et al. 2007), but they fundamentally are usually examining patterns among populations to look for genetic loci that deviate from patterns of neutrality. There are even new bioinformatic programs that are able to process large amount of data, and conduct genome-wide tests to identify markers associated with environmental variables (Joost et al. 2007).

9.1.4 An Example of Landscape Genomics

There have only been a few studies that we would consider landscape genomic studies published thus far. One of the best examples is the use of both neutral genes and genes under selection to understand both the evolution of humans and the ecology of pathogens (Prugnolle et al. 2005a, b). In the first paper these authors use landscape genetics to provide support for the "recent African origin" model of human evolution by showing that geographic distance from East Africa along probable colonization routes is the best predictor for neutral genetic diversity in human populations. Subsequently, these authors examined patterns in MHC loci (associated with resistance against pathogens; also called Human Leukocyte Antigen or HLA) across 61 human populations to test a hypothesis regarding the high genetic diversity found in MHC loci. Prugnolle et al. (2005b) showed, using landscape resistance models, that the MHC loci had greater variation in areas with high pathogen diversity, while accounting for the fact that the contemporary pattern of diversity at this locus worldwide was influenced by human colonization. While still working with relatively few markers, Prugnolle et al. (2005a, b) demonstrate the power and potential associated with contrasting neutral patterns with those under selection. The pattern of neutral genetic diversity allowed these authors to disentangle the effects of past colonization history from patterns of natural selection on a particular locus with important function. We expect more efforts in the near future on both humans and wildlife that combine both marker types in a fully integrated landscape genomics study.

9.2 Identifying Evolutionary Significant Units Using Genes Under Selection

The field of landscape genomics offers important insights for conservation biologists as well as evolutionary biologists. One fundamental issue in conservation biology is the defining of Evolutionary Significant Units (ESUs), which are population units of conservation interest often below the species level (Fraser and Bernatchez 2001). Defining ESUs has been hotly debated over the past two decades (e.g., Moritz 1994; Waples 1995; Crandall et al. 2000; Fraser and Bernatchez 2001; Palsboll et al. 2007), but the fundamental definition usually suggests that groups of individuals must show adaptive (or ecological) divergence and historical isolation from other groups to be considered ESUs (Allendorf and Luikart 2007).

Historical isolation can be readily analyzed through standard neutral genetic approaches, whereas genetic analysis of adaptive divergence requires the discovery and novel evaluation of genes under selection (see above). Biologists have therefore often looked at historical isolation through molecular genetic studies, but have relied on detailed ecological studies to determine adaptive divergence. Because genetic analysis of isolation can be done quickly and is relatively inexpensive, the designation of ESUs has relied on a plethora of molecular studies and resulting information on contemporary and historical population isolation, but little ecological information. Not surprisingly, ESU designation has been heavily criticized for the overemphasis on reproductive isolation and under-emphasis on ecological data that suggest adaptive differences among populations (Crandall et al. 2000; Pearman 2001).

The heavy reliance on isolation rather than adaptation may bias what we choose to conserve. Consider the species that has multiple populations with high gene flow but selection for different traits on the landscape of each population. Examining neutral genetic markers across the landscape would reveal high gene flow and suggest for us to lump these populations into one ESU. However, we know that substantial functional divergence and reproductive isolation can take place despite high levels of gene flow (Smith et al. 1997; Crandall et al. 2000; Dieckmann and Doebelli 1999). Here is one place where landscape genomics may help ESU designation. Following the approaches pioneered by Prugnolle et al. (2005a, b) discussed above, we could genetically evaluate both isolation and adaptation, and use the gene flow rates to scale the degree of adaptation present in various sub-populations. Not only would this approach be much more powerful and quantitative than current methods, but it would remove the time and cost differential for obtaining measures of isolation versus adaptation.

9.3 Conclusion

Population genomic approaches can facilitate landscape genetics in three main ways. First, genotyping numerous loci provides high statistical power to quantify gene flow, genetic differentiation (F_{ST}), and diversity. Second, analyzing many loci can help reduce biases when measuring gene flow using methods that require the assumption that loci are neutral (e.g., N_e, F_{ST} and migration rates), because analyzing many loci helps identify and exclude loci that are under selection. Third, the measurement of adaptive genes and detection of locus-specific effects could help detect important selection gradients in the landscape. With these tools we can hopefully move away from delineating ESUs based primarily on isolation and move towards ESUs that will conserve based both isolated populations and adaptive differences across space (see Crandall et al. 2000).

The main barriers to the use of genomics approaches for population monitoring are the current expense and, in some taxa, the lack of availability of numerous markers (including markers in genes). Fortunately costs are decreasing and genomic information is rapidly increasing for most species. As pointed out by Hauser and Seeb (2008), these barriers are decreasing exponentially over time. We are therefore on the cusp of answering long-standing ecological and evolutionary questions in secretive and elusive species, thanks to improved noninvasive sampling of elusive species (see Long et al. 2008 for noninvasive methods to sample carnivores) and new technologies for SNP genotyping and sequencing short DNA fragments (Morin and McCarthy 2007). This includes questions about the genetic basis of local adaptation that can be addressed by using genome-wide scans and population genomic approaches (Luikart et al. 2003) to identify and characterize patterns of adaptive genetic variation. It also includes questions about how landscape features influence gene flow and dispersal in natural populations.

It is exciting time to conduct landscape genetic/genomic studies. The recent boom in genetic technological advances and computational approaches in landscape ecology (i.e. Garroway et al. 2008) and molecular biology will lead to rapid advances changing the relatively new field of landscape genomics.

References

Allendorf FW, Luikart G (2007) Conservation and the Genetics of Populations. Blackwell, Oxford

Antao T, Lopes A, Lopes RJ, Beja-Pereira A, Luikart G (2008) LOSITAN: A workbench to detect molecular adaptation based on an Fst-outlier method. BMC Bioinformatics 9:323

Balkenhol N, Gugerli F, Cushman S, Waits L, Coulon A, Arntzen J, Holderegger R, Wagner H (2007) Identifying future research needs in landscape genetics: where to from here? Landscape Ecology 24:455–463

Beaumont MA (2005) Adaptation and speciation: what can Fst tell us? Trends Ecol Evol 20:435–440

Bonin A, Taberlet P, Miaud C, Pompanon F (2006) Explorative genome scan to detect candidate loci for adaptation along a gradient of altitude in the common frog (*Rana temporaria*). Mol Biol Evol 23:773–783

Collins FS, Morgan M, Patrinos A (2003) The human genome project: Lessons from large-scale biology. Science 300:286–290

Crandall KA, Bininda-Evans ORP, Mace GM, Wayne RK (2000) Considering evolutionary processes in conservation biology. Trends Ecol Evol 15:290–295

Da Silva A, Gaillard J-M, Yoccoz NG, Hewison AJM, Galan M, Coulson T, Allainé D, Vial L, Delorme D, Van Laere G, Klein F, Luikart G (2009) Heterozygosity-fitness correlations revealed by neutral and candidate gene markers in roe deer from a long-term study. Evolution 63:403–417

Dieckmann U, Doebeli M (1999) On the origin of species by sympatric speciation. Nature 400:354–357

Eid J et al (2009) Real-time DNA sequencing from single polymerase molecules. Science 323:133–138

Foll M, Gaggiotti O (2008) A genome-scan method to identify selected loci appropriate for both dominant and codominant markers: a Bayesian perspective. Genetics 180:1–17

Fraser DJ, Bernatchez L (2001) Adaptive evolutionary conservation: towards a unified concept for defining conservation units. Mol Ecol 10:2741–2752

Garroway CJ, Bowman J, Carr D, Wilson PJ (2008) Applications of graph theory to landscape genetics. Evol Appl 1:620–630

Gonzalez-Martinez SC, Ersoz E, Brown GR, Wheeler NC, Neale DB (2006) DNA sequence variation and selection of tag singlenucleotide polymorphisms at candidate genes for droughtstress response in *Pinus taeda* L. Genetics 72:1915–1926

Hauser L, Seeb JE (2008) Advances in molecular technology and their impact on fisheries genetics. Fish and Fisheries 9:473–486

Hodges E et al (2007) Genome-wide in situ exon capture for selective resequencing. Nat Genet 39:1522–1527

Holderegger R, Wagner HH (2008) Landscape genetics. Bioscience 58:199–207

Joost S, Bonin A, Bruford MW, Despres L, Conord C, Erhardt G, Taberlet P (2007) A spatial analysis method (SAM) to detect candidate loci for selection: towards a landscape genomics approach to adaptation. Mol Ecol 16:3955–3969

Kim S, Misra A (2007) SNP genotyping: Technologies and biomedical applications. Annu Rev Biomed Eng 9:289–320

Kohn MH, Murphy WJ, Ostrander EA, Wayne RK (2006) Genomics and conservation genetics. Trends Ecol Evol 21:629–637

Lawson Handley LJ, Manica A, Goudet J, Balloux F (2007) Going the distance: human population genetics in a clinal world. Trends Genet 23:432–439

Long R, MacKay P, Ray J, Zielinski W (2008) Noninvasive survey methods for carnivores. Island Press, Washington, DC

Luikart G, England PR, Tallmon D, Jordan S, Taberlet P (2003) The power and promise of population genomics: from genotyping to genome typing. Nature Reviews Genetics 4:981–994

Luikart G, Pilgrim K, Visty J, Ezenwa VO, Schwartz MK (2008) Candidate gene microsatellite variation is associated with parasitism in wild bighorn sheep. Biol Lett 4:228–231
Manel S, Schwartz M, Luikart G, Taberlet P (2003) Landscape genetics: combining landscape ecology and population genetics. Trends Ecol Evol 18:189–197
Mardis ER (2008) The impact of next-generation sequencing technology on genetics. Trends Genet 24:133–141
Meyer M, Stenzel U, Hofreiter M (2008) Parallel tagged sequencing on the 454 platform. Nature Protocols 3:267–277
Morin PA, McCarthy M (2007) Highly accurate SNP genotyping from historical and low-quality samples. Mol Ecol Notes 7:937–946
Morin PA, Luikart G, Wayne RK SNP-workshop group (2004) Applications of single
Moritz C (1994) Defining 'evolutionary significant units' for conservation. Trends Ecol Evol 9:373–375
Palsboll PJ, Berube M, Allendorf FW (2007) Identification of management units using population genetic data. Trends Ecol Evol 22:11–16
Pearman PB (2001) Conservation value of independently evolving units: sacred cow or testable hypothesis? Conserv Biol 15:780–783
Perkel J (2008) SNP genotyping: six technologies that keyed a revolution. Nat Meth 5:447–453
Porreca GJ, Zhang K, Li J, et al (2007) Multiplex amplification of large sets of human exons. Nat Meth 4:931–936
Prugnolle F, Manica A, Balloux F (2005a) Geography predicts neutral genetic diversity of human populations. Curr Biol 15:R159–R160
Prugnolle F, Manica A, Charpentier M, Guegan JF, Guernier V, Balloux F (2005b) Pathogen driven selection and worldwide HLA Class I diversity. Curr Biol 15:1022–1027
Shendure J, Ji H (2008) Next-generation DNA sequencing. Nat Biotechnol 26:1135–1145
Smith TB, Wayne RK, Girman DJ, Bruford MW (1997) A role for ecotones in generating rainforest biodiversity. Science 276:1855–1857
Storfer A et al (2007) Putting the "landscape" in landscape genetics. Heredity 98:128–142
Storz JF, Nachman MW (2003) Natural selection on protein polymorphism in the rodent genus *Peromyscus*: evidence from interlocus contrasts. Evolution 57:2628–2635
Thurston MI, Field D (2005) Msatfinder: detection and characterisation of microsatellites. Distributed by the authors at http://www.genomics.ceh.ac.uk/msatfinder/. CEH Oxford, Mansfield Road, Oxford OX1 3SR
van Tienderen P, de Haan A, van der Linden C, Vosman B (2002) Biodiversity assessment using markers for ecologically important traits. Trends Ecol Evol 17:577–582
Vasemägi A, Primmer CR (2005) Challenges for identifying functionally important genetic variation: the promise of combining complementary research strategies. Invited Review in Mol Ecol 14:3623–3642
Vasemägi A, Nilsson J, Primmer CR (2005) Expressed sequence tag (EST) linked microsatellites as a source of gene associated polymorphisms for detecting signatures of divergent selection in Atlantic salmon (*Salmo salar L.*). Mol Biol Evol 22:1067–1076
Vera JC et al (2008) Rapid transcriptome characterization for a non-model organism using 454 pyrosequencing. Mol Ecol 17:1636–1647
Waples RS (1995) Evolutionarily significant units and the conservation of biological diversity under the Endangered Species Act. In: Nielsen JL, Powers DA (eds) Evolution and the aquatic ecosystem: defining unique units in population conservation. American Fisheries Society Symposium No. 17. Bethesda, MD
Wilding CS, Butlin RK, Grahame J (2001) Differential gene exchange between parapatric morphs of *Littorina saxatilis* detected using AFLP markers. J Evol Biol 14:611–619

Chapter 10
Spatial Information Management in Wildlife Ecology: Adding Spatially Explicit Behaviour Data to the Equation?

Kim Jochum and Falk Huettmann

10.1 Overview

The implementation of spatial data to wildlife management is not really new as such, although it has not been explicitly demanded and practiced (Braun 2005). By adding spatial information to the discipline, resounding success has already been achieved elsewhere improving specific science goals (e.g. Kadmon et al. 2004; Hirzel et al. 2006).

When developing wildlife management strategies, traditionally, only ecological/biological baseline wildlife monitoring data are used (Caughley and Sinclair 1994). Classic components of the wildlife management formula consist for instance of demography, basic telemetry data, body dimensions, physiology, and environmental information such as vegetation type and climatic factors. Disease information just got added recently. And including human dimensions becomes virtually mandatory now for Wildlife Managers, too (Braun 2005). However, behaviour data is still missing in the majority of cases. We find this surprising because Behaviour data have been collected for a long time, have fascinated mankind (e.g. Lorenz 1966; Goddall 1988; Montgomery 1991) and often deal with charismatic animals of high management interest. Further, wildlife behaviour data can present a 'missing link' between success and failure of sustainable population management (Festa-Bianchet and Appollonio 2003). As these authors state, especially, social status and dominance hierarchy systems in species play a key role for animal survival and for a successful species management. So what are the key reasons for not including behaviour information in wildlife management, and how can it be overcome; e.g. with the help of digital approaches widely common elsewhere by now? Together with a review of existing behaviour data that can be used for wildlife management, we will address the following three questions.

K. Jochum (✉) and F. Huettmann
EWHALE lab- Biology and Wildlife Department, Institute of Arctic Biology, University of Alaska-Fairbanks, 419 IRVING I, Fairbanks, AK 99775-7000, USA
e-mail: kimjochum@gmx.de

S.A. Cushman and F. Huettmann (eds.), *Spatial Complexity, Informatics, and Wildlife Conservation*
DOI 10.1007/978-4-431-87771-4_10, © Springer 2010

1. Where can we find online available behaviour data, supporting global data sharing policies (digital and online behaviour databases, spatial and online behaviour databases, inclusion of behaviour data in online biology databases, behaviour journals)?
2. In which case studies contribute spatially explicit behaviour data to wildlife management and where is such data accessible in the first place?
3. How can one improve future wildlife management by adding behaviour data to the equation?

In the behaviour sciences, it is widely believed that detailed information on researched individuals has to exist foremost in a high detail long-term observational format; it basically can only be obtained through years of minuscule field work (Drickamer et al. 2002). This assumed requirement makes behaviour research in the wild for the majority of species, even in the twenty-first century and with high-tech, nearly impossible though. Mainly such circumstances account for the vast lack of behaviour data of free ranging animals across populations and ecosystems (Insley et al. 2003). Most examples for successful behaviour data collections in the wild consist of highly time- and money-consuming long-term studies (Van Lawick-Goodall 1973; Zucker and Clarke 1998; Whiten et al. 1999). Considering that research data can take up to 20 years until they are implemented in relevant governmental management strategies, and that results from behaviour studies are usually provided in outlets and (hardcopy) formats not readily available and accessible for managers in time, such situation lend itself for behavioural data virtually left out of the current wildlife management equation. Consequently, wildlife does not receive the best possible management. Results of these situations we can find virtually everywhere.

However, the possibility to collect, and use opportunistic short-term behaviour data, analysed through an algorithmic modelling and data mining approach for generalizing patterns, and to make the implementation of behaviour data accessible online, reveals completely new possibilities for wildlife management (Jochum 2008).

Not only has the cumbersome collection of long-term data limited the successful implementation of behaviour data for wildlife management decisions, but further is the immediate accessibility of these long- and short-term data from wild populations also extremely rare. One would think the limited amount of data existing would at least be spread across researchers to make the best management implications possible. But from our experience and inquiries we found that this is not the case, yet. At first, spatial behaviour data gets not collected in a standardized fashion (e.g. by using widely accepted protocols) and thus suffers from subjectivity (Drickamer et al. 2002; Alcock 2005). It's highly indicative that none of these text books make any reference to databases and how to design, store and work up behaviour data and Metadata (Martin and Bateson 2005, for basic details on computer data entry). Secondly, especially the sharing, online exchange of data in times of global digital initiatives, remains still virtually unutilized by behaviour researchers and wildlife managers alike.

Data sharing in general has not been an easy task and culture to get implemented into the sciences and beyond, yet (Huettmann 2005, 2007a). Across research topics, researchers were, and still are, afraid of other people steeling their data and information; a fear from the past and somewhat a heritage of the Watson and Crick era (Edelson 1998). Such type of thinking fails to understand the way data sharing works and what their own benefit, as well as the benefit of global sustainable management, could be; it's the wider good that matters; reciprocity (Esanu and Uhlir 2004; Huettmann 2007a, b). When sharing information about data in Metadata Clearinghouses, only the data description such as background information, including a project abstract, project leader contact and data type for columns in data sheets on the gathered data, will be available in a standardized fashion (e.g. NBII Clearinghouse http://mercdev3.ornl.gov/nbii/; see also Chapter 12, this volume). Offering this information in a standardized manner is highly important for successful global online search results and comparability as such. Whereas, the raw data will only be exchanged after both-sides agree with the data exchange. In these federated concepts, and as globally promoted and supported by the Rio Convention and GBIF (see Table 10.1) for instance, the data holder remains in complete control of his/her data.

Table 10.1 List of major online databases searched for behavioural databases and information

Name of database	Details	Relevant amount of behaviour data	URL
GCMD	Global Change Master Directory by NASA (also Metadata portal for the International Polar Year IPY)	No	http://gcmd.nasa.gov/
MaNIS	Mammal Information Networked System	No	http://manisnet.org/
ORNIS	Ornithological Information Network System	No	http://olla.berkeley.edu/ornisnet/
FishBase	A global Fish Information Database	No	http://www.fishbase.org/HOME.HTM
OBIS	Ocean Biogeographic Information System	No	http://www.iobis.org/
OBIS-Seamap	Ocean Biogeographic Information System- SeaMap	No	http://seamap.env.duke.edu/
Tracking Ocean Wanderer by BirdLife International	Database for tracking seabird migration	No	http://www.birdlife.org/action/science/species/seabirds/tracking.html
Mercury	Distributed Metadata Management, Data Discovery and Access System	No	http://mercury.ornl.gov/
NBII Clearinghouse	National Biological Information Infrastructure	Yes	http://mercdev3.ornl.gov/nbii/
GBIF	Global Biodiversity Information Facility	No	www.gbif.org

The data sharing policy gets either stated in the Metadata, or can be set specifically when using the popular DIGIR protocols, that allow for a federated network of databases providing the data owner with full control of the data to be contributed and shared.

Another important term should be considered when talking about data sharing: Open Access (OA). Open Access does not really describe the sole exchange of the data, but it more or less starts where traditional data sharing ends – with peer-reviewed publications and their pre-prints. OA stands for literature being digital, online, free of charge, and free of most copyright and licensing restrictions (Peter Suber OA http://www.earlham.edu/~peters/fos/overview. htm). We found that most behaviour journals and their publishers are still not involved in being on the forefront of supporting this fast, general education-supportive way of exchange via OA. Up to now, many databases are not perceived and set up as an OA tool, yet, for facilitating the global use of behaviour data. Databases in general are defined as collections of information organized in such a way that a computer program can quickly select desired pieces of data (University of Queensland 2008).

In addition to the actual lack of behavioural databases, most of them are also short of geo-referencing. This is a special problem when one wants to map those in time and space, and for doing specific and sophisticated analysis such as autocorrelation or neighbourhood relationships (see Chapter 15, this volume). Using such aspatial data makes for incomplete science to be overcome. Items of concern are usually the spatial resolution of the coordinates, and as well the technical format, and method to obtain these data, and the Metadata (see Jonker 2006 for a good example how this can be achieved and delivered online).

10.2 Digital and Online Behaviour Databases

Behaviour databases are difficult to find online, and even more sadly, they are rarely accessed and used. The user community does not have a culture to search and use for such data. This claim can easily be assessed when looking at publications of the Journal of wildlife management. Many Wildlife Databases in general are rarely online accessible in the first place, and secondly, tend to charge a fee for access, based on a wrong underlying business model that actually constrains the use of data, instead of promoting their use widely (Esanu and Uhlir 2004; Curry and Humphries 2007). Overall, when searching for behaviour databases in the World Wide Web, some larger wildlife databases were found to exist in a digital form, but they have no direct online-accessibility, charge a fee for access, lack high-quality Metadata, and do not belong into the traditional animal behaviour database field.

For instance, the State Behaviour Database for Pure Liquids and Data Correlation (Cibulka and Takagi 2004) is designed for storage, maintenance and evaluation of published experimental data. But at the same time, it's not online accessible and overly commercialized. How do these statements fit together? When searching the Web, more information service companies selling databases, animal agencies and computational behaviour portals can be found, exposing the tiny amount of Wildlife Behaviour databases. The Cell Behaviour Database (CBDB) of Japan for instance collects data of molecular level information on biological phenomena. The database itself can not be found in the web and only exists in Japanese (Iwazume et al. 2005), and thus, is not widely available to the global village.

The only 'real' wildlife behaviour database we located, and being accessible online, so far, is the Marine Wildlife Behaviour Database (MWBD) (Vigness Raposa 2008) from the University of Rhode Island, USA. But this database just compiles Metadata, not the underlying raw data. It deals with diving and movement pattern of marine mammals and has the goal to predict and minimize environmental impacts. As the publication by Mann et al. (2000) shows, we can learn a lot from marine behaviour biologists, despite the fact that only marginal data can be collected for marine mammals due to their subsurface life cycle.

10.3 Spatial and Online Behaviour Databases

Online accessible, but still not part of the research field, is for example the Group Behaviour Database. This database is part of the SWARMS (Scalable sWarms of Autonomous Robots and mobile Sensors) project created to provide a guided source for papers useful to engineers, applied mathematicians, robotocists and others interested in using the biological literature to create artificial networks (SWARMS 2006). Similar applies to the Australian Spatial Data Directory (ASDD). ASDD aims to improve access to Australian spatial geospatial dataset descriptions for all stakeholders (ASDD 1999). No separate online behaviour databases exist with geo-referenced zoological information. Therefore, our next step is to take a closer look at biological online databases to see what behaviour related information can be found.

10.4 Review – Online Biology Databases

Overall, the amount and usage of biological databases grows steadily. But does the amount of accessible behaviour metadata grow in a similar fashion? Is geo-referenced well documented behaviour data included and made available in large biological databases in the first place? This review includes large and world wide used biological databases such as the following: GCMD, MaNIS, ORNIS, FishBase, OBIS and OBIS-SEAMAP, BirdLife International, Mercury, NBII Clearinghouse and GBIF (see Table 10.1 for details). Although we are aware that this review does not cover the entire discipline and all efforts globally, we believe it allows for a representative state-of-the-art of online Behaviour Data. If we did not find specific online Behaviour Databases, this simply speaks for their hidden existence, not readily offering and marketing themselves to a global audience of users that eagerly wants to include them into wildlife management applications and elsewhere. We strongly believe that science products should be 'used', and especially when it comes to data and adding behavioural data for a better wildlife management. We believe this is the only way to strengthen science-based management (e.g. Huettmann 2007c).

The Global Change Master Directory (GMCD) (National Aeronautics and Space Administration (NASA) 2008) was established to discover earth science and environmental data. Although the GCMD database holds more than 25,000 descriptions of earth science data sets, no behaviour data is included.

The Mammal Networked Information System (MaNIS) (Wieczorek 2001) was established by seventeen North American institutions and their collaborators; it has a global focus and includes terrestrial as well as marine mammals. The original objectives of MaNIS are to include the facilitation of Open Access to combined specimen data from a web browser and to enhance the value of specimen collections. However, when searching for the general term 'behaviour' (as well as 'behaviour') no relevant results could be found. MaNIS has still the main focus on geo-referenced specimen collections as such.

Similar behaviour-less data results were revealed in the ORNIS database (Ornithological Information System) (ORNIS 2004), which was established as a cooperation between the USA, Canada, Belgium, France and the United Kingdom to gather world wide bird information. This database expands the American Mammal Network MaNIS into the ornithological field. One should point out that birds have been among the most studied subjects by behavioural ecologists.

The American database FishBase (Froese and Pauly 2008) combines Metadata on fish species across the world focussing on dispersal and movement pattern. Existing search possibilities do not allow the search for behaviour data separately. The search engine offers only to search by name, family, country, ecosystem and topics. The term behaviour is only used in connection with physiology terms of the senses such as 'vision'. Through FishBase one can learn a lot about the distribution of certain species and their movement pattern, but nothing about the actual behaviour of fish (a traditional stronghold in behaviour research).

The OBIS (Ocean Biogeographic Information System) database (OBIS 2002) holds data on locations of marine animals and plants for documenting the oceans' diversity, distribution, and abundance of life. Currently, 16.7 million records of 104.000 species from 501 databases are gathered in OBIS, but these data do not show any type of behaviour data associated with them, nor spatial behaviour data. OBIS presents an associated member of GBIF (Global Biodiversity Information Facility) implying data translocation from many small databases to a major large web portal and database. The OBIS database was established to evolve a strategic alliance of people and organizations sharing a vision to make marine biogeographic data freely available over the World Wide Web. However, behaviour data, urgently needed for an improved management, are not yet included in this vision, and make it less convincing.

OBIS-SEAMAP (Read et al. 2009) is part of OBIS, representing the Ocean Biogeographic Information System Spatial Ecological Analysis of Megavertebrate Populations. Spatially and temporally interactive online archives are built for marine mammal, sea turtle and seabird data. Joining data from providers all over the world make such work possible. Despite it's specific focus on spatial and transect data and direct sightings of marine animals in the wild, this database also lacks relevant behaviour data.

BirdLife International presents another international organization, distributed in over one hundred countries, hosting the BirdLife International Telemetry database (Tracking Ocean Wanderers http://www.birdlife.org/action/science/species/seabirds/tracking.html). This global partnership of conservation organisations striving to conserve birds, their habitats and global biodiversity, includes considerable data files on spatial bird behaviour, e.g. for endangered species, but makes them not readily available to the public online. However, in achieving their goal of '*working with people towards sustainability in the use of natural resources*' sharing Open Access policies would be highly supportive of this goal, that otherwise would not be possible to meet.

Mercury is defined as the Distributed Metadata Management, Data Discovery and Access System (Mercury 2008) presenting a web-based system to search for metadata and retrieve associated data. This system is used by many databases including NBII, representing the only biological and therefore relevant search machine for behaviour data inside the Mercury information system. Mercury incorporates a number of important features for successful data sharing including providing a single portal to information contained in distributed data management systems and supporting international metadata standards.

In the National Biological Information Infrastructure (NBII Clearinghouse) (NBII 2008) some behaviour data is finally revealed. Searching with the keyword 'behaviour', 769 compliant datasets can be found. When searching for 'behaviour' even 2,808 metadata sets are encountered. It might be rather useful for more successful searches, and for promoting behaviour-related online data, to present an efficient solution for this language problem, and so that either record is shown. However, when combining the words 'behaviour' and 'wildlife' in the search engine, only nine datasets were found, all offered by one university (University of Alaska Fairbanks). The search for the terms 'behaviour' and 'wildlife' still reveals 683 dataset matches. By including the word 'mammal' to 'behaviour' and 'wildlife', 34 datasets matches are shown. These results represent probably the best free and Open Access online behaviour datasets existing in the descriptive form of Metadata.

The easily misleading thought of all the discovered datasets being directly focussed on behaviour studies, makes it difficult to judge the capability of the database engines existing so far to work with behaviour data. Merging protocols, and what exact study protocols were used, is hardly mentioned, neither. Probably, at least the search possibilities would have to be adjusted for a more precise behaviour data discovery for the following reasons: For example, when looking at the 34 datasets showing in the search of 'behaviour', 'wildlife' and 'mammal', nearly all studies represent environmental monitoring or datasets describing specific locations. The number of real and descriptive behaviour studies are actually fewer than 10 datasets. Further, half of the so-called behaviour data sets (17 of the 34) originate from the USGS (United States – Geological Survey), six of the remaining 17 datasets are placed again by one institution (UAF). Conclusively, few relevant online behaviour Metadata sets really exist so far, and the ones existing originate from only few data providing collector-organisations.

GBIF stands for the Global Biodiversity Information Facility database (GBIF 2008), representing the largest data portal for Open Access data sharing across the world. Here, using a federated database network based on the DIGIR protocol (Distributed Generic Information Retrieval; http://digir.net), major sources of online data are gathered from many smaller databases to a major search engine and data body. The goal is to combine as much biodiversity data as possible from all over the world and make it freely available to everyone through Open Access (e.g. Huettmann 2005). This data portal is a service providing access to millions of scientific data records being shared via the GBIF webportal and its underlying network with the global citizens (the biggest growing audiences are currently found in China, India, Brazil and Africa). This data structure seems to be highly promising, but when searching for behaviour data, virtually no entry can be found. This global biodiversity database hosted in Copenhagen includes mainly species occurrence records, names and classifications of organism, so far. It would be of great promise to propose the inclusion of behaviour data to their disseminated data-fields.

In addition to DIGIR as a globally dominating data exchange protocol favoring specimen records, there are no direct links to Metadata, and specific Metadata profiles that support behaviour data as such. Behavioural Ecologists and Wildlife Managers alike are poorly represented in GBIF due to the heritage of specimen collection exchanges among museums and their technical features, which have received most attention, so far. Changes are supposed to occur with a new protocol called ABCD BioCase and TAPIR (TDWG (Biodiversity, Information Standard) Access Protocol for Information Retrieval) (www.tdwg.org/activities/tapir). However, we are not aware of any relevant applications in the disciplines of wildlife management and Behaviour Sciences, yet.

10.5 Review – Wildlife Management and Behaviour Journals

A variety of well-established Behaviour Journals exist with many of them overlapping in a comprehensive fashion between social sciences, environmental sciences and zoological sciences (Springer 2009; Thomson Reuters 2009). At the same time, the majority of behaviour publications is aligned with specific behaviour journals and is less often found to be published in their focus field of research, e.g. wildlife; for wildlife behaviour research, no separate journal can be described.

Overall, not many of the well-established behaviour journals are in favour of complete Open Access policies. However, new initiatives in journal publications (especially online journals) follow and support the OA ideas, and hold very high citation factors at the same time. A key example provides the PLoS Journal Line. *PLoS Biology* as an open-access peer-reviewed general biology journal published by the Public Library of Science (PLoS). It is compliant with OA policies and holding a certified OA license. The impact factor of this journal lies above 14, much higher than journals charging fees (see Table 10.2). PLoS is proud to be

Table 10.2 List of major behaviour and wildlife journals and their comparison

Name of peer-review journal	Details of international journals	Relevant amount of behaviour data	Open access	Impact factor	URL
PLoS Biology	General biology journal	Yes	Yes	14.1	http://journals.plos.org/
Behaviour	Comparative and experimental ethology	Yes	No	1.604	http://www.brill.nl/beh
Animal Behaviour	All aspects of animal behaviour	Yes	No	2.752	http://ees.elsevier.com/anbeh/
Applied Animal Behaviour Science	Application of ethology to animals managed by humans	Yes	No	1.404	http://ees.elsevier.com/applan/
Adaptive Behaviour	Adaptive behaviour in living organism and autonomous artificial systems	Yes	Optional	1.024	http://adb.sagepub.com/
Journal of Wildlife Management	Scientific and management foundations of wildlife management	Rare	No	1.528	http://www.bioone.org/loi/wild
Wildlife Biology	Theoretical, empirical and practical studies	Rare	No	0.724 (2005)	http://www.wildlifebiology.com/
Wildlife Research	Ecology and management of wild animals in natural and modified habitats	Rare	Optional	?	http://www.publish.csiro.au/nid/144.htm
European Journal of Wildlife Research	All aspects of wildlife biology	Rare	No	0.979 (2007)	http://www.springer.com/life+sci/zoology/journal/10344
Journal of Animal Ecology	Original research on all aspects of animal ecology	Rare	No	3.747	http://www.wiley.com/bw/journal.asp?ref=0021–8790

"a nonprofit organization of scientists and physicians committed to making the world's scientific and medical literature a public resource" (PLoS 2009).

The well-established behaviour journal generation includes the following major journals:

The journal *Behaviour* (Albers and Wisenden 2009) is in existence since 1948 publishing original research reports on behaviour of all animals on substantial contributions to the biological analysis of the causation, ontogeny, function and evolution. *Animal Behaviour* was first published in 1953 and describes itself as a 'leading international publication containing critical reviews, original papers, and research articles on all aspects of animal behaviour' (Elsevier 2009) and is published for the Association for the Study of Animal Behaviour in collaboration with the Animal Behaviour Society. The Journal of *Applied Animal Behaviour Science* is published in charge of the International Society for Applied Ethology (ISAE) publishing information on the behaviour of domesticated and utilized animals. No separate internet webpage exists to provide background information on these journals; only information accessible is provided through the publishers (e.g. Elsevier for *Animal Behaviour*, www.elsevier.com).

The journal *Adaptive Behaviour* is published by SAGE, an independent scholarly publisher, and focuses on research and review articles on adaptive behaviour in living organisms and autonomous artificial systems (SAGE 2009). SAGE Publications, is offering their authors the option to make their primary research articles freely available on publication via Open-Access. It is up to the author to decide who will pay for the publication, the author himself or the reader. The same possibility have authors of the journal *Wildlife Research* (CSIRO 2009). *Wildlife Research's* publishing focuses on research from the ecology and management field of wild animals in natural and modified habitats.

Non-supportive Open Access data of research on the scientific and management foundations of wildlife management are provided through T*he Journal of Wildlife Management* (JWM).This journal is a publication of The Wildlife Society (TWS) presenting a knowledge database for wildlife science research and management (The Wildlife Society 2009).

In 1994, the journal *Wildlife Biology* was initiated by the Nordic Council for Wildlife Research (NKV) supporting research out of the theoretical, empirical and practical field from throughout the world. The publisher's primary focus is to create the scientific basis for the enhancement of conservation and management practices for wildlife species and their environments focussing on game species (Wildlife Biology 2009). All aspects of wildlife biology are welcome to be submitted to the *European Journal of Wildlife Research* (Lutz 2009). This journal promotes publications from wildlife ecology, diseases, toxicology, conservation, management, and the sustainable use of natural resources. The *Journal of Animal Ecology* is fostered by the British Ecological Society and supports mainly long-term ecological research in our changing world. The publication focus includes specifically Behavioural Ecology (Hays et al. 2009). These journals do not provide an Open-Access publication. The Journal *Game and Wildlife Science* does not maintain a website.

Overall, good publication possibilities across disciplines and journals do exist, and at least three journals support already Open-Access, with numbers increasing. A certain change and rethinking can be experienced already and will probably stay in everybody's minds driving the support of Open-Access and data-sharing with the public. It is hoped that this new culture will also be taken on by the Behavioural Sciences and with their rich data.

10.6 Two Case Studies for the Successful Use of Spatial Explicit Behaviour Data

10.6.1 Humpback Whale (Megaptera Novaeangliae) Studies

Research on humpback whales gets conducted for a long period of time (Chapman 1974). But since spatial behaviour data got included, it became possible for researchers to conclude on the existence of social structures and social systems between humpback whale populations (Valsecchi et al. 2002; Clapham 1996; Mann et al. 2000) even although behaviour data of whales can only get collected in a patchy style due to whales spending their majority of time under the water surface. The variations of vocal calls between group sizes and group competitions support these research findings significantly (Silber 1986). Analysis of tissue samples (DNA) and spatial distributions of these animals has already been used to describe group competitions in more detail. Clapham (1996), Valsecchi (2002) and Marino (2002) described social pattern, all leading to the same conclusion: Humpback whales are social animals living in a social fission–fusion system with a dominance hierarchy exhibited. Conclusively, with the help of spatial behaviour data, and in combination with genetic information and ecological factors, researchers have been able to describe the humpback whale's social structure in a detailed fashion. Similar results have been found by Jochum (2008). However, only data by the author are described with Metadata, globally findable online and accessible for the global audience.

10.6.2 Study Across Mammal Species

A lack of behaviour data on many mammal species exists due to their inaccessibility in the wild and due to difficulties of long-term data collection, e.g. due to funding, widely believed to be essential for conclusion drawing in the social structure and dominance hierarchy research field. Trying to overcome this situation, a new approach to this discipline 'behaviour modeling' was tested, concentrating on the use of short-term opportunistic samples (Jochum 2008). Frequently, such data present the only quickly obtainable behaviour data to be accomplished in the wild and with 'real world' budgets.

Marginal behaviour data for six species were investigated and compared: Howling monkeys (*Alouatta palliata*), humpback whales, muskoxen (*Ovibos moschatus*), spotted seals (*Phoca largha*), brown bears (*Ursus arctos*) and polar bears (*U. maritimus*). The non-invasive instantaneous scan sampling methodology was used to estimate spatial proximities between individuals. Exploratory graphs from data mining show indicative distances kept between adult male howling monkeys (Fig. 10.1) in comparison to adult female howling monkeys (Fig. 10.2).

Making strategic decisions, adult male howlers stay further apart from each other than female howling monkeys do; adult howlers never come closer to each other than 2 m and their median distance kept from each other is 17 m. Female adult howlers on the other side spend about 26% of the observed time close to each other (2 m or less) and their median distance kept between each other is only 8 m. Here, specific spatial differences become recognizable even in an opportunistic data collection scheme of less than 2 weeks of observation.

Overall, findings in this study indicated non-consistent patterns for different animal strata, as expected for social species with a well-structured and individual-based dominance hierarchy. Most anthropologists have not agreed yet in defining the overall structure of existing social networks in howling monkeys.

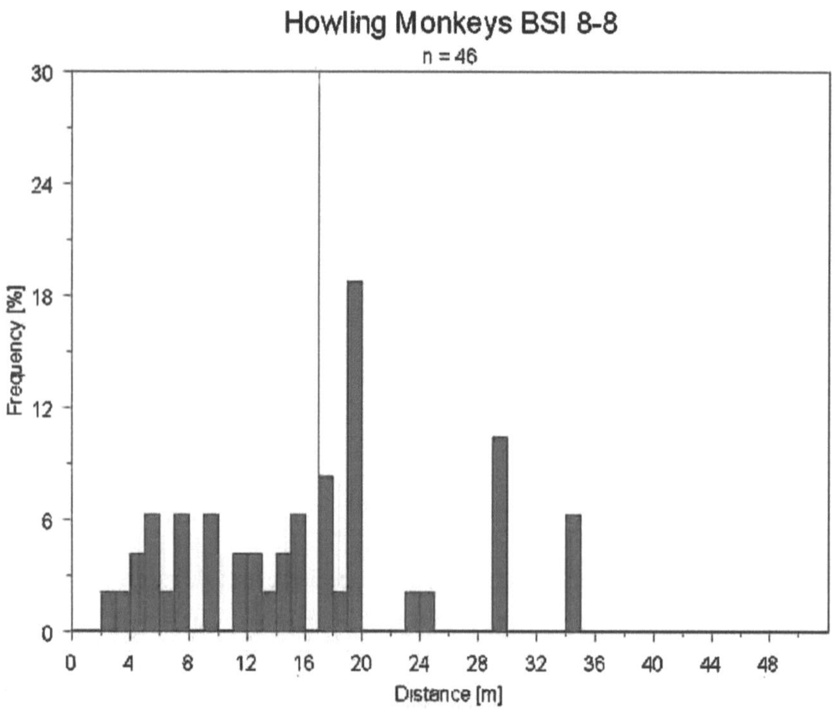

Fig. 10.1 Descriptive example for the applicability of spatial behaviour data in Howling Monkeys of Nicaragua; Body Size Index (BSI) eight (large adult) implementing the distance kept between two adult males. The vertical line shows the expected mean (Taken from Jochum 2008)

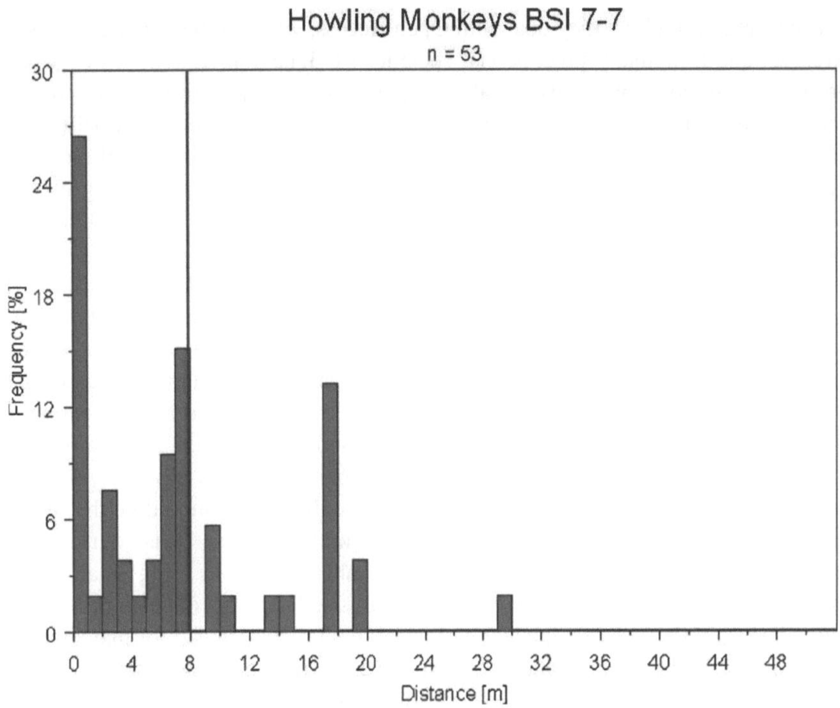

Fig. 10.2 Descriptive example for the applicability of spatial behaviour data in Howling Monkeys of Nicaragua; Body Size Index (BSI) 7 (adult female) implementing the distance kept between two adult females. The vertical line shows the expected mean (Taken from Jochum 2008)

But all recent studies confirm that the howling monkey truly is a social species with a sophisticated social structure built upon dominance hierarchy, often described as fission-fusion system (Clarke et al. 1998; Zucker and Clarke 1998; Dias and Luna 2006; Bezanson et al. 2007).

By implementing a new approach for geo-referenced behaviour data and analysis, many new opportunities and perspectives come to bear; a wealth of synergies forms. Data investigating spatially explicit social traits in animals can be collected for many species world-wide and where no long-term studies are applicable, as the case for many endangered animal species. Further, and when made publicly available, already existing marginal data sets can now be re-investigated for underlying patterns and get complemented to test new theories, fill data gaps, and expand data collections.

10.7 Conclusion and Outlook

Our findings are not in support of a current practice of best possible and holistic wildlife management; geo-referenced behaviour-related information is widely left out. These concerns should be addressed immediately by a number of behaviour journals, regarding how they deal with Open Access, whether they have an (online) Archive, and how many professional societies of Behaviour Ecologists exist and promote these concepts. It would also be good to identify institutions where one actually can study 'wildlife behaviour', whether these programs are certified for data sharing, and whether the curriculum and the used textbooks of these institutions include databases and digital principles, as required for making elevant contributions in the year 2008 and beyond, as well as for a successful wildlife management as such (see Huettmann 2007c for an example).

We conclude that in reality, a vast majority of behaviour data is still kept in widely locked up paper boxes and folders, hidden from the outer world, serving only a few. Also unpublished data or yet non peer-reviewed data could be important to put online. It is obvious that a sole focus on hardcopy data summaries and locally done statistical analysis is insufficient for contributing to an efficient wildlife management that actually achieves global sustainability. It is further surprising to us to learn that the push to include, and make behaviour data available, does not come from the Wildlife Managers themselves; we think they should be the first to know what constitutes good wildlife management. Only when behaviour data are readily available researchers can find out relevant details and bring small pieces together, for example in the form of Meta-Analysis. That way, one can bring them into a large picture, provide context and make them relevant, also for management decisions, and when these data are all we have. In the meantime, and as a proxy, one can datamine and model short-term datasets to find out about their predictability and help to describe management implications (as shown in Jochum 2008 for example).

The exchange of research background information - on what, when and where research had been conducted - should be a role model and an additional key focus for the research community as a whole. Not doing so presents a vast loss of effort and of the possibility to achieve major goals due to having to start fresh each time again caused by the unknown existence or poor description of certain data. Public access of (behaviour) data sets will remain a key factor for successful research and management world-wide.

References

Alcock J (2005). Animal Behavior. Sinauer, Sunderland, MA.

ASDD (1999). ASDD - Australian Spatial Data Directory. Australian Spatial Data Infrastructure (ASDI). Accessed 11.01.2009 *from* http://asdd.ga.gov.au/asdd/.

Bezanson M, Garber PA, Murphy JT, Premo LS (2007). Patterns of subgrouping and spatial affiliation in a community of mantled howling monkeys (*Alouatta palliata*). Amer J Primatol 69: 1–17.

Braun CE (2005). Techniques for Wildlife Investigations and Management. The Wildlife Society (TWS), Bethesda, MD. ISBN 0–933564–15–5

Caughley G, Sinclair ARE (1994). Wildlife Ecology and Management. Blackwell, Cambridge, MA.

Chapman DG (1974). Status of Antarctic rorqual whale stocks. *In* The whale problem: A status report. *(W.B. Schevill)*, Harvard University Press. MA. 218–238.

Cibulka I, Takagi T (2004). State behavior database for pure liquids and data correlation. Internat J Thermophys 25: 361–369.
Clapham PJ (1996). The social and reproductive biology of humpback whale: An ecological perspective. Mammal Rev 26(1): 27–49.
Clarke MR, Glander KE, Zucker EL (1998). Infant-nonmother interactions of free-ranging manteled howlers (*Alouatta palliata*) in Costa Rica. Int J Primatol 19(3): 451–472.
CSIRO (2009). Wildlife Research. CSIRO Publishing. Accessed 05. February 2009 *from* http://www.publish.csiro.au/nid/144.htm.
Curry GB, Humphries CJ (2007). Biodiversity Databases: Techniques, Politics and Applications. CRC. Boca Raton, FL.
Dias PAD, Luna ER (2006). Seasonal changes in male associative behavior and subgrouping of *Alouatta palliata* on an Island. Int J Primatol 27: 1635–1651.
Drickamer LC, Vessey SH, Jakob EM (2002). Animal Behavior, Mechanisms, Ecology, Evolution 5th edn., McGraw–Hill, New York.
Edelson E (1998). Francis Crick and James Watson: And the Building Blocks of Life. Oxford Portraits in Science. Kindle Book.
Elsevier (2009). Animal Behavior. Elsevier New York. Accessed 05. February 2009 *from* http://ees.elsevier.com/anbeh/.
Esanu JM., Uhlir PF (eds.) (2004). Open Access and the Public Domain in Digital Data and Information for Science: Proceedings of an International Symposium. U.S. National Committee for CODATA, National Research Council.
Festa-Bianchet M, Appollonio M (eds.) (2003). Animal Behavior Wildlife Conservation. Island, Washington, DC.
Froese R, Pauly D (2008). FishBase. World Wide Web electronic publication. Accessed 12.11.2008 *from* www.fishbase.org.
GBIF (2008). Global Biodiversity Information Facility. Biodiversity occurrence data provided by: GBIF Data Portal. Accessed 11.01.2009 *from* www.gbif.net.
Goodall J (1988). My Life with the Chimpanzees. Aladdin, New York.
Hays G, Norris K, Boots M, Coulson T (2009). J Anim Ecol. Wiley. New York. Accessed 05. February 2009 *from* http://www.wiley.com/bw/journal.asp?ref=0021–8790.
Hirzel AH, Lay GL, Helfer V, Randin C, Guisan A (2006). Evaluating the ability of habitat suitability models to predict species presences. Ecol Modelling 199: 142–152.
Huettmann F (2005). Databases and science-based management in the context of wildlife and habitat: towards a certified ISO standard for objective decision-making for the global community by using the internet. J Wildlife Manage 69: 466–472.
Huettmann F (2007a). Constraints, suggested solutions and an outlook towards a new digital culture for the oceans and beyond: experiences from five predictive GIS models that contribute to global management, conservation and study of marine wildlife and habitat, in: Vanden Berghe E. et al. (Ed.) Proceedings of 'Ocean Biodiversity Informatics': an international conference on marine biodiversity data management Hamburg, Germany, 29 November –1 December, 2004. IOC Workshop Report, 202, VLIZ Special Publication 37: pp. 49–61. http://www.vliz.be/vmdcdata/imis2/imis.php?module=ref&refid=107201
Huettmann F (2007b). The digital teaching legacy of the International Polar Year (IPY): Details of a present to the global village for achieving sustainability. Eds M. Tjoa and R.R. Wagner. Proceedings 18th International Workshop on Database and Expert Systems Applications (DEXA) 3–7 September 2007, Regensburg, Germany. IEEE Comput Soc, Los Alamitos, CA. pp. 673–677.
Huettmann F (2007c). Modern adaptive management: Adding digital opportunities towards a sustainable world with new values. Forum Public Policy: Clim Change Sustainable Develop 3: 337–342.
Insley SJ, Phillips AV, Charrier I (2003). A review of social recognition in pinnipeds. Euro Ass Aquat Mammals 29: 181–201.
Iwazume M, Kaneko H, Dan M (2005). CBDB: Cell Behavior Database – Toward understanding high order level biological phenomena-. *In* Proceedings. Chemo-Bio Informatics. (*C.-B.I. Society*), Japan, p. 115.

Jochum K (2008). Benefits of using marginal opportunistic wildlife behavior data: Constraints and applications across data - a dominance hierarchy example relevant for wildlife management. German Diploma Thesis, University of Veterinay Medicine Hannover Foundation. Hannover, Germany.

Jonker R (2006). Quantifying influences of disturbances on foraging behavior and movement pattern of Caribou. Unpublished M.Sc. thesis. University of Wageningen, Holland.

Kadmon R, Farber O, Danin A (2004). Effect of roadside bias on the accuracy of predictive maps produced by biclimatic models. Ecol Appl 14: 401–413.

Lorenz K (1966). On Aggression. Hartcourt, Brace & World, New York.

Lutz W (2009). European Journal of Wildlife Research. Springer. Accessed 05. February 2009 *from* www.springer.com/life+sci/zoology/journal/10344.

Mann J, Connor RC, Tyack PL, Whitehead H (2000). Cetacean societies: Field studies of dolphins and whales. University of Chicago Press. Chicago, IL.

Marino L (2002). Convergence of complex cognition abilities in cetaceans and primates. Brain Behav Evol 59: 21–32.

Martin P, Bateson P (2005). Measuring Behavior: An introductory guide, 5th edn. Cambridge University Press.

Mercury (2008). Mercury (Distributed Metadata Management, Data Discovery and Access System). Oak Ridge National Laboratory. Accessed 11.01.2009 *from* http://mercury.ornl.gov/.

Montgomery S (1991). Walking With the Great Apes: Jane Goodall, Dian Fossey, Birute Galdikas. Houghton Mifflin, Co Publishers, New York.

National Aeronautics and Space Administration (NASA) (2008). GCMD – Global Change Master Directory. Retrieved 24 October 2008.

NBII (2008). NBII Clearinghouse. National Biological Information Infrastructure. Accessed 13.03.2008 *from* http://www.nbii.gov/portal/community/Communities/NBII_Home/.

OBIS (2002). Ocean Biogeographic Information System (OBIS v3.1). Census of Marine Lifestrives. The State University of New Jersey. Accessed 11.01.2009 *from* http://www.iobis.org/.

ORNIS (2004). ORNIS – ORNithological Information System. Accessed 29.11.2008 *from* http://olla.berkeley.edu/ornisnet/.

PLoS (2009). PLoS Biology. Public Library of Science. Accessed 05. Februry 2009 *from* http://biology.plosjournals.org/.

Read AJ, Halpin PN, Crowder LB, Best BD, Fujioka E (2009). OBIS-SEAMAP: mapping marine mammals, birds and turtles. World Wide Web electronic publication. Accessed 10.01.2009 *from* http://seamap.env.duke.edu.

SAGE (2009). Adaptive Behavior. Sage Journals Online. Accessed 05. February 2009 *from* http://adb.sagepub.com.

Silber GK (1986). The relationship of social vocalisation to surface behavior and aggression in the Hawaiian humpback whale (*Megaptera novaeangliae*). Canad J Zool 64: 2075–2080.

Springer (2009). SpringerLink. Springer Science and Buisness Media. Accessed 05 February 2009 *from* http://www.springerlink.com/home/main.mpx.

SWARMS (2006). Group Behavior Database – SWARMS (Scalable Swarms of Autonomous Robots and Mobile Sensors). Yale University School of Forestry and Environmental Studies. Accessed 11.01.2009 *from* http://alliance.seas.upenn.edu/~ swarms/dynamic/index.php.

The Wildlife Society (2009). The Journal of Wildlife Management. BioOne. Accessed 05 February 2009 *from* http://www.bioone.org/loi/wild.

Thomson Reuters (2009). Scientific. Retrieved 05 February 2009 from http://scientific.thomson.com/cgi-bin/jrnlst/jloptions.cgi?PC=D.

University of Queensland (2008). What is a Database? Library, University of Queensland. Accessed 29.11.2008 *from* www.library.uq.edu.au/training/skills/what_dbase.html.

Valsecchi E, Hale P, Corkeron P, Amoss W (2002). Social structure in migrating humpback whales (*Megaptera novaengliae*). Mol Ecol 11: 507–518.

Van Lawick-Goodall J (1973). The behavior of chimpanzees in their natural habitat. Amer J Psychiatry. 130: 1–12.

Vigness Raposa KJ (2008). Marine Wildlife Behavior Database (MWBD). University of Rhode Island. Accessed 11 January 2009 *from* http://mwbd.edc.uri.edu/default.htm.

Whiten A, Goodall J, McGrew WC, Nishida T, Reynolds V, Sugiyama Y, Tutin CEG, Wranhgham RW, Boesch C (1999). Cultures in Chimpanzees. Nature 399: 682–685.

Wieczorek J (2001). MaNIS - Mammal Networked Information System. University of California, Berkeley. Accessed 29.11.2008 *from* http://manisnet.org/.

Wildlife Biology (2009). Wildlife Biology. Nordic Counsil for Wildlife Research (NKV). Accessed 05 February 2009 *from* http://www.wildlifebiology.com/.

Zucker EL, Clarke MR (1998). Agonistic and affiliative relationships of adult female howlers (*Alouatta palliata*) in Costa Rica over a 4-year period. Int J Primatol 19: 433–449

Chapter 11
Critical Remote Sensing Contributions to Spatial Wildlife Ecological Knowledge and Management

Gregory J. McDermid, Nicholas C. Coops, Michael A. Wulder, Steven E. Franklin, and Nicole E. Seitz

11.1 Introduction

A spatial information management approach to applied wildlife ecology will rely on our capacity to link animal-based data sets – observations related to a species' distribution, abundance, health, or genetics, for example – to a variety of spatially explicit environmental variables. This idea is based on the general concept that an organism's characteristics and behaviors at both the individual and population levels are inextricably linked to the physical habitat in which it occurs (Guisan and Zimmermann 2000; Braun 2005). While the investigation of these links must be well-grounded by solid field observations, the multiple scales and extent over which information must be compiled suggests a key role for remote sensing instruments and related technologies. For example, it is becoming increasingly evident that the health of wild species is adversely affected by human activities and landscape change (e.g. Daszak et al. 2001; Farnsworth et al. 2005). Ongoing research may reveal a direct link between human-induced habitat changes and long-term physiological stress, leading to damaging health consequences in

G.J. McDermid (✉)
Department of Geography, University of Calgary, Calgary, Alberta, Canada T2N 1N4

N.C. Coops
Department of Forest Resource Management, University of British Columbia, Vancouver, British Columbia, Canada V6T 1Z4

M.A. Wulder
Canadian Forest Service (Pacific Forestry Centre), Natural Resources Canada, Victoria, British Columbia, Canada V8Z 1M5

S.E. Franklin
Department of Geography, Trent University, Peterborough, Ontario, Canada K9J 7B8

N.E. Seitz
Canadian Forest Service (Pacific Forestry Centre), Natural Resources Canada, Victoria, British Columbia, Canada V8Z 1M5

S.A. Cushman and F. Huettmann (eds.), *Spatial Complexity, Informatics, and Wildlife Conservation*
DOI 10.1007/978-4-431-87771-4_11, © Springer 2010

individual animals (i.e. impaired reproduction, diminished growth, suppressed immune function) and subsequent negative effects at the population level (i.e. low natality and survival rates, diminished abundance). An approach to understanding these relationships, based on sensitive and reliable measures of health, stress, and landscape change, is both urgently needed and impossible to conceive without remote sensing.

For decades, remote sensing has been acknowledged as a critical data source of environmental information, theoretically capable of supporting a broad range of ecological applications (Robinson 1985; Greegor 1986; Graetz 1990; Franklin 2001; Lewis 2003). The promise of the technology lies in its ability to deliver high-quality, spatially explicit observations over large areas with regular revisit intervals, in a format that is both well-suited for ecological interpretation and readily integrated with data from other components of a modern spatial information management system (e.g., GPS, GIS modules). Recent reviews by Cohen and Goward (2004), Leimgruber et al. (2005), McDermid et al. (2005), and Gottschalk and Huettmann (2006), for example, document more than 30 years of success applying remote sensing technology to the analysis and modeling of wildlife–habitat relationships. However, as with all such multi-disciplinary partnerships, effectiveness can be hindered by miscommunication between the *methods* experts – practitioners of remote sensing, GIS, and other spatial technologies – and *applications* personnel with expertise in wildlife and ecology.

The linkages between ecologists and remote sensing scientists can be improved by addressing gaps in understanding through the establishment of widely accepted standards. As a data source and analysis tool, remote sensing is still relatively new, and its data are often not well known and improperly handled. However, in the few decades that the data have been available, solid synthesis applications have emerged. For example, land-cover and change-detection analyses are two instances of remote sensing products that have become widely accepted in the various user communities (Franklin and Wulder 2002; Wulder et al. 2003). Overall user satisfaction with these products can be partially attributed to an increased understanding of development strategies and characteristics, leading in turn to realistic user expectations surrounding results. However, remote sensing rarely replaces traditional field work. Instead, the technology represents a powerful suite of data sets and methodological procedures capable of *complementing and extending* ground observations accurately and efficiently over large geographic areas, and, as a result, provides an effective foundation for performing wildlife-ecological studies.

In this chapter, we explain the basic operation of remote sensing devices, and provide an overview of information products and major distribution centers. We then review the critical targets that remote sensing can address in support of wildlife research and modeling, with specific reference to key advancements and sample applications of interest to researchers in applied wildlife ecology. We conclude with the presentation of an application framework designed to link ecological information needs with the correct remote sensing imagery and information–extraction strategies in order to improve wildlife research and management.

11.2 Remote Sensing Background

For our purposes, remote sensing devices can be thought of as instruments, typically mounted on air- or space-borne platforms, which are designed to measure the electromagnetic radiation that is reflected and/or emitted by the surface of the Earth. For the most part, these earth-observing (EO) instruments are adapted to exploit portions of the electromagnetic spectrum that are not strongly impacted by the atmospheric gases and particles through which this radiation must pass. These *atmospheric windows* occur primarily in the visible, infrared, and microwave portions of the spectrum, and represent the wavelengths within which the vast majority of relevant sensors operate. *Passive* remote sensing devices rely primarily on the visible and infrared light provided by the sun, and constitute the familiar optical systems most commonly applied to wildlife studies. *Active* sensors such as lidar (light detection and ranging) and radar (radio detection and ranging) provide their own sources of illumination, and are increasingly employed in ecological work. Together, they represent a valuable emerging component of the discipline that is especially adept at difficult applications, such as characterizing the detailed vertical structure of vegetation (e.g., Hyde et al. 2006). Here we focus primarily on the use of passive EO instruments, a selection of which are presented in Table 11.1. To augment the background summary on remote sensing provided here, readers are advised to consider general text books by Campbell (2007), Richards and Jia (2006), and Lunetta and Elvidge (1998). We do not provide information on the aerial photography aspects of remote sensing, which have been reviewed in a related context elsewhere (Hall 2003).

Lefsky and Cohen (2003) review the background and considerations for the selection of remotely sensed data, describing in detail the various resolutions used to describe remote sensing systems: *spatial, spectral, temporal,* and *radiometric*. Gaining an understanding of these characteristics helps users select the data source that is best-suited to meet a given information need. *Spatial resolution* is the image characteristic that is typically of most interest to users. While the true concept is more nuanced, spatial resolution is often considered analogous to the pixel size of a digital image. A given image's pixel size indicates the instantaneous field of view of the sensor, or the surface area across which radiance measures have been generalized; 30 × 30 m in the case of Landsat Thematic Mapper, for example. This generalization occurs for each spectral band that is acquired for a given image type, and different bands may have different spatial resolutions.

The *extent* of an image is also linked to the spatial resolution, through sensor optics and altitude, among other factors. Images with smaller pixels typically cover a smaller portion of the Earth's surface, while those with larger pixels generally cover larger areas.

The *spectral resolution* defines the specific electromagnetic wavelengths across which an individual instrument acquires measurements. Multispectral sensors commonly have a relatively modest number of bands (up to 20) placed at locations where vegetation reflectance or absorption features are known to be present,

Table 11.1 Characteristics of low-, medium-, and high-spatial-resolution optical sensors (after Coops et al. 2006)

Sensor	Footprint (km × km)	Spatial resolution[a] (m)	Spectral resolution (nm)
Low spatial resolution sensors			
NOAA 17 (AVHRR)	2,940	1,100	500–1,250
SPOT 4 (VGT)	2,250	1,000	430–1,750
Terra (MODIS)	2,330	500	366–14,385
Medium spatial resolution sensors			
Landsat-5 (TM)	185	30	450–2,350
Landsat-7 (ETM+)	185	30 (MS/SWIR); 15 (pan)	450–2,350
SPOT 2 (HRV)	60	20 (MS); 10 (pan)	500–890
SPOT 4 (HRVIR)	60	20	500–1,750
SPOT 5 (HRG)	60	10 (MS); 20 (SWIR)	500–1,730
IRS (RESOURCESAT-1)	141	23.5	520–1,700
Terra (ASTER)	60	15	530–1,165
EO-1 (HYPERION)	37	30	433–2,350
High spatial resolution sensors			
Orbview-3	8	4 (MS); 1 (pan)	450–900
WorldView-1	17.6 km swath	0.5 (pan)	450–900
QuickBird-2	16.5	2.44 (MS); 0.8 (pan)	450–900
IKONOS	13.8	4 (MS); 1 (pan)	450–850

[a]*MS* multispectral, *SWIR* shortwave infrared, *pan* panchromatic

and that also coincide with clear atmospheric windows. Hyperspectral systems are characterized by having a large number (greater than 20) of near-contiguous narrow spectral bands. The quality of spectral resolution is established not by the *number* of bands of data collected, but rather by their width. Narrow spectral bands are better able to characterize detailed reflectance characteristics, without unnecessarily generalizing the response (Wulder et al. 2004).

Temporal resolution is the frequency at which a given location on the Earth's surface is imaged, or can be imaged. A high temporal resolution would indicate that a location is imaged frequently. The *temporal extent* of a remote sensing data source is an aspect for additional consideration in wildlife ecology studies, indicating the historical holdings and archives of a given sensor. Efforts to quantify changes in habitat through change detection, for example, require imagery that 'bookends' the targeted time interval, and can be limited by the temporal extent of the relevant data sets. Coops et al. (2006) provide insights regarding image selection and processing options for monitoring ecological disturbance and change.

Radiometric resolution refers to the precision with which radiance measurement is possible from a given data type, as indicated by the range of digital numbers recorded for each pixel. Inevitably, users must consider the trade-offs between such characteristics as image extent, desired local detail, temporal revisit frequency, and spectral information when selecting imagery for a given application. The 'perfect' remote sensing data source does not exist.

11.2.1 Remote Sensing Data Products and Sources

A description of the technical details surrounding the processing of raw sensor measurements into useable information products is beyond the scope of this chapter; interested readers are referred instead to the general remote sensing text books cited previously for the basics on this topic. However, users of remote sensing technology seeking to acquire appropriate data products require (1) a basic understanding of the various outputs available, and (2) some general guidelines regarding image sources and data distribution centers.

While remote sensing outputs are typically sensor- and organization specific, the general categories tend to be consistent across agencies and organizations. The Committee on Earth Observation Satellites categorizes EO data products according to their level of processing (Table 11.2). Level 0 represents *raw* data, which are generally not suitable for use in applied wildlife ecological studies. Level 1 products, by contrast, have been radiometrically calibrated, are often geometrically registered, and represent the familiar *unprocessed* imagery that supports large numbers of contemporary wildlife initiatives. However, while L1 is certainly the most flexible and widely used brand of EO imagery, the data remain unrefined, and typically require significant investments in order to generate useable information layers. Less well-known to the wildlife community is the wealth of *derived* L2 and L3 products available from many sensors: biophysical and geophysical information attributes generated by science teams eager to add value to EO data for the benefit of researchers and managers in other fields. For example, the National Aeronautics and Space Administration's (NASA's) Earth Observing System of satellites support hundreds of L2 and L3 data products designed to contribute to the measurement and monitoring of our planet (Parkinson and Greenstone 2000). While the global focus of EOS data emphasizes spatial scales that are often too coarse for detailed wildlife research, regional studies would be well-advised to take advantage of these underutilized – and often free – sources of information.

The strategy and cost of acquiring EO imagery and products depends primarily on the sensor and platform of interest. Sensors are managed by a wide range of corporate and government entities that operate under a variety of price structures and copyright arrangements. In general, public organizations such as NASA and the European Space Agency provide relatively low-priced outputs under generous copyright conditions, and archived image databases are often freely available to the public through various on-line geoportals. The International Center for Remote Sensing Education maintains a useful list of Internet clearinghouses and data distribution portals for satellite imagery at http://www.r-s-c-c.org. Of particular interest to ecologists is the recent release of the full 30-plus-year archive of the Landsat mission into the public domain (Woodcock et al. 2008), and available through the United States Geological Survey at http://landsat.usgs.gov.

Unlike the widely available low- and medium-spatial-resolution satellite sensors operated by government organizations, most high-spatial-resolution sensors are controlled by private corporations with more restrictive copyright policies and market-driven data

Table 11.2 Summary of standard remote sensing data products. Product levels are generally consistent across earth-observing agencies and organizations throughout the world

Data product	Description	Example
L0	Reconstructed but unprocessed 'raw' data with all available supplementary information (ephemeris, calibration) appended	Landsat 7 L0R
L1	Radiometrically corrected data converted to units of absolute radiance; often scaled to integers (digital numbers) for storage efficiency. Some L1 products have also been corrected for systematic geometric errors, and re-sampled to a user-specified map projection	Landsat 7 L1R (radiometrically calibrated) and L1G (radiometrically calibrated and geolocated) data; MODIS L1A (radiometrically calibrated) and L1B (radiometrically calibrated and geolocated)
L2	Derived biophysical and geophysical information products, distributed at the same locations and resolutions as the L1 source data	MODIS Surface Reflectance, LAI, Evapotranspiration, Land Cover, Sea, Sea Ice Cover, Chlorophyll Fluorescence, and Surface Temperature products
L3	Derived biophysical and geophysical information products that have been resampled in space and/or time, often for the purpose of completeness and consistency. Resampling may include aspects organization of averaging and compositing	MODIS gridded land, ocean, and atmosphere products. Polar grids and climate modeling grids are also provided to facilitate use by those research communities

prices. Complicating the issue of data availability in these cases (and in some medium-spatial-resolution systems as well) is the need to *task* the sensor in order to acquire imagery over a given area of interest. This obviously requires forethought and communication with the satellite operator. Projects seeking to purchase these types of data after the fact will often come away frustrated. Up-to-date information on pricing, availability, and tasking requirements is normally available from official sources on-line.

11.3 Linking Information Needs with Remote Sensing Strategy

The widespread availability of digital imagery and sophisticated software packages has greatly enhanced the accessibility of remote sensing technology in wildlife ecology studies. However, the data and processing strategies are both varied and complex, and, as a result, the potential for misuse by the uninformed is high. Chief among all concerns regarding the use of remote sensing data sources and processing strategies is the selection of the *proper* tools and techniques, and a subsequent correct interpretation of findings. The ecological literature contains multiple references to disappointing experiences with remote sensing data products (e.g. Plummer 2000; Thogmartin et al. 2004; Gottschalk and Huettmann 2006), and while some have asserted that the technology has failed to deliver consistently on its initial ecological promise, it seems clear that the larger issue revolves around a lack of solid understanding regarding the use of the remote sensing in an ecological setting (McDermid et al. 2005; Fassnacht et al. 2006). As with all technical collaborations, there is an on-going need for users and producers of remotely sensed information to seek common ground with respect to the capabilities of the tools, and the wildlife community would benefit substantially from the development and adoption of a methodological framework that links ecological information needs with appropriate remote sensing strategy.

The past decade has witnessed a tremendous increase in the number of publicly-accessible imaging platforms designed to deliver data at ever-increasing spatial, spectral, radiometric, and temporal resolutions. In addition to the familiar optical systems, there are the newly emerging technologies of lidar and various types of radar. While these recent choices have greatly enhanced our ability to conduct ecological modeling and monitoring activities, they also present complex challenges surrounding the selection of appropriate data and processing techniques. However, since the characteristics of ecosystems are generally determined by the primary tropic level – vegetation (Graetz 1990) – the discussion of information–extraction strategies for use in ecological studies must begin with a review of the remote sensing scene model, and how it relates to vegetation as a hierarchical, multi-scale phenomenon.

11.4 Multi-Scale Vegetation Structure

Complex systems theory describes the behavior of ecological systems characterized by a large number of components interacting in a non-linear fashion and exhibiting adaptive properties through time (Kay 1991; Hay et al. 2002). An important characteristic of these systems – or at least, our perception of them – is that they intuitively take on the form of a nested hierarchy, in which finer objects (leaves, branches, trees) are nested within broader ones (stands, forests, cover types). These ideas are important, since they help to define conceptual models that categorize vegetated landscapes amongst scale domains, and express the spatial dimensions at which selected information occurs. These considerations are the foundation of a variety of classification systems (e.g. Anderson et al. 1976; Woodcock and Harward 1992; Franklin and Woodcock 1997) that are designed to organize information hierarchically, and provide a helpful basis for linking ecological information with remote sensing data sources and appropriate information–extraction techniques.

11.5 The Remote Sensing Scene Model

Strahler et al. (1986) described the remote sensing model as one composed of three distinct elements: the sensor, the atmosphere, and the scene. The scene comprises the area of interest, which, in the context of a terrestrial ecology application, might consist of a forested landscape viewed at a specific scale. A model of this landscape can normally be constructed as a series of two- or three-dimensional objects distributed on a homogeneous background (Jupp et al. 1988, 1989). In real imagery, these objects would appear as groups of similar-looking pixels, and could take several different forms depending upon scale. For example, a conifer forest could be modeled at fine scales as a series of two-dimensional objects representing trees, shadows, and patches of understory, or, at a broader scale, as collections of structurally homogeneous forest stands. In these two cases, the groups of pixels in the imagery might appear similar, but they would represent markedly different ground objects.

One of the keys to understanding the nature of remote sensing imagery is to know the relationship between the objects of interest in the scene and the pixels in the image. Generally speaking, this relationship can be described as belonging to one of two distinct types: H-resolution or L-resolution (Strahler et al. 1986). The H-resolution case occurs when the pixels are *smaller* than the objects under investigation; in other words, when there are many pixels per object. L-resolution imagery, on the other hand, occurs when the pixels are *larger* than the objects, or there are many objects per pixel. This designation is important, since it neatly summarizes the physical relationship between pixels and objects: knowledge that can be exceptionally useful in selecting the appropriate image-processing techniques.

H-resolution imagery tends to display high amounts of local variability, and, as a result, contains large amounts of spatial information. In general, H-resolution scenes are best suited for classification; particularly strategies involving object-based approaches (e.g. Burnett and Blaschke 2003; Benz et al. 2004), texture variables (e.g. Cohen and Spies 1992; Franklin and McDermid 1993; Carr and de Miranda 1998), contextual decision rules (e.g. Gong and Howarth 1992; Groom et al. 1996; Sharma and Sarkar 1998), and other forms of spatial analysis (e.g. Atkinson and Lewis 2000; Csillag and Kabos 2002). By definition, classification involves the placement of pixels into distinct categories, and works best, therefore, on H-resolution entities that fit cleanly into information classes of interest defined at a specific scale.

L-resolution imagery, on the other hand, typically has low amounts of local variability, and, consequently, small quantities of spatial information. These scenes tend to be better-suited for a variety of sub-pixel analyses, such as mixture modeling (e.g. Hall et al. 1996; Atkinson et al. 1997; Heinz and Chang 2001), or empirical techniques that operate on a per-pixel basis and relate multi- or hyper-spectral response patterns to various intra-pixel properties (e.g. Michaelsen et al. 1994; Cohen et al. 2001, 2003). The selection of an inappropriate processing strategy – texture analysis in an L-resolution scene, for example, or per-pixel regression modeling in an H-resolution case – leads almost invariably to frustration to many users.

Since natural systems are comprised of a hierarchy of objects nested one inside another at different scales, a single image can be H-resolution with respect to some objects and L-resolution with respect to others. For example, Landsat imagery would be considered L-resolution with respect to tree objects, since a single 30 m pixel contains several individual tree crowns (top part of Fig. 11.1). However, at the stand scale, the same imagery would be considered H-resolution, since a 500-ha forest stand can contain many different 30 m pixels (bottom part of Fig. 11.1). As a result, the *correct* information–extraction strategy is obtained not through blind devotion to a single image type or processing routine, but depends on a rather more sophisticated evaluation of (1) the scale of the information desired and (2) the spatial resolution of the imagery available.

11.6 Critical Targets for Remote Sensing in Wildlife Ecology

The diversity of remote sensing systems and their varying spectral, spatial, temporal, and radiometric resolutions allows for an impressive suite of wildlife ecology and habitat attributes to be modeled, predicted, and monitored through

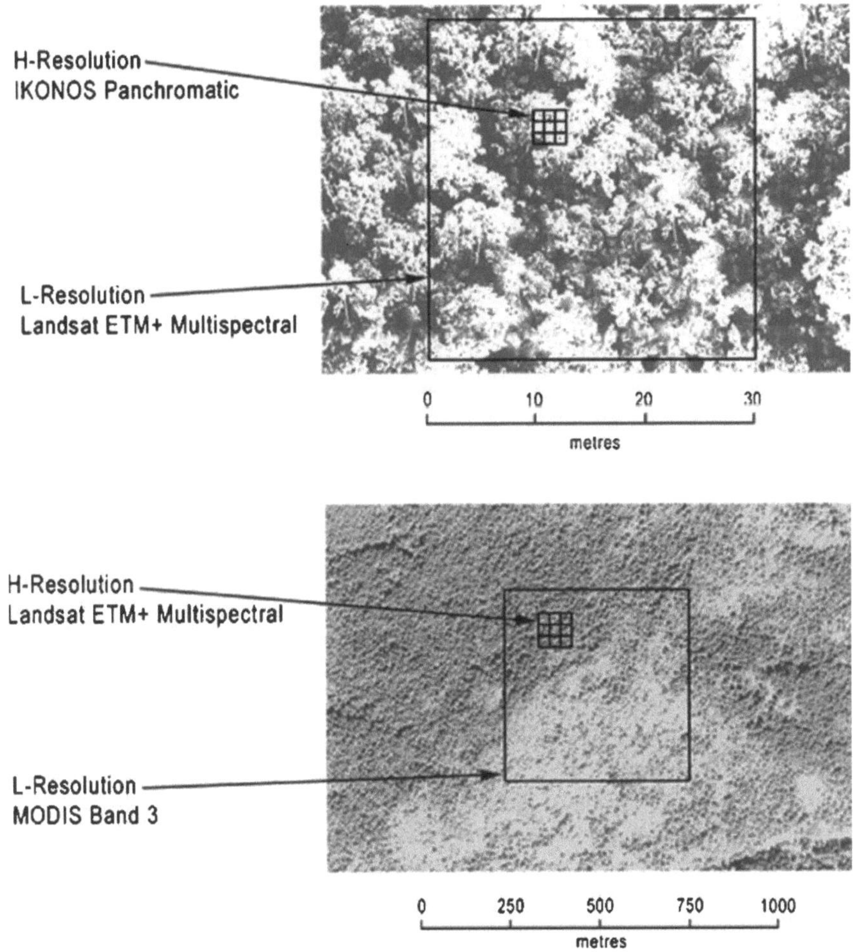

Fig. 11.1 Examples of H- and L-resolution imagery for a forested scene. At the tree level (*top*), Landsat ETM+ pixels are L-resolution, while IKONOS panchromatic pixels are H-resolution. At the stand level (*bottom*) Landsat pixels are H-resolution, while visible-band MODIS pixels are L-resolution

time. Principally, these attributes provide information on such factors as land cover physiognomy, vegetation structure and condition, forage characteristics, specific nutrient concentrations, overall productivity, and biomass. From these attributes, information on the distribution of shelter, shade, and nesting resources for wildlife, as well as potential to meet their metabolic needs, can be assessed. We surveyed the literature to investigate the role remote sensing has played in contemporary wildlife research (Table 11.3). In the following section, we highlight a number of the critical attributes that have been successfully predicted using remote sensing technology and have high relevance for wildlife ecology studies. Remote sensing of animals – another key remote sensing contribution to wildlife ecology – is reviewed elsewhere (Gillespie 2001; Ramanujan 2004; Majumdar et al. 2005).

11.7 Land Cover, Condition, and Change

A large number of key issues for wildlife management, including habitat suitability, land clearing and conversion to other land uses, and regional conservation planning, all require information on contemporary land cover condition and change. In most cases, remote sensing technology is used to predict *land cover* which is then used to infer spatially explicit habitat suitability for a wide range of wildlife species. One of the most common approaches to predicting land cover information from remotely sensed data is through the use of image classification, which involves the categorization of pixels a number of land cover classes, based on their similar spectral and/or spatial properties. When the user supervises this process by locating optimum examples of the land cover classes based on prior knowledge, field plots, or other information, the process is known as *supervised classification* (Richards and Jia 2006).

Cannon et al. (1982) was one of the first researchers to utilize Landsat Multi-Spectral Scanner (MSS) (80 m spatial resolution) imagery in a wildlife context, using 1978 MSS data to classify shinnery oak rangelands in Western Oklahoma, US. A strong positive correlation was found between percentage of grassland habitat and the population density of male lesser prairie chicken (*Tympanuchus pallidicinctus*). Even at this early adoptive stage, wildlife managers were encouraged to consider the use of timely satellite-based remote sensing observations as a cost-effective means of supplementing ground surveys. In a similar approach, Palmeirim (1988) undertook a supervised classification to generate a land cover map using Landsat Thematic Mapper (TM) imagery (30 m spatial resolution) to produce habitat suitability estimates for a number of avian species in northeastern Kansas. Seven land cover classes were predicted, and from this, patch statistics of different key land cover types such as the ratio of forest to grasses were calculated and related to different bird populations.

Also using Landsat TM data, Luoto et al. (2004) performed a supervised classification to produce a five-class land cover map of the boreal forests of Finland, and again, combined the land cover information with patch statistics and topographic

Table 11.3 Examples from literature showing the use of a variety of remotely sensed inputs, over a range of scales, to aid in studies of wildlife, habitat, and environmental management

	Focal species	Core habitat attributes	Methodology	Sensor platform	Location	Resolution	Author/year
	Lesser prairie chicken	Land cover	Supervised classification	Landsat MSS	Oklahoma, US	80 m	Cannon et al. 1982
	Various bird populations	Land cover	Supervised classification	Landsat TM	Kansas, US	30 m	Palmeirim 1988
	Boreal bird populations	Land cover	Supervised classification	Landsat TM	Finland	30 m	Luoto et al. 2004
	Mexican spotted owl	Land cover	Supervised classification	Landsat TM	Mexico	30 m	Peery et al. 1999
	Kirtland's warbler	Land cover	Supervised classification	SPOT XS	Bahamas	20 m	Miller and Conroy 1990
	29 Terrestrial mammals	Land cover	Supervised classification	Landsat TM	Great Britain	10 km	Cardillo et al. 1999
	Various bird populations	Land cover	Supervised classification	Landsat TM	Great Britain	30 m	Fuller et al. 2005
	Grizzly bear	Land cover	Supervised classification	Landsat TM	Montana, US	30 m	Waller and Mace 1997
	Elk	Land cover	Supervised classification	Landsat TM	Colorado, US	30 m	Huber and Casler 1990
[Au4]	Greater crane	Land cover	Supervised classification	Landsat TM	Northwestern Minnesota	30 m	Herr and Queen 1993
	Various bird/butterfly populations	Land cover	Unsupervised classification	Landsat TM	Yellowstone NP, US	30 m	Debinski et al. 1999
[Au5]	Buzzard	Land cover	Unsupervised classification	Landsat TM	Argyll, Scotland	30 m	Austin et al. 1996
	Grizzly bear	Land cover	Combined classifications	Landsat TM	Alberta, CA	30 m	Franklin et al. 2001
	Southwestern fly-catcher	Land cover	Unsupervised classification	Landsat TM	Arizona, US	30 m	Hatten and Paradzick 2003
	Waterfowl	Land cover	Combined classifications	Ikonos	Saskatchewan, CA	4 m	Dechka, et al. 2002
	Various bird/butterfly populations	Biomass/primary production	NDVI	Landsat TM	Western North America	30 m	Bailey et al. 2004
	African weaver-bird	Biomass/primary production	NDVI	NOAA AVHRR	Sub-Saharan Africa	1 km	Wallin et al. 1992
	Amphibian richness	Biomass/primary production	GVI	NOAA AVHRR	Europe	1 km	Rodriguez et al. 2005

(continued)

Table 11.3 (continued)

Focal species	Core habitat attributes	Methodology	Sensor platform	Location	Resolution	Author/year
Impala	Biomass/primary production	NDVI	Landsat TM	Botswana	30 m	Van Bommel et al. 2006
Ungulates and ostrich	Biomass/primary production	NDVI	NOAA AVHRR	Kalahari of Botswana	1 km	Verlinden and Masogo 1997
Barren-ground caribou	Biomass/primary production	NDVI	NOAA AVHRR	Arctic Tundra	1 km	Stow et al. 2004
12 Carnivore species	Biomass/primary production	NDVI/fPAR	MODIS	North Hemisphere	1 km	Nilsen et al. 2005
Greater gliders	Leaf pigments	Reflectance spectroscopy	Ground-based	Australia	Leaf-level	McIlwee et al. 2001
Possum, greater glider/koala	Leaf pigments	Reflectance spectroscopy	Ground-based	Canberra Australia	Leaf-level	Dury et al. 2001
Possum/greater glider/koala	Leaf pigments	Reflectance spectroscopy	HYMAP	Australia	5 m	Huang et al. 2002
Potential forage quantity	Non-photosynthetic vegetation	Reflectance spectroscopy	Ground-based	Nevada	Leaf-level	Elvidge 1990
Potential forage quantity	Non-photosynthetic vegetation	Reflectance spectroscopy	MODIS	Mongolia	1 km	Kawamura et al. 2005
Potential forage quantity	Non-photosynthetic vegetation	Reflectance spectroscopy	AVIRIS	Colorado	20 m	Roberts et al. 1992
Sky lark and the great/blue tits	Structure	lidar height metrics	lidar	Oxfordshire, UK	3 m	Mason et al. 2003
Fox squirrel	Structure	lidar height metrics	lidar	Delaware, US	25 cm	Nelson, et al. 2005
Bird communities	Structure	Radar backscatter	radar	Kakadu National Park, Australia	25 m	Imhoff et al. 1997
Potoroo/large wallabies	Structure	Spatial statistics	Digital camera	Nadgee nature reserve, Australia	2 m	Coops and Catling 1996
Chinook salmon	Energy	Water temperature	FLIR (forward looking infra-red)	North-Eastern Oregon, US	20–60 cm	Torgersen et al. 1999.
Schistosoma/Fasciola	Energy	Land surface temperature/NDVI	NOAA AVHRR	Ethiopia	1 km	Kristensen et al. 2001

data to predict the distribution of selected bird species. Areas of high predicted-bird-species richness in the boreal agricultural-forest mosaic were found mainly concentrated along river valleys with steep topography. Analysis indicated that the explanatory power of the topography-moisture models increased when the composition and land cover information, derived from remote sensing, were included.

Rather than undertaking a broad land cover classification, Peery et al. (1999) classified TM imagery into seven classes, using a supervised classification approach which captured a range of conifer age classes, as well as aspen stands, to predict Mexican spotted owl (*Strix occidentalis lucida*) distributions. Raw spectral bands and image band ratios were combined with topographic data in the final classification. Habitat composition and suitability was then assessed by overlaying the mapped home ranges of the species. Results indicated that Mexican spotted owls occurred in sites with more mature, mixed conifer species, and with proportionally less pinyon pine, than random sites on the landscape.

Miller and Conroy (1990) employed slightly finer SPOT High Resolution Visible (HRV) imagery (20 m spatial resolution) and a supervised classification to predict seven land cover classes ranging from agriculture and native grasses to a variety of broadleaf successional stages. Maps of the early seral stages were then linked to potential habitat for the Kirtland's warbler (*Dendroica kirtlandii*), an endangered species wintering in the Bahamas. In their conclusions, Miller and Conroy (1990) highlighted that remotely sensed data provided critically important information due to its timeliness and ability to provide information on vegetation composition in areas which were logistically difficult to access.

In a number of countries, remote sensing-based land cover maps have been generated over extensive areas, and are publicly available (for example, the 2001 National Land Cover Database of the United States – http://www.mrlc.gov; the Land Cover Map of Great Britain – http://www.ceh.ac.uk/data; and the land cover map of the forested region of Canada – http://www.pfc.forestry.ca/EOSD). Cardillo et al. (1999) assessed the benefit of a remote sensing-derived land cover map of Great Britain based on a supervised classification of land cover with 25 classes. These land cover maps were aggregated to a 1 × 1 km grid and then related to the richness and occurrence of 29 terrestrial mammals from four regions of Britain. Results indicated that over the entire country, the predictive power of the land cover information was poor and explained less than half of the variation in mammal species richness and occurrence. This predicative ability was considerably stronger however when the country was stratified into regions and analyzed separately (R^2 = 0.21–0.55 for the stratified regions, versus R^2 = 0.05–0.29 for the aggregated data). The same land cover dataset was also used by Fuller et al. (2005) to examine bird habitat preferences across south-eastern England. They found the land cover products to be an effective way to link predictions of bird species richness to habitat, and that the maps captured strong regional patterns associated with distinctive habitat assemblages. The authors concluded that remote sensing was an excellent tool to assess habitats, comprehensively, over large areas.

In addition to birds, remote sensing technology has also been used to map land cover in support of other wildlife studies. Waller and Mace (1997) utilized a

Landsat TM supervised classification to obtain seven land cover classes to predict grizzly bear (*Ursus arctos*) populations in Montana, US. Results indicated the bears utilized avalanche chutes and slab rock (classified as soil and bare ground) all year round, whereas shrub and timber harvest areas were selected relative to availability in summer and fall. Clear patterns of movement were also differentiated using a combination of the land cover maps with data from radio-collared bears. Recent work with grizzly bears in Alberta (Nielsen et al. 2006) has examined relationships between spatial landscape structure, human-caused landscape change, grizzly bear health and population performance through combined use of remote sensing technology, Global Positioning System (GPS) radio-telemetry, wildlife health evaluation, and molecular techniques.

Elk (*Cervus canadensis*) habitat mapping was undertaken by Huber and Casler (1990) in Colorado, US. Again, a land cover map was produced using a supervised classification with 13 classes of forest, grass, and shrub communities. Results showed that the large number of detailed forest classes resulted in significant misclassification of some of the key classes critical for elk habitat mapping. The study concluded caution should be exercised when attempting to classify highly-detailed land cover types, which may not appear spectrally distinct from other classes in satellite data.

In the above cases, satellite remote sensing imagery was classified using a supervised approach, wherein the user guides the land cover classes being created through training data. *Unsupervised classification* involves the statistical examination of a sample of image pixels with the goal of dividing the image into spectral classes based on the inherent spectral clusters present within the image. Unlike supervised classification, unsupervised approaches do not require the user to specify training data to initiate the process, rather classes are labeled after the classification process based on ground truth information. Debinski et al. (1999) used an unsupervised classification approach on Landsat TM imagery to produce an initial separation of 50 classes, which allowed for subsequent discrimination of gross land cover types, both individually and along a hydrological gradient. Each spectral class was subsequently identified and labelled using aerial photography and personal knowledge of the study area, resulting in a detailed and locally-specific vegetation map. Six non-forested meadow classes, representing a distinct xeric-to-hydric gradient, were mapped and correlated to the abundance of a number of butterfly and bird species. Results indicated that the sites of highest species richness coincided for both birds and butterflies, and could be found predominately in the mesic meadows. Hatten and Paradzick (2003) employed a similar unsupervised classification routine to map vegetation and floodplain land cover types, developing a relationship with southwestern flycatcher habitat in Arizona, US. The method provided a basis for predicting landscape configuration at both the local site location and environments 200 km further away.

In the tropics, Ortega-Huerta and Medley (1999) performed an unsupervised classification on Landsat TM imagery to derive a range of land cover classes for input into a jaguar (*Panthera onca*) habitat suitability model for sites in Mexico. Land cover classes were combined with topography and hydrological layers to map optimum jaguar habitat and provide management criteria to maintain and further enhance available habitat. Similarly, Cua on (2000) compared land cover and land cover change data from 1974 to 1986 derived from Landsat MSS data to a number of mammal abundance records in Neotropical regions of South America. Results indicated that there was a declining trend in the amount of habitat suitable for approximately 59% of the tropical species examined.

Image classification techniques can also be applied to high-spatial-resolution imagery from satellite or airborne platforms. Dechka et al. (2002), for example, applied both supervised and unsupervised techniques on two Ikonos satellite images (4 m spatial resolution) to map a number of wetland habitat classes and vegetation communities in southern Saskatchewan, Canada. A number of image processing approaches were used, resulting in a range of accuracies of land cover classes, including a key wetland habitat class used to map waterfowl and migratory bird patterns across central Canada.

11.8 Biomass and Primary Production

There is strong evidence that contemporary climate drives broad-scale species richness gradients of both plants and animals (Hawkins et al. 2003). For plants, it is widely accepted that energy and water together drive diversity and form (Currie and Paquin 1987; Leathwick et al. 1998; Francis and Currie 2003). For animals, energy either alone or in combination with water has been linked to large-scale variation in diversity, depending largely on the location in the world the study is focused (Hawkins et al. 2003). Primary production can be derived using remotely sensed data by examining the spectral reflectance centered on the near-infrared and visible red bands of the electromagnetic spectrum. Within-leaf scattering is high in the near-infrared region, thus the reflectance signal returned from the canopy is also high in this spectral region. In contrast, the red wavelengths of the electromagnetic spectrum are selectively absorbed by leaf pigments, which results in low reflectance in these bands (Coops et al. 2007). As a consequence, foliage area and increasing photosynthetic activity are correlated to the contrast in reflectance between near-infrared and red wavelengths. The Normalized Difference Vegetation Index (NDVI) is the most common and widely-applied of these *vegetation indices*. NDVI has been used as an estimator of 'greenness' (Stow et al. 2007) and a surrogate for large number of vegetation attributes, including biomass, leaf area index, phytomass, amount of green cover, productivity, photosynthetic activity, and leaf nitrogen content (Turner et al. 1992; Huete et al. 1994; Asner and Wessman 1997).

Bailey et al. (2004) computed the annual maximum value of Landsat NDVI to provide an estimate of maximum annual primary productivity and the relationship between this measure of productivity and its spatial heterogeneity and bird and butterfly species richness was then examined. Positive linear relationships between the Landsat imagery and the number of functional guilds of birds and species richness of neotropical migrant birds were found. Spatial variation in NDVI however was negatively correlated with number of functional guilds, and species richness,of resident birds. In a more species-specific study, Wallin et al. (1992) analyzed a time sequence of 1-km NDVI data from the NOAA Advanced Very High Resolution Radiometer (AVHRR) sensor to relate vegetation dynamics to potential breeding habitat of the African weaver-bird. The very large continental spatial scale of the imagery, combined with the highly mobile nature of the species, made the scale of the datasets well-matched. The results demonstrated that coarse-spatial-resolution satellite data could be effectively used to monitor potential breeding habitat through time. Relationships between avian species diversity and annual vegetative biomass were also found in Senegal using broad-scale satellite information on vegetation greenness (Jorgensen and Nohr 1996). Similarly, Skidmore et al. (2003) predicted mammal and bird species richness using broad-scale (1 km spatial resolution) satellite indicators of vegetation greenness. However, they noted that climate parameters were better predictors of species richness than the satellite data alone. Bonn et al. (2004) used a similar approach to investigate the relationship between species richness and productivity, and found that higher productivity levels do relate to higher levels of species richness.

Rodriguez et al. (2005) utilised a global vegetation index (GVI – an indicator of standing plant biomass), obtained from the AVHRR at 1 km spatial resolution to predict amphibian richness. Results indicated that while potential evaporation was the best explanatory variable overall, plant biomass derived from GVI was almost as good as potential evaporation at predicting amphibian richness, suggesting that plant productivity plays a role in determining diversity for this group. Van Bommel et al. (2006) used NDVI from a number of seasonal Landsat TM scenes to develop subclasses of broad physiognomic vegetation types in Botswana and related them to the occurrence of impala at landscape scales, under seasonally varying conditions. Results indicated that impala displayed selectivity for vegetation subclasses assigned on the basis of NDVI characteristics only, and not climate, allowing simple population models to be developed. Given the recent progress in the fields of satellite tracking of animals in the field (Amstrup et al. 2004) and vegetation condition assessment using the NDVI and other indices, a strong link between animal movements and functional vegetation analysis using remote sensing was proposed. Earlier, Verlinden and Masogo (1997) utilized AVHRR-derived NDVI and found excellent relationships between NDVI and green grass conditions in the Kalahari of Botswana. Based on these predictions of green cover, the density and distributions of a number of species, including wildebeest, hartebeest and ostrich, were made. In general, relationships between NDVI and animal distribution were difficult to test using the available presence/absence data. However, results indicated that both ostrich and wildebeest were associated with areas with higher NDVI. The authors concluded that NDVI from AVHRR data could be used to monitor suitable habitat in the wet season for some abundant species that preferentially select green patches in their environment.

In a study which attempted to employ phenological information from NDVI time series, Stow et al. (2004) used the annual median and rate of change of NDVI to estimate the quality and quantity of green forage that was available to the Porcupine caribou herd in northern Alaska and the Yukon Territory of Canada. At the broad scale, females were found to select annual calving grounds with a high proportion of easily-digestible forage (characterized by a high rate of NDVI increase), and then locally select concentrated calving areas with relatively high plant biomass. The amount of forage available at peak lactation times of the year provided the best model of calf survival. The authors concluded the timing of snowmelt and vegetation phenology influenced both the annual selection of calving areas and subsequent survival rate of the calves. Similarly, Nilsen et al. (2005) linked satellite measured greenness with measures of fauna diversity by comparing variations in the mean and seasonal greenness over a two year period with the home ranges of 12 carnivore species in the northern hemisphere, testing the hypothesis of Harestad and Bunnell (1979) that species' home ranges should decrease as a function of increasing productivity. Results indicated that the accuracy of prediction of eight of the 12 species' home range sizes was improved through the inclusion of satellite estimated greenness.

11.9 Energy Relations

In grasslands in particular, information on vegetation productivity and biomass can be combined with remotely sensed estimates of land surface temperature. As NDVI increases over grasslands, a linear decrease in surface temperature can be observed. This pattern is ascribed to an increase in latent heat flux away from the surface due to transpiration by grassland plants. While this relationship changes based on land cover type and season, it can be exploited to describe the moisture availability of the landscape (Hill 2004). This type of analysis has been applied in epidemiology studies such as estimating risk for the snail-borne diseases caused by *Schistosoma* spp. and *Fasciola* spp. Models using NDVI and temperature data derived from the AVHRR over regions of Ethiopia explained over 90% of the variance in observed snail sample sites (Kristensen et al. (2001).

Torgersen et al. (1999) utilized remotely-sensed stream temperature data to evaluate changes in the riverine habitat and thermal spatial structure that may influence the distribution patterns of Chinook salmon (*Oncorhynchus tshawytscha*). Using airborne thermal imagery, at 20–60 cm spatial resolution in north-eastern Oregon, results indicated that cool-water temperature patterns were strongly related to salmon distributions. The authors concluded that the heterogeneity of thermal properties in streams should be recognized for their biological potential to provide habitat information for species existing near the margin of their environmental tolerances.

11.10 Chemical/Pigment Constituents

An important application for remote sensing technology, in addition to the classification of land cover and habitat types and the prediction of biomass, is the detection of different vegetation characteristics based on underlying pigment and chemical constituents. These differences in the composition of vegetation are often difficult to detect, since the changes are often subtle and gradual, with species having many similar spectral characteristics. These surveys however can be successfully undertaken using data with very fine spectral resolution, allowing finer detail in the spectral signatures of vegetation to be examined. This type of remote sensing imagery is available predominately from airborne sensors, such as the Airborne Visible/Infrared Imaging Spectrometer (AVIRIS) and the Compact Airborne Spectrographic Imager (CASI). Most recently, however, the Hyperion sensor onboard the Earth Observer (EO1) satellite platform, launched in 2000, provides similar imagery over larger areas in a more cost-effective manner. Finally, spectra can also be obtained from hand-held instruments, using near-infrared reflectance spectroscopy, allowing for in situ estimation of detailed vegetation properties.

Initial research in this area was undertaken by Norris et al. (1976) who utilized near-infrared-reflectance spectra of 87 samples of ground dry forages such as alfalfa and tall fescue. Multiple-linear-regression techniques were used to determine the optimum wavelengths for predicting each of the chemical concentrations, with the authors concluding that infrared reflectance has the potential to assist in the rapid evaluation of forage quality.

McIlwee et al. (2001) investigated the potential of hand-held reflectance spectroscopy as to assess the chemical constituents of Eucalyptus leaves in Australia, and thus directly assess the quality of intake of foliage by gliders and possums. The authors concluded that concentrations of foliar nitrogen, tannins, and phenolics could be estimated using regression approaches correlating near-infrared reflectance spectra of foliage samples. Spectral-based models of food intake were found to be highly accurate for both species, and these models were then used to assess the food intake for gliders and examine the relationship between leaf palatability and food preferences of the mammals. Differences in leaf palatability for a number of tree species were consistent with the known food preferences of greater gliders, and it was concluded that field-based spectroscopy provides a powerful tool for predicting the foraging behavior of herbivores in situations where forage choices are determined by the compositional attributes of food. In a follow-on study, Dury et al. (2001) assessed the feasibility of extending the results spatially across larger areas using airborne remotely-sensed canopy biochemistry. Using laboratory spectra, they first reported on developing calibration equations for key concentrations based on the reflectance spectra, then scaled-up the relationships using airborne hyperspectral imagery (HYMAP). Strong correlations were found to exist between canopy-level pigment contractions and mean spectra. The study also identified several other spectral bands that corresponded to other key chemical concentrations, with the authors concluding that airborne HYMAP data could be used to estimate selected foliage chemical concentrations at the canopy level with acceptable accuracy (Huang et al. 2002).

11.11 Non-Photosynthetic Vegetation

In addition to live vegetation components, information on dry plant materials such as dry leaves, dry reproductive structures, bark, and woody debris can also provide important insights to wildlife populations and abundance, due to their key roles as both food source and shelter. Elvidge (1990) acquired very fine, hand-held spectra of green leaf (or green stem), senesced leaf, decayed leaf, brown wood, grey wood and bark for eight tree species including sagebrush, pinyon pine, white peppermint and sycamore. Results showed that the spectral features of dry plant materials bear little resemblance to those of green leaves. Green leaf spectra are dominated by the spectral features of chlorophyll and water, but when leaf senescence occurs these two compounds are lost and a host of previously-masked spectral features emerge. Elvidge (1990) concluded that all previous vegetation indices had been based on the spectral features of green vegetation. The development of vegetation indices for dry plant materials will be of major utility in assessing the biomass and biochemistry of dormant plant communities, vegetation stress, and measuring fuel loadings in areas subject to fire, and may also have a key role to play in biodiversity assessment.

Following on these ideas, Kawamura et al. (2005) employed AVHRR and Moderate Resolution Imaging Spectroradiometer (MODIS) sensors to detect seasonal vegetation changes (phenology) with regard to forage quantity and quality, focusing on, amongst others, attributed dead standing biomass. The dynamic range of the MODIS NDVI was analyzed and its sensitivity in discriminating between vegetation differences was evaluated across sparsely- and densely-vegetated areas. Results suggested that the MODIS NDVI can reliably detect the phenology and attributed dead biomass forage quantity and quality of grassland steppe areas.

In research on understory fuel loads associated with fire potential, Roberts et al. (1992) developed techniques to map non-photosynthetic vegetation (NPV), shade, and soil from a hyperspectral AVIRIS airborne image, and then interpreted these maps in an ecological context. Results confirmed the maps of NPV were distinguished from soil through spectral variations attributed to lignin and cellulose. These types of spatial predictions, in addition to being useful for fire fuel estimation, have direct relevance on wildlife habitat, wherein these components provide food, shelter, and grazing (Wessman et al. 1997). Similarly Jia et al. (2006) used airborne AVIRIS imagery to map major forest components in montane coniferous forests in Colorado, US. Again, mapping the fractional covers of NPV and bare soil proved critical, with high fractions of NPV and bare soil found in areas with recent disturbance such as fire or insect infestation.

11.12 Vertical Vegetation Structure

Thus far, all the techniques discussed in this chapter have employed *passive* satellite or airborne data. These systems, such as Landsat, collect data by sensors operating in the visible and infrared regions of the spectrum, designed primarily to detect reflected light and temperature (such as weather or meteorological satellites). Alternatively, *active* remote sensing systems are those that emit energy, in one form or another, and then measure the rate or amount of return back to the instrument. Active sensors can therefore operate under expanded meteorological conditions, since solar illumination is not required. The choice of active versus passive systems for vegetation structural mapping will depend primarily on the information need. Since active sensors can operate regardless of weather, they may be most effectively used in areas where there is perpetual cloud cover (e.g. tropical rainforests) (Lefsky and Cohen 2003). Synthetic aperture radar (SAR) is one form of radar remote sensing that utilizes microwave wavelengths many times longer than that of visible light. Terrestrial lidar sensors typically capture data at a single spectral band, often between 900 and 1,064 nm using cohesive laser beams operating primarily on airborne platforms.

Information on vertical vegetation structure is difficult to quantify from passive remote sensing technology, and yet is a key mechanism underlying many wildlife–habitat models. Manual survey of vegetation structure becomes prohibitive in terms of time and cost if sampling needs to be of sufficient density to characterize fine-grained heterogeneity at a landscape extent (Bradbury et al. 2005). Zimble et al. (2003) characterized a suite of vertical and horizontal forest attributes at fine scales for inclusion in decision-support systems in central Idaho. Analysis of field-derived tree height variance demonstrated that this metric could accurately distinguish between single-storey and multi-storey vertical structural classes.

Mason et al. (2003) used a combination of lidar and high-spatial-resolution optical data to extract information on landscape and vegetation structure at a spatial scale fine enough to match the fine-grained predictor variables used in most wildlife–habitat models. Results for the predicted abundance of two bird species, the sky lark and the great/blue tits, indicated that the derivation of vegetation structure data from airborne lidar has several clear advantages over field surveys in the construction of habitat models. First, the vertical resolution and sampling density of the data is equivalent or better than that which can be achieved by field measurement; second, the vertical and horizontal resolution available from LIDAR is scalable to landscape scales; and finally, the predicted attributes from these data allow heterogeneity in vegetation structure to be expressed at a variety of spatial scales, ranging from the foraging patch or territory to landscape (Bradbury et al. 2005).

Nelson et al. (2005) utilized an extensive lidar dataset over the state of Delaware to identify forested sites that potentially could support Delmarva fox squirrel (DFS; *Sciurus niger cinereus*) populations, an endangered species endemic to mature forests with open understories. The study indicated that, based on site characteristics of known habitat, a systematic airborne lidar data can be used to screen extensive areas of forest to locate potential DFS habitat, and that over 70% of locations met the canopy structural criteria thus supporting DFS populations, according to a habitat suitability model. The authors recommended also that a systematic survey across the county and state could be developed and undertaken at regular intervals to monitor changes to the areal extent of potential habitat over time.

In contrast to lidar data, radar remote sensing employs microwave energy emitted from airborne or spaceborne antennas, rather than laser bursts. These microwave pulses are emitted and received at wavelengths from 1 cm to 1 m and are configured on either a single vertical or horizontal plane (known as polarization; Coops 2002). Operationally designated wavelengths for radar are 3 cm (X-band), 5.5 cm (C-band), 24 cm (L-band) and 68 cm (P-band). Surfaces inclined towards the radar will have a stronger reflection (termed backscatter) than those which slope away from the radar. As a result, smooth flat surfaces will reflect little or no microwave energy and thus will appear dark in radar images. Vegetation, which is usually moderately rough at the scale of most radar wavelengths, appears gray.

A key attribute affecting the response of vegetation on radar backscatter is vegetation structure, in particular the vertical distribution of biomass through the canopy. As a result, the natural and anthropogenic processes that affect vegetation structure, such as regeneration and succession, can be readily-apparent in radar imagery. Kasischke et al. (1997) undertook a detailed review of the application of radar in ecological studies and highlighted a wide range of radar applications, including land cover classification (Henebry and Kux 1995; Hoekman and Quinones 2000), measurement of above ground woody biomass (Bergen and Dobson 1999), and delineation of wetland inundation (Falco et al. 1996). Imhoff (1995) compared forest canopy biometric data from a variety of tropical and sub-tropical forests with varying structural differences to simulated scenes of radar backscatter, showing that the structure of forest stands can have a considerable effect on backscatter amount, even when the amount above ground biomass remains equivalent. Beaudoin et al. (1994) utilized multi-polarized P-band data to document significant correlations with forest biomass, and found backscatter amount to be sensitive to a variety of ground surface attributes such as plant undergrowth, relief, and soil conditions. Yanasse et al. (1997) showed the ability of L-band, HV-polarized data to detect biomass changes occurring during tropical forest succession, and found backscatter to provide an accurate measure of biomass when soil conditions were dry. Imhoff et al. (1997) integrated aerial photography and field data with P-, L-, and C-band SAR data obtained from NASA's airborne AIRSAR system to study landscape spatial heterogeneity and bird community ecology at Kakadu National Park in Australia. Results indicated that SAR data were able to discern structural vegetation differences, and that multispectral sensors successfully identified floristic differences relevant to bird habitat quality. The authors concluded that the developed approaches advanced the use of SAR data for three-dimensional mapping of animal habitats from remotely sensed data (Imhoff et al. 1997).

Dobson et al. (1995) used multi-polarized SAR data and ancillary terrain information to estimate a range of forest structural attributes in boreal forests of Northern Michigan, including basal area, height, and dry crown biomass. Results indicated that biophysical attributes could be estimated with relatively small errors from SAR data, and that the combination of shorter wavelengths (X and C) yielded substantial improvements in estimates of crown biomass. Despite this success, a number of other studies have found SAR to be unresponsive to biomass when a certain threshold has been reached. For example Rauste et al. (1994) found that L-band SAR reached saturation at 70 mg/ha. The recent launch of advanced satellite-based SAR platforms, including the Canadian 3-m-resolution, fully polarimetric RADARSAT-2 instrument, potentially promises future advances on this front.

Demonstrating the exciting potential of *integrated* passive and active remote sensing data sets, Hyde et al. (2006) investigated the estimation of structural information from lidar, SAR, and optical sensors including Landsat to combine the highly-accurate vertical information available from active remote sensing with the broad-scale capabilities of optical imagery. The results indicated that while lidar was the best single dataset for estimating stand height and biomass, the use of Landsat metrics in addition, improved the prediction of large tree structures. Results also showed that high-spatial-resolution Quickbird imagery improved estimates only marginally when compared to the lidar datasets, however the combination of all sensors combined was better than lidar alone, however only slightly better than the dual combination of lidar and Landsat.

Despite the overriding message here that active remote sensing instruments are better-suited for extracting vegetation structural information, a number of authors have reported considerable success using passive high-spatial-resolution optical imagery from either satellite or airborne systems (Wulder et al. 2004). These include estimation of individual crown closure (Cohen et al. 1995, 2001), prediction of stem density and stand height (Franklin and McDermid 1993), and classification of relative stand age or stage of development (Cohen et al. 1995, 2001; Franklin et al. 2001; Nelson et al. 2005). Employing high-spatial-resolution imagery to extract structural attributes also lends itself to the use of textural attributes, which can provide information on stand crown gaps (Blackburn and Milton 1997) and foliage estimation (Wulder et al. 1998). Additionally, the derivation of image variance and semivariance to provide measures of stand structure (St-Onge and Cavayas 1997; Wulder et al. 1998; Levesque and King 2003), and the fitting of spatial statistical models such as semivariograms to represent forest structure (Levesque and King 2003), has also been successful. In this latter approach, the semivariogram range, sill, and nugget are fitted to image objects and subsequently interpreted. For example, Coops and Catling (1996) used a modified local variance method that assessed the changes in the standard deviation of a moving 3×3 window on successively spatially degraded images. A relationship was then developed that related the maximum level of variance observed with the vertical distribution of biomass within a forested canopy. These assessments were then applied to estimate mammalian distributions and abundances over large areas (Coops and Catling 2002).

11.13 An Application Framework

An application framework is a strategy for optimizing the appropriate remote sensing data and methods in an ecological context (Phinn et al. 2003). The process consists of a number of steps, including: (1) identifying the information requirements for the project (e.g. McDermid et al. 2005); (2) organizing the information needs into an ecological hierarchy (e.g. Franklin and Woodcock 1997); (3) conducting an exploratory analysis using existing digital data (e.g. Franklin 2001); (4) identifying the ideal remote sensing data, considering spatial, spectral, radiometric, and temporal resolutions (e.g. Lefsky and Cohen 2003; Coops et al. 2006); (5) selecting and applying a suitable set of processing strategies to extract the required information (e.g. Wulder 1998; Campbell 2007); and (6) conducting a cost-benefit analysis (e.g. de Bruin and Hunter 2003). In developing this application framework for ecological studies, Phinn et al. (2003) stressed the importance of the (typically) interdisciplinary analysis team understanding that the specifications for the ideal remote sensing data can vary, depending on vegetation conditions, study area size, and available image processing techniques. The choice of data should dictate – at least initially – the subsequent image processing techniques to be pursued: in general, classification for H-resolution data and per-pixel modeling for L-resolution imagery. Assessing the benefits of the resulting investment should take into account, among other things, the accuracy of the information products generated, the value of the resulting habitat maps, and the utility of the vegetation database for other resource management applications.

Acknowledgements The authors are grateful for support from the Natural Sciences and Engineering Research Council of Canada, the Canadian Forest Service, the Alberta Innovation and Science Fund, and the many partners and colleagues in the Foothills Model Forest Grizzly Bear Research Program.

References

Amstrup SC, McDonald TL, Durner GM (2004) Using satellite radiotelemetry data to delineate and manage wildlife populations. Wildl Soc Bull 32:661-679.

Anderson JR, Hardy EE, Roach JT, Witmer E (1976) A land use and land cover classification system for use with remote sensor data. U.S. Geological Survey Professional Paper 964. Washington, DC.

Asner GP, Wessman CA (1997) Scaling PAR absorption from the leaf to landscape level in spatially heterogeneous ecosystems. Ecol Model 103:81-97.

Atkinson PM, Lewis P (2000) Geostatistical classification for remote sensing: an introduction. Comput Geosci 26:361-371.

Atkinson PM, Cutler EJ, Lewis H (1997) Mapping sub-pixel proportional land cover with AVHRR imagery. Int J Remote Sens 18:917-935.

Austin GE, Thomas CJ, Houston DC, Thompson DBA (1996) Predicting the spatial distribution of buzzard Buteo buteo nesting areas using a Geographical Information System and remote sensing. Journal of Applied Ecology 33(6): 1541-1550.

Bailey SA, Horner-Devine MA, Luck G, Moore LA, Carney KM, Anderson S, Betrus C, Fleishman E (2004) Primary productivity and species richness: relationships among functional guilds, residency groups and vagility classes at multiple spatial scales. Ecography 27:207-217.

Beaudoin A, Letoan T, Goze S, Nezry E, Lopes A, Mougin E, Hsu CC, Han HC, Kong JA, Shin RT (1994) Retrievel of forest biomass from SAR data. Int J Remote Sens 15:2777-2796.

Benz UC, Hofmann P, Willhauck G, Lingenfelder I, Heynen M (2004) Multi-resolution, object-oriented fuzzy analysis of remote sensing data for GIS-ready information. Int J Photogram Remote Sens 58:239-258.

Bergen KM, Dobson MC (1999) Integration of remotely sensed radar imagery in modeling and mapping of forest biomass and net primary production. Ecol Model 122:257-274.

Blackburn GA, Milton EJ (1997) An ecological survey of deciduous woodlands using airborne remote sensing and geographical information systems (GIS). Int J Remote Sens 18:1919-1935.

Bonn A, Storch D, Gaston KJ (2004) Structure of the species-energy relationship. Proc R Soc Lond B Biol Sci 271:1685-1691.

Bradbury RB, Hill RA, Mason DC, Hinsley SA, Wilson JD, Balzter H, Anderson GCA, Whittingham MJ, Davenport IJ, Bellamy PE (2005) Modelling relationships between birds and vegetation structure using airborne LiDAR data: a review with case studies from agricultural and woodland environments. Ibis 147:443-452.

Braun CE (2005) Techniques for wildlife investigations and management. The Wildlife Society (TWS), Bethesda, MD.

Burnett C, Blaschke T (2003) A multi-scale segmentation/object relationship modelling methodology for landscape analysis. Ecol Model 168:233-249.

Campbell JB (2007) Introduction to remote sensing. The Guilford Press, New York.

Cannon RW, Knopf FL, Pettinger LR (1982) Use of Landsat data to evaluate lesser prairie chicken habitat in western Oklahoma. J Wildl Manage 46:915-922.

Cardillo M, Macdonald DW, Rushton SP (1999) Predicting mammal species richness and distributions: testing the effectiveness of satellite-derived land cover data. Landsc Ecol 14:423-435.

Carr JR, de Miranda FP (1998) The semivariogram in comparison to the co-occurrence matrix for classification of image texture. IEEE Transact Geosci Remote Sens 36:1945-1952.

Cohen WB, Goward SN (2004) Landsat's role in ecological applications of remote sensing. Bioscience 54:535-545.

Cohen WB, Spies TA (1992) Estimating structural attributes of douglas-fir western hemlock forest stands from Landsat and SPOT imagery. Remote Sens Environ 41:1-17.

Cohen WB, Spies TA, Fiorella M (1995) Estimating the age and structure of forests in a multi-ownership landscape of western Oregon, USA. Int J Remote Sens 16:721-746.

Cohen WB, Maiersperger TK, Spies TA, Oetter DR (2001) Modelling forest cover attributes as continuous variables in a regional context with Thematic Mapper data. Int J Remote Sens 22:2279-2310.

Cohen WB, Maiersperger TK, Gower ST, Turner DP (2003) An improved strategy for regression of biophysical variables and Landsat ETM+ data. Remote Sens Environ 84:561-571.

Coops NC (2002) Eucalypt forest structure and synthetic aperture radar backscatter: a theoretical analysis. Trees Struct Funct 16:28-46.

Coops NC, Catling PC (1996) Utilising airborne multispectral videography to predict habitat complexity in eucalypt forests for wildlife management. Wildl Res 24:691-702.

Coops NC, Catling PC (2002) Prediction of the spatial distribution and relative abundance of ground-dwelling mammals using remote sensing imagery and simulation models. Landsc Ecol 17:173-188.

Coops NC, Wulder MA, White J (2002) Identifying and describing forest disturbance and spatial pattern: data selection issues and methodological implications. Forest disturbance and spatial pattern: remote sensing and GIS approaches. Taylor and Francis, Boca Raton, FL.

Coops NC, Wulder M, White JC (2006) Identifying and describing forest disturbance and spatial pattern: data selection issues and methodological implications. In Wulder M, Franklin SE Forest Disturbance and Spatial Pattern: Remote Sensing and GIS Approaches. Taylor and Francis. CRC Press. pp. 33-60.

Coops NC, Black TA, Jassal, RPS, Trofymow JAT, Morgenstern K (2007) Comparison of MODIS, eddy covariance determined and physiologically modelled gross primary production (GPP) in a Douglas-fir forest stand. Remote Sens Environ 107:385-401.

Csillag F, Kabos S (2002) Wavelets, boundaries, and the spatial analysis of landscape pattern. Ecoscience 9:177-190.

Cuaŕon AD (2000) Effects of land-cover changes on mammals in a neotropical region: a modelling approach. Conserv Ecol 14(4):1676-1692.

Currie DJ, Paquin V (1987) Large-scale biogeographical patterns of species richness of trees. Nature 329:326-327.

Daszak P, Cunningham AA, Hyatt AD (2001) Anthropogenic environmental change and the emergence of infectious diseases in wildlife. Acta Trop 78:103-116.

de Bruin S, Hunter GJ (2003) Making the trade-off between decision quality and information cost. Photogram Eng Remote Sens 69:91-98.

Debinski DM, Kindscher K, Jakubauskas ME (1999) A remote sensing and GIS-based model of habitats and biodiversity in the Greater Yellowstone Ecosystem. Int J Remote Sens 20:3281-3291.
Dechka JA, Franklin SE, Watmough MD, Bennett RP, Ingstrup DW (2002) Classification of wetland habitat and vegetation communities using multi-temporal Ikonos imagery in southern Saskatchewan. Can J Remote Sens 28:679-685.
Dobson MC, Ulaby FT, Pierce LE (1995) Land-cover classification and estimation of terrain attributes using synthetic aperture radar data. Remote Sens Environ 51:199-214.
Dury SJ, Turner BJ, Foley WJ (2001) Can hyperspectral data be used to map koala and possum habitat? Pages 1648-1650 IGARSS 2001.
Elvidge CD (1990) Visible and near-infrared reflectance characteristics of dry plant materials. Int J Remote Sens 11:1775-1795.
Falco T, Francis F, Lovejoy S, Schertzer D, Kerman B, Drinkwater M (1996) Universal multifractal scaling of synthetic aperture radar images of sea-ice. IEEE Trans Geosci Remote Sens 34:906-914.
Farnsworth ML, Wolfe LL, Hobbs NT, Burnham KP, Williams ES, Theobald DM, Conner MM, Miller WM (2005) Human land use influences chronic wasting disease prevalence in mule deer. Ecol Appl 15:119-126.
Fassnacht KS, Cohen WB, Spies TA (2006) Key issues in making and using satellite-based maps in ecology: a primer. For Ecol Manage 222:167-181.
Francis AP, Currie DJ (2003) A globally consistent richness-climate relationship for angiosperms. Am Naturalist 161:523-536.
Franklin SE (2001) Remote sensing for sustainable forest management. Lewis Publishers, New York.
Franklin SE, McDermid GJ (1993) Empirical relations between digital SPOT HRV and CASI spectral response and lodgepole pine (Pinus contorta) forest stand parameters. Int J Remote Sens 14:2331-2348.
Franklin J, Woodcock CE (1997) Multiscale vegetation data for the mountains of southern California. Pages 141-168 *in* D. A. Quattrochi and M. F. Goodchild, editors. Scale in Remote Sensing and GIS. Lewis Publishers, New York.
Franklin SE, Wulder MA (2002) Remote sensing methods in medium spatial resolution satellite data land cover classification of large areas. Prog Phys Geog 26:173-205.
Franklin SE, Wulder MA, Gerylo GR (2001) Texture analysis of IKONOS panchromatic data for Douglas-fir forest age class separability in British Columbia. Int J Remote Sens 22: 2627-2632.
Fuller RM, Cox R, Clarke RT, Rothery P, Hill RA, Smith GM, Thomson AG, Brown NJ, Howard DC, Stott AP (2005) The UK land cover map 2000: planning, construction and calibration of a remotely sensed, user-oriented map of broad habitats. Int J Appl Earth Obs Geoinform 7:202-216.
Gillespie TW (2001) Remote sensing of animals. Prog Phys Geogr 25:355-362.
Gong P, Howarth PJ (1992) Frequency-based contextual classification and gray-level vector reduction for land-use identification. Photogram Eng Remote Sens 58:423-437.
Gottschalk T, Huettmann F (2006) Thirty years of analyzing and modeling avian habitat relationships using satellite imagery data: a review. J Ornith 147:175-175.
Graetz RD (1990) Remote sensing of terrestrial ecosystem structure. *in* R. J. Hobbs and H. A. Mooney, editors. Remote Sensing of Biosphere Functioning. Springer-Verlag, New York.
Greegor DH (1986) Ecology from space. Bioscience 36:429-432.
Groom GB, Fuller RM, Jones AR (1996) Contextual correction: techniques for improving land cover mapping from remotely sensed images. Int J Remote Sens 17:69-89.
Guisan A, Zimmermann NE (2000) Predictive habitat distribution models in ecology. Ecol Model 135:147-186.
Hall RJ (2003) The role of aerial photographs in forestry remote sensing image analysis. *in* MA Wulder and SE Franklin, editors. Remote Sensing of Forest Environments: Concepts and Case Studies. Kluwer Academic Publishers, Norwell, MA, pp 47-77.
Hall FG, Peddle DR, Ledrew EF (1996) Remote sensing of biophysical variables in boreal forest stands of Picea mariana. Int J Remote Sens 17:3077-3081.

Harestad AS, Bunnell FL (1979) Home range and body-weight – re-evaluation. Ecology 60:389-402.

Hatten JR, Paradzick CE (2003) A multiscaled model of southwestern willow flycatcher breeding habitat. J Wildl Manage 67:774-788.

Hawkins BA, Field R, Cornell HV, Currie DJ, Guegan JF, Kaufman DM, Kerr JT, Mittelbach GG, Oberdorff T, O'Brien EM, Porter EE, Turner JRG (2003) Energy, water, and broad-scale geographic patterns of species richness. Ecology 84:3105-3117.

Hay GJ, Dube P, Bouchard A, Marceau DJ (2002) A scale-space primer for exploring and quantifying complex landscapes. Ecol Model 153:27-49.

Heinz DC, Chang CI (2001) Fully constrained least squares linear spectral mixture analysis method for material quantification in hyperspectral imagery. IEEE Trans Geosci Remote Sens 39:529-545.

Henebry GM, Kux HJH (1995) Lacunarity as a texture measure for SAR imagery. Int J Remote Sens 16:565-571.

Herr AM, Queen LP (1993) Crane habitat evaluation using GIS and remote sensing. Photogrammetric Engineering and Remote Sensing 59:1531-1538.

Hill M (2004) Grazing agriculture: managed pasture, grassland, and rangeland. Pages 449-529 *in* S. Ustin, editor. Remote Sensing for Natural Resource Management and Environmental Monitoring. John Wiley & Sons, Hoboken, NJ.

Hoekman DH, Quinones MJ (2000) Land cover type and biomass classification using AirSAR data for evaluation of monitoring scenarios in the Colombian Amazon. IEEE Trans Geosci Remote Sens 38:685-696.

Huang ZXJ, Turner B, Foley W, Dury S (2002) Use of HYMAP image data to estimate sideroxylonal-A concentration of eucalypt foliage. Pages 1652-1654 IGARSS 2002.

Huber TP, Casler KE (1990) Initial analysis of Landsat TM data for elk habitat mapping. Int J Remote Sens 11:907-912.

Huete A, Justice C, Liu H (1994) Development of vegetation and soil indexes from MODIS-EOS. Remote Sens Environ 49:224-234.

Hyde P, Dubayah R, Walker W, Blair JB, Hofton M, Hunsaker C (2006) Mapping forest structure for wildlife habitat analysis using multi-sensor (LiDAR, SAR/InSAR, ETM plus, Quickbird) synergy. Remote Sens Environ 102:63-73.

Imhoff ML (1995) A theoretical analysis of the effect of forest structure on synthetic aperture radar backscatter and the remote sensing of biomass. IEEE Trans Geosci Remote Sens 33:341-352

Imhoff ML, Sisk TD, Milne A, Morgan G, Orr T (1997) Remotely sensed indicators of habitat heterogeneity: use of synthetic aperture radar in mapping vegetation structure and bird habitat. Remote Sens Environ 60:217-227.

Jia GJ, Burke IC, Goetz AFH, Kaufmann MR, Kindel BC (2006) Assessing spatial patterns of forest fuel using AVIRIS data. Remote Sens Environ 102:318-327.

Jorgensen AF, Nohr H (1996) The use of satellite images for mapping of landscape and biological diversity in the Sahel. Int J Remote Sens 17:91-109.

Jupp DL, Strahler AH, Woodcock CE (1988) Autocorrelation and regularization in digital images: I. Basic theory. IEEE Trans Geosci Remote Sens 26:463-473.

Jupp DL, Strahler AH, Woodcock CE (1989) Autocorrelation and regularization in digital images: II. Simple image models. IEEE Trans Geosci Remote Sens 27:247-258.

Kasischke ES, Melack JM, Dobson MC (1997) The use of imaging radars for ecological applications – a review. Remote Sens Environ 59:141-156.

Kawamura K, Akiyama T, Yokota H, Tsutsumi M, Yasuda T, Watanabe O, Wang G, Wang S (2005) Monitoring of forage conditions with MODIS imagery in the Xilingol steppe, Inner Mongolia. Int J Remote Sens 26:1423-1436.

Kay JJ (1991) A nonequilibrium thermodynamic framework for discussing ecosystem integrity. Environ Manage 15:483-495.

Kristensen TK, Malone JB, McCarroll JC (2001) Use of satellite remote sensing and geographic information systems to model the distribution and abundance of snail intermediate hosts in Africa: a preliminary model for Biomphalaria pfeifferi in Ethiopia. Acta Trop 79:73-78.

Leathwick JR, Burns BR, Clarkson BD (1998) Environmental correlates of tree alpha-diversity in New Zealand primary forests. Ecography 21:235-246.
Lefsky MA, Cohen WB (2003) Selection of remotely sensed data. Pages 13-46 *in* M. A. Wulder and S. E. Franklin, editors. Remote Sensing of Forest Environments: Concepts and Case Studies. Kluwer Academic Publishers, Norwell, MA.
Leimgruber P, Christen CA, Laborderie A (2005) The impact of Landsat satellite monitoring on conservation biology. Environ Monit Assess 106:81-101.
Levesque J, King DJ (2003) Spatial analysis of radiometric fractions from high-resolution multispectral imagery for modelling individual tree crown and forest canopy structure and health. Remote Sens Environ 84:589-602.
Lewis M (2003) Inventing global ecology: tracking the Biodiversity Ideal in India 1945-1997. Orient Longman, New Delhi.
Luckman A, Baker J, Kuplich TM, Yanasse CDF, Frery AC (1997) A study of the relationship between radar backscatter and regenerating tropical forest biomass for spaceborne SAR instruments. Remote Sens Environ 60:1-13.
Lunetta RS, Elvidge C (1998) Remote sensing change detection: enviromental monitoring methods and applications. Ann Arbor Press, Chelsea, MI.
Luoto M, Virkkala R, Heikkinen RK, Rainio K (2004) Predicting bird species richness using remote sensing in boreal agricultural-forest mosaics. Ecol Appl 14:1946-1962.
Majumdar SK, Huffman JE, Brenner F, Panah AI (2005) Wildlife diseases: landscape epidemiology, spatial distribution, and utilization of remote sensing technology. Pennsylvania Academy of Sciences, Easton, PA.
Mason DC, Anderson GCA, Bradbury RB, Cobby DM, Davenport LJ, Vandepoll M, Wilson DJ (2003) Measurement of habitat predictor variables for organism-habitat models using remote sensing and image segmentation. Int J Remote Sens 24:2515-2532.
McDermid GJ, Franklin SE, LeDrew EF (2005) Remote sensing for large-area habitat mapping. Prog Phys Geogr 29:449-474.
McIlwee AM, Lawler IR, Cork SJ, Foley WJ (2001) Coping with chemical complexity in mammal-plant interactions: near-infrared spectroscopy as a predictor of Eucalyptus foliar nutrients and of the feeding rates of folivorous marsupials. Oecologia 128:539-548.
Michaelsen J, Schimel DS, Friedl MA, Davis FW, Dubayah RC (1994) Regression tree analysis of satellite and terrain data to guide vegetation sampling and surveys. J Veg Sci 5:673-686.
Miller KV, Conroy MJ (1990) SPOT satellite imagery for mapping Kirtland's Warbler wintering habitat in the Bahamas. Wildl Soc Bull 18:252-257.
Nelson R, Keller C, Ratnaswamy M (2005) Locating and estimating the extent of Delmarva fox squirrel habitat using an airborne LiDAR profiler. Remote Sens Environ 96:292-301.
Nielsen SE, Stenhouse GB, Boyce MS (2006) A habitat-based framework for grizzly bear conservation in Alberta. Biol Conserv 130:217-229.
Nilsen EB, Herfindal I, Linnell JDC (2005) Can intra-specific variation in carnivore home-range size be explained using remote-sensing estimates of environmental productivity? Ecoscience 12:68-75.
Norris KH, Barnes RF, Moore JE, Shenk JS (1976) Predicting forage quality by infrared reflectance spectroscopy. J Anim Sci 43:889-897.
Oindo BO, Skidmore AK, De Salvo P (2003) Mapping habitat and biological diversity in the Maasai Mara ecosystem. Int J Remote Sens 24:1053-1069.
Palmeirim JM (1988) Automatic mapping of avian species habitat using satellite imagery. Oikos 52:59-68.
Parkinson CL, Greenstone R (2000) EOS Data Products Handbook. NASA/Godard Space Flight Center, Greenbelt, MD.
Peery MZ, Gutierrez RJ, Seamans ME (1999) Habitat composition and configuration around Mexican spotted owl nest and roost sites in the Tularosa Mountains, New Mexico. J Wildl Manage 63:36-43.

Phinn SR, Stow DA, Franklin J, Mertes LAK, Michaelsen J (2003) Remotely sensed data for ecosystem analyses: combining hierarchy theory and scene models. Environ Manage 31:429-441.

Plummer SE (2000) Perspectives on combining ecological process models and remotely sensed data. Ecol Model 129:169-186.

Ramanujan K (2004) New tools for conservation. Earth Observatory. http://earthobservatory.nasa.gov/Study/Conservation/printall.php accessed January 6, 2008.

Rauste Y, Hame T, Pulliainen J, Heiska K, Hallikainen M (1994) RADAR-based forest biomass estimation. Intl J Rem Sens 15:2797-2808.

Richards JA, Jia X (2006) Remote sensing digital image analysis: an introduction. Fourth Edition. Springer-Verlag, Berlin.

Roberts DA, Smith MO, Sabol DE, Adams JB, Ustin S (1992) Mapping the spectral variability in photosynthetic and non-photosynthetic vegetation, soils, and shade using AVIRIS. Summaries of the Third Annual JPL Airborne Geoscience Workshop, Pasadena, CA, 92-14, 38-40.

Robinson IS (1985) Satellite oceanography, Chinchester, England.

Rodriguez MA, Belmontes JA, Hawkins BA (2005) Energy, water and large-scale patterns of reptile and amphibian species richness in Europe. Acta Oecologica Int J Ecol 28:65-70.

Rotunno OC, Treitz PM, Soulis ED, Howarth PJ, Kouwen N (1996) Texture processing of synthetic aperture radar data using second-order spatial statistics. Comput Geosci 22:27-34.

Sharma KMS, Sarkar A (1998) Modified contextual classification technique for remote sensing data. Photogram Eng Remote Sens 64:273-280.

Skidmore AK, Oindo BO, Said MY (2003). Biodiversity assessment by remote sensing. Proceedings of the 30th International Symposium on Remote Sensing of the Environment: Information for Risk Management and Sustainable Development.

St-Onge BA, Cavayas F (1997) Automated forest structure mapping from high resolution imagery based on directional semivariogram estimates. Remote Sens Environ 61:82-95.

Stow DA, Hope A, McGuire D, Verbyla D, Gamon J, Huemmrich J, Houston S, Racine C, Sturm M, Tape K, Hinzman L, Yoshikawa K, Tweedie C, Noyle B, Silapaswan C, Douglas D, Griffith B, Jia G, Epstein H, Walker S, Daeschner S, Petersen A, Zhou LM, Myneni R (2004) Remote sensing of vegetation and land-cover change in Arctic Tundra Ecosystems. Remote Sens Environ 89:281-308.

Stow D, Petersen A, Hope A, Engstrom R, Coulter L (2007) Greenness trends of Arctic tundra vegetation in the 1990s: comparison of two NDVI data sets from NOAA AVHRR systems. Int J Remote Sens 28:4807-4822.

Strahler AH, Woodcock CE, Smith JA (1986) On the nature of models in remote sensing. Remote Sens Environ 20:121-139.

Thogmartin WE, Gallant AL, Knutson MG, Fox TJ, Suarez J (2004) Commentary: a cautionary tale regarding use of the National Land Cover Dataset 1992. Wildl Soc Bull 32:970-978.

Torgersen CE, Price DM, Li HW, McIntosh BA (1999) Multiscale thermal refugia and stream habitat associations of chinook salmon in northeastern Oregon. Ecol Appl 9:301-319.

Turner CL, Seastedt TR, Dyer ML, Kittel TGF, Schimel DS (1992) Effects of management and topography on the radiometric response of a tallgrass prairie. J Geophys Res Atmos 97:18855-18866.

van Bommel FPJ, Heitkonig IMA, Epema GPF, Ringrose S, Bonyongo C, Veenendaal EM (2006) Remotely sensed habitat indicators for predicting distribution of impala (*Aepyceros melampus*) in the Okavango Delta, Botswana. J Trop Ecol 22:101-110.

Verlinden A, Masogo R (1997) Satellite remote sensing of habitat suitability for ungulates and ostrich in the Kalahari of Botswana. J Arid Environ 35:563-574.

Waller JS, Mace RD (1997) Grizzly bear habitat selection in the Swan Mountains, Montana. J Wildl Manage 61:1032-1039.

Wallin DO, Elliott CCH, Shugart HH, Tucker CJ, Wilhelmi F (1992) Satellite remote sensing of breeding habitat for an African weaverbird. Landsc Ecol 7:87-99.

Wessman CA, Bateson CA, Benning TL (1997) Detecting fire and grazing patterns in tallgrass prairie using spectral mixture analysis. Ecol Appl 7:493-511

Woodcock C, Harward VJ (1992) Nested-hierarchical scene models and image segmentation. Int J Remote Sens 13:3167-3187.

Woodcock CE, Allen R, Anderson M, Belward A, Bindschadler R, Cohen W, Gao F, Goward SN, Helder D, Helmer E et al. (2008) Free access to Landsat imagery. Science 320:1011-1011.

Wulder M (1998) Optical remote-sensing techniques for the assessment of forest inventory and biophysical parameters. Prog Phys Geogr 22:449-476.

Wulder MA, LeDrew EF, Franklin SE, Lavigne MB (1998) Aerial image texture information in the estimation of northern deciduous and mixed wood forest leaf area index (LAI). Remote Sens Environ 64:64-76.

Wulder MA, Dechka JA, Gillis MA, Luther JE, Hall RJ, Beaudoin A (2003) Operational mapping of the land cover of the forested area of Canada with Landsat data: EOSD land cover program. For Chron 79:1075-1083.

Wulder MA, Hall RJ, Coops NC, Franklin SE (2004) High spatial resolution remotely sensed data for ecosystem characterization. Bioscience 54:511-521.

Yanasse CCF, Sant'Anna SJS, Frery AC, Renno CD, Soares JV, Luckman AK (1997) SIR-C dependence on tropical forest regeneration stages. Remote Sens Environ 59:180-190.

Zimble DA, Evans DL, Carlson GC, Parker RC, Grado SC, Gerard PD (2003) Characterizing vertical forest structure using small-footprint airborne LiDAR. Remote Sens Environ 87:171-182.

Ortega-Huerta MA, Medley KE (1999) Landscape analysis of jaguar (*Panthera onca*) habitat using sighting records in the Sierra de Tamaulipas, Mexico. Environ Conserv 26(4):257-269.

Chapter 12
Spatial Data Management Through Metadata: Global Concepts, Formats, Tools and Requirements

Vivian Hutchison

In an age of increasing technological abilities and access to information, metadata represents a critical element in information-sharing environments. Data management is greatly enhanced by the use of metadata standards. It provides for the foundation for data to be used by the public. Metadata assure a consistent way to display information describing datasets and other research. Consequently, the use of metadata standards in spatial data management environments also greatly facilitates data exchange processes. Metadata repositories serve as valuable tools to research data that has already been collected, analyzed, and reviewed because they give users an opportunity to evaluate existing information in order to make informed decisions. Repositories also offer an invaluable opportunity for collaboration between organizations on research projects. Furthermore, the creation of metadata provides a systematic documentation of valuable resources developed by an organization, which benefits the organization by capturing institutional knowledge to be preserved. Records provide a method for maintaining information for future use, and a standard way to share information with data catalogs and clearinghouses.

12.1 Metadata Standards for Spatial Data Management

A metadata standard is a common set of terms and definitions that are presented in a structured format. In essence, metadata is documentation that can answer questions – who, what, when, where, why, and how – that describe every facet of the data or resource being documented. The metadata record captures content, quality, accessibility, collection methods, processing, and availability of a dataset for use by others. It is extremely important to use established standards in the documentation of

V. Hutchison
National Biological Information Infrastructure, United States Geological Survey, Seattle, WA, USA
USGS Western Fisheries Research Center, 6505 NE, 65th Street, Seattle, WA, USA
e-mail: vhutchison@usgs.gov

S.A. Cushman and F. Huettmann (eds.), *Spatial Complexity, Informatics, and Wildlife Conservation*
DOI 10.1007/978-4-431-87771-4_12, © Springer 2010

geospatial datasets because the more standardized the structure and content of the information, the more efficiently it can be used by both computers and humans.

Four important standards have been established to describe datasets with a geospatial focus. When President Clinton signed Executive Order 12906 in 1994 requiring all Federal government agencies in the United States to create and publish metadata, the Federal Geographic Data Committee (FGDC) was assigned to develop a standard. It was titled "Content Standard for Digital Geospatial Metadata" and has subsequently been adopted by many other organizations beyond the Federal government interested in data sharing and interoperability with spatial data resources developed by the government. Further, it has been adopted as the required standard by over 200 national and international catalogs and clearinghouses worldwide.

A metadata profile contains additional elements that are incorporated into the main metadata standard to fill the need of a particular community. These profiles are officially approved by the FGDC. For example, biologists needed elements in the standard that allow frequently used information in biology such as taxonomy and methodologies to be documented. The Biological Data Profile (BDP) was developed by the National Biological Information Infrastructure (NBII) to fill this need. It was approved in 1998. Other profiles have been developed to serve other types of communities such as the Shoreline Data Profile, developed by the National Oceanic and Atmospheric Administration (NOAA), and the Remote Sensing Profile, created by the FGDC.

Another standard, Ecological Metadata Language (EML) was developed by a collaboration of individuals at National Center for Ecological Analysis and Synthesis (NCEAS), the Long Term Ecological Research Program, and the Joseph W. Jones Ecological Research Center. The standard was developed to better document projects in ecology and for using sensors. The standard was built in a modules and extensible fashion; such that users of the standard can determine which modules are most pertinent to describing their data, literature, and software resources. These modules can then be linked through a crosswalk to other established standards. EML is implemented in eXtensible Markup Language (XML) format.

The National Aeronautic Space Administration (NASA) Global Change Master Directory (GCMD) program developed a metadata standard in 1987 that preceded the establishment of the FGDC standard in 1994. It is called the Directory Interchange Format, or DIF standard, and it concentrates on the documentation of Earth Science data. When the FGDC standard was established as the official Federal standard in 1994, elements were added to the DIF to make it more compatible with the FGDC standard. The DIF exists today because it is a standard that provides more rigorous rules for information entry than other standards demand, making search and retrieval more efficient. GCMD provides a technical crosswalk from the DIF to other major standards, including the FGDC standard.

Finally, the International Organization for Standardization (ISO) has completed a metadata standard for documenting geospatial datasets that was approved by the American National Standards Institute (ANSI) in 2004. ISO is a non-governmental industrial organization that establishes standards for the international exchange of goods and services. It's widely supported by OECD (Organisation for the Economic and Commercial Development). Development of elements included in the ISO 19115 standard show influence primarily from the FGDC Content Standard for Digital Geospatial Metadata and the Australia ANZLIC Spatial Land Information Council standard.

ISO 19115 adds functionality not found in the FGDC standard, for example, in areas such as multi-lingual data sharing, topic categories for high-level metadata classification, unique identifiers for metadata records, roles and responsibilities for a dataset, and descriptions of geospatial service metadata. Countries are encouraged to create profiles to the ISO standard for which the conditionality of elements can be established (mandatory, mandatory if applicable, optional) and elements can be added to reflect areas of importance to a country or region. For example, Canada, the United States, and Mexico have produced the "North American Profile," or NAP, to the ISO standard. It is expected to be released in the late fall of 2008. The NAP will essentially become the next version of the FGDC Content Standard, currently used in the production of metadata records in the United States. In addition, organizations such as the NBII will be gathering the biological community together to produce a biological extension to the NAP that can be employed if a dataset is biological. Other organizations will produce extensions in a similar manner for remote sensing, marine data, and shoreline data. These extensions will be approved by the FGDC before their release.

The following Table 12.1 outlines the major standards and profiles for geospatial data description and websites where more information can be found. (Note that there are many other metadata standards that describe different types of data. Those, such as Darwin Core [for museum specimens] and Dublin Core [Primarily for web resources] are not listed here.)

12.2 Metadata Clearinghouses

Metadata clearinghouses are repositories for metadata records that provide effective search mechanisms for users. These clearinghouses are essential means for data sharing, and are rapidly improving in functionality. In the recent past, metadata clearinghouses were primarily useful for mere search and discovery purposes; however, they are progressing rapidly into dynamic tools that provide users with a more robust research experience and for linking Metadata into workflows for assessing, using and interpreting data in-time and online. There are many examples of such repositories.

GeoSpatial One Stop (GOS) is a major repository for metadata records on a variety of topics, and from a variety of contributors. Through a series of data portals, geodata.gov aims to become "Your One Stop Shop for Federal, State, and Local Geographic Data." The system currently has hundreds of metadata providers and offers a variety of activities that can be performed by a user that include finding data or map services, making a map, and publishing data and map services.

Table 12.1 List of major standards and profiles for geospatial data description

Standard name	Creator	Profile, extension or a main standard	Used by	Website
Content Standard for Digital Geospatial Metadata (CSDGM)	Federal Geographic Data Committee (FGDC)	Standard	United States: Federal agencies State agencies Universities Non-profits	www.fgdc.gov
Biological Data Profile	National Biological Information Infrastructure (NBII)	Profile (to FGDC CSDGM)	United States: Federal agencies State agencies Universities Non-profits	www.nbii.gov
Shoreline Metadata Profile	National Oceanic and Atmospheric Administration (NOAA)	Profile (to FGDC CSDGM)	United States: Federal agencies State agencies Universities Non-profits	http://www.csc.noaa.gov/metadata/sprofile.pdf
Extension for Remote Sensing Metadata	Federal Geographic Data Committee (FGDC)	Extension (to FGDC CSDGM)	United States: Federal agencies State agencies Universities Non-profits	http://www.fgdc.gov/standards/projects/FGDC-standards-projects/csdgm_rs_ex/MetadataRemoteSensingExtens.pdf

Ecological Metadata Language	National Center for Ecological Analysis and Synthesis (NCEAS), University of California at Santa Barbara	Standard	LTER program and various ecology organizations	http://knb.ecoinformatics.org
Directory Interchange Format	NASA Global Change Master Directory	Standard	NASA GCMD and various other organizations	http://gcmd.nasa.gov/index.html
ISO 19115 and ISO 19139	International Organization for Standardization	Standard	International community	http://www.iso.org/iso/home.htm
North American Profile	FGDC	Profile (to ISO)	Canada United States Mexico	www.fgdc.gov
ANZLIC Metadata Profile	ANZLIC Spatial Land Information Council	Profile (to ISO)	Australia and New Zealand	http://www.anzlic.org.au/

Other clearinghouses specialize in certain topical areas. The NBII, for example, hosts a metadata Clearinghouse that focuses on biological metadata records. This is perhaps not surprising since the NBII, a collaborative effort coordinated by the U.S. Geological Survey, provides access to data and information on the nation's biological resources. The NBII Clearinghouse presents an opportunity to discover metadata, create an RSS (Really Simple Syndication) feed for pre-defined searches, bookmark a page, and email a link to a particular search result. Conservation Commons is yet another example of a portal that hosts metadata records specifically oriented around conservation and geographic information. Many more repositories exist. States each host metadata clearinghouses, and many counties do as well.

Many Clearinghouses collect metadata records from a variety of partners because they harvest from web accessible folders. The NBII, for example, has established links with partner folders where FGDC compliant metadata records, displayed in XML, reside. Each week, the NBII Clearinghouse discards all records it has collected the week before and harvests all partner metadata again, thus capturing any changes that may have occurred in the folders. Records may have been added, updated, or deleted from a partner's collection. In this way, the partner is the sole possessor of its own metadata records, on its own server. The NBII is merely making a copy of the partner's records to display in the Clearinghouse. These federated systems have received a great appreciation, and use, by the global data user community because they are flexible, frequently updated, and keep the data owner in charge of information delivery and content.

Clearinghouses often share metadata resources as well. For example, the GeoSpatial One Stop system harvests specified metadata records from the NBII. To prevent duplication of records on any one clearinghouse, the NBII assists its new partners in also registering with the GOS system. Additionally, some repositories might request that certain types of records are forwarded into their systems. For example, the NBII provides a subset of forestry records to the Global Forestry Information System (GFIS). This partner then combines these records with a larger repository of global forestry records, thus increasing the exposure of all of the metadata records pertaining to forestry. The sharing of resources by clearinghouses allows greater, global, exposure for each metadata record in general, thus increasing the chances that the records will be used in beneficial ways to scientists, managers and the global public as a whole.

Metadata repositories are progressing quickly past their original intent – they are no longer merely containers for static metadata records. Clearinghouses now offer more tools to the user beyond discovery, such as the ability to download map layers and utilize map services. The virtually endless prospects that these types of Clearinghouses present are invaluable. Scientists have the opportunity to create collaborations that may not have been possible before, based purely on their use of an innovative avenue to discover new colleagues who are working on similar scientific questions to their own. Organizations benefit from these new collaborations, as creative scientists can find ways to share resources, funding, and datasets, potentially saving an organization a great deal of money on research and data management and delivery even. Clearinghouses offer the scientific community a chance to use technology to share results of research, some of which has taken decades to gather, analyze, and publish. With such resources, more broadly based analyses can be conducted that better explain our scientific world.

12.3 Metadata Editor Tools

A plethora of tools for the creation of metadata records have been developed by a variety of organizations. These editors shown in Table 12.2 all perform similar functions in creating metadata records; however, there has yet to be developed a tool that "does it all." For example, a tool might be proficient in capturing information from an ESRI Arc Catalog program, but it is limited in that it does not include the option to use the Biological Data Profile, and import species lists and taxonomies, e.g. from ITIS. Or, a tool might be stand-alone software and include Profiles but may not be able to extract information out of a Geographic Information System program like ArcCatalog. Not surprisingly, tools are generally produced with a bias to the organization producing the tool; so although they are offered generously for use by other organizations, they often contain predisposed elements that reflect producer needs. Finally, it is rather astonishing that online tools are relatively new, but that is the reality, and changing. Most tools are downloadable as a desktop feature or must be purchased.

Table 12.2 Prominent metadata editors (Please note this list represents a small sample of metadata creation tools)

Tool name	Creator	Comments
ArcCatalog	ESRI	Automatically populates metadata fields from Arc products. Does not include Profiles
Spatial Metadata Management System	Intergraph	Stand-alone software backed by database. Includes BDP
Metavist	USDA Forest Service	Stand-alone software for Windows only. Incorporates the BDP
NPS Metadata Tools and Editor	National Park Service	Contains powerful tools to be used with ESRI products or as a stand-alone tool. BDP included
MetaScribe	NOAA Coastal Services Center	Template driven web application to aid in creating metadata for collections of a similar data type that have similar metadata record content
Metadata Enterprise Resource Management Aid (MERMAID)	NOAA National Coastal Data Development Center (NCDDC)	Internet based tool with ability To define user roles and permissions, version control, and validation. No Profiles
EPA Metadata Editor	Environmental Protection Agency	A 3-tab approach offering a simplified interface. Works with ESRI products
Morpho	Knowledge Network for Biocomplexity	Creates Ecological Metadata Language (EML) metadata records

12.4 Moving Forward with Metadata

Strides are being taken to extend the usefulness of metadata records. Web service catalogues are being developed that capture mapping services referenced in metadata records that can then be utilized on a dynamic map. Researchers can use a map to find out where specific research is taking place. Analyses can be performed using metadata records to create historical trend data. There is much work to be done, and user needs will dictate future technological advances in metadata functionality beyond data management utility.

An extensive amount of outreach has occurred in the form of metadata workshops and websites that help people understand how to create metadata records. Workshops are offered throughout the United States by government organizations such as the NBII and the FGDC. State agencies have offered training, as have non-profit organizations. Requests for workshops are numerous, and they are well-attended events. In a typical workshop, participants learn what metadata is, why it is valuable, how to interpret the standard, how to use a Clearinghouse, and how to use software to create a record.

Workshops, however, can only reach a limited number of people each year. Funding for travel has dried up in recent years for all types of organizations, thus making it difficult for employees to attend. Websites have been created to help navigate the steps to creating a metadata record, but independent online learning is not always appealing to some. Further, it is difficult to monitor results of a workshop such as the number of records that results in the future because of a workshop event.

Regulation of metadata requirements has been a weak point for the metadata world in the past. Legislation basically exists, but does not get enforced. The concept of ‘carrots and sticks’, as used in the International Polar Year, is still weak on both ends, resulting still in an avoidance of documenting online data. Despite the existence and interest in training workshops, many people and funding bodies are still unaware of metadata requirements for geospatial datasets or, when made aware, remain unwilling to fund and participate in the creation of records. Thus, much data exists that has not been documented or archived even though standards and Federal requirements have existed for over a decade. There is no “metadata cop,” as people in the field like to say. Metadata are a true community responsibility; they work as a ‘bottom up’ effort.

However, as grassroots work succeeds and agencies and organizations begin to recognize the importance of creating metadata records, better support will develop from upper level management. This type of support is critical for the benefits of metadata to be realized and put into practice. Further, metadata practices need to begin early. Just think about the benefits of introducing metadata practices into the university setting. The more students of GIS who are made aware of data management techniques such as the creation of metadata records, the better chance the practice would carry forward into professional settings. Further, in the Federal government, scientists are promoted in similar ways as university professors are tenured – based on publications. Supporters of metadata advise policy changes in the Federal government to require that metadata records are produced and published in Clearinghouses *before* a promotion can occur. If metadata records were a requirement along with publications, we would likely see a dramatic increase in records. What a boon this would be to both organizations and scientific collaboration!

For Further Information:

Website Resources Used in this Paper:
Ecological Metadata Language: http://knb.ecoinformatics.org/software/eml/
Federal Geographic Data Committee: www.fgdc.gov
National Biological Information Infrastructure: www.nbii.gov
National Oceanic and Atmospheric Administration: www.noaa.gov
ANZLIC Spatial Land Information Council: http://www.anzlic.org.au/
International Organization for Standardization: www.iso.org
GeoSpatial One Stop: http://www.geodata.gov
Conservation Commons: www.conservationcommons.org
Wisconsin State Cartographer's Office http://sco.wisc.edu/wisclinc/metatool/mtools.htm
Environmental Protection Agency: http://geodata.epa.gov/eme.html
Socio Economic Data and Applications Center (CIESIN): http://sedac.ciesin.columbia.edu/metadata/

Chapter 13
Free Database Availability, Metadata and the Internet: An Example of Two High Latitude Components of the Census of Marine Life

Bodil Bluhm, David Watts, and Falk Huettmann

13.1 The Need for Open Access Biodiversity Databases

Our understanding of science is based on data. For hundreds of years, natural scientists have collected observations and compiled measurements. Historically, the data have been recorded and archived in notebooks, typed reports and more recently in electronic format. Most projects had a small and somewhat opportunistic scope, and lacked a wider international coordination and strategy; this is specifically the case for databases and their dissemination. Only in the last decades has the culture of data sharing and open access databases developed, and the focus changed to international collaborations and questions of global relevance (Esanu and Uhlir 2004; Huettmann 2005; see Graham et al. 2004 for Natural History Collections). In times of globalization, such an approach is urgently required. In the field of biodiversity, the need for open access databases has grown tremendously in the last decades because of the growth of humankind to the extent where man directly explores resources within and beyond their sustainability and directly and indirectly affects biota to extinction (Wilson 1999). Inventories represent the essence for sustainable management, for assessing and understanding changes and to inform educated decision-making (Braun 2005).

Analyses of large marine biodiversity data sets in the last decades have resulted in publications with direct management implications documenting the status quo on issues such as global overfishing (Pauly et al. 2003; Worm et al. 2005, 2006),

B. Bluhm (✉)
School of Fisheries and Ocean Sciences, PO Box 757220, University of Alaska-Fairbanks, Fairbanks, AK 99775-7220, USA
e-mail: bluhm@ims.uaf.edu

D. Watts
Australian Antarctic Data Centre, Australian Antarctic Division, Kingston, Tasmania, Australia

F. Huettmann
EWHALE lab- Biology and Wildlife Department, Institute of Arctic Biology, University of Alaska-Fairbanks, 419 IRVING I, Fairbanks, AK 99775-7000, USA
e-mail: fffh@uaf.edu

S.A. Cushman and F. Huettmann (eds.), *Spatial Complexity, Informatics, and Wildlife Conservation*
DOI 10.1007/978-4-431-87771-4_13, © Springer 2010

effects of climate changes on biomass and community structures (Brodeur et al. 2002; Grebmeier et al. 2006), occurrence and effects of invasive species (Mooney and Hobbs 2000) etc. Such analyses are necessary for developing the sustainable use of living resources, addressing conservation goals and conducting reasonable predictions, as well as for a critical evaluation of the magnitude of the human footprint (Costanza et al. 1998; Pauly et al. 1998). They are not possible without the tedious compilation of data from various sources, a process often including the need to overcome data release blockage, to spend many hours in libraries and often re-enter published data because of the lack of access to the original electronic source or the lack of an electronic source. Slowly, the scientific community at large including scientists, academic institutions, funding agencies, the public and editors/publishers is beginning to transition from a 'my-data'-attitude to a 'data sharing' culture (www.earlham.edu/~peters/fos/overview.htm; Esanu and Uhlir 2004). In the biodiversity realm, researchers are beginning to join forces to promote on-line availability of data with metadata (www.vliz.be/events/obi/statement.php) and stress the increased efficiency and reduction of duplication of research when open access databases are in place. For ocean-related data, the formation of the Global Ocean Observing System (GOOS) in 1989 demonstrates the awareness of the value and long-term benefits of online data to science and stakeholders on an international scale.

In high latitudes where the authors conduct their research, the observed and predicted changes in global climate were estimated early on to have the earliest and most pronounced effects (Manabe and Stouffer 1995). Measurable temperature shifts have been recorded for the Arctic shelf shallow water bodies (Hunt and Stabeno 2002), Arctic sea ice cover and thickness have declined in recent decades (Serreze et al. 2007; Comiso et al. 2008), and water masses have changed their characteristics in the central Arctic (Shimada et al. 2004). Temperature and salinity changes in the deepest parts of the Southern Ocean are now apparent (Rintoul 2008). With increased surface UV and ocean acidity, these abiotic changes may lead to substantial and profound changes in the ocean's ecosystem (Thomson et al. 2008), with global repercussions. The fauna in polar areas is typically highly adapted in their life history, ecology and physiology to the extreme and highly seasonal conditions of their environment (Thiel et al. 1996; Clarke 1998; Pörtner and Playle 1998). It has evolved a unique character over millions of years with high endemism particularly in the Southern Ocean, making it a special contribution to global biodiversity. Any change in the environmental conditions will have direct effects on the marine biota on multiple scales, from communities and populations to individuals (Schumacher et al. 2003).

Detecting these effects over a wider area, and responding to them appropriately, requires long-term time-series which are scarce in polar regions and which might well require a new science culture to evolve. In the Arctic, with the exception of the subarctic Bering and Barents Seas (e.g. Hamre 1994), few species are harvested commercially and, hence, the need for long-term monitoring of any particular species or biodiversity as a whole has largely been ignored. The US National Marine Fisheries Service has been working to develop a fisheries management plan for the US EEZ of the Arctic which may spur enhanced monitoring activities of selected species in the future. In the Antarctic, the Commission of the Conservation of Antarctic Marine Living Resources (CCAMLR) has built a framework for the continual monitoring of stocks and the setting of catch limits, with a particular focus on toothfish (e.g. *Dissostichus eleginoides*). However, the majority of non-fish species remain unmonitored also in the Southern Ocean. With the recent focus on climate change research in the polar oceans and the focus on the poles during the International Polar Year (IPY), the glaring gaps in our knowledge of biodiversity and climate change effects on biota have clearly emerged to the surface.

13.2 Global Efforts of Biotic Inventories: The Examples of the Census of Marine Life (CoML), Arctic Ocean Diversity (ArcOD) and the Census of Antarctic Marine Life (CAML)

Most data collections on biotic inventories are not as easily automated or remotely organized as physical measurements are. This is for instance because of access problems, the need for mostly costly and time-consuming manual species identification, animal movements etc. Challenges of access to the targeted biota, and the high costs related to the collections resulted in much of the biodiversity data collected to date originating from relatively short, funding cycle-driven individual projects with widely varying methods, investigators and target species. Data collections often lack compatibility and international coordination, in the Arctic amplified by a long-standing and unfortunate lack of communication between western and Russian scientists. Few notable exceptions to short projects on a global scale include, for example, the time series of phyto- and zooplankton off the North Sea island of Heligoland (e.g. Hickel et al. 1993), the Southern Ocean Continuous Phytoplankton Recorder series (Hosie 2005) since 1990, and 20 years of ship-based seabird observations in the eastern hemisphere of Antarctic waters (Woehler et al. 2003, 2006). For decades, lack of interest in, and funding for, geo-referenced taxonomic work and monitoring of biological systems with, instead a strong focus on ecological mechanisms and processes resulted in an obvious gap in our knowledge of 'what is there' for a given habitat (Yarincik and O'Dor 2005). This shortcoming has been recognized by the Rio Convention and its inventors (e.g. Strong 2001). In the marine realm, the realization of this gap led a team of visionary scientists to organize the Census of Marine Life, a global network of about 2,000 researchers in more than 80 nations engaged in a 10 year scientific initiative (years 2000–2010) to assess and explain the distribution, diversity and abundance of life in the world oceans (O'Dor and Gallardo 2005). Within this ambitious effort, the authors are involved in two of the 14 field projects, the Arctic Ocean Diversity project (www.arcodiv.org; Bluhm et al. 2008) and the Antarctic Census of Marine Life (www.caml.org; CAML 2005), reporting on efforts to overcome problems outlined.

ArcOD is a 6 year international collaborative effort to inventory biodiversity in the Arctic Ocean's three realms, sea ice, water column and sea floor, from the shallow shelves to the deep basins. CAML is a 5 year census program investigating the distribution and abundance of Antarctica's vast marine biodiversity to develop a benchmark and assess what prioritized taxa can tell us about the known and as yet unknown aspects of marine biodiversity and ocean change. Within these and other CoML projects, historic data are compiled and new data collected focusing on regional, taxonomic and temporal gaps. Examples of such activities in ArcOD and CAML include a collaborative digitization of the vast high latitude collections of the Zoological Institute, Russian Academy of Sciences, in St. Petersburg Russia, new comprehensive collections of under-sampled areas such as the Canada Basin (e.g. Gradinger and Bluhm 2005) and the area under the recent Larsen shelf break-up, beginning of a time series in the Chukchi Sea (www.arctic.noaa.gov/aro/russian-american/cruise2-objectives.htm), descriptions of new species (e.g. Rogacheva 2007; Gagaev 2008; Piraino et al. 2008) and predictive modeling to overcome data gaps (e.g. Rutzen 2008), and to learn more about the ecosystem as a whole and inform management and conservation efforts. ArcOD and CAML were selected as lead projects for the Arctic and Antarctic Marine Biodiversity clusters, respectively, of the IPY (www.ipy.org). Clusters were identified around topical themes by the IPY Joint Committee based on the initial submissions of over 1,000 letters of intent (http://classic.ipy.org/development/eoi/). As cluster lead projects, ArcOD and CAML offer to serve as archives for data collected by cluster projects, and all released data are made publicly available via a website.

13.3 Biodiversity Data Bases and Database Set-Up and Design: Ocean Biogeographic Information System (OBIS), Global Biodiversity Information Facility (GBIF), Scientific Committee on Antarctic Research and Marine Biodiversity Information Network (SCAR-MarBIN)

Biodiversity data bases generally deal with species information and increasingly also with their environment. The concept of global biodiversity data sharing goes back to the visionary demands brought forward by the Antarctic Treaty in 1956 (http://en. wikipedia.org/wiki/Antarctic_Treaty_System) and the Rio Convention in 1992 (http://www.cbd.int/). It was the Antarctic Treaty and its progressive information sharing vision for peaceful purposes (see JCADM – Joint Committee on Antarctic Data Management for digital data policies http://www.jcadm.scar.org/) that set the stage of the Rio Convention. After the initiation of the internet, public data delivery became technically possible through the development of Species Analyst (http://xml.coverpages.org/speciesAnalyst.html). Developed by D. Viglais and colleagues, it presented the first foundation for a federated data structure allowing to exchange and access data in-time and internationally. Thousands of data bases with millions of records worldwide and focusing on a particular taxon have been set up as centralized databases in the last decades, for example, in the marine world, FishBase (Froese and Pauly 2000), AlgaeBase (Nic Donnchan and Guiry 2002), CephBase (Wood et al. 2000), NeMys (Deprez et al. 2004) and Obis-Seamap (http://seamap.env.duke.edu/). In recent years, integrated networked databases have been established that use common schemas and protocols (see below) and, thereby, are able to harvest data hosted by data providers (Costello and Vanden Berghe 2006). Examples of such web portals include the database of the Census of Marine Life, the Ocean Biogeographic Information System (Zhang and Grassle 2003; www.iobis.org) and the Global Biodiversity Information Facility (Edwards et al. 2000, www.gbif.org). A commonly used access protocol is the Distributed Generic Information Retrieval (DiGIR; Zhang and Grassle 2003) in conjunction with the Darwin Core schema, which makes all the data sharing features technically possible (see Fig. 13.1).

The ArcOD data base for Arctic marine biodiversity data and SCAR-MarBIN, the data portal for Antarctic biodiversity marine data, have evolved within different administrative cultures, contexts and funding schemes. SCAR-MarBIN was established in 2005, under the auspices of the Scientific Committee on Antarctic Research (SCAR) and supported by the Belgian Science Policy office willing to spearhead these efforts. It utilizes the same well-established data exchange

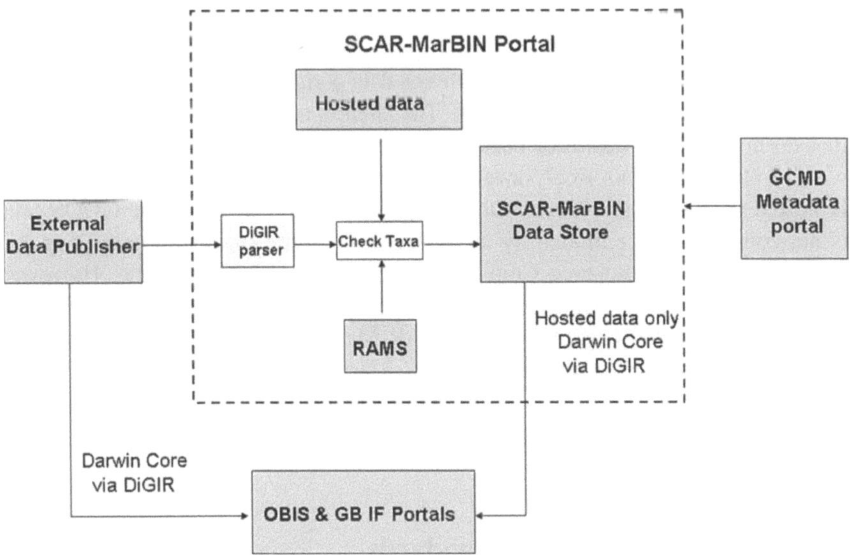

Fig. 13.1 SCAR-MarBIN workflow. Acronyms explained in text. The external data provider directly informs SCAR-MarBIN of any relevant taxa occurrence data within the SCAR-MarBIN region of interest. After harvesting these new records, RAMS provides a taxa validation and filter service for this data and any hosted data. Hosted data is then published to OBIS and GBIF portals under the assurance it is unique and a new dataset. There are links from each dataset to the metadata record within GCMD

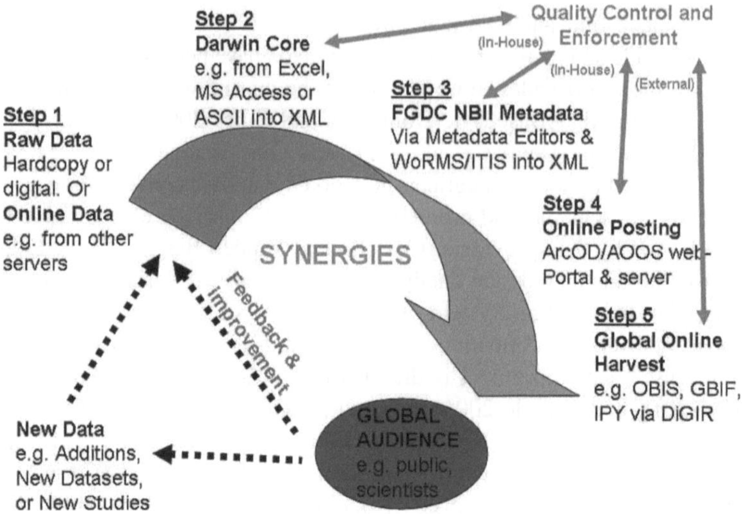

Fig. 13.2 ArcOD data base workflow. Acronyms explained in text. Arctic species occurrence data is delivered to ArcOD in electronic format or hardcopy and transformed into Darwin Core format. Species/taxon names are checked against WoRMS and adjusted if necessary. FGDC NBII metadata are produced in collaboration with the data provider if available and based on publications associated with the data files. Posting online is through the AOOS and ArcOD web portals where data get harvested by OBIS and GBIF

protocols used by OBIS and GBIF, both as a data portal reading other data providers and as a data provider in its own right. The latter is a service for those institutions that do not have the desire or capacity to become active data providers and still wish to make their data freely available. The ArcOD database in its current form is hosted by the Alaska Ocean Observing System (AOOS), part of GOOS, but is not connected to or funded by an overarching Arctic organization such as the International Arctic Science Committee and is, hence, smaller scale. However, their goal is virtually identical, and data are to be fully compatible. Workflows presented for ArcOD (Fig. 13.1) and SCAR-MarBIN (Fig. 13.2) need to be seen and assessed in this light. Currently, SCAR-MarBIN includes close to one million records, the ArcOD database is approaching 100,000 with more in preparation for both systems.

13.4 Data and Metadata Standards

With the advent of large on-line databases including those that get used internationally, global standards facilitating interoperability and smooth data exchange, uniform data set descriptions and common nomenclature have become even more necessary when compared to pre-internet times. ArcOD and SCAR-MarBIN use the Darwin Core schema, a data exchange standard intended to support the retrieval and integration of primary data that documents the occurrence of organisms in space and time and the occurrence of organisms in biological collections (TDWG 2007; http://www.tdwg./org/activities/darwincore/). Elements of the Darwin Core include, for example, record level elements such as a global unique catalogue identifier, taxonomic elements such as a species name and the associated taxonomic hierarchy, locality elements, collecting event entries such as the sampling gear, and biological elements such as sex and reference elements. As a minimum, a taxon name with latitude and longitude of the specimen collected at a given date is needed.

Every data set needs to be associated with metadata describing the data, thereby preserving the usefulness of the data over time. One can refer to metadata as book keeping of scientific content needed each time a data point is collected and assessed. ArcOD uses the Federal Geographic Data Committee (FGDC) standard, specifically the biological profile by the National Biological Information Infrastructure (www.nbii.gov/portal/server.pt; see also Braun 2005 for wildlife in general). This includes over 400 fields such as for example, a data set title and abstract, geographic location, point of contact, taxonomic information, references, geographic projection information etc. CAML, along with all other Antarctic activities, uses the Antarctic Master Directory (AMD) hosted by the Global Change Master Directory (GCMD) facility, which uses the Directory Interchange Format (DIF). This discovery metadata format is designed to store all types of data metadata and therefore, only has very generalized taxonomic group elements. Great detail in the metadata is of crucial importance for appropriate use and interpretation of the data by multiple user groups including management agencies and stakeholders for reaching sustainability decisions. Many governmental institutions and funding agencies today have made detailed metadata a part of their reporting requirements (see also Data Quality Act in the USA). It is part of 'best professional practices'. As an example, the Australian Antarctic Data Centre has developed a specific compliance framework within the Australian Antarctic science program to ensure all possible data is captured, described and archived for re-use (Finney 2008). Metadata records are reviewed and updated and there are appropriate linkages to projects, publications, datasets or databases. This framework also provides methods and feedback to data providers to ensure the supplied data can be accessed by contemporary software, it can be mapped to the real world via either place names or coordinates and all abbreviations and terminology used are clarified.

Taxonomy as an active research field is a discipline constantly in flux where new research from morphological and molecular analyses and disagreement result in synonyms and different names used for the same species. This problem necessitates the use of a common naming reference. One such reference is the World Register of Marine Species (WoRMS), which aims at providing an authoritative list of names of all marine species worldwide including synonyms (SMEBD 2008). Content is managed and advised by taxonomic experts each of whom is responsible for quality control of a particular taxon. The interface at www.marinespecies.org provides access to an underlying database called 'Aphia', which is continuously being updated. Beyond its convenient online use and strict maintenance, WoRMS is fully compatible to another extensive reference system, the Integrated Taxonomic Information System (ITIS; www.itis.org), provides Taxonomic Serial Numbers (TSN) and taxonomic hierarchy, and links with Genbank. SCAR-MarBIN is built around a subset of WoRMS called the Register of Antarctic Marine Species (RAMS). Only those data associated with species validated in RAMS are subsequently exposed via mapping or download services, and thereby improving the data quality of the SCAR-MarBIN portal. ArcOD contributes to, and uses WoRMS whenever possible to update species names in older publications and standardize species names in newer ones before posted data sets.

13.5 Web-Based Data Delivery

The ArcOD data base and SCAR-MarBIN are inherently online sources. They deliver their data, metadata and information in a digital format free of charge to a global audience. Both portals and their data are available through OBIS and GBIF ('one stop shopping'). These data can be globally harvested by everybody satisfying the technical requirements. This is one attempt to reduce the digital divide (Stiglitz 2006) and provide information for all citizens of the world. Currently, it is planned to track downloads and data set uses to assess performance and success of our data sets offered to the global community. In addition, we are trying to add more functionality to the web delivery such as for instance links with data mining tools, OpenModeler (http://openmodeller. sourceforge.net/), GeophyloBuilder (https://www.nescent.org/wg_ EvoViz/GeoPhyloBuilder) and other websites and services.

13.6 Outlook

Geography and remoteness apart, previously fragmented polar research communities were based on the cold-war, and national and continental (digital) divides that are now in the process of being (partially) overcome with a new focus on open access data deliveries and opportunities. The move towards digital data, and a change in data sharing culture, administration, business models and the increased focus on polar regions during IPY and its digital data (IPY Data Information Service http://ipydis.org/) have brought the polar communities together and initiated global communications and information exchanges that were impossible in earlier times. It is hoped that the data, experience and culture, provided by ArcOD and CAML will contribute to wise decision-making in terms of resource use and conservation. We hope that these initiatives will trigger the release of many more datasets and synergies, and present a global role model for data delivery and public information on the polar regions that allows for a fair data exchange and could support sustainable resource use. As a next step, it is relevant that the global public and decision-makers or advisors make the best possible use of the provided information, and implement a new peaceful and sustainable culture that fully uses these substantial opportunities provided.

Acknowledgments This publication is a contribution to the Census of Antarctic Marine Life and the Arctic Ocean Diversity Census of Marine Life projects. We thank all of our team members and contributors world-wide.

References

Bluhm BA, Gradinger R, Hüttmann F, Hopcroft RR, Gebruk A, Sirenko BI (2008) Arctic Ocean Diversity – A Census of Marine Life project in the Arctic. BIUZ (in German) 38:40–49

Braun CE (2005) Techniques for Wildlife Investigations and Management. The Wildlife Society (TWS), Bethesda, MD

Brodeur RD, Sugisaki H, Hunt GL Jr (2002) Increases in jellyfish in the Bering Sea: implications for the ecosystem. Mar Ecol Prog Ser 233:89–103

CAML (2005) The Census of Antarctic Marine Life (CAML). http://www.caml.aq/education-outreach/documents/20061114_CAMLSciStatement_000.pdf

Clarke A (1998) Temperature and energetics: an introduction to cold ocean physiology. In: Pörtner H-O, Playle RC (eds.) Cold ocean physiology. Soc. Exp. Biol. Seminar Series. Cambridge University Press, Cambridge, pp. 3–30

Comiso JC, Parkinson CL, Gersten R, Stock L (2008) Accelerated decline in the Arctic sea ice cover. Geophys Res Lett 35: L01703, doi:01710.01029/02007GL031972

Costanza R, Andrade F, Antunes P, van den Belt M, Boersma D, Boesch DF, Catarino F, Hanna S, Limburg K, Low B, Molitor M, Pereira JG, Rayner S, Santos R, Wilson J, Young M (1998) Principles for sustainable governance of the oceans. Science 281:198–199

Costello, M.J. and E. Vanden Berghe (2006) Ocean Biodiversity Informatics: enabling a new era in marine biology research and management. Mar Ecol Prog Ser 316:203–214

Deprez T, Vanden Berghe E, Vincx M (2004) NeMys: a multidisciplinary biological information system. In: Vanden Berghe E, Brown M, Costello MJ, Heip C, Pissierssens P (eds), Proceedings of ‘The Colour of Ocean Data’ Symposium, Brussels 25–27 November 2002, IOC Workshop Report 188 (UNESCO), Paris), [and VLIZ Special Publication 16], pp. 57–63

Edwards JL, Lane MA, Nielsen ES (2000) Interoperability of biodiversity databases: biodiversity information on every desktop. Science 289:2312–2314

Esanu JM, Uhlir PF (eds.) (2004) Open Access and the Public Domain in Digital Data and Information for Science: Proceedings of an International Symposium. U.S. National Committee for CODATA, National Research Council.

Finney KT (2008) Australian Antarctic Program (AAP) Data Policy. Accessed at http://data.aad.gov.au/aadc/about/ on 3-Feb-2009.

Froese R, Pauly D (eds) (2000) FishBase 2000: concepts, design and data sources. ICLARM Contribution, 1594. ICLARM: Los Baños, Laguna, Philippines. XVII, 344 pp.

Gagaev SY (2008) *Sigambra healyae* sp. n., a new species of polychaete (Polychaeta: Pilargidae) from the Canadian Basin of the Arctic Ocean. Russian J Mar Biol 34:73–75

Gradinger R, Bluhm B (2005) Arctic Ocean Exploration 2002. Polar Biol (Special Issue) 28:169–170

Graham CH, Ferrier S, Huettmann F, Moritz C, Peterson AT (2004). New developments in museum-based informatics and applications in biodiversity analysis. Trends Ecol Evol 19: 497–503.

Grebmeier JM, Overland JE, Moore SE, Farley EV, Carmack EC, Cooper LW, Frey KE, Helle JH, McLaughlin FA, McNutt SL (2006) A major ecosystem shift in the northern Bering Sea. Science 311:1461–1464

Hamre J (1994) Biodiversity and exploitation of the main fish stocks in the Norwegian - Barents Sea ecosystem. Biodivers Conserv 3: 473–492
Hickel W, Mangelsdort P, Berg J (1993) The human impact in the German Bight: Eutrophication during three decades (1962–1991)) Helgoland Mar Res 47:243–263
Hosie G (2005) Southern Ocean CPR Survey. Sir Alistair Hardy Foundation for Ocean Science, Annual Report 2004, pp. 16–17
Huettmann F (2005) Databases and science-based management in the context of wildlife and habitat: towards a certified ISO standard for objective decision-making for the global community by using the internet. J Wildlife Manage 69: 466–472
Hunt GL, Stabeno PJ (2002) Climate change and the control of energy flow in the southeastern Bering Sea. Prog Oceanogr 55:5–22
Manabe S, Stouffer RJ (1995) Simulation of abrupt climate change induced by freshwater input into the North Atlantic Ocean. Nature 378:165–167
Mooney HA, Hobbs RJ (eds) (2000) Invasive species in a changing world. Island, Washington, DC
Nic Donnacha E, Guiry MD (2002) AlgaeBase: documenting seaweed biodiversity in Ireland and the world. Biol Envir Proc R Ir Acad 102B:185–188
O'Dor R, Gallardo VA (2005) How to census marine life: ocean realm field projects. Sci Mar 69:181–199
Pauly D, Christensen V, Dalsgaard J, Froese R, Torres F Jr (1998) Fishing down marine food webs. Science 279:860–863
Pauly D, Alder J, Bennett E, Christensen V, Tyedmers P, Watson R (2003) The future for fisheries. Science 302:1359–1361
Piraino S, Bluhm BA, Gradinger R, Boero F (2008) *Sympagohydra tuuli* gen. nov. et sp. nov. (Cnidaria: Hydrozoa) a cool hydroid from the Arctic sea ice. J Mar Biol Ass UK 88: 1637–1641
Pörtner H-O, Playle RC (eds.) (1998) Cold ocean physiology. Soc Exp Biol Seminar Series. Cambridge University Press, Cambridge
Rintoul, S (2008) Antarctic route highlights new ocean-climate links – the SURVOSTRAL project, CSIRO Media release 08/24. http://www.csiro.au/news/Ocean-ClimateLinks.html
Rogacheva AV (2007) Revision of the Arctic group of species of the family Elpidiidae (Elasipodida, Holothuroidea). Mar Biol Res 3: 367–396
Rutzen I (2008) Predictive GIS-modelling of arctic mesozooplankton distribution based on available online data for four copepod species (*Calanus glacialis, Calanus hyperboreus, Metridia longa* and *Metridia pacifica*). Unpublished M.Sc. Thesis, University of Hannover, Germany
Schumacher J D, Bond NA, Brodeur RD, Livingston PA, Napp JM, Stabeno PJ (2003) Climate change in the southeastern Bering Sea and some consequences for biota. In: Hempel G, Sherman K (eds.) Large marine ecosystems of the world. Elsevier, Amsterdam, pp 17–40
Serreze MC, Holland MM, Stroeve J (2007) Perspectives on the Arctic's shrinking sea-ice cover. Science 315:1533–1536
Shimada K, McLaughlin F, Carmack E, Proshutinsky A, Nishino S, Itoh M (2004) Penetration of the 1990s warm temperature anomaly of Atlantic Water in the Canada Basin. Geophys Res Lett: 31 L20301, doi:10.1029/2004GL020860.
Stroeve JC, Holland MM, Meier W, Scambos T, Serreze M (2007) Arctic sea ice decline: Faster than forecast. Geophysl Res Lett 34, L09501, DOI 10.1029/2007GL029703
SMEBD (2008). World Register of Marine Species. Accessed at http://www.marinespecies.org on
Stiglitz J (2006). Making Globalization work: The next steps to global justice. Penguin Books, New York
Strong M (2001). Where on Earth are we going? Texere, New York
TDWG (2007) Taxanomic Database Working Group Darwin Core wiki site. http://wiki.tdwg.org/twiki/bin/view/DarwinCore/WebHome
Thiel H, Pörtner HO, Arntz WE (1996) Marine life at low temperatures - a comparison of polar and deep-sea characteristics. In: Uiblein F, Ott J, Stachowitsch M (eds.) Deep-sea

and extreme shallow-water habitats: affinities and adaptations. Biosystematics and Ecology Series11:183–219
Thomson PG, Davidson AT, Cadman N (2008) Temporal changes in effects of ambient UV radiation on natural communities of Antarctic marine protists. Aquat Microb Ecol 52:131–147
Wilson, EO (1999) The Living Planet in Crisis. 1st edn. Columbia University Press
Woehler EJ, Raymond B, Watts DJ (2003) Decadal-scale seabird assemblages in Prydz Bay, East Antarctica. Mar Ecol Prog Ser 251:299–310
Woehler EJ, Raymond B, Watts DJ (2006) Convergence or divergence: where do short-tailed shearwaters forage in the Southern Ocean? Mar Ecol Prog Ser 324:261–270
Wood JW, Day CL, Lee P, O'Dor RK (2000) CephBase: testing ideas for cephalopod and other species-level databases. Oceanography 13: 14–20.
Worm B, Sandow M, Oschlies A, Lotze HK, Myers RA (2005) Global patterns of predator diversity in the open oceans. Science 309:1365–1369
Worm B, Lotze HK, Myers RA (2006) Ecosystem effects of fishing and whaling in the North Pacific and Atlantic Ocean. In: Estes JA, Brownell RL, DeMaster DP, Doak DF, Williams TM (eds) Whales, whaling and ocean ecosystems. University of California Press, Berkley, CA, pp. 333–341
Yarincik K, O'Dor R (2005) The Census of Marine Life: goals, scope and strategy. Sci Mar 69 (suppl 1) 2005:201–208
Zhang Y, Grassle JF (2003) A portal for the Ocean Biogeographic Information System. Oceanol Acta 25:193–197.

Chapter 14
Components of Spatial Information Management in Wildlife Ecology: Software for Statistical and Modeling Analysis

Hawthorne L. Beyer, Jeff Jenness, and Samuel A. Cushman

14.1 Introduction

Spatial information systems (SIS) is a term that describes a wide diversity of concepts, techniques, and technologies related to the capture, management, display and analysis of spatial information. It encompasses technologies such as geographic information systems (GIS), global positioning systems (GPS), remote sensing, and relational database management systems (RDBMS), which were once relatively independent but increasingly have been merging into cohesive, integrated systems.

Technological developments are facilitating the collection of unprecedented volumes of spatially referenced ecological data. The historical trend towards improved accuracy, resolution, precision and frequency of spatial data is likely to continue in the future. However, what have been lacking are the advances in statistical methods and analytical software that are needed to facilitate our use of this information to answer ecological questions in a rigorous manner. In Chapter 1 we presented a conceptual model for the interdependence between data, method and ecological theory. The collection of spatial data has exceeded the pace by which statistical methods and analytical software has advanced.

The task of collecting and using spatial data presents us with three general problems: how do we manage this data, how do we visualize, inspect, and process this data

S.A. Cushman (✉)
US Forest Service, Rocky Mountain Research Station, 800 E Beckwith, Missoula, MT 59801, USA
e-mail: scushman@fs.fed.us

H.L. Beyer
Faculty of Biomedical & Life Sciences, Ecology & Evolutionary Biology, Graham Kerr Building, University of Glasgow, Glasgow G12 8QQ, Scotland, UK
e-mail: hawthorne@spatialecology.com

J. Jenness
Jenness Enterprises, 3020 N. Schevene Blvd. Flagstaff, Arizona, 86004, USA
e-mail: jeffj@jennessent.com

S.A. Cushman and F. Huettmann (eds.), *Spatial Complexity, Informatics, and Wildlife Conservation*
DOI 10.1007/978-4-431-87771-4_14, © Springer 2010

efficiently, and how do we use this data to inform the decision making process and improve our understanding of ecological systems? We have three goals for this chapter. First, we will describe a range of existing kinds of software for spatial ecological analysis, and review their functionality. Second, we will discuss a number of important issues facing researchers who wish to develop their own custom software tools, and provide some general suggestions to assist their efforts. Third, we look forward to discuss the future of analytical computing, and identify key technological developments that are driving current and future progress in spatial analytical software development.

14.2 Data Storage and Management

There are several important motivations for adopting formal data management strategies. Given the increasing rate of collection of spatially referenced data it becomes imperative to automate the storage and organization of that data, enforce quality control and consistency standards, and provide a method of disseminating data to stakeholders. Also, the large volumes of data typical of spatial datasets require systems that can efficiently query and process the data, ideally by multiple concurrent users. Relational database management systems (RDMBS) are key data administration devices as they are designed to handle very large amounts of data (terabytes), efficiently process queries on huge databases, facilitate the sharing and distribution of data over networks, provide user access control and other security measures, permit simultaneous querying and editing of data by multiple users (including the ability to resolve editing conflicts), and drive network based data dissemination services (web portals and data servers). Most statistical and spatial analysis software is able to access data directly from a RDBMS.

The expense of setting up and maintaining a RDBMS is offset by the long term indirect cost savings of having a centralized, consistent, updated, permanent data warehouse that requires relatively minimal human management but that can drive a wide range of services and uses. Database management systems also help to ensure that data is appropriately archived so that it can be accessed in the future (possibly by the broader scientific community). Permanent archiving of data is something that is often neglected in the field of ecology and wildlife management. In Chapter 6 we presented a conceptual model for flexible, multi-variate, multi-scale, spatial databases to serve as the foundation for ecological monitoring and spatial and temporal analysis of ecological pattern-process relationships across scale. The construction of these databases will require the use of RDMBS within an explicitly spatial template.

Examples of RDBMS software include PostGRESQL, SQL Server, and Oracle (see Table 14.1 for details).

Table 14.1 Examples of spatial information system, statistical, modeling and spatial analysis software listed alphabetically within each section (this list is not comprehensive)

General purpose statistical and modeling software	
R[a]	http://www.r-project.org
Mathmatica	http://www.wolfram.com
MATLab	http://www.mathworks.com
WinBUGS[a]	http://www.mrc-bsu.cam.ac.uk/bugs
Software specializing in quantitative spatial analysis	
CrimeStat[a]	http://www.icpsr.umich.edu/CRIMESTAT
FRAGSTATS[a]	http://www.umass.edu/landeco/research/fragstats/fragstats.html
GUESS[a]	http://graphexploration.cond.org
PASSAGE[a]	http://www.passagesoftware.net
General purpose Geographic Information System (GIS) software	
ArcGIS	http://www.esri.com
ArcReader[a]	http://www.esri.com
GeoDa	https://www.geoda.uiuc.edu
GeoMedia	http://www.intergraph.com
Google Earth[a]	http://earth.google.com
GRASS[a]	http://grass.itc.it
IDRISI	http://www.clarklabs.org
Landserf[a]	http://www.landserf.org
Manifold	http://www.manifold.net
MapInfo	http://www.mapinfo.com
Quantum GIS[a]	http://www.qgis.org
RAMAS	http://www.ramas.com
SAGA[a]	http://www.saga-gis.org
TNTMips	http://www.microimages.com
uDIG[a]	http://udig.refractions.net
Spatial analysis software specifically designed for wildlife ecology	
ade4[a], adehabitat[a]	http://pbil.univ-lyon1.fr/ADE-4/home.php?lang=eng
CDPOP	http://LandguthResearch.dbs.umt.edu
Corridor Design[a]	http://www.corridordesign.org
Geospatial Modelling Environment[a]	http://www.spatialecology.com
HRT: Home Range Tools For ArcGIS[a]	http://blue.lakeheadu.ca/hre
Marine Geospatial Ecology Tools[a]	http://code.env.duke.edu/projects/mget
Tools for Graphics and Shapes[a]	http://www.jennessent.com/
Relational Database Management System (RDBMS) software	
Oracle	http://www.oracle.com
PostGRESQL[a]	http://www.postgresql.org
SQL Server	http://www.microsoft.com
Remote sensing software	
eCognition	http://www.definiens.com
ENVI	http://www.ittvis.com/envi/index.asp
ER Mapper	http://www.ermapper.com

(continued)

Table 14.1 (continued)

ERDAS Imagine	http://www.erdas.com
SPRING[a]	http://www.dpi.inpe.br/spring/english
PCI	http://www.pci.on.ca
Simulation software	
AnyLogic	http://www.xjtek.com/
CelLab[a]	http://www.fourmilab.ch/cellab
MASON	http://cs.gmu.edu/~eclab/projects/mason
NetLogo[a]	http://ccl.northwestern.edu/netlogo
SWARM[a]	http://www.swarm.org

[a]Open source or free software

14.3 Mapping and Visualization

A wide range of commercial and free software solutions exist for displaying and exploring spatial data. The principal features this software usually offers include: (1) on-the-fly reprojection of spatial data so that datasets with different projections are appropriately aligned without having to manually reproject the dataset, (2) the ability to access datasets stored on remote spatial data servers, (3) the ability to perform tabular and spatial queries on the data (e.g. select all features with a specified attribute value, or select all features from one dataset that overlap the features from another dataset), (4) the ability to directly read or import a wide variety of data formats, and (5) options for graphical display and map making. Although we do not offer a comparative review of mapping and visualization software here (because we focus on quantitative spatial analysis), many examples are provided in Table 14.1 (see the General Purpose GIS Software section).

14.4 Quantitative Spatial Analysis

Spatial analysis is a nebulous term that covers a wide range of statistical and numerical techniques. It ranges from incorporating spatial attributes (e.g. coordinates or distance measures) into standard statistical and modeling analyses, which handle the spatial data in a spatially implicit manner, to using techniques that have been specifically designed for application to spatial data, such as geostatistics (Cressie 1993; Diggle and Ribeiro 2007), pattern analysis, graph theory and network analysis, spatial randomization tests, and spatial stochastic simulation (Dale 1999; Fortin and Dale 2005). A wide diversity of software has been developed to facilitate these types of analysis. We briefly describe each of these categories of analysis and provide examples of software that can be used to implement them.

14.4.1 Geostatistics

Geostatistics involves modeling continuous spatial phenomena based on a set of spatially referenced samples. There are a variety of geostatistical analytical approaches; however, they all share a basic focus on predicting a quantitative response variable across space based on the pattern and autocorrelation of that variable among spatially referenced sample locations. Typically the sample data are point samples arranged in space using a regular, random, or stratified random design (e.g. point locations at which elevation is measured) and the goal of the modeling process is often spatial prediction. *Model-based geostatistics* (Diggle and Ribeiro 2007) is the term that describes the general, formal statistical framework of which commonly used techniques like simple kriging are one example.

Geostatistics often involves simple data structures: coordinates describing the spatial location of point samples, and sometimes also a boundary (polygon) that describes the area of interest. Predictions resulting from geostatistical models are often represented as raster data structures.

Examples of software that implement geostatistics include the geoR and geoRglm packages in R, GSTAT, and Geostatistical Analyst (ESRI ArcGIS) (see Table 14.1 for details).

14.4.2 Pattern Analysis

Spatial pattern analysis is a method of quantitatively describing heterogeneously distributed (patchy) phenomena. In studying pattern, we hope to learn something about the processes that have generated these patterns, i.e. to help us to develop a mechanistic understanding of ecological processes. Pattern analysis covers a broad range of techniques (see Dale 1999; Fortin and Dale 2005) and employs a variety of data structures, for instance: points (nearest neighbour analysis, point pattern analysis), quadrats and transects (quadrat variance analysis), rasters (spectral analysis and wavelets, boundary detection, landscape metrics), and polygons (polygon change analysis).

Examples of relevant software include several R packages (e.g. spatial, spatstat, splancs, waveslim, wavelets), PASSaGE, and FRAGSTATS (see Table 14.1 for details).

14.4.3 Randomization Tests

Statistical significance is quantified using a theoretically justifiable reference distribution. In the case of spatial patterns and processes we often have no preconceived notion about what a suitable reference distribution would be. Randomization techniques (bootstrapping, Monte Carlo simulation) allow us to generate appropriate reference distributions that can be used to quantify significance, and can be applied to many different sorts of problems. This approach forms the basis of techniques like Mantel's test.

Examples of relevant software include PASSAGE, R, and the Geospatial Modelling Environment (see Table 14.1 for details).

14.4.4 Stochastic Simulation

Like randomization tests, stochastic simulation helps us to quantify how likely our observed data is given a proposed process model. They can also be used to make predictions about the state of a system at a given time based on the process model, which captures the spatial and temporal dynamics of a system and range from mechanistic models to phenomenological models.

Process models tend to be specific to a particular system/question (e.g. animal movement models, landscape fire models) and it is therefore difficult to write software that can accommodate all possible model forms and variations. Most stochastic simulation involves writing simulation programs, either in a 'high-order' statistical language like R, or any of the lower level programming languages (C, C++, Java,.NET, Fortran, Pascal, etc.). However, for certain types of simulations (like cellular automata and individual based models) there is generic software available that may be useful in implementing a process model.

Examples of relevant software include R, MATLab, Mathmatica, SWARM, and CelLab (see Table 14.1 for details).

14.4.5 Graph Theory and Network Analysis

A graph is a collection of nodes connected by edges to form a network. Although graphs do not necessarily represent spatial systems (e.g. the nodes may represent producers and consumers at different trophic levels in an ecosystem and the edges represent the trophic relationships between them), here we focus on the spatial applications (cities connect by a network of roads, or foraging patches connected by a network of trails). One of the premises of network analysis is that it allows us to identify the emergent properties of complex systems that might not otherwise be apparent by direct observation. Furthermore, because many ecological processes are well represented by networks (e.g. disease transmission) they provide a powerful and often more realistic/mechanistic method of modeling the process.

Examples of relevant software include R (igraph package), GUESS, and Network Analyst (ESRI ArcGIS) (see Table 14.1 for details).

14.5 Developing Custom Geospatial Analysis Tools

Implementing a novel analysis often involves software customization. Many software providers provide a framework for adding custom functionality to their software. In the case of open source software everyone has access to the source code so we have complete freedom to customize the application (subject to open source license agreements). But many commercial products also provide powerful interfaces for customization.

Customization not only allows us to add new functionality, it allows us to improve the efficiency of existing functionality. This can be important in the context of stochastic simulations and Monte Carlo applications where a procedure must be run a large number of times, potentially requiring long processing times (weeks or months). It is not uncommon to reduce processing times by an order of magnitude or more following code optimization for a particular analysis, making it more feasible to use.

There are many different development environments that can be used to develop customized geospatial analyses and there are several issues that should be considered when selecting one. One very important consideration is to what extent you can re-use existing code. The ability to call functionality in existing tools (so that you do not have to program them yourself) is the factor that is most likely to reduce development time for an application. Understanding what functionality you require, and where it is available in existing software is a necessary first step.

Another consideration is what level of control and access you have to this existing code. With open source software the potential for code re-use is very high. With commercial environments, you are limited to the objects and methods that the company chooses to expose to you, and are somewhat at that company's mercy if they choose to change that code.

In the case of quantitative spatial analysis one must also decide whether to approach customization from the geospatial or the statistical side of the problem. Many tools exist for manipulating and exporting geospatial data, so implanting customization using statistical and modeling software (e.g. R) may be the most efficient approach.

Familiarity with a programming language is the other factor that most affects development time. If you are new to programming it is not advisable to begin with a 'low-level' programming language (e.g. C, C++, Java, .NET). Geospatial applications tend to be quite complex, involving a wide array of programming techniques and technologies. The higher order scripting languages will are easier to learn and be productive with (e.g. R, Python) because they do not require the same level of technical programming knowledge.

Examples of customized geospatial analysis tools pertaining to wildlife ecology include "CDPOP" (Landguth and Cushman in press), "Corridor Design" (Majka et al. 2007), "HRT: Home Range Extension for ArcGIS" (Rodgers et al. 2007), and the "Geospatial Modelling Environment" (see Table 14.1 for details).

14.5.1 Software Transparency, Documentation and Certification

Reporting a methodology in enough detail that it could be independently repeated is an important aspect of the scientific method. For quantitative analyses it is important to understand exactly what software is doing so that we are able to document our methodology and understand what the implications of a software program or algorithm might be on the results we obtain. However, there is little consensus on what an appropriate level of documentation should be for software. Commercial software, for which the source code is not available and algorithms are often considered proprietary and are therefore sparsely described or not described at all, is particularly problematic; we input data and get new data as output but we have little or no definitive information about how the algorithm works, hence the term "black box" software. In such cases many authors will report the name and version number of the software used, which in some sense would allow a method to be independently repeated if other people had access to the same software. However, we argue that black box software does little to improve our understanding of the systems we study and should be avoided where well documented alternatives exist.

A simple example illustrates this problem. Often an analytical technique can be implemented in a number of different ways that can result in qualitatively different solutions. For instance, bivariate kernel density estimation (KDE) is a commonly used technique in ecology. There are a number of kernels that can be used (Gaussian, quartic, etc.) and in the case of a 2D problem the bandwidth parameter is a 2×2 covariance matrix (Silverman 1986). Some commonly used GIS software packages make simplifying assumptions in the implementation of KDE by using a particular kernel and assuming an equal bandwidth for both dimensions with no covariance, but none of these details are documented. This black box approach to KDE can be misleading as these simplifying assumptions can have a profound influence on the results obtained.

In a discipline in which peer review plays such an important role in advancing understanding it is perhaps surprising that no formal peer review process has been adopted for software. In the case of "open source" software the source code is freely available so can be reviewed for logical consistency. It is therefore completely transparent, well documented (because there is no more precise documentation than the code itself) and open to peer review. In such cases a somewhat informal peer review process often takes place whereby a number of developers will review and improve an algorithm (the statistical software R is a good example of this).

However, it would be impractical to suggest that we should review source code or documentation every time we wish to use software. It would be more efficient to identify cross-disciplinary standards for scientific software and establish a system of peer review. Software certification would help to ensure that software was bug free, that appropriately detailed documentation existed, and that adequate testing had been performed. These standards would apply equally well to open source and commercial software. Although we are moving towards such a system, in the mean time we encourage users to favour software that is well documented and tested.

14.5.2 *The Future of Analytical Computing*

In the past spatial software was designed for large, independent workstations and used proprietary data formats that prohibited convenient software interoperability. Computing technology is now moving towards virtualization (operating systems and programs are not bound to a particular instance of computer hardware), distributed and cloud computing (data processing using remote systems, software distributed via the network or internet), and widely adopted standards that ensure software interoperability. The spatial data, processing and exchange standards being defined by the Open Geospatial Consortium (OGC; http://www.opengeospatial.org), which has broad support across many industries, are particularly important for ensuring interoperability in the future.

The ramifications of seamless software interoperability on spatial analysis are likely to be that we use a wider array of software, developed by a broad community of developers, and often accessed via the internet. Web Processing Services (WPS) are a particularly promising approach to making analyses available via the internet. WPS is a set of standards that facilitates access to an analysis algorithm or model that uses spatial data. It defines the required input data, the output data that is generated, runs the process and returns the results. The provider of the WPS service is able to maintain and improve the service independently of any software that might be installed on the client machine. The client can run the WPS process using data that is stored locally or on a network data server, and processing is performed locally.

The importance of providing transparent, peer-reviewed and high quality quantitative analysis solutions is likely to be greatly facilitated by R (R Development Core Team 2008), which has had a profound impact on the field of statistical computing. Several GIS applications already use R as the statistical and modeling engine that drives spatial analyses.

References

Cressie NAC (1993) Statistics for spatial data. Wiley, New York

Dale MRT (1999) Spatial pattern analysis in plant ecology. Cambridge University Press, Cambridge

Diggle PJ, Ribeiro PJ (2007) Model-based geostatistics. Springer, New York

Fortin M, Dale MRT (2005) Spatial analysis: a guide for ecologists. Cambridge University Press, Cambridge

Landguth EL, Cushman SA (in press) CDPOP: a spatially explicity, individual-based, population genetics model. Mol Ecol Resour. doi:10.1111/j.1755-0998.2009.02719.x

Majka D, Jenness J, Beier P (2007) CorridorDesigner: ArcGIS tools for designing and evaluating corridors. http://corridordesign.org

R Development Core Team (2008) R: a language and environment for statistical computing. R Foundation for Statistical Computing, Vienna, Austria. ISBN 3–900051–07–0, http://www.R-project.org

Rodgers AR, Carr AP, Beyer HL, Smith L, Kie JG (2007) HRT: home range tools for ArcGIS. Version 1.1. Ontario Ministry of Natural Resources, Centre for Northern Forest Ecosystem Research, Thunder Bay, Ontario, Canada

Silverman BW (1986) Density estimation for statistics and data analysis. Chapman & Hall/CRC, London, New York

Chapter 15
Spatial Analysis of Wildlife Distribution and Disease Spread

Marie-Josée Fortin, Mark R.T. Dale, and Stefania Bertazzon

15.1 Introduction

Many of the interactions between organisms depend on the distance or the ease of movement (accessibility) between them which can be based on the concept of the neighbors or of the neighborhoods of given individuals. A number of different statistical approaches have been developed (Fortin and Dale 2005; Perry 1995) to address the definitions of neighbors and neighborhoods in order to implement measures of those characteristics that are most important to the interactions under study. In particular, the numbers of neighbors (however defined) and their distances can be combined into measures of aggregation, dispersion or crowding (Lloyd 1967), which can have clear effects on important demographic processes, such as the spread of disease, beyond the simple effect of distance to the nearest organisms of the same or different kinds.

These aggregation measures (Downing 1991) are not able to differentiate among different intensities and types of spatial patterns. This is why spatial statistics (Haining 2003; Fortin and Dale 2005) are more appropriated to analyze species spatial distribution. Specifically, spatial statistics allow us to determine the spatial distributions of species and their interactions by: (1) mapping their spatial distributions; (2) characterizing and quantifying the spatial patterns; and (3) relating these spatial distributions to underlying environmental conditions (e.g., species–

M-.J. Fortin (✉)
Department of Ecology and Evolutionary Biology, University of Toronto,
25 Harbord Street, Toronto, ON, Canada M5S 3G5
e-mail: mariejosee.fortin@utoronto.ca

M.R.T. Dale
Department of Environmental Science and Engineering, University of Northern British Columbia, Prince George, BC, Canada V2N 4Z9

S. Bertazzon
Department of Geography, University of Calgary, Calgary, AB, Canada T2N 1N4

S.A. Cushman and F. Huettmann (eds.), *Spatial Complexity, Informatics, and Wildlife Conservation*
DOI 10.1007/978-4-431-87771-4_15, © Springer 2010

environment relationship) and to intra- and interspecies interactions (Dormann 2007; Segurado et al. 2006). Mapping plants by species is relatively easy because plants are mostly sessile, yet this task is not feasible for most animals because they are mobile. Thus, to study the spatial distribution of animal species, surrogate measures of space occupancy need to be used, such as: (1) nest and den locations; (2) telemetry/GPS data through time; (3) territory, home range, defoliated area and disease regions; (4) presence/absence or quantitative survey or atlas data; and (5) infected individuals or disease's cluster locations. Each of these surrogate data types can be analyzed using different spatial statistical methods: (1) point pattern methods for location data (Diggle 2003; Ripley 1981); (2) line or "fibre" pattern methods (Ripley 1981) or graph theory (Fall et al. 2007) for telemetry/GPS path data; (3) categorical polygons analysis (McGarigal and Marks 1995) or polygon change analysis (Robertson et al. 2007; Sadahiro and Umemura 2001) for delineated regions; (4) surface pattern methods (Fortin and Dale 2005) for presence/absence or quantitative survey data; and (5) spatio-temporal connectivity for disease data (Real and Biek 2007).

One logical way to proceed in analyzing animal spatial pattern is therefore according to the data type and according to assumptions about the process: (1) whether we have a complete census or just a sample, and (2) whether the animals can be assumed to be "resident" in a territory or to be mobile, motile, wandering, or migratory. If the animals move, the question is whether their full movement paths are known or just intermittent locations on those paths. Here we summarize an array of spatial statistical approaches (Fig. 15.1) that are most useful for characterizing animal spatial occupancy data and disease spread. We emphasize the assumptions on which the application and interpretation of each method are based, rather than their technical aspects; the mathematical details are available elsewhere (among others Dale et al. 2002; Fortin and Dale 2005; Haining 2003; Rangel et al. 2006; Waller and Gotway 2004).

15.2 Spatial Analysis of Animal Census Points

15.2.1 Stationarity

Almost all spatial statistics are based on the assumptions of *stationarity* and *isotropy*. *Stationarity* requires that the parameters of the distributions of interest (e.g., mean and variance) have the same expected values independent of position in the study area. *Isotropy* requires that the characteristics of the spatial pattern do not depend on direction (i.e., they are the same in all directions). If these assumptions are violated, statistical inference and significance tests are no longer valid. Another common assumption is that organisms surveyed can be treated all as equivalent (univariate) or divided into a small number of categories of equivalent entities (bivariate if only two categories and multivariate otherwise).

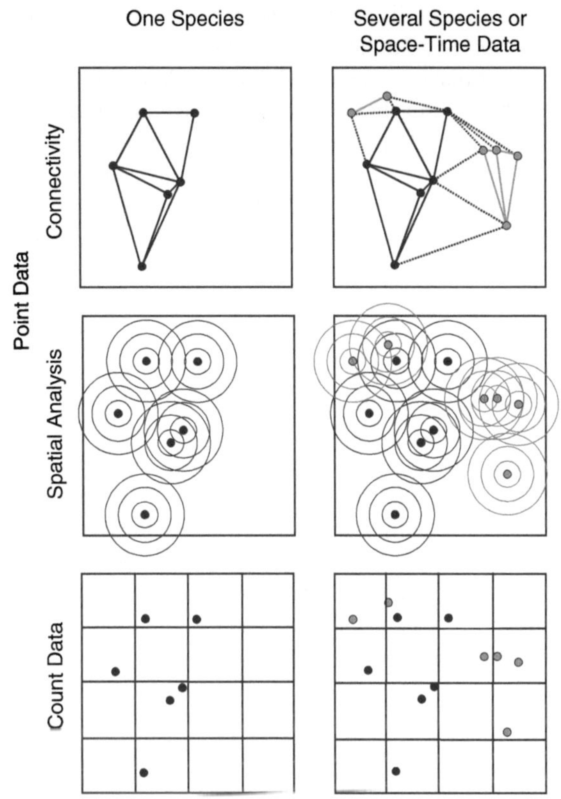

Fig. 15.1 Data types can be point locations of species present, measurements of characteristics of species at point locations, or number of individuals per quadrat. Such univariate (one species) or multivariate data (several species or one species analyzed through time) can be analyzed by determining the connectivity among points, by quantifying the spatial aggregation or the degree of spatial autocorrelation

15.2.2 Point Data: Spatial Aggregation

If the spatial distribution of an animal can be represented by point data such as den or nest locations, point pattern analysis methods can be used. This set of spatial analysis methods includes many of the familiar methods from the plant ecology or geographical literature. These point pattern analyses can be univariate, bivariate or multivariate and as such they can be used to analyze species spatial interactions or disease spread between infected vs. "at risk" populations. For such point pattern analyses, the data are assumed to be a complete census of the positions of zero-dimension events or points in a two dimensional plane.

One obvious starting place for characterizing point patterns is to consider the nearest neighbor of each event, and to base any inferences on the distribution of these event-to-nearest-event distances. In some cases, we will want to compare the distribution of observed event-to-nearest-event distances to the distribution of random event-to-nearest-event distances expected from random points in a plane. Because these nearest neighbor approaches essentially use average distances, they are called first-order methods. Second-order methods compare the observed and expected numbers of events as a function of distance from other events in the pattern. The most commonly used second-order method is Ripley's *K*-function (Ripley 1981), and its transformations (e.g. *L*-function). In these methods, events in circles of radius *t* centered on events are counted, and the average observed count is compared with the expected count for circles of the same radius and randomly distributed points (complete spatial randomness, CSR). Radii for which the observed counts exceed expected indicate scales of clumping in the spatial pattern, and radii for which expected count exceeds observed indicate scales of overdispersion. To visualize these methods, consider expanding circles centered on events: the nearest neighbor distance method asks how big an expanding circle, centered on event *i*, must be before it first encounters another event; the Ripley's *K* approach compares the observed count of events in circles of given size with the expected number.

This second-order approach is easily expanded from the univariate case in which all the events are of the same kind to the bivariate case, in which the events belong to two categories (e.g. juveniles and adults, males and females, or diseased and healthy). If the two categories are labeled A and B, we can count the number of events of type A in circles centered on events of type B and the number of events of type B in circles centered on events of type A, comparing the observed with the expected for both combinations as a function of circle size. For multivariate analysis, with more than two categories, there are a number of possible approaches to analysis. For example the categories A, B, C, D,… can be combined into fewer, such as A and ~A (everything that is not A) for one analysis, then B and ~B, and so on, or just "same" (X and X, whatever X is) vs. "different" (X and ~X, for all X). For more details, see Fortin and Dale (2005, section 2.1.5). Finally, Wiegand and Moloney (2004) modify Ripley's *K* by using a "ring" instead of a circle to avoid the cumulative effects of increasing radii, which may make results easier to interpret.

To test the significance of aggregation statistics, the most basic null hypothesis is that the events are randomly and independently distributed in the plane, according to Complete Spatial Randomness (Diggle 2003; Ripley 1977). Other null hypotheses can be formulated using restricted spatial randomization procedures which better reflect prior knowledge about the behavior of the species and the regionalized properties of the study area (Fortin and Dale 2005; Wiegand and Moloney 2004). Regardless of the null hypothesis used, the point pattern methods described here assume that all points within the study area have been enumerated in a complete census (Fig. 15.2a, c). This assumption is needed because the degree of spatial aggregation is proportional to the number of points in the study area and in the randomization procedure all the points have the same probability of occurrence within the study area. Therefore, if these methods are used with sampled points (Fig. 15.2b, d) the degree of spatial aggregation computed is biased and significance values are wrong (Fig. 15.2c–d).

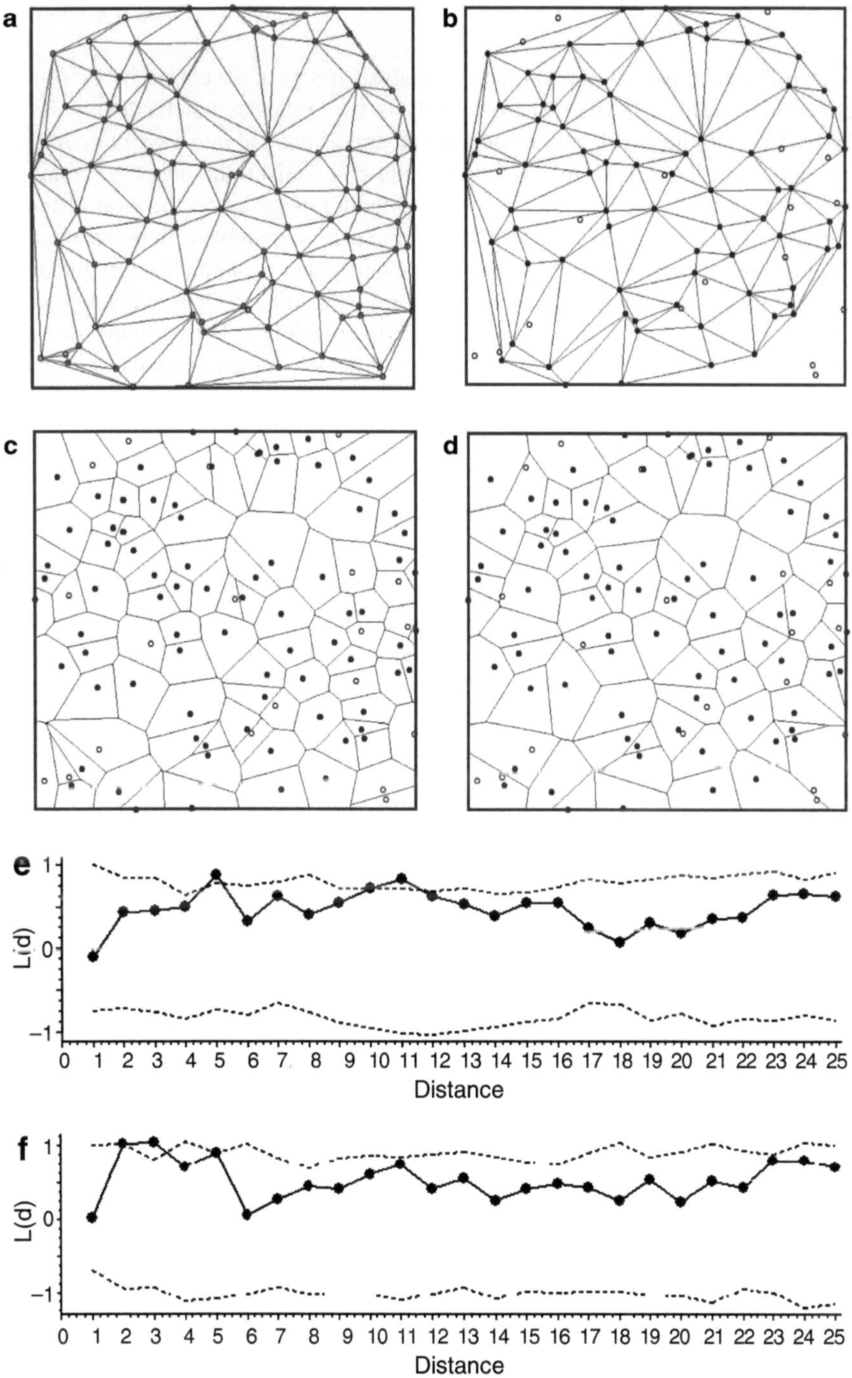

Fig. 15.2 Census (all circles, n = 100 points) versus sampled data (open circles, n = 80 points) effects on Delauney connectivity (**a**, **b**), area occupancy per point based on Voronoi polygons (**c**, **d**) and Ripley's K statistic here plotted using its corresponding L function (solid line with circles) based on 100 points (**e**), and 80 points (**f**) where the dashed lines are the confidence envelop of the null hypothesis of complete spatial randomness at the probability level of 0.05

A different approach to point pattern analysis has been developed by Perry and co-workers, called SADIE (Spatial Analysis by Distance IndicEs; Perry 1995, 1996, 1998). In this method, the background area is divided into a grid of contiguous subunits, like a field divided into sample plots. Given the observed count of events per sample unit, the measure of aggregation is based on the total distance individuals would have to be moved in order to achieve complete "regularity" with the same count in each unit ("distance to regularity") or a complementary measure of the total distance events would have to be moved to have them in a single unit, thus achieving maximal clustering ("distance to crowding"). Perry (1995) provides an algorithm to calculate these distances and a randomization procedure evaluates the significance of the result.

15.2.3 Point Data: Animal Movement Analysis

There are at least two kinds of data sets to be considered as representing a "complete census" of animal positions: (1) the case in which full continuous paths of movement are known (Fig. 15.3a), and (2) the case in which the paths of movement are not known but rather a series of locations (e.g. GPS points) through time. When only a series

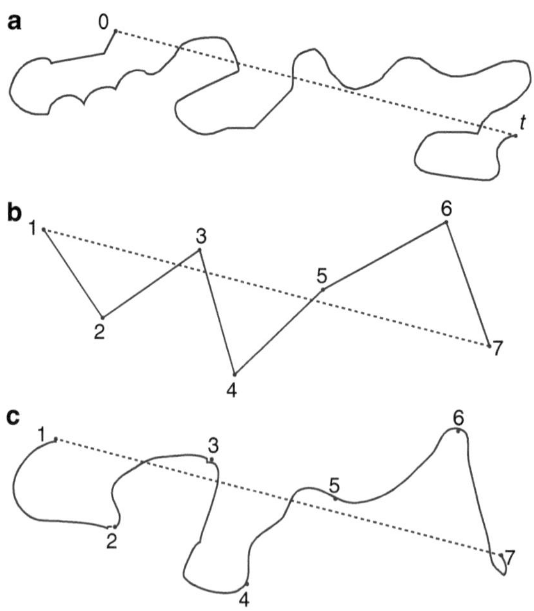

Fig. 15.3 Animal movement methods based on the data type: (**a**) known path (solid line) and displacement (dotted line); (**b**) known positions in order with an assumed path of straight line-segments and displacement; and (**c**) known positions in order with a derived least-cost path and displacement

of locations are available, we might join these known positions as approximations of the unknown paths. Animal movement analysis can then be determined in performing a point pattern analysis of the sequence of points, in which the analysis preserves the order of the locations (Fig. 15.3b), or in performing a connectivity analysis based on Euclidean distances between points, or based on the least-cost distance as weighted by the matrix quality between the points (Fall et al. 2007). Often we will want to evaluate the tortuosity of a path, which is a measure of the path complexity (Laube et al. 2007; Wiens et al. 1993). For a curving path the simplest index might be the ratio of path length to net displacement: the longer the total distance traveled for a given net movement, the more tortuous the path taken. Where the "path" is approximated by straight line segments, as where only the positions are known for given intervals, which are then joined in sequence, tortuosity can also be measured by the ratio of the total apparent path length to the "size" of the complex hull for the entire path, with the recommendation that "size" is the diameter of the convex hull, thus giving a measure of path "compactness" (Fortin and Dale 2005). Tortuosity measures can also be determined using least-cost distance rather than Euclidean distances but their interpretation requires careful thought.

15.2.4 Line Pattern

We have described the analysis of objects in space that are reduced to dimensionless events, often called points. In wildlife ecology, we may encounter spatial structures that are not sensibly reduced to dimensionless points, but are more easily considered as lines, whether straight (e.g., dead woody debris) or curving, of one dimension. Such linear objects (straight or possibly curved, but smooth and differentiable) can be studied using fibre process analysis (Benes and Rataj 2004; Stoyan et al. 1995). A large number of ecologically interesting phenomena can be treated as fibre processes, including invertebrate burrows in sediment, animal migration routes, and foraging search patterns. Characteristics of interest in such systems include: the aggregation or overdispersion of fibres; anisotropy of pattern or the tendency for the fibres to be parallel; spatial autocorrelation of segment lengths; and similarity of shape. The scope of analysis can be expanded beyond fibres to include both dimensionless points and one-dimensional linear structures, with and without an essential structural relation (flowering shoots that emerge from a plant's spreading rhizomes; duck nests in relation to watercourses). Fibre analysis (Stoyan et al. 1995) is a relatively new area of spatial analysis with direct relevance to wildlife ecology, and it is definitely worth pursuing as it develops new techniques and new applications.

15.2.5 Connectivity

Animal movement ability in a landscape is affected by the spatial configuration of land cover types (Taylor et al. 1993). The realized connectivity of a landscape for an animal, given the ability of the animal to move across the landscape, is known as functional connectivity (Taylor et al. 1993; With et al. 1997). Hence, to determine how animal movement is affected by a spatially heterogeneous landscape, functional connectivity can be measured by first creating a "cost" surface based on the matrix quality (composition) and on how the matrix quality affects individual movement (neutral, high energetic cost, avoidance, barrier, etc.). Then the functional connectivity is determined by finding the least-cost path between sampled points obtained by telemetry or GPS (Fig. 15.3c) as the shortest path measured in cost units rather than Euclidian distance (Adriaensen et al. 2003). Analysis of functional connectivity using graph theory is a growing field of research with application to animal movement and reserve design (Fall et al. 2007; O'Brien et al. 2006).

15.3 Polygon Change Analysis

Animal point data gathered either by sampling, telemetry or GPS can be used to determine animal territory, home range or geographical range (i.e., polygon data). Methods to delineate home ranges from point locations are usually based on minimum convex hulls (St-Louis et al. 2004; Fortin et al. 2005) or kernel functions (a weighted averaging method; Gitzen et al. 2006). Once home range polygons are delineated, polygon patterns can be characterized using landscape metrics (McGarigal and Marks 1995). Changes in polygon patterns through time can be analyzed using polygon change analysis (Sadahiro and Umemura 2001), which quantifies (Fig. 15.4) overlap, contraction, retraction, appearance, and disappearance of polygon areas over time. To assess the significant differences among landscape metrics form different landscapes or time periods, a stochastic spatial modelling approach should be used as the one suggested by Fortin et al. (2003).

15.4 Spatial Analysis of Sampled Animal Data

15.4.1 *Spatial Autocorrelation*

Spatial autocorrelation is the correlation of among values of a quantitative variable (e.g. count data) that are separated by some distance from one another (Cliff and Ord 1981; Haining 2003; Fortin and Dale 2005). The intensity of spatial autocorrelation at a given distance, d, is based on the squared difference between pairs of quantitative values that are separated by the given distance. Measures of spatial autocorrelation (Moran's I, Geary's c) are closely related to measures of spatial variance (semivariance) (Fortin and Dale 2005). These methods have been used to study the spatial pattern of a number of species including birds (Betts et al. 2006;

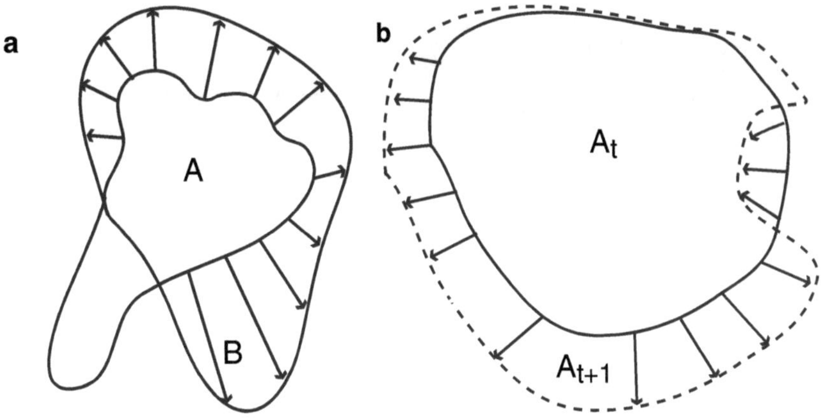

Fig. 15.4 Polygon change analysis to determine the type of spatial overlap and amount of area: (**a**) home ranges of two individuals; and (**b**) home range of the same individual through time

Koenig 1998; Ludovisi et al. 2005; Villard and Maurer 1996). Measures of spatial autocorrelation are typically estimated at a series of increasing distances (lags), and the plot of spatial autocorrelation against lag is referred to as a spatial correlogram (or variogram, for spatial variance). The shape of these plots provides useful information about the intensity of spatial autocorrelation, and the distance at which the spatial autocorrelation is no longer affecting the observed spatial pattern (known as the range). The statistical significance of an autocorrelation coefficient at a given lag can be computed using either an approximation test or a randomization procedure (Fortin and Dale 2005). Because autocorrelation coefficient values at different lags are not independent of one another, to account for lack of independence, a Bonferroni correction or progressive Bonferroni correction should be used (Legendre and Legendre 1998).

Global analyses of spatial autocorrelation assume stationarity (Fig. 15.5). However, many (or most) landscapes are heterogeneous. To address this reality, local spatial autocorrelation coefficients (e.g., local Moran's I, local Getis statistics) have been developed (Anselin 1995; Getis and Ord 1992) to characterize spatial autocorrelation at each sampling location. The advantage of local indicators of spatial association (LISA) is that values can be mapped, and hence can be used to identify subregions where spatial autocorrelation is unusually high (hotspots) or low (coldspots). The identification of such hotspots and coldspots is a way to determine whether or not the entire study area is stationary. The disadvantage of LISA methods is that a coefficient is calculated for each sampling location. Given the very large number of coefficients, the significance of each coefficient cannot be reliably assessed (since Bonferroni correcting for a very large number of tests ensures that nothing remains significant). That said, LISA methods are useful as indicator of local spatial pattern in the data.

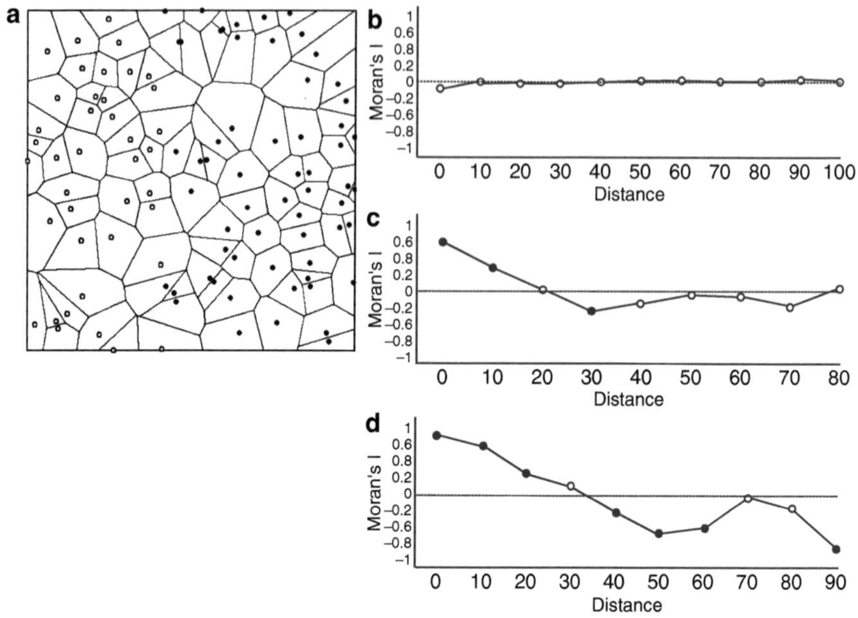

Fig. 15.5 Spatial autocorrelation and the importance of the assumption of stationary. (**a**) Map of the 100 "nest" points given two type of strata (stratum A indicated by open circles: $n = 40$; stratum B indicated by filled circles: $n = 60$) where the range of eggs found in the nest of the stratum A is from 0 to 2 (open circles) and for the stratum B from 0 to 5 (filled circles). Moran's I correlograms based on: (**b**) the 100 nest points which comprise the two strata (i.e., violating the stationarity assumption) where no significant spatial autocorrelation was found; (**c**) the 40 nest points of stratum A and (**d**) the 60 nest points of stratum B, where significant spatial autocorrelation was found in both strata separately

15.4.2 Spatial Interpolation

Often sampled data are used to map species distributions. Information about the spatial autocorrelation (spatial variance) can be used to fill in missing values (Fig. 15.6). The most commonly used spatial interpolation technique is a geotatistical method known as kriging (Cressie 1993) which uses parameters from a variogram and its model: the "range" is the distance at which values are not longer autocorrelated if there is one; the "sill" is the value of the variance at the range; and the "nugget effect" is the variance at distance 0 (due to sampling design and random noise). Values at unsampled locations are "kriged" (estimated) from values at sampled locations and the model of how autocorrelation changes with distance. Since a single set of variogram's parameters is estimated, the method assumes spatial stationarity (i.e. mean and variance are constant over the entire study area). The advantage of kriging is that for each interpolated value (location) a measure of uncertainty or error can be calculated. By examining mapped errors, it is possible to identify regions where more sampling would be useful (errors are high) (Lin et al. 2008). Kriging has been used frequently in animal ecology (Gribko et al. 1995; Villard and Maurer 1996) to map species distributions.

Most landscapes are highly spatially heterogeneous, thus violating the stationarity assumption required for kriging. One option is to stratify the landscape into subareas that are approximately stationary, and calculate a separate variogram for each stratum or subarea. Stratified kriging can then be performed using separate variogram parameters for each stratum (Fig. 15.6, Burrough and McDonnell 1998).

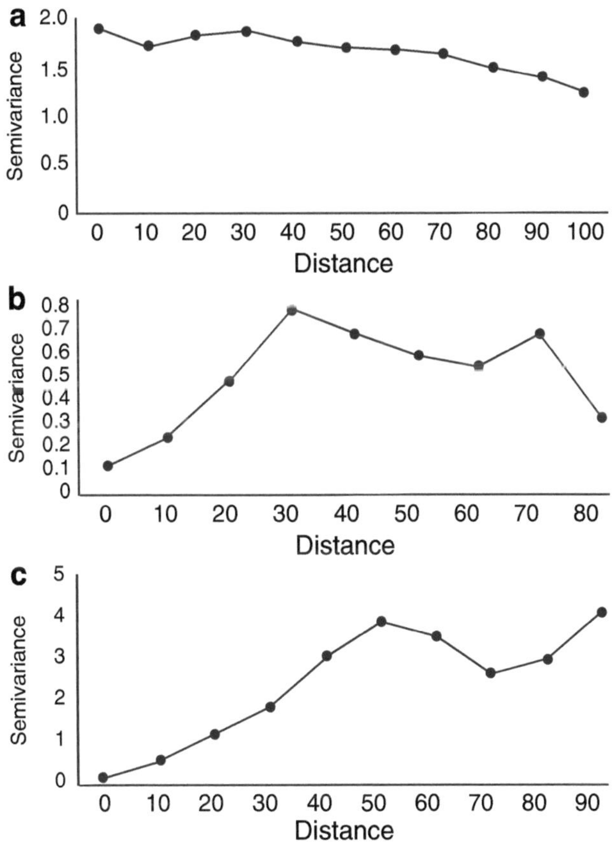

Fig. 15.6 Spatial variance and the importance of the assumption of stationary. Using the same data as in Fig. 15.5, the variograms are based on: (**a**) the 100 nest points which comprise the two strata (i.e., violating the stationarity assumption) where no spatial variance was found as indicated by the "flat" plot of the semivariance values; (**b**) the 40 nest points of stratum A and (**c**) the 60 nest points of stratum B, where spatial variance was found at short distance lags in both stratum separately

15.5 Disease Spread and Dynamic Modeling

Numerous epidemics affecting animals and humans have dominated the headlines over the last few years. Examples include West Nile virus, avian flu and BSE (mad cow disease), with serious implications for animal and human populations, as well as considerable economic concerns. With the emergence of numerous epidemics, diseases present complex spatial patterns and spread mechanisms, with intricate space–time dynamics, which make their modeling and prediction an increasingly difficult task.

Spatial epidemiology, also referred to as geographical epidemiology or medical geography, is concerned with the study of the spatial patterns of disease and mortality, with the determinants of disease and their dynamic interaction in space and time (Waller and Gotway 2003; Elliot et al. 2000; Elliott and Wartenberg 2004). Models are often application-specific, and there is wide variation in the sophistication of methods used to integrate space and time statistically (Crabbe et al. 2000; Burra et al. 2006; Buzzelli et al. 2006; Hooten et al. 2007; Law et al. 2006; Ohta et al. 2006). Here we will focus on dynamic models, which can be temporal or spatio-temporal; implicit (incidence, risk, rates) or explicit (diffusion and contagion).

Dynamic and spatio-temporal modeling presents several conceptual and methodological challenges (Sterner and Smith 2006; Tobin et al. 2007; Wikle 2007). Spatio-temporal processes are multidimensional, and the development of spatio-temporal analytical models is hampered by intricate patterns of correlation and non-stationarity in space and time (Egenhofer and Golledge 1998; Raper 2000; Peuquet 2002).

15.5.1 Spatial and Spatio-Temporal Dependencies and Stationarities

Spatial dependencies and spatial non-stationarities are present, simultaneously, in most observed spatial data. Similarly, dependencies and non-stationarities are common in temporal data, where they also tend to appear simultaneously. When space and time are considered jointly, spatial and temporal dependencies and non-stationarities become dynamically intertwined. As noted by Griffith and Amrhein (1991), most of the multivariate analytical methods used by geographers can be (re)formulated as regression analyses. Regression analyses require independence of the observations and stationarity of the process; violation of either assumption inflates the variance of regression estimates, resulting in less efficient and less reliable models (Anselin 1988; Legendre et al. 2002). Advanced spatial analytical methods are typically designed to correct the negative effects of spatial dependencies or non-stationarities. For example, spatial autoregressive methods (Anselin 1988) correct for spatial dependence, and local, or geographically weighted, methods (Fotheringham et al. 2002) address non-stationarity. In spatial autoregressive models, a model of spatial dependency is contextually estimated, while in geographically weighted models, non-stationarities are addressed by the specification of a multitude of local models, each one meeting the stationarity assumption. As spatial and temporal dependencies and non-stationarities interact in space, time, and space–time, the stationarity assumption becomes harder to meet even for the local models. Furthermore, because all these spatial and temporal dependencies, it is harder to disentangle them. As a result, only partial solutions are typically sought, and the resulting models are either spatial, with limited consideration of time, or temporal, with limited consideration of space.

15.5.2 Temporal Models and Temporal Extensions to Spatial Models

Several types of model have been developed either as spatial extensions to temporal models or as temporal extensions to spatial models. Examples of these are spatial versions of SUR (Seemingly Unrelated Regression) models (Anselin 1988). SUR models are systems of regression equations, whose parameters are estimated jointly to account for interactions among the various parts of the system. The system of regression equations lends itself to a treatment of space and time, as the data can be organized into pools of time series, each one pertaining to a spatial location, or pools of spatial series, each one pertaining to a time period. The latter instance allows for the explicitly spatial specification of the individual models, i.e., spatial autoregressive or geographically weighted models. When the leading dimension is time, the model may include an autoregressive or moving average component, such as STARIMA (Space–Time AutoRegressive Integrated Moving Average) models (Pfeifer and Deutsch 1980). These models can effectively analyze spatial and temporal dynamics, modeling the evolutionary trend of phenomena, such as disease spread and retreat. While they can provide efficient estimates in either space or time, they are not truly spatio-temporal analytical, and they can hardly deal with truly spatio-temporal dependencies.

15.5.3 Spatial Diffusion Models

Diffusion models, also referred to as spatial diffusion models or innovation diffusion models (Okubo and Levin 2000), are largely based on the theoretical work of Swedish geographer Torsten Hagerstrand (1973). Spatial diffusion is a good framework for simulating and predicting diffusion processes (Rodgers 1995). Any diffusion process is composed of a number of typical stages: (1) *penetration*: initial agglomerations, with a few adopters, usually in small areas; (2) *expansion*: radial dissemination and new agglomerations, with a significant increment in the number of new adopters, this stage ends with the first significant shrinkage in the number of adopters; and (3) *saturation*: only small and isolated increments, the adoption rate is always decreasing. In the case of disease spread, two additional stages can be considered:

(4) *decay*: a long period of rapidly declining intensity; and (5) *extinction*: the last throes of the epidemic waves (Cliff and Haggett 2006).

This simple, descriptive model is a useful conceptual tool to describe an observed process and it can aid some prediction of its future developments. Due to the complexities of spatial representation, the temporal dimension is better developed than the spatial one; therefore, the model is traditionally complemented by various forms of spatial interaction models (Hyndman et al. 2002), usually based on the classic gravity model (Hua and Porell 1979): in analogy with the physics concept of gravity, the expected interaction among places is directly proportional to the mass of their power of attraction, reduced by the effect of distance. The gravity model rests on the definition of an indicator of mass (e.g., susceptible population), a measure or model of distance, and the subsequent calibration of three parameters: an exponent indicating the attraction effect, an inverse exponent, representing the effect of distance, and a multiplicative constant. In spatial diffusion analysis, the gravity model constitutes a powerful tool to analyze the transmission mode of the diffusion (e.g. contagion). A large and significant distance decay coefficient indicates the prevalence of a contagious transmission mode, where the prevalent interaction takes place among nearby units, and the mass effect tends to be negligible. In contrast, a large and significant attraction coefficient expresses the prevalence of a hierarchical transmission mode, where the process "jumps" among large nodes of an urban hierarchy or a transportation network, eventually trickling down to smaller nodes, and where the distance effect is secondary.

15.5.4 Spatio-Temporal Modelling of Epidemic Waves

Within the general framework of spatial diffusion analysis, several specific models have been defined, with various applied and analytical foci. One such recent development is a model aimed at analyzing the spatial dynamic form of epidemic waves, as they advance in the early epidemic stages and eventually retreat in response to public health measures (Cliff and Haggett 2006). While a large literature deals with modeling epidemic waves in the time domain, this work by quantitative geographers borrows geomorphology concepts to create a model that focuses on the spatial dimension and shape of the waves, defining a method to identify their leading and trailing edges over time.

The representation of space in this model differs from that in the traditional epidemiological literature (Elliott and Wartenberg 2004), where the typical units of observation tend to be individual infected cases; here, instead, the units of observation are spatial units of interest, such as health regions, management units, or user-defined areas. It is assumed that these units are discrete, and that time can also be split into discrete intervals. The spatial extent of a wave is measured, in each time period, as the area enclosed between its leading and trailing edges; the study region thus is divided, in each time period, into areas defined as susceptible, infected, and recovered. Statistical analysis of the wave characteristics and temporal analysis of the transition phase allows a classification of the epidemic in its fundamental characteristics, including speed of the epidemic wave, spatial distribution (i.e., level of clustering), and transmission mode. Owing to these characteristics, the model has potential for use in forecasting and the definition of early warning systems.

Cliff and Haggett (2006) argue that one of the main advantages of this type of model is its relatively low data-intensity, because it only requires data on presence or absence of disease in each sub-region at given time intervals. The model is therefore particularly useful in countries or situations characterized by paucity or discontinuity of health records and data. We argue that for the same reasons the model constitutes an effective tool for the analysis of animal disease. The problem of spatial standardization, i.e., comparisons across geographical scales or between different countries, is likely to have a relatively low impact on wildlife disease studies.

15.6 Conclusion

Given the brevity of this presentation, our intent here was not to provide a complete description of the methods, but to provide guidance that will allow the reader to begin exploring available approaches with greater ease and confidence. Also, it is useful to note that some general principles apply to all spatial and spatio-temporal analyses. The first important principle is that, for any analysis, the implications of the underlying assumptions of process stationarity (independence of position in time and space) and isotropy (independence of direction) must be understood, and these assumptions must be critically evaluated. Second, data type will determine the methods that can be used. Data types are classified according to geometry (point, line, or polygon), completeness (sample or census), and dynamics (temporally static or changing over time). Once data type has been identified, the next step in analysis is to determine spatial parameters from the spatial structure that can be used into spatial modeling approaches.

Acknowledgement We are grateful to Josie Hughes for her very useful comments and to Stephanie Melles for her help with the figures.

References

Adriaensen F, Chardon JP, De Blust G, Swinnen E, Villalba S, Gulinck H, Matthysen E (2003) The application of 'least-cost' modelling as a functional landscape model. Landscape and Urban Planning 64:233–247

Anselin L (1988) *Spatial Econometrics: Methods and Models*. New York: Kluwer

Anselin L (1995) Local indicators of spatial association —LISA. Geographical Analysis 27:93–115

Benes V, Rataj J (2004) *Stochastic Geometry: Selected Topics*. New York: Springer

Betts MG, Diamond AW, Forbes GJ, Villard M-A, Gunn JS (2006) The importance of spatial autocorrelation, extent and resolution in predicting forest bird occurrence. Ecol Modell 191:197–224

Burra T, Elliott S, Eyles J, Kanaroglou P, Wainman B, Muggah H (2006) Effects of residential exposure to steel mills and coking works on birth weight and preterm births among residents of Sidney, Nova Scotia. Canad Geographer 50:242–255

Burrough PA, McDonnell RA (1998) *Principles of Geographical Systems*. Oxford: Oxford University Press

Buzzelli M, Su J, Le N, Bache T (2006) Health hazards and socio-economic status: A neighbourhood cohort approach, Vancouver, 1976–2001. Canad Geographer 50:376–391

Cliff AD, Haggett P (2006) A swash-backwash model of the single epidemic wave. J Geograph Syst 8:227–252

Cliff AD, Ord JK (1981) Spatial Processes: Models and Applications. London: Pion

Crabbe H, Hamilton R, Machin N (2000) Using GIS and dispersion modelling tools to assess the effect of the environment on health. Trans GIS 4:235–244

Cressie NAC (1993) *Statistics for Spatial Data*. 2nd edition. New York: Wiley

Dale MRT, Dixon P, Fortin M-J, Legendre P, Myers D, Rosenberg MS (2002) The conceptual and mathematical relationships among methods for spatial analysis. Ecography 25:558–577

Diggle PJ (2003) *Statistical Analysis of Spatial Point Patterns*. 2nd edition. London: Hodder Arnold

Dormann CF (2007) Effects of incorporating spatial autocorrelation into the analysis of species distribution data. Global Ecol Biogeography 16:129–138

Downing JA (1991) Biological and physical heterogeneity in lakes. Pp. 160–180. in: Kolasa J, Pickett S (eds.), Ecological Heterogeneity. New York: Springer

Egenhofer MJ, Golledge RG (1998) *Spatial and Temporal Reasoning in Geographical Information Systems*. Oxford: Oxford University Press

Elliott P, Wakefield JC, Best NG, Briggs DJ (2000) *Spatial Epidemiology: Methods and Applications*. Oxford: Oxford University Press

Elliott P, Wartenberg D (2004) Spatial epidemiology: Current approaches and future challenges. Environ Health Persp 12: 998–1006

Fall A, Fortin M-J, Manseau M, O'Brien D (2007) Spatial graphs: Principles and applications for habitat connectivity. Ecosystems 10:448–461

Fotheringham S, Brunsdon C, Charlton M (1992) *Geographically weighted regression. The analysis of spatially varying relationship*. Wiley

Fortin M-J, Boots B, Csillag F, Remmel TK (2003) On the role of spatial stochastic models in understanding landscape indices in ecology. Oikos 102: 203–212

Fortin M-J, Dale MRT (2005) *Spatial Analysis: A Guide for Ecologists*. Cambridge: Cambridge University Press

Fortin M-J, Keitt TH, Maurer BA, Taper ML, Kaufman DM, Blackburn TM (2005) Species ranges and distributional limits: pattern analysis and statistical issues. Oikos 108:7–17

Getis A, Ord JK (1992) The analysis of spatial association by use of distance statistics. Geographical Analysis 24:189–206

Gitzen RA, Millspaugh JJ, Kernohan BJ (2006) Bandwidth selection for fixed-kernel analysis of animal utilization distributions. J Wildl Manage 70:1334–1344

Gribko LS, Liebhold AM, Hohn ME (1995) A model to predict gypsy moth *(Lepidoptera: Lymantriidae)* defoliation using kriging and logistic regression. Environ Entomol 24:529–537

Hagerstrand T (1973) *Innovation Diffusion as a Spatial Process*. Chicago, IL: University of Chicago Press

Haining R (2003) *Spatial Data Analysis: Theory and Practice*. Cambridge: Cambridge University Press

Hooten MB, Wikle CK, Dorazio RM, Royle JA (2007) Hierarchical spatiotemporal matrix models for characterizing invasions. Biometrics 63: 558–567

Hua C, Porell H (1979) A critical review of the development of the gravity model. Int Regional Sci Rev 4:97–126

Hyndman RJ, Koehler AB, Snyder RD, Grose S (2002) A state space framework for automatic forecasting using exponential smoothing methods. International J. Forecasting 18:439–454

Koenig WD (1998) Spatial autocorrelation in California land birds. Conservation Biol 12:612–620

Laube P, Dennis T, Forer P, Walker M (2007) Movement beyond the snapshot – Dynamic analysis of geospatial lifelines. Comput Environ Urban Syst 31:481–501

Law J, Haining R, Maheswaran R, Pearson T (2006) Analyzing the relationship between smoking and coronary heart disease at the small area level: A Bayesian approach to spatial modeling. Geograph Anal 38:140–159

Legendre P, Dale MRT, Fortin M-J, Gurevitch J, Hohn M, Myers D (2002) The consequences of spatial structure for the design and analysis of ecological field surveys. Ecography 25:601–616

Legendre P, Legendre L (1998) *Numerical Ecology*. 2nd English edition, Amsterdam: Elsevier Science

Lin YP, Yeh MS, Deng DP, Wang YC (2008) Geostatistical approaches and optimal additional sampling schemes for spatial patterns and future sampling of bird diversity. Global Ecol Biogeography 17:175–188

Lloyd M (1967) Mean crowding. J Anim Ecol 36:1–30

Ludovisi A, Minozzo M, Pandolfi P, Taticchi MI (2005) Modelling the horizontal spatial structure of planktonic community in Lake Trasimeno (Umbria, Italy) using multivariate geostatistical methods. Ecol Modell 181:247–262

McGarigal K, Marks BJ (1995) *FRAGSTATS: spatial pattern analysis program for quantifying landscape structure*. USDA For Serv Gen Tech Rep PNW-351

O'Brien D, Manseau M, Fall A, Fortin M-J (2006) Testing the importance of spatial configuration of winter habitat for woodland caribou: An application of graph theory. Biol Conserv 130:70–83

Ohta K, Kobashi G, Takano S, Kagaya S, Yamada H, Minakami H, Yamamura E (2006) Analysis of the geographical accessibility of neurosurgical emergency hospitals in Sapporo city using GIS and AHP. Int J Geograph Inform Sci 21:687–698

Okubo A, Levin, SA (2000) *Diffusion and Ecological Problems: Modern Perspectives*. New York: Springer

Perry JN (1995) Spatial-analysis by distance indexes. J Anim Ecol 64:303–314

Perry JN (1996) Simulating spatial patterns of counts in agriculture and ecology. Comput Electron Agric 15:93–109

Perry JN (1998) Measures of spatial pattern for counts. Ecology 7:1008–1017

Peuquet DJ (2002) *Representations of Space and Time*. New York: Guilford

Pfeifer PE, Deutsch SJ (1980) Identification and interpretation of first order space-time ARMA models. Technometrics 22:397–408

Rangel TFLVB, Diniz-Filho JAF, Bini LM (2006) Towards an integrated computational tool for spatial analysis in macroecology and biogeography. Global Ecol Biogeography 15:321–327

Raper, J. (2000) *Multidimensional Geographic Information Science*. London: Taylor & Francis

Real LA, Biek R (2007) Spatial dynamics and genetics of infectious diseases on heterogeneous landscapes. J Roy Soc Interf 4: 935–948

Ripley BD (1977) Modeling spatial patterns. Journal of the Royal Statistical Society Series B-Methodological 39:172–212

Ripley BD (1981) *Spatial Statistics*. New York: Wiley-Interscience

Robertson C, Nelson TA, Boots B, Wulder MA (2007) STAMP: Spatial-temporal analysis of moving polygons. J Geograph Syst 9:207–227

Sadahiro Y, Umemura M (2001) A computational approach for the analysis of changes in polygon distributions. J Geograph Syst 3:137–154

Segurado P, Araujo MB, Kunin WE (2006) Consequences of spatial autocorrelation for niche-based models. J Appl Ecol 43:433–444

Sterner RT, Smith GC (2006) Modelling wildlife rabies: Transmission, economics, and conservation. Biol Conserv 131:163–179

St-Louis, V, Fortin M-J, Desrochers A (2004) Association between microhabitat and territory boundaries of two forest songbirds. Landsc Ecol 19:591–601

Stoyan D, Kendall WS, Mecke J (1995) *Stochastic Geometry and Its Applications*. New York: Wiley

Taylor PD, Fahrig L, Henein K, Merriam G (1993) Connectivity is a vital element of landscape structure. Oikos 68:571–572

Tobin, PC, Whitmire SL, Johnson DM, Bjornstad ON, Liebhold AM (2007) Invasion speed is affected by geographic variation in the strength of Allee effects. Ecol Lett 10:36–43

Villard M-A, Maurer BA (1996) Geostatistics as a tool for examining hypothesized declines in migratory songbirds. Ecology 77:59–68

Waller LA, Gotway CA (2003) *Applied Spatial Analysis of Public Health Data*. New York: Wiley
Waller LA, Gotway CA (2004) *Applied Spatial Statistics for Public Health Data*. Hoboken, NJ: Wiley
Wiegand T, Moloney KA (2004) Rings, circles and null-models for point pattern analysis in ecology. Oikos 104:209–229
Wiens JA, Crist TO, Milne BT (1993) On quantifying insect movements. Environ Entomol 22:709–715
Wikle CK (2003) Hierarchical Bayesian models for predicting the spread of ecological processes. Ecology 84: 1382–1394
With KA, Gardner RH, Turner MG (1997) Landscape connectivity and population distributions in heterogeneous environments. Oikos 78:151–169

Chapter 16
Current State of the Art for Statistical Modelling of Species Distributions

Troy M. Hegel, Samuel A. Cushman, Jeffrey Evans, and Falk Huettmann

16.1 Introduction

Over the past decade the number of statistical modelling tools available to ecologists to model species' distributions has increased at a rapid pace (e.g. Elith et al. 2006; Austin 2007), as have the number of species distribution models (SDM) published in the literature (e.g. Scott et al. 2002). Ten years ago, basic logistic regression (Hosmer and Lemeshow 2000) was the most common analytical tool (Guisan and Zimmermann 2000), whereas ecologists today have at their disposal a much more diverse range of analytical approaches. Much of this is due to the increasing availability of software to implement these methods and the greater computational ability of hardware to run them. It is also due to ecologists discovering and implementing techniques from other scientific disciplines. Ecologists embarking on an analysis may find this range of options daunting and many tools unfamiliar, particularly as many of these approaches are not typically covered in introductory university statistics courses, let alone more advanced ones. This is unfortunate as many of these newer tools may be more useful and appropriate for a particular analysis depending upon its objective, or given the quantity and quality of data available (Guisan et al. 2007; Graham et al. 2008; Wisz et al. 2008). Many of these new tools represent a paradigm shift (Breiman 2001) in how ecologists approach data analysis. In fact,

T.M. Hegel (✉)
Yukon Government, Environment Yukon, 10 Burns Road, Whitehorse, YT, Canada Y1A 4Y9
e-mail: troy.hegel@gov.yk.ca

S.A. Cushman
US Forest Service, Rocky Mountain Research Station, 800 E Beckwith, Missoula, MT 59801, USA

J. Evans
Senior Landscape Ecologist, The Nature Conservancy, NACR – Science

F. Huettmann
EWHALE lab- Biology and Wildlife Department, Institute of Arctic Biology, University of Alaska-Fairbanks, 419 IRVING I, Fairbanks, AK 99775-7000, USA
e-mail: fffh@uaf.edu

S.A. Cushman and F. Huettmann (eds.), *Spatial Complexity, Informatics, and Wildlife Conservation*
DOI 10.1007/978-4-431-87771-4_16, © Springer 2010

for a number of these approaches, referring to them as *new* is a misnomer since they have long been used in other fields and only recently have ecologists become increasingly aware of their usefulness (Hochachka et al. 2007; Olden et al. 2008).

The purpose of this chapter is to introduce and provide an overview of the current state of the art of tools for modelling the distribution of species using spatially explicit data, with particular reference to mammals. We include statistical approaches based on data models (e.g. regression) and approaches based on algorithmic models (e.g. machine learning, data mining). Breiman (2001) refers to these approaches as the *two cultures*. Our goal is not to recommend one approach over another, but rather to provide an introduction to the broad range of tools available, of which many ecologists may not be familiar. Our descriptions of these approaches are admittedly brief due to the necessity of space, and indeed a complete review would require an entire book itself. We hope that our overview provides sufficient information for a starting point to search out more detailed information for an analysis. Indeed, we strongly recommend those interested in using any of the tools described here to become familiarized with additional resources, which we have attempted to provide as references. We avoid a detailed discussion of animal and environmental data, as this is covered at length elsewhere in this book (Part III); nor do we delve in depth into the theory of animal–habitat relationships which is also previously discussed (Part II). We begin by outlining some basic concepts and definitions providing an ecological context for SDMs. Following this we briefly describe the types of data used for SDMs and how this affects model interpretation. Subsequently, we outline statistical modelling tools within the *data model* realm, followed by tools grouped under *algorithm models*. Finally, we provide an overview of a number of approaches for model evaluation.

16.2 Species Distribution Models in Their Ecological Context

16.2.1 The Ecological Niche

Ecological theory suggests that species exhibit a unimodal response to limiting resources in n-dimensional ecological space (Whittaker 1975; Austin 1985; ter Braak 1986). The volume of this ecological space in which an organism can survive and reproduce defines its *environmental niche* (Hutchinson 1957). Many SDMs are based on this niche concept, in which the niche is as an n-dimensional hypervolume, where axes represent n resources limiting an organism's fitness. The niche is defined by the boundaries of these resources, with the volume itself representing the total range of resources providing for the average fitness of an organism to be zero or greater. That is, these boundaries identify the range in which a species can physiologically persist. Hutchinson (1957) proceeded to differentiate between the fundamental niche, described above, and the realized niche in which these resource boundaries are reduced due to inter-specific interactions (e.g. competition, predation). The fundamental niche can be thus viewed as the theoretical limits of resources allowing an organism to persist, whereas the realized niche is the actual limits of resources in which an organism exists. For a sample of additional resources on the niche concept readers are referred to Chase and Leibold (2003), Kearney (2006), Pulliam (2000), Soberón (2007), and Soberón and Peterson (2005).

Quantification of niche space at the species level is a first step toward predicting the distribution, occurrence, or abundance of wildlife species with SDM approaches. Often, the large number of factors which compose the niche can be reduced to a relative few that explain much of the variance in species responses. This technique is heuristically powerful, but it can often obscure relationships between mechanism and response. Importantly, without clear linkages between cause and effect, reliable extrapolation to new conditions (e.g. different study areas, future predictions) is problematic. Therefore, it is preferable to identify limiting factors, which are key variables associated with species tolerances that explain substantial proportions of variance and make sense in terms of well-understood mechanisms.

16.2.2 Scale and Spatial Complexity

Biophysical gradients are clines in *n*-dimensional ecological space. In geographical space these gradients often form complex patterns across a range of scales. The fundamental challenge of using SDMs to predict habitat suitability and occurrence in complex landscapes is linking non-spatial niche relationships with the complex patterns of how environmental gradients overlay heterogeneous landscapes (Cushman et al. 2007). By establishing species optima and tolerances along environmental gradients, researchers can quantify the characteristics of each species' environmental niche. The resulting statistical model can be used to predict the biophysical suitability of each location on a landscape for each species. This mapping of niche suitability onto complex landscapes is the fundamental task required to predict individualistic species responses.

High levels of spatial and temporal variability are typically found in ecological systems. This variability in environmental conditions strongly affects the distribution and abundance of species and the structure of biological communities across the landscape. Details of the spatial and temporal structure of ecosystems are important at a range of scales. There is no single correct scale of analysis for SDM. The fundamental unit of ecological analysis is the organism (Schneider 1994) and fundamental scales are those at which the organism strongly interacts with critical or limiting resources in its environment. Each species will respond to factors across a range of scales in space and time based on its life history strategy and ecological adaptations (Cushman et al. 2007). Ecological responses to environmental gradients must be quantified at scales that match the biological interactions of individual organisms. Analyses at inappropriate scales risk missing or misconstruing relationships between mechanisms and responses (Wiens 1989; Cushman and McGarigal 2003). Accounting for multiple interactions across ranges of spatial and temporal scales is the fundamental challenge to understanding relationships between species distributions and environmental variables in complex landscapes (Levin 1992;

Turner et al. 2003). Where data allow, it is advantageous to quantitatively measure the relationships among driving factors across a range of scales simultaneously to identify these dominant scales and quantify interaction of factors across scale (e.g. Cushman and McGarigal 2003). Ideally, ecological analysis will therefore not be between hierarchical *levels*, such as populations, communities, or ecosystems, but instead will focus on relationships among organisms and driving processes across continuous ranges of scale (Levin 1992; Cushman et al. 2007).

The literature surrounding SDM consists of a myriad of confusing terminology (Hall et al. 1997; Mitchell 2005; Kearney 2006). There has historically been two classes of SDMs: distribution models (DMs; Soberón and Peterson 2005) that predict the broad scale occurrence (i.e. range) of a species over large spatial extents (e.g. globally or continentally), and habitat or resource selection models (RSMs) which predict the occurrence of an organism at finer scales, such as within a population's range. In many instances the analytical process of developing either class of model is similar. Ecologically, the difference between the two is one of scale (Johnson 1980; Wiens 1989). The inferences gained and the type of variable used for each class of model may differ substantially because of the different ecological processes acting on the organism at each scale.

16.2.3 Non-equilibrium Dynamics and Disturbance

Disturbance is central to ecology (Cooper 1913; Leopold 1933; Watt 1947; Reiners and Lang 1979; Turner et al. 2003). Many ecosystems and populations are inherently non-equilibrial or depend on disturbances to maintain community structure and ecosystem function (White 1979; Mooney and Godron 1983; Sousa 1984; Glenn and Collins 1992; Collins et al. 1998). Linking niche relationships of species to environmental variables at appropriate spatial scales is complicated by changing biophysical conditions through time, species invasion and succession, and the legacy of past disturbances (Cushman et al. 2007). Current environmental conditions will not fully explain existing population distribution and abundance due to lack of equilibrium or time lags in establishing new equilibria following disturbances. For example, Baker (1995) argued that the time required for fire regimes to adjust to climatic change may often exceed the time that climate is stable, creating a perpetual temporal disequilibrium between climate, fire regimes, fuel loads, and forest structure, and thus wildlife habitat related to these ecological processes. Integrating spatial and temporal complexity into SDMs requires mechanistic understanding of the key drivers that limit species distributions and abundances, including the spatial scales at which they operate and temporal lags in their effects (Cushman et al. 2007).

16.2.4 Population Ecology

Often of interest to researchers and managers is the relationship between habitat/ environmental variables and population abundance or density predicted via the probability of occupancy or occurrence (e.g. Boyce and McDonald 1999; He and Gaston 2000). For instance, in the case of endangered species managers are often tasked with ensuring population persistence. Habitat improvement or protection is one tool to achieve this. While intuitively appealing, in practice the relationship between habitat quality and population density is not well defined, and demographic mechanisms linking habitat to density are needed (Van Horne 1983; Hobbs and Hanley 1990). For example, areas that are attractive to dispersers yet are not suitable for reproduction may have high density yet do not provide the resources for a self-sustaining population. Pulliam (2000) notes that species may be present in unsuitable habitat and absent in suitable habitat and thus without a mechanistic understanding of what is occurring in a system, simple relationships between habitat and density can lead to incorrect inference regarding the importance of environmental resources to a species. Management based on this information could result in the enhancement or protection of habitat entirely unsuitable for a population's (or species) long-term persistence.

With the recent development of individual-based RSMs, the mechanistic relationship between habitat and population density may be more forthcoming. How an individual uses the landscape (i.e. selects resources) should influence its individual fitness (e.g. Buskirk and Millspaugh 2006). Focusing on individuals allows for the development of the mechanistic relationships between environmental variables (habitat) and the demographic parameters affecting population growth. For example, McLoughlin et al. (2006) related lifetime reproductive success of red deer (*Cervus elaphus*) in Scotland to the selection coefficients from a RSM and McLoughlin et al. (2005) report a relationship between predation mortality and resource selection in woodland caribou (*Rangifer tarandus caribou*) in Canada. Studies such as these enable us to understand how habitat influences populations through its effect on demographic parameters. Boyce and McDonald (1999) suggested that population density could be predicted from environmental variables using RSMs. However, recent work by Johnson and Seip (2008) indicates this is tenuous and requires a number of assumptions (e.g. population at equilibrium) and that a range of factors can confound predictions. For example, they found that when populations were well below ecological carrying capacity, model predictions of density based on habitat were overestimated since there was ample suitable habitat with no animals present. Hence, caution is warranted when making any assumption regarding population density or abundance based on occupancy or occurrence measures.

16.3 Data Types, Model Types, and Interpretation

16.3.1 Data Terminology and Application

In developing SDMs, data can come in a number of forms and can represent spatially referenced occurrences (e.g. single locations) or abundances (e.g. animal counts). Counts, or densities, of animals within some defined area (e.g. 1 km^2 pixel) can be modelled as a function of environmental characteristics within that area (e.g.

see Section 16.4.1.3). More traditionally however it is the binary variable, presence/absence, which is modelled. Identifying the presence of a species, or individual, at a location is relatively straight forward. Of much greater issue is the identification or confirmation of absence (MacKenzie 2005). How does one know that a species truly is absent from an area and that its absence is not due to sampling issues? For example, can absence be inferred from a lack of sighting in just one sample? Indeed, true absences may be very difficult to detect in nature. Thus, if one has a sample of used (or presence) locations (we use the terms use and presence interchangeably) and a sample of absence locations in which the surety of absences is questionable, modelling approaches to differentiate and predict the two may be problematic.

If a random sample of presences and true absences are used as data, model predictions can be inferred as absolute probabilities (i.e. the absolute probability that a species will occur on some unit based on its environmental characteristics). Manly et al. (2002) classify such a model as a resource selection probability function (RSPF). Often, true absences cannot realistically be assumed and yet many modelling approaches require a binary dependent variable. One approach to deal with these situations is to use *pseudo-absences*. Pseudo-absences most often represent the habitat *available* for an organism to select. Thus, it is assumed that pseudo-absences inherently include both used and unused locations. While model estimation using use and pseudo-absence data (other terms include use/availability) occurs similar to a use/non-use approach, model predictions cannot be considered as absolute probabilities. Rather, model predictions are inferred as relative probabilities and hence different landscape units can be compared relative to one another, but not in absolute terms. Manly et al. (2002) term such models as resource selection functions (RSF) and note that a RSF is directly proportional to a RSPF by some unknown constant. Whereas predictions for landscape or habitat units can be quantified probabilistically from a RSPF, these same units should only be ranked or indexed based on predictions from a RSF. A third situation applicable to some modelling approaches is termed *presence-only*. These types of models (e.g. see ENFA and Maxent below) do not require the user to explicitly provide a binary response variable; rather, only the used locations are entered. The modelling procedure then generates an availability sample for comparison. However, model predictions and output must still be interpreted appropriately because used locations were not compared to true absences. Specific requirements and assumptions of each modelling approach are further described in subsequent sections. A number of resources are available describing data collection and study designs. Manly et al. (2002) and Thomas and Taylor (1990, 2006) are good resources for those interested in RSMs. Hirzel and Guisan (2002) discuss sampling strategies for habitat models (i.e. DMs) and Graham et al. (2004) discuss the use of museum-based informatics.

16.3.2 Model Types

For local or regional natural resource managers, RSMs are typically the type of model of interest, particularly for wide-ranging, generalist species [e.g. elk (*C. elaphus*), coyotes (*Canis latrans*)], where landscape use in a specific area (e.g. for one population) is of management concern (e.g. Sawyer et al. 2006). SDMs have been used to investigate patterns of geographic invasion (Peterson 2003) and changes in species ranges due to climate change (e.g. Beaumont et al. 2007), and Guisan and Thuiller (2005) provide a thorough review of SDMs and their potential for broader ecological insight. Although ecologically the differences between SDMs and RSMs may be substantial, the types of data (i.e. use/availability, use/non-use) used for either class of model can be similar. Hence, the tools we describe here can often be applied to either class of model. However, based on the published literature ecologists developing SDMs seemingly use a much broader array of analytical tools from both the data- and algorithm-based approaches (e.g. Elith et al. 2006), than have ecologists developing RSMs who have adopted data-based models more frequently (e.g. Huzurbazar 2003).

The tools we discuss here are most applicable for correlative models (Soberón and Peterson 2005) rather than a physiological mechanistic approach (Guisan and Zimmermann 2000) since the data are typically animal observation data (e.g. presence/absence) in which occurrence is correlated with environmental conditions which we infer has some relation to fitness, although this relationship is unspecified. Indeed, the tools we describe here are more appropriate for general distributional models (RSMs or DMs) rather than formal ecological niche models (Peterson 2006). Ultimately, it is up to the researcher to define the objectives of a study prior to any data analysis. While these analytical tools can certainly help with completing these objectives, they most certainly cannot define them.

16.4 Data-Based Models

16.4.1 Generalized Linear Models

Many of the first statistical approaches to species distribution modelling used generalized linear models (GLM; McCullagh and Nelder 1989; Guisan and Zimmermann 2000). Through the use of a link function (e.g. logit, log), GLMs allow a non-Gaussian response variable to be modelled as a linear function of some predictor variable(s). A linear predictor $\mathbf{X}\boldsymbol{\beta}$, where $\mathbf{X}$ represents a vector of predictors and $\boldsymbol{\beta}$ represents a vector of estimated parameters plus an intercept (α), is transformed via the link function to predict a response. For example, the log link function is often used with count data and the Poisson distributional family, and model predictions (μ) are estimated by:

$$\mu = \exp(\mathbf{X}\beta)\left(\text{which is equivalent to } \log_e(\mu) = \mathbf{X}\beta\right), \tag{16.1}$$

where $\exp(\mathbf{X}\beta) = e^{\mathbf{X}\beta}$.

Nonlinear relationships can be modelled through the use of quadratic, cubic, or higher-order terms. Rather than assuming that the error term of the model is normally distributed, as in ordinary least-squares linear regression, GLMs allow errors to be distributed following a number of other distributional families including binomial, Poisson, or negative-binomial; hence GLMs are parametric models. Model parameters (coefficients) are estimated via maximum likelihood and represent the change in the response following a one-unit change in predictor. Parameters in models with multiple predictors are interpreted the same way, but with all other predictors held constant.

Selecting between two, or among multiple, models with differing specifications of predictor variables is termed *model selection*. Two nested models (i.e. when one model represents a subset of another) can be compared using a likelihood-ratio test (LRT). Information-theoretic (IT) criteria (e.g. AIC; Burnham and Anderson 2002) allow for simultaneous multi-model comparison of both nested and non-nested models and avoids the need to carry out many pairwise LRTs, which are only valid for nested models. Further, IT criteria such as AIC allow models to be weighted and ranked relative to the entire set of models under consideration. From these weights, models can be averaged and the relative influence of individual predictors compared. Model selection using IT criteria is based on the principle of parsimony. Models with higher log-likelihoods indicate they fit the data better than those with lower log-likelihoods. However, any increase any predictors will also increase the log-likelihood, even if only by a minute amount. Therefore, IT criteria penalize a model based on its number of estimated parameters. Thus balancing between an overfit model (i.e. too many predictors) with one having too few predictors to be meaningful and not explaining sufficient variability in the data. Burnham and Anderson (2002) discuss a small sample size version, AIC*c*, when the ratio of the number of observations (data) to the number of estimated parameters is <40, and argue that this should always be used. They also describe a quasi-AIC (QAIC) which they suggest should be used when overdispersion (i.e. greater variance than expected) in the data is suspected, common in count data (Gelman and Hill 2007).

Models not meeting the assumptions of GLMs (McCullagh and Nelder 1989) may result in biased or overly optimistic (i.e. deflated standard errors) estimates. Perhaps the most critical assumption is that of independent observations. Roughly, this refers to each observation contributing *equally* to the model. If two or more observations are not independent (i.e. correlated in some way) due to, for example, temporal sampling issues (e.g. telemetry observations collected close together in time), or repeated observations from the same individual in which those data are correlated to a greater degree than data across individuals, parameter estimates may be biased and measures of precision may be biased low (Dormann 2007). For instances in which data can be grouped, say within individuals or groups, there are methods to deal with this lack of independence such as robust sandwich estimators (e.g. Nielsen et al. 2002), however mixed models (see below) may be a better option (Gillies et al. 2006), particularly when these groups are unbalanced (i.e. unequal sizes). Collinearity between predictor variables can also seriously affect estimates through biased parameters and deflated standard errors. Prior to inclusion of predictor variables in a model, their correlation should be assessed and if collinearity is detected, one of the variables removed. The choice of collinearity threshold is somewhat arbitrary.

A correlation of $r > 0.9$ should be a serious concern, and a correlation of $r > 0.7$ should be examined closely. Variance inflation factors (VIF; VIF = $1/1-R^2$), or tolerance (1/VIF), can be used after model fitting to assess how much the variance of an estimate is inflated due to collinearity among variables in the model. One rule of thumb indicating strong collinearity is VIF > 10 (Chatterjee and Hadi 2006), which is equivalent to $r = 0.95$.

16.4.1.1 Logistic Regression

Ordinary logistic regression (OLR; Hosmer and Lemeshow 2000) has been the traditional workhorse for estimating SDMs (e.g. Guisan and Zimmermann 2000, Guisan et al. 2002). Logistic regression is an intuitive approach as it uses binary (e.g. presence/absence) data as the dependent variable. The mean of these binary data is modelled as a binomial probability distribution and the relationship between some predictor(s) (e.g. landscape variables) and the probability of an event (e.g. an occurrence) is linearized through the use of the logit link function (16.2), which ensures the transformation of $\mathbf{X}\boldsymbol{\beta}$ to a probability of occurrence (μ) ranges between 0 and 1.

$$\mu = \exp(\mathbf{X}\beta)/(1+\exp(\mathbf{X}\beta)) \tag{16.2}$$

By definition, OLR assumes the use of the logit link function; however in a GLM framework other link functions are available such as the probit, complimentary log–log, and log–log. The majority of SDMs have used the logit link function and hence we restrict our discussion to logistic regression.

When a random sample of true presence/absence data are used, μ is an absolute probability (i.e. the model is a RSPF) and inferences and predictions based on estimated model parameters are straight forward. In situations where a sample of used and a separate sample of unused (or available) locations are the dependent variable, a case-control approach is appropriate (Manly et al. 2002; Keating and Cherry 2004). Since the dependent data were not collected as one random sample representative of the overall population, an adjustment must be made to Eq. (16.2) to account for sampling fractions (i.e. the proportion of used locations in the sample relative to the total number of used locations in the population). Without this adjustment probabilities will be biased (Keating and Cherry 2004). In practice, knowing these sampling fractions in an ecological setting may be virtually impossible. Without knowing these sampling fractions the intercept parameter cannot be estimated (Manly et al. 2002), or more specifically, the intercept parameter estimated in a GLM framework is biased and should not be used.

When presence/pseudo-absence (use/availability) data are used, μ is a relative value (i.e. the model is a RSF). Manly et al. (2002) suggest the use of an exponential model (16.1) as a recommended form of the RSF. They suggest estimating model parameters ($\boldsymbol{\beta}$) by logistic regression and using them in the exponential RSF rather than the logistic model (16.2). This assumes $\mathbf{X}\boldsymbol{\beta} \le 0$ to yield a valid probability model. However, because sampling fractions are not known, the estimated intercept parameter is not valid and therefore not included in the RSF. Keating and Cherry (2004) critiqued the use of OLR for use/availability RSMs and noted in their simulation study that: RSFs were not always directly proportional to RSPFs, RSFs did not always rank resource units properly, parameter estimates obtained via OLR do not always result in $\mathbf{X}\boldsymbol{\beta} \leq 0$, and even small levels of contamination (i.e. numbers of used locations in the availability sample) can yield invalid parameter estimates. Johnson et al. (2006) demonstrated empirically that: the likelihood for a use/availability model is valid and is closely related to a logistic discriminate function and hence yields valid log-likelihoods and parameter estimates, RSFs are generally proportional to RSPFs with relatively high (>50%) levels of overlap (i.e. when a used location is found in both the used and available samples), and parameter estimates were generally stable with high (~50%) levels of data contamination.

The decision to adopt a use/non-use versus a use/availability approach, and hence develop a RSPF or RSF respectively, is not trivial and has both biological and analytical consequences. Non-use implies that a site or location was sampled and an organism was not present. The justification of assuming the non-use of a site is important as the absence of an organism may be due to sampling effort or bias rather than true lack of occurrence. Indeed, particularly for RSMs, it is difficult to envision many instances in which true absence can definitively be assumed (MacKenzie 2005; see Section 16.4.1.4 below for methods to deal with detection). The use of logistic regression assumes that the detection of an organism is perfect, and hence the differentiation between RSFs and RSPFs does have implications. Since predictions from a RSF are relative, maps, for example, can only identify relative rankings of suitable habitat. To conceptualize this, imagine a hypothetical RSF with the only predictor being categories from a landcover map. If category z is chosen as the reference category during maximum likelihood estimation, the parameter estimates for all other categories are interpreted relative to it (Hosmer and Lemeshow 2000). After removing the intercept from the linear predictor and calculating μ in category z, the result is zero, or more specifically there is no result as there would be nothing to calculate (i.e. it is nonidentifiable; Lele and Keim 2006). Treating this value as a true probability could have serious implications from a management perspective.

Parameters estimated from logistic regression are simple to interpret and can be described in two ways. First, the parameter value (β_i) can be used to identify the change in μ (16.2) with a one-unit change in the value of the independent variable i. Parameters can also be interpreted as an odds ratio, $\exp(\beta_i)$. The odds ratio is the factor by which the odds of an event occurring changes with a one unit change in the independent variable i. The choice of how one interprets these parameters is largely based on the research objectives.

We have devoted substantial space to the discussion of logistic regression models largely because they have been, and still are, used frequently by ecologists. As will be evident in the proceeding sections, many newer approaches accommodate many of the limiting issues with OLR. Further, many of the details presented here are applicable for other data-based models, such as collinearity, independence, and model selection.

16.4.1.2 Conditional Logistic Regression

An alternative to OLR is the use of conditional logistic regression (CLR). CLR models use data that are matched, in which a used (presence) location is specifically matched to a number of unused, or available, locations to create a group (stratum), and results of the model are conditional upon each group. These approaches have been used to deal with situations in which habitat availability changes during the course of a study (Arthur et al. 1996), and to deal with potential temporal autocorrelation arising from data collected from GPS radio-telemetry data (Johnson et al. 2004). Such models are known as discrete choice models (DCM; Manly et al. 2002), conditional fixed-effects logistic regression (CFE; Johnson et al. 2004), or case-control models (Pearce and Boyce 2006). Johnson et al. (2004) used CFE to develop a RSM for caribou in northern British Columbia, Canada. They created a group for each telemetry observation, with 10 random locations representing available resources. These ten random locations were generated within a circle whose area was determined by the location of the previous observation. While they used ten locations, this is somewhat of an arbitrary decision, as is any sample size selection for pseudo-absence data. This approach would be most useful when some inherent lack of independence in the data is present; common with GPS radio-telemetry data. Interpretation of model coefficients is the same as for OLR, yet may be viewed as more reliable given that the natural clustering in the data is accounted for. In CFE, there is no intercept estimated since the model is conditioned on each stratum. Arthur et al. (1996) developed a RSM for polar bears (*Ursus maritimus*) that was essentially a DCM (Manly et al. 2002) which accounted for changing sea ice availability over time. The application of these types of models in which availability is matched to a specific used location may be especially well suited for modelling species distribution in changing environments. Some detailed references for those interested in CLR models include Cooper and Millspaugh (1999), Craiu et al. (2008), Manly et al. (2002), McCracken et al. (1998), and McDonald et al. (2006).

16.4.1.3 Count-Based Models

An alternative to modelling a binary response, as in logistic regression, the abundance (counts) of animals or observations can be modelled as a function of some environmental features (Pearce and Ferrier 2001; Potts and Elith 2006). GLMs readily provide for this by using either a Poisson or negative-binomial distribution, typically with a log link function (McCullagh and Nelder 1989; Long 1997). Negative-binomial models are often used over Poisson models because they estimate an overdispersion parameter. Manly et al. (2002) describe a log-linear modelling approach in which counts of observations are recorded in categorical habitat types. The probability of use of (μ) an individual habitat type (i) can be predicted by:

$$\mu_i = B_i \exp(X_i \beta_i), \quad (16.3)$$

where B_i is the overall proportion of habitat type *i* available in the study area. Counts of observations can be related to continuous environmental variables also. Millspaugh et al. (2006) used counts of observations from GPS radio-collared elk within randomly placed units (circles or squares) on the landscape and used a Poisson regression model to estimate an RSF. They used an offset term (a variable with a coefficient fixed to 1.0) of the total number of observations across the entire landscape to ensure that model predictions were probabilities rather than counts of observations. In many ecological datasets there can be a high frequency of zeros (Martin et al. 2005). Because these data may not conform to the assumptions of the Poisson or negative-binomial distributions, a number of approaches have been developed to deal with these extra zeros including zero-inflated and hurdle models (Welsh et al. 1996; Fletcher et al. 2005; Potts and Elith 2006). Nielsen et al. (2005) compared logistic regression predictions of occupancy with predictions from a zero-inflated model of moose (*Alces alces*) abundance and found that probability of occurrence predicted abundance well at low densities. However, they concluded that different environmental factors were involved in predicting distribution and abundance at higher densities. Potts and Elith (2006) provide a good introduction of count-based models in an ecological context and Long (1997) and Cameron and Trivedi (1998) provide good statistical overviews.

16.4.1.4 Resource Utilization Functions

A relatively new approach for modelling species distributions is the resource utilization function (RUF; Hepinstall et al. 2004; Marzluff et al. 2004; Millspaugh et al. 2006). The RUF incorporates more broad-scale space use patterns (e.g. home range) into the development of a RSM. Its benefits include not requiring an unused/availability sample which may be difficult to adequately identify, and the spatial accuracy of animal locations may not be as critical because of the smoothing nature of space-use estimators (Rittenhouse et al. 2008). With this approach a utilization distribution (UD; Kernohan et al. 2001) must first be estimated which uses the spatial location of animal observations to estimate the intensity (or probability) of use within, for example, an animal's home range. Hence, the location data for this approach must be suitable to adequately estimate a UD. The UD estimates this intensity of use based strictly on the spatial distribution of the observations and not any underlying environmental features. Two methods for estimating UDs include kernel approaches (Worton 1989; Seaman and Powell 1996) or local convex hulls (Getz and Wilmers 2004; Getz et al. 2007).

A grid is created with each pixel value representing intensity. With the UD estimated, resource (landscape) variables under each pixel are extracted, much in the same way in which the values of these variables would be extracted from point data. These resources can subsequently be related to intensity of use in a number of ways (Millspaugh et al. 2006). Multiple linear regression can be used with the intensity values for each pixel as the dependent variable and the underlying resources as predictors and the spatial autocorrelation among pixels accounted for (Marzluff et al. 2004). Jachowski (2007) used this approach to develop a RUF for endangered black-footed ferrets (*Mustela nigripes*) in relation to prey distribution. Another approach is to categorize the UD values into ranks (i.e. an ordinal measure) and use polytymous logistic regression to model these ranks, rather than the actual UD values (Rittenhouse et al. 2008).

Another approach is to use the estimated UD in a compositional analysis (Aitchison 1986; Aebischer et al. 1993; Millspaugh et al. 2006) which is a multivariate analysis of variance. Under this approach, habitat use is based on the proportion of each habitat type within the UD, with each habitat type weighted by the intensity of use as determined by the individual values of the UD pixels. Millspaugh et al. (2006:391) term this a "weighted compositional analysis" and provide an example using radiocollared female elk. These animals exhibit non-random space use within their home range which emphasizes the rationale for a weighted compositional analysis in which even use within a home range is not assumed, which is the case with the original compositional analysis approach described by Aebischer et al. (1993). Each habitat type within the UD (i.e. home range) is weighted by the overall sum of their associated UD intensity values.

16.4.1.5 Occupancy Models

An assumption with models using presence/absence data is that both presences and absences are detected perfectly. This is a rather naïve assumption though, that virtually gets violated in nature every single time (Buckland et al. 2001; MacKenzie et al. 2002). While presence of a species or individual at a site is relatively easy to confirm (e.g. it was directly observed), absence is much more difficult to assume because non-detection may be a result of either the sampling approach (i.e., false negatives) or that absence was indeed true (MacKenzie 2005). Failure to account for imperfect detection of absences can bias model parameter estimates and subsequent inferences (Gu and Swihart 2004; MacKenzie 2006; Tyre et al. 2003). MacKenzie et al. (2002) introduced an approach (occupancy models) to model occupancy rates when detection probability is less than perfect, which is closely related to mark-recapture models. While OLR models the probability that a site will be occupied based on some covariate(s) assuming perfect detection, occupancy models model presence as well as the probability of detection (MacKenzie et al. 2005) thus reducing bias induced by false-negatives. Logistic regression can be used to estimate a RSM of species presence or absence while taking into account the detectability of a certain resource unit (MacKenzie 2006). Of note is that to estimate detection probability, a site must be visited (i.e. sampled) more than once and that for this approach a designed study may be necessary rather than using existing observational data in which it may rarely be the case that a site has been visited multiple times. MacKenzie and Royle (2005) and Bailey et al. (2007) outline sampling strategies and recommendations for designing a site occupancy study.

MacKenzie et al. (2005) provide a comprehensive overview of occupancy models including study design, statistical estimation, and ecological inferences, and is a necessary resource for anyone undertaking this modelling approach. Donovan and Hines (2007) provide an on-line tutorial and instruction manual for implementing occupancy models. MacKenzie (2006) provides a detailed description of using occupancy models to develop RSMs with specific reference to pronghorn antelope (*Antilocapra americana*). Ball et al. (2005) use an occupancy modelling approach to evaluate an existing habitat model for Palm Springs ground squirrel (*Spermophilus tereticaudus chlorus*). Occupancy models can be estimated using the wide-ranging mark-recapture software Program MARK (Cooch and White 2007) or the specially designed program Presence (Mackenzie et al. 2002).

16.4.2 Quantile Regression

Another approach for SDMs not having received much attention (Austin 2007) is quantile regression, which has its roots in econometrics (Koenker 2005). Rather than assuming that the relationship (i.e. slope) between observations and predictors is constant across the range of predictor values (i.e. environmental space), quantile regression assumes this relationship changes over different ranges (i.e. quantiles) of the predictors (Cade and Noon 2003, Cade et al. 2005). Vaz et al. (2008) used this approach in the upper boundaries of organism–environment relationships to identify limiting factors for distribution and provide a modelling strategy for using quantile regression. These semi-parametric models estimate unique parameters relating environmental features to a response across different quantiles of the data. Thus, for each quantile of the data a separate coefficient is estimated; having specific fits for specific data sections, this can be somewhat equated with approaches used by MARS and even mixed-models (see further below). Austin (2007) discusses how comparison of these quantile-specific coefficients can be used to assess limiting factors and environmental gradients. These models are an extension of linear least-squares models and are most readily fit to continuous response data; however Koenker (2005) describes approaches to fit binary and count data, as well as nonlinear models. Additionally, because the data are partitioned into quantiles, a sufficient number of observations must be present in each quantile to adequately estimate its respective parameter and therefore quantile regression models may be difficult to fit with sparse data. Model parameters can be interpreted as in least-squares linear models. More applications are needed to assess the validity and specific niche where these model types can be used for SDMs.

16.4.3 Generalized Additive Models

An approach similar to GLMs, yet more flexible, are generalized additive models (GAM; Hastie and Tibshirani 1990, Wood 2006), which have also been used extensively for developing SDMs (Guisan et al. 2002). While GLMs are somewhat restrictive in their assumption of a linear relationship between response and predictor variables, GAMs allow for more complex relationships to be estimated, using cubic or thin-plate regression splines, and do not assume a linear relationship. Tensor-product splines can be used to estimate smooth functions of interactions of predictors. Like GLMs, GAMs can use binary data (for SDMs they may use presence/absence or use/non-use data) within the binomial family, and link functions such as the logit. GAMs can also be used to fit non-binary data and can be used with Gaussian continuous data, Poisson distributed count data, and zero-inflated data (Barry and Welsh 2002). However, unlike GLMs which have a single parameter for each predictor variable, GAMs estimate a non-parametric, smooth *function* for one or more of the predictors in a model. A GAM with no smooth functions of any predictor variables is identical to a GLM. GAMs can include both linear and smooth terms in a model, and selection among models with differing specifications of predictors can be used as with GLMs (e.g. AIC).

When fitting GAMs, the primary decision by the analyst is how *smooth* should the relationship be between outcome and predictor. The smoothest relationship is a simple straight line (df = 1) and as the response tracks values of the predictor variables more closely, the function becomes less smooth (i.e. more complex) and the degrees of freedom rise. The main caution with using GAMs is to avoid over-fitting the data. This essentially becomes a model selection problem of choosing the effective degrees of freedom for each smooth term which can be computationally intensive (Wood and Augustin 2002). Hastie and Tibshirani (1996) describe an approach called BRUTO which uses a back-fitting procedure using cross-validation to identify not only the variables to remain in the final model, but also the effective degrees of freedom (i.e. smoothness) for each variable. Wood and Augustin (2002) discuss an approach using penalized regression splines to carry out integrated model selection through generalized cross-validation to *select* the appropriate degree of smoothing. Interpreting and communicating the results and parameter estimates from a GAM is more challenging than GLMs. If inference is the primary objective of the SDM, graphical displays of the smooth function relating predictor to response(s) may be the most useful approach as this visual representation is likely easier to understand than would a written description, particularly for complex relationships. GAMs do not provide coefficient estimates and standard errors for smoothing functions. Rather, the significance of the function itself is evaluated. Spatial predictions for use in a GIS are most readily obtained within the statistical package used to estimate the GAM, and subsequently exported to the GIS, rather than attempting to apply the GAM model directly within a GIS as is readily carried out with GLM model parameters.

16.4.4 Bayesian Analysis

Bayesian statistical approaches (e.g. Gelman et al. 2004) are becoming increasingly more common in ecological research (Clark 2005; Ellison 2004) and species distribution modelling more specifically (e.g. Gelfand et al. 2006; Latimer et al. 2006). All of the more traditional frequentist-based approaches described thus far in this section can be analyzed using a Bayesian approach. For example, Wintle and Bardos (2006) used a Bayesian autologistic regression model to model habitat relationships of the yellow-bellied glider (*Petauroides australis*) in the presence of spatially autocorrelated animal observation data, Thomas et al. (2006) developed a Bayesian discrete-choice model to investigate caribou resource selection with cow-calf pairs treated as random effects (see Section 16.4.5), and La Morgia et al. (2008) examined potential impacts of human land use to roe deer (*Capreolus capreolus*) habitat using GAMs in a Bayesian setting. Martin et al. (2005) describe a Bayesian approach, and provide code, for a zero-inflated count model. Software BayesX (Brezger et al. 2008) provides a software environment for analysing structured additive regression models such as GAMS, semiparametric splines, and geographically weighted regression (GWR), in a Bayesian framework.

One of the difficulties in Bayesian data analysis is in empirically defining the posterior distribution of the model parameters. The improvement of Monte Carlo techniques have made the posterior distributions of Bayesian models possible to approximate from flat (i.e. uniform) prior distributions. Historically, the primary criticism of Bayesian approaches is that a model converges on a distribution defined by expert opinion, thus returning an answer that was pre-defined. By starting a Bayesian model with flat priors and converging on an optimal distribution given the data, we can ask the question; what is the probability that the resulting model will arrive at the observed result? A flexible approach for this type of analysis is Markov Chain Monte Carlo (MCMC) with Gibbs Sampling (Casella & George 1992). Gibbs Sampling generates samples from the joint probability of two or more variables, thus approximating the joint distribution. This approach is useful when the joint distribution is not explicitly known. This algorithm is well adapted to sampling the posterior distribution along a MCMC. This approach is implemented in software WinBUGS (Lunn et al. 2000).

Bayesian approaches can be valuable because of their flexibility in building complex (e.g. hierarchical) models with multiple random effects, their ability to incorporate multiple probability distributions, and their ability to incorporate prior information about factors influencing a species distribution (Gelman et al. 2004). Kynn (2005) developed software (Elicitor) to generate prior distributions from expert opinion for use in Bayesian species distribution modelling. While there are many advantages to adopting a Bayesian approach, it can be challenging and likely not recommended for those new to species distribution modelling or those without a particularly strong background in statistics. Indeed, there are important and non-trivial differences both philosophically and inferentially between frequentist and Bayesian approaches that should be recognized prior to embarking on this path (Ellison 2004). McCarthy (2007) provides a good introductory text with an ecological focus, while Gelman et al. (2004) provide a thorough technical overview.

16.4.5 Random Effects Models

For data collected opportunistically or in a field setting where few variables can be controlled for, unobserved heterogeneity (i.e. variability) can be present (Gelman and Hill 2007). This may result from variability across year in which some source of annual, but unknown, variability may affect the occurrence of a species. In a radio-collar telemetry study, the individual may be the source of this heterogeneity and while some features of the sampled individuals may be possibly included in the modelling framework, other unmeasurable factors (e.g. genetic) may be at play and be influencing resource selection by that individual. In these situations where some factor may be affecting the process being measured, random effects (or mixed models) may be employed (Pinheiro and Bates 2000; Skrondal and Rabe-Hesketh 2004; Gelman and Hill 2007). In these models, the unmeasurable factors are treated as a random variable in which the individual levels of that variable (e.g. specific years) are randomly drawn from the overall distribution of "year". It is possible to have >1 random effect which can either be independent of one another (i.e. cross-classified) or can be hierarchically organized (e.g. individual within group within the population). Models are termed "mixed" when they incorporate both random and fixed effects. Fixed effects are those that are unchanging across different levels of the random effect (i.e. the slope of the relationship between predictor and response is constant). Random effects models are particularly valuable for identifying the source of unobserved variability and subsequently accounting for it, thus reducing the overall variance of the model. Random effects can be specified as a random intercept so that the relationship between predictor and response is constant, but the y-intercept is shifted up or down for different levels of the random effect. Random effects can also be specified as random coefficients such that for each level of the random effect, the slope of the relationship between response and predictor is different. Mobæk et al. (2009) used a mixed modelling approach to examine habitat selection by sheep (*Ovis aries*) across different levels of forage productivity and sheep stocking density. In their analysis they considered two random effects; individual sheep hierarchically nested within an enclosure.

Random effects models can be used to explicitly account for lack of independence, for example within telemetry points from the same radio-collared individual. They can also deal with unbalanced sample designs, for example where some years have more data collected than others. Gillies et al. (2006) describe how random effects models can be used to deal with these issues. Further, because random effects are assumed to be drawn from a probability distribution, these models can be generalized beyond the data used to train them. That is, by treating year as a random effect, the model can be used for inference and prediction to years not included in the modelling process. This also saves degrees of freedom for parameter estimation since each random effect (e.g. individual) does not have a unique parameter estimated for it, which in the presence of numerous random effects could prove exceptionally large. Given that most data used for SDMs are collected in the field, often opportunistically, random effects models may be a valuable tool to account for and quantify unobserved variability in the data. Further, examining the random effects parameters themselves can be useful for gaining greater ecological insight into processes driving these patterns and can be useful for designing future study to investigate these in a more detailed fashion (Afshartous and Wolf 2007).

16.4.6 Ecological Niche Factor Analysis (ENFA)

In situations where species absence data are unavailable, poor, or irrelevant, and data consist of presence-only records, Hirzel et al. (2002) developed a multivariate approach, and the Biomapper software, called ecological niche factor analysis (ENFA) to estimate habitat-suitability. Conceptually, the principle behind ENFA is relatively straight forward. The distribution of some environmental, non-categorical, predictor variables where the species is present is compared to, and is a subset of, the distribution of those same predictors across the entire (i.e. available) study area (see Hirzel et al. 2002:Fig. 1). Predictor distributions are normalized prior to analysis through a Box-Cox transformation. Four parameters are required: ms and σs are the mean and standard deviation of the predictor distribution where the species was present respectively, and mG and σG are the mean and standard deviation of the predictor distribution for the overall (global) region respectively. Marginality (M) is the absolute difference between the means of the two distributions for a predictor. It is divided by $1.96^{*}\sigma G$ to remove any bias from the global distribution and generally ensures M ranges between 0 and 1. Specialization (S) is the ratio of σG:σs and measures the range of use of a predictor relative to the overall range of that predictor. That is, a species that used some environmental feature(s) within a very narrow range of its values is considered specialized. An important consideration is that M and S are dependent upon how the global area is defined and will change with different definitions of the study area, thus inferences are relative to how this study area is defined. Sample sizes are dependent on the variability of the predictors in the area since it is most important to obtain accurate distributions of predictors. With highly variable environments this may require substantial data. For more homogeneous environments, fewer data may be required.

To incorporate multiple predictors in multivariate space, principal components analysis (PCA) is used in which the first axis (factor) accounts for M across all predictors, and subsequent axes maximize S. Suitability for a pixel (cell) is estimated by comparing its location on the PCA factors (axes) to the total number of cells that are as far or farther from the median value of the axes, and then normalized to range between 0 and 1 (Hirzel et al. 2002). A good example of the application of ENFA is provided by Sattler et al. (2007) in their model of two bat (*Pipistrellus* spp.) species. Hirzel et al. (2004) describe an approach, which is implemented in Biomapper, to convert categorical data into quantitative distance and frequency metrics required by ENFA. Biomapper also produces spatially explicit data layers for visualization and use in a GIS. Additionally, the R software package 'adehabitat' (Calenge 2006) also carries out ENFA.

16.4.7 Generalized Dissimilarity Modelling

A relatively recent approach to SDM is the use of generalized dissimilarity modelling (GDM; Ferrier et al. 2007) which was initially used to model turnover in species diversity in community ecology (Ferrier 2002). Observed rates of turnover (change) between individuals units (e.g. pixels, land units) are compared and modelled as a function of environmental characteristics of those units. Because many pairwise comparisons can be made, GDM uses a matrix regression approach in which an n × n matrix of pairwise dissimilarity measures, for n units, represents the environmental predictor variables (non-categorical), and a matrix of the pairwise difference in species turnover the response variable. The GDM approach does not assume the relationship between response and predictor variables is linear and models two types of non-linearity. First, GDM assumes that the separation between the response dissimilarity between pixels and the environmental variables is curvilinear, rather than linear as in a traditional regression, and uses appropriate link and variance functions to model this. Second, GDM assumes that the rate at which change occurs is not constant and in fact changes along some environmental gradient. To model this non-linearity GDM uses splines to fit transforming functions to each environmental variable. Elith et al. (2006) extend this approach beyond modelling species diversity to modelling species distribution. To do this, a kernel regression is used on the output of the transformed environmental variables from the GDM to predict the likelihood of species occurrence. Using presence/pseudo-absence data with GDM should be considered cautiously (Ferrier et al. 2007). Statistical significance testing for choosing when to include (or drop) a predictor is carried out using matrix permutations (see Ferrier et al. 2007). Ferrier et al. (2007) have also developed the R software package 'gdm' for estimating GDMs.

16.4.8 Multivariate Adaptive Regression Splines (MARS)

Friedman (1991) introduced multivariate adaptive regression splines (MARS) as a flexible regression approach to model nonlinear relationships and can be used for both binary and continuous responses, and continuous and categorical predictors. It also bridge's Breiman's (2001) two modelling cultures, as it is a data mining tool, yet still uses a regression-based framework. It has recently been used to develop SDMs but is not yet commonly applied (Leathwick et al. 2005, 2006). MARS builds piecewise-linear relationships between the response and predictors, with differing slopes in different regions of the predictor variable space, with changes in slope at estimated knots. For each knot there is a pair of basis functions which describe the slope of the relationship on either side of it. Multiple knots can be present for a single predictor. To estimate the model, MARS first fits an overly complex model (forward-selection) with too many 'knots', including all predictors, which minimizes the residual error. MARS also can estimate parameters for interactions between two or more predictors, which can provide for an exceptionally complex model, especially considering that interactions may occur only over certain ranges of the predictors. These knots are subsequently removed (i.e. pruned) such that those knots providing little information to the model are removed (backward-selection). Variable (predictor) selection occurs here such that predictors providing little information are removed from the model. The final model has the lowest generalized cross-validation measure of mean squared error. In this way MARS differs from GLMs and GAMs in that model estimation and selection occur simultaneously, and a multi-model comparison need not be undertaken by the modeller, as MARS itself carries this out. Due to the underlying optimization algorithm, MARS also has a great, but yet rarely applied, capability to assess model selection questions with linear regression settings. Details of predictor interactions can also be specified such that, for example, only pairwise interactions may be considered, or that certain predictors cannot interact. Therefore, MARS is also a effective tool to assess, and describe specific interactions; a problem that greatly plagues the GLM models. Specific tuning parameters that must be specified include the maximum number of basis functions initially allowed during forward-selection, which should be high (Salford Systems 2001) and the minimum number of observations between knots which controls the level of smoothing. Larger spaces between knots results in smoother models. Leathwick et al. (2006) compared MARS and GAM models and reported similar results between the two approaches, and Elith et al. (2006) reported that MARS, with and without predictor interactions, performed modestly compared to other approaches. The complexity of a MARS model is related to the amount of data used for training it. With sparse data, complex relationships will not be able to be estimated (Wisz et al. 2008). MARS appear to be less strong when used to predict out of its training space (Huettmann, unpublished). MARS's strength appears to lie in its ability to model multi-responses for community-type data in which it shows very strong performance (Elith et al. 2006; Elith and Leathwick 2007; Leathwick et al. 2006) and a wide array of applications (Huettmann 2007 for marine applications).

16.5 Algorithm-Based Approaches

16.5.1 Neural Networks

Artificial Neural Networks (ANN) is a family of statistics that consists of an interconnected group of processing nodes (neurons) that can be used in non-linear statistical computation. ANN's are adaptive in nature, changing structure based on input that flows through the network. Each node in the network performs a simple task and weights between nodes lend structure to the entire network. Learning is based on cost constraint and gradient-decent which is an optimization approach to find the local minimum. Learning algorithms within ANN's are highly variable and include Evolutionary models, simulated annealing, and nonparametric models.

Neural Networks have been around in the machine learning community since the 1960s onwards. With mathematical and software progress, a push for ANNs came in the 1980s. Some ecological applications have used Neural Networks already for SDMs and elsewhere (Özesmi and Özesmi1999, Yen et al. 2004). However An unsupervised version of a neural net, Self Organizing Maps (SOM) is gaining popularity in cluster analysis and visualization (Kohonen 1995). SOM's utilize vector weights and distances between neurons to project the n-dimensional space of the input data to a 2 or 3 dimensional map, while maintaining the topology. SOM's are akin to multidimensional scaling and are a robust approach to both unsupervised clustering and visualization of high dimensional data.

16.5.2 Classification and Regression Trees

Classification and Regression Trees (CART) are nonparametric greedy algorithms that recursively partition the training data into a set of rules (Breiman et al. 1984). They have been among the leading algorithms in the machine learning community (Breiman 2001). Theoretically, the predicted value may be anything for which a function can be defined that returns a measure of impurity and a distance measure between impurities. CARTs are a powerful approach that can deal with mixed data types and do not violate assumptions of parametric statistics. CARTs can handle well correlated data structures. These approaches have gained popularity in ecology and elsewhere due to the easy interpretability of the resulting model (De'ath and Fabricius 2000). There are a large numbers of variants in CART models that apply different splitting criteria, pruning approaches, and validation. Here we focus on binary trees (multi-split criteria and mixed multi-split criteria also exist). Two common statistics used in splitting are the Gini Impurity Index (Breiman et al. 1984) and information gain based on entropy (Quinlan 1993). A critical parameter in controlling how far a tree grows is complexity. Unfortunately, this is a difficult parameter to define and resulting models are quite sensitive to it. High complexity will potentially overfit the model and low values can create too simple a tree, not adequately explaining the variation. Some prefer to grow a very complex tree and then apply an algorithm to prune the tree, thus avoiding over fit. Pruning is carried out through a cost-complexity approach. The cost can be defined by a metric such as sum of squared errors or deviance with a complexity penalty based on the size of the tree. Cross-validation can then be used to choose from the candidate trees. CART models are very attractive due to the interpretability of the node splits (rules), avoidance of parametric assumptions (i.e. distribution, independent residuals), and its ability to handle noisy data. It however, should be noted that although node splits are easily interpretable, inference from an entire tree, such as it occurs in complex ecological applications, is difficult. Pruning makes such inferences easier, and even more so, if one simply focuses on the predictions, their underlying tree model and the accuracy assessment from such approaches. Overfit has also proven to be a problem (Hastie et al. 2001). For a review of CART induction see for instance Bell (1996), O'Connor and Jones (1997), De'ath and Fabricius (2000) or Kothari and Dong (2001).

16.5.2.1 Bagging

Bagging is an ensemble technique (see entry below) that averages models across bootstrap aggregates (Breiman 1996). Bagging as such can virtually be applied to any number of machine learning or algorithmic models. Bagging generates n bootstrap samples (with replacement), builds a model for each and, then averages the resulting models across bootstrap aggregates. The idea behind this method is that the rule of large numbers demonstrates that a series of weak learners will reduce variance and converge on the true classification without overfitting the data (Breiman 1996). A classic example of bagging are Random Forests trees (Breiman 2001); see below. Bagging gets even more powerful when combined with boosting.

16.5.2.2 Boosting

Boosting is another ensemble method (see entry below) that uses the previous model in the ensemble to improve error, thus boosting accuracy. Commonly the CART algorithm is used in Boosting, however many machine learning/algorithmic approaches may be implemented. Boosting assigns a weight to each model based on classification error. At each iteration, weights are increased on the incorrectly classified classes to focus the algorithm on these cases. Together with bagging, boosting can get very powerful of an approach to modelling, e.g. in the TreeNet algorithm. For applications using the TreeNet algorithm implication see for instance Popp et al. (2007) and Craig and Huettmann (2008).

16.5.2.3 Random Forests

Random Forest (Breiman 2001) is a CART method based on Bagging that is generating considerable interest in the ecological community (Cutler et al. 2007; Magness et al. 2008). The Random Forest algorithm is executed by bootstrapping (with replacement) 63% of the data and generating a weak learner based on a CART for each bootstrap replicate. Within the pre-set specification (e.g. node depth and number of samples per node) each CART is unconstrained (grown to fullest) and prediction is accomplished by taking the 'majority votes' across all nodes in all random trees. At each replicate the data not used [out of bag (OOB)] to construct the tree are used for validation, providing a quasi-independent validation of model fit. Independent variables are randomly selected at each node and variable importance is assessed using the mean decrease in accuracy (MDA) by dividing the standard error by the misclassification rate. The number of variables randomly selected at each node is defined by m [commonly defined as sqrt(number of independent variables)].

The number of bootstrap replicates should be defined by the stabilization of OOB error, however it should be noted that the interaction of variables stabilizes at a slower rate than the error (pers. comm. A. Cutler), stressing the necessity of a large number of bootstrap replicates. Murphy et al. (accepted) demonstrated that models may be improved by a more parsimonious set of variables utilizing a model selection criteria based on the variable importance measures. In addition, they observed that multi-collinearity can affect model performance. Imbalance between classes has proven to reduce the power of ensemble learners (Chawla 2006). There have been few suggestions for addressing this problem (Chawla 2006; Chen et al. 2004) based on generating synthetic data. Evans and Cushman (2009) developed an iterative down-sampling method that proved robust in SDM.

Due to its unique approach to modelling, the use of Random Forest in SDM has proven robust and stable. It can handle abundance information, multicategorical or binary, e.g. presence and absence data. Prasad et al. (2006) and Rehfeldt et al. (2006) both demonstrated the utility of Random Forest for large scale prediction of tree species. Random Forest can easily be projected into new variable space, making it an appropriate algorithm for projective modeling such as climate change (Rehfeldt et al. 2006). Evans and Cushman (2009) predicted the probability of four tree species in northern Idaho at a landscape scale, demonstrating the utility of Random Forest as a model for SDM at fine scales.

16.5.3 Entropy Maximization

Entropy in information theory is a quantitative measure of randomness (uncertainty) in a random variable. The focus in this algorithm is on 'information'. The loss of information in a signal due to randomness is a relevant measure of the information contained in the signal. Recently, entropy maximization has shown to be a powerful tool in ecology (Harte et al. 2008; Phillips et al. 2006). Maximum entropy is indicated when partial information contained in a random variable exhibits the least biased (most uniform) distribution given all distributions of independent information or constraints. Phillips et al. (2004, 2006) introduced entropy maximization as an approach to modelling species distributions and developed associated software called Maxent. In the Maxent model a target probability distribution is identified based on an optimal convergence on the maximum entropy (Phillips et al. 2006). Due to the lack of interaction between independent variables, it is possible to elucidate how an independent variable is influencing the focal species. However, since the derived probability distribution is dependent on the observed data, sample bias can add error to the resulting predictions (Phillips et al. 2006). It is also not clear how to control the error component and overfit is an issue.

The Maxent model explicitly assumes that used locations are compared to a sample of available locations across the landscape. As such, presence-only data can be used in the analysis. Maxent can incorporate both continuous and categorical predictors and the functional relationship between predictor and response can take a variety of forms including linear and quadratic relationships, as well as threshold and hinge (i.e. piecewise) forms, and interactions between two variables (Phillips and Dudík 2008). Categorical variables are treated as indicators. The relationship between response and predictor can change at different ranges of the predictor values. For example, a single predictor could be related to the response via a linear relationship in its lowest range, via an interaction with another predictor in its middle range, and via a threshold function within it upper range. For each unique relationship between predictor and response, a unique weight is estimated, akin to a regression coefficient. Variables providing no information for predicting occurrence are weighted at zero. Hence, these models have the potential to become highly complex, particularly with increasing numbers of predictors.

Maxent creates a probability distribution across the entire study area, with each location (e.g. pixel) receiving a non-negative value, such that the sum of values of all locations in the study area equals one. This probability distribution is modelled by the predictors, and their various relationships with the response. Maxent estimates this model, and variable weights, by finding the combination of predictors, and their various functional relationships and interactions, which maximizes the log-likelihood of the model. To avoid overfitting and estimating a highly and overly complex model, the log-likelihood is penalized by a regularization parameter which increases as the complexity of the model increases (Phillips and Dudík 2008). Model selection is automated within Maxent and all predictors specified by the user, and various functional relationships, are considered; however not all may be retained in the final model. In a broad comparative analysis, modelling via entropy maximization has been demonstrated to provide models with often higher predictive ability than other approaches (Elith et al. 2006). However, Haegeman and Loreau (2008) argue that the mechanistic requirements for entropy maximization may not be met in all cases thus invalidating the resulting model. They further demonstrate that the specification of scales can also dramatically affect the validly of the model. Entropy maximization is an immature yet very powerful approach in ecology that needs further evaluation. However, care should be used when interpreting results, utilizing predictions, and extrapolating results into new variable space.

16.5.4 Genetic Algorithms

Genetic algorithms (GA) are an iterative optimization approach based on evolutionary theory. The GA generates random populations by encoding individuals as binary finite length vectors. A fitness, f(x), is assigned to each individual. The algorithm seeks to use selective combinations of individuals with optimal f(x) to generate an optimization. Through a stochastic process, mutation and cross-over are applied at every step to prevent the population from stagnating at a local optima and providing optimal combinations, respectively. Cross-over is achieved by selecting two finite length vectors with optimal f(x), randomly selecting bits from each and passing the result to the next iteration of the model. Mutation occurs when a low probability criteria is met, causing bits within the vector to be flipped.

One widely used GA in niche modelling is the Genetic Algorithm for Rule-Set Prediction (GARP; Stockwell and Peters 1999). GARP searches for relationships between presence locations and environmental features compared to random locations within the overall study area. GARP uses a genetic algorithm, a stochastic optimization tool, to uncover these relationships, and because the algorithm is stochastic, each run of the model can yield a unique solution. The stochasticity derives from the random selection of testing and training data used in each iteration of the algorithm. Olden et al. (2008) provide a detailed description of genetic algorithms in an ecological context. GARP identifies the relationship between presence locations and environmental features of the landscape via rule sets. The first step of the modelling process involves GARP choosing a method from a suite of possibilities. These include logistic regression, atomic, bioclimatic envelopes (i.e. range rules), or negated range rules.

Briefly, logistic regression is as described in Section 16.4.1.1 whereby the probability of presence is estimated for a location on the landscape, envelope (range) rules specify the range of some bioclimatic variables within which presence is recorded (or conversely for negated range rules non-presence is recorded), and atomic rules are those specifying categories or upper/lower thresholds of a variable in which presence (or absence) is recorded. Once a method is selected, a rule (i.e. functional relationship) is identified and the predictive accuracy of that rule is assessed. This process is iterated by identifying new rules until either 2,500 iterations is reached or the predictive accuracy of the overall rule-set converges (Anderson et al. 2003). Rules are retained if they increase the predictive accuracy of the rule-set. If not, the rule is omitted. The final rule-set is then the one used to predict and map the final species distribution model. The multiple methods used and potentially large number of rules can make interpretation of the final model difficult. Further, recent comparative work by Elith et al. (2006) indicates that GARPs predictive performance may not be as strong as other methods such as maximum entropy (but see Peterson et al. 2007).

16.5.5 Support Vector Machines (SVM)

A relatively new method in machine learning approaches are support vector machines (SVM; Kecman 2005), also known as kernel machines, which have recently been applied to SDMs (Drake et al. 2006; Guo et al. 2005; Shan et al. 2006). Guo et al. (2005) provide a good overview of SVMs, which can be used to model presence-only data (one-class SVMs) or presence/absence data (two-class SVMs). For one-class models in which only occurrences are predicted, SVMs use the predictor variables to find the smallest possible multivariate hypersphere in which all observations are included. In this sense SVMs produce models conceptually in line with Hutchinson's (1957) niche concept (Drake et al. 2006). Some outlying points are permitted to be outside the sphere, which is a trade-off (v) between its volume and rejected observations. The predictor data are assumed to be multivariate spherical and any categorical data must be converted to some numerical format prior to analysis, such as binary dummy coding. It is also recommended that data be scaled to range between 0 and 1 or +1 prior to analysis (Hsu et al. 2009). To account for the unlikely occurrence of the data being spherically distributed, kernel functions are applied to the data. In two-class SVMs, presence and absence are predicted by finding a hyperplane, which tries to maximize the separation between the two classes. This is termed the margin which is the distance between the nearest points in each class. The points on the boundaries of the classes are termed the support vectors. For one-class SVMs v is the only tuning parameter and for two-class SVMs C (a penalty parameter for misclassification) and γ (a parameter for the kernel) are the tuning parameters. Tuning parameters can either be assessed manually, or via a grid search over a vast possible number of parameter combinations and the performance measured by cross-validation (Hsu et al. 2009; Dimitriadou et al. 2009). Drake et al. (2006) demonstrated that SVMs are insensitive to collinearity in the predictors and perform equally as well as other SDM modelling approaches, even with relatively sparse data (e.g. 40 observations). However, Shan et al. (2006) commented that the predictive accuracy of SVMs may be offset by their relatively low comprehensibility in their SDM of the southern brown bandicoot (*Isoodon obesulus*). The output of SVMs, rather than parameters or weights, is the hypersphere or hyperplane. The data lying on the boundaries of these features are the support vectors, and thus predictions are made on new data by assessing whether or not they fall within the hypersphere, or in a specific category on either side of the hyperplane. Kecman (2005) notes that in SVMs the concept of parsimony refers to output having few support vectors as opposed to fewer model parameters or weights. In low dimensions, these features could be visualized but in higher dimensions visualization would be nearly impossible.

16.5.6 Ensemble Models

Ensemble models can offer a complex suite of models and optimization approaches for obtaining best possible predictions, and for each pixel. Starting out with high-performance algorithms, they have received attention as being currently among the best possible modeling techniques. However, their underlying software and approaches are diverse, and here we can simply present them on a general scheme. Ensemble models try to optimize for each case, row, pixel, a prediction that helps to improve the overall model accuracy (see the section below on Model Evaluation). Having such a 'truth' at hand can then allow to reach this goal. A relatively simply and straight forward approach to ensemble modeling is to have a 'suite of model algorithms', an ensemble, compute the best possible predictions for each pixel. The ensemble model would then pick the best possible algorithm that minimizes the deviation from 'truth' to the prediction for each pixel provided by the algorithms used in the suite of models. In that case, the absolute performance of the ensemble model depends a lot on the algorithms used in the 'suite'. Such methods are employed for instance in the Bayesian Model Averaging (BMA) in R.

Another, and conceptually even more powerful approach would be to start with a powerful suite of modeling algorithms. Then, break the data in an optimized set and apply to each section the best possible algorithm that achieves the best prediction for the nature of the data. In a certain way, this resembles the approach used in MARS, but allowing for many more modeling algorithms than the linear ones used there. Besides the question what algorithms may enter the suite, another crucial question is what optimization is used to obtain the optimal section breaks in the data (Elder 2003). Many of these questions enter the realm of hyperdimensional optimizations (similar to the Traveling Salesman problem), which are not well resolved in mathematics, yet. Resolving it though would offer much progress on the SDM approach, and to Ecology and global sustainability overall, and that tries to find the best possible prediction!

16.6 Model Evaluation

A final and critical stage of any modelling and predictive application is to evaluate the final model (Fielding and Bell 1997). Without an assessment of the predictive ability of a model, its utility for either ecological inference, generalization or management application is in doubt. One should note that one easily can produce and present a prediction surface. But without a relevant accuracy assessment it has virtually no scientific value (apart from having compiled and build up the GIS data and statistical analysis, helping to form a modelling infrastructure and culture). Here we provide a brief overview of a number of approaches which may be of use for evaluating the performance of SDMs. It should be noted that almost all true spatial model evaluations still have to be performed manually, and that no coherent tool exist to do it automatically all in one approach. Depending on the type of data used to estimate a model, different evaluation approaches may required. One of the first and classic approaches used is the confusion matrix (Fielding and Bell 1997). It is a simple table classifying evaluation data into one of four categories; presence and absence correctly predicted or incorrectly predicted. The criterion for determining whether a test observation was correctly or incorrectly predicted is problematic however as the confusion matrix is a binary approach to model evaluation although many modelling approaches generate predictions on a continuous scale (e.g. probability of occurrence). Thus, the arbitrary choice of cut-off (threshold) for determining occurrence can change the assessment of a model substantially (Manel et al. 2001). For example, if a probability of 0.5 were used, locations with predicted values of 0.49 and 0.51 would be classified differently, regardless of their ecological similarity. Further, this approach is usually not recommended for use-availability data, as availability data inherently include used locations, thus a confusion matrix would underestimate the performance of the model. A number of accuracy measures can be derived from the confusion matrix including sensitivity/specificity, omission/commission, Kappa, and prevalence, and are described in detail in Fielding and Bell (1997).

One approach used to avoid the arbitrary choice of threshold is the receiver operating characteristic curve (ROC; Fielding and Bell 1997; Pearce and Ferrier 2000; Fawcett 2006). The Kappa statistic and ROC are the two most commonly used evaluation metrics in machine learning classification problems. Rather than using a single arbitrary threshold, a range of thresholds are used and the proportions of correctly and incorrectly classified test data are tabulated. The ROC curve is defined as sensitivity plotted against 1-specificity. Sensitivity is a measure of the proportion of true positives whereas specificity is the proportion of false negatives (commission error). The inherent nature of the correlation between these terms gets used to describe model accuracy. An assessment of the balance between sensitivity and specificity can lend considerable insight into model performance at a class level. A curve is developed and the area under the ROC curve (AUC) is used as the global metric predicting the overall discriminatory ability of the model, where a value of 1.0 indicates a perfectly predicting model and a value of 0.5 implies the model predicts no better than chance. However, the ROC approach also requires true presences and absences and will also underestimate model performance when using use-availability data (Boyce et al. 2002). Phillips et al. (2006) comment that the use of the ROC for presence-only modelling approaches (e.g. Maxent) is valid if one interprets the AUC metric appropriately, as discriminating between use and random use. This is also noted by Anderson et al. (2003) for interpretation of confusion matrices.

Manel et al. (2001) and Lobo et al. (2007) critique the ROC as a measure of model performance. Reasons for this are: 1) the error components are weighted equally (Peterson et al. 2008); 2) can over-value a models of rare species (Manel et al. 2001); and 3) certain modelling techniques do not predict across the spectrum of probabilities thus, violating the assumption that the specificity spans the entire range of probabilities (0–1). Manel et al. (2001) recommended using the Kappa statistic to avoid these problems. However, Peterson et al. (2008) proposed modifying ROC by formulating a partial ROC, thus limiting the x-axis to the domain given the model under assessment. The two major limitations of the ROC is that it is only suited towards discrete data and few strategies exist for validating more than two classes. The Kappa statistic (Cohen 1960; Monserud and Leemans 1992) is a flexible metric for multi-class models that evaluates the agreement between classes and then adjust for amount of agreement that could be observer by random chance. The primary criticisms of Kappa are that it is not truly chance constrained because it does not account for the expected frequency of a class, and it does not make distinctions among various types and sources of disagreement. To overcome limitations in the Kappa a weighting function (Cohen 1968) has been implemented to account for near agreement and adjust for expectation in the frequency of observations.

Boyce et al. (2002) describe the k-fold classification procedure in which only the presence data are used to assess model performance. The entire dataset is partitioned into k folds (groups) and the model estimated k times with each run of the model using k-1 folds to train the model and the remaining fold to evaluate it, such that each fold is used once for model evaluation. For each run of the model predictions are generated and subsequently classified into a user-defined number of bins.

In each bin the frequency of presence locations from the testing fold are tallied, and a Spearman rank correlation coefficient estimated to assess model performance. A coefficient of +1.0 implies the model predicts presences perfectly. This approach can only assess the ability of the model to predict presences, not absences. A version of this approach has been included in the Biomapper software for ENFA (Hirzel et al. 2006). Wiens et al. (2008) recently introduced a three-way k-fold cross-validation to evaluate model performance over space and time.

Elith et al. (2005) introduced an "evaluation strip" as a method to evaluate the predictive ability of SDMs in relation to specific predictors. This more or less graphic method involves adding an evaluation strip to each environmental predictor's raster grid. This strip consists of values generated by the user to represent the full range of values present in that variable. The strip is placed in an area where no observations are found, and hence these added data are not included in model training. Once the model has been estimated, predictions are made onto these evaluation strips and the predictive performance of the model can be visualized graphically in relation to either one or two variables. This approach may be particularly valuable for assessing where greater certainty (or uncertainty) in model predictions may lie, in terms of predictor variable values.

In a famous assertion Chrisman (1989) referred to a spatial random field as "a distribution of possible realizations in which the true value lies". The uncertainty in a spatial random field that Chrisman (1989) referred to exists in any random field (measured variables) as a function of the statistical precision and accuracy. Uncertainty is an extremely relevant issue in spatial models where there is an error around our measurement, in either the dependent and/or independent variables, or we are predicting to an entire landscape. Validation methods for testing the accuracy of spatial prediction where no independent validation data exists has plagued the modelling community. Monte Carlo simulation is a useful tool for assessing both spatial and model uncertainties. By iteratively assessing the model given an error distribution, the uncertainty can readily be quantified and visualized. Conceptually, uncertainty can be quantified as either "error" or "a distribution of possible values" given an observation. Random fields may have measurement error that varies throughout the range of variability yielding higher uncertainty in a portion of the distribution (i.e. measurement error is not constant across all values of the predictor). By assessing the uncertainty of a model the error component of a prediction can be quantified and thus controlled for. When model predictions are used for management applications, for instance, knowledge of the spatial distribution of uncertainty may aid in decision-making. Sensitivity analysis is the measurement of the contribution of individual input parameters to the uncertainty in model predictions (Lilburne et al. 2006). This contribution is typically measured by making small adjustments to an input parameter and assessing the subsequent changes in model predictions. Knowledge of the sensitivity of model predictions to input parameters can highlight variables contributing greater uncertainty to model predictions and where additional resources may be required to obtain more accurate measurements. Sensitivity analysis is also useful in machine learning approaches where multiple interactions may be modelled.

Here we conclude our brief description of SDMs. Many more points can be made, and will develop over time. We envision for instance the required training of managers and the legal community in SDMs and their interpretation. We further promote 'Best Professional Practices', such as outlined in this book, and look with interest into concepts such as cloud computing (as basically implemented in Open Modeler already). We hope that the community interested in global sustainability will pick up these concepts and push them forward towards best possible applications.

References

Addicott JF, Aho JM, Antolin MF, et al (1987) Ecological neighborhoods: scaling environmental patterns. Oikos 49:340–346.

Aebischer NJ, Robertson PA, Kenward RE (1993) Compositional analysis of habitat use from animal radio-tracking data. Ecol 74:1313–1325.

Afshartous D, Wolf M (2007) Avoiding 'data snooping' in multilevel and mixed effects models. J R Stat Soc A 170:1035–1059.

Aitchison J (1986) The statistical analysis of compositional data. Chapman & Hall, London.

Anderson RP, Lew D, Peterson AT (2003) Evaluating predictive models of species' distributions: criteria for selecting optimal models. Ecol Model 162:211–232.

Arthur SM, Manly BFJ, McDonald LL, Garner GW (1996) Assessing habitat selection when availability changes. Ecol 77:215–227.

Austin, MP (1985) Continuum concept, ordination methods, and niche theory. Annu Rev Ecol Systemat 16:39–61.

Austin M (2007) Species distribution models and ecological theory: a critical assessment and some possible new approaches. Ecol Model 200:1–19.

Bailey LL, Hines JE, Nichols JD, MacKenzie DI (2007) Sampling design trade-offs in occupancy studies with imperfect detection: examples and software. Ecol Appl 17:281–290.

Baker WL (1995) Longterm response of disturbance landscapes to human intervention and global change. Landsc Ecol 10:143–159.

Ball LC, Doherty PF Jr, McDonald MW (2005) An occupancy modeling approach to evaluating a Palm Springs ground squirrel habitat model. J Wildl Manag 69:894–904.

Barry SC, Welsh AH (2002) Generalized additive modeling and zero inflated count data. Ecol Model 157:179–188.

Beaumont LJ, Pitman AJ, Poulsen M, Hughes L (2007) Where will species go? Incorporating new advances in climate modeling into projections of species distributions. Glob Chang Biol 13:1368–1385.

Bell JF (1996) Application of classification trees to the habitat preference of upland birds. J Appl Stat 23, 349–359.

Betts MG, Ganio L, Huso M, Som N, Huettmann F, Bowman J, Wintle JA (2008). The ecological importance of space in species distribution models: A comment on Dormann et al. Ecography 32:1–5.

Boyce MS, McDonald LL (1999) Relating populations to habitats using resource selection functions. Trends Ecol Evol 14:268–272.

Boyce MS, Vernier PR, Nielsen SE, Schmiegelow FKA (2002) Evaluating resource selection function. Ecol Model 157:281–300.

Breiman L, Friedman JH, Olshen RA, Stone CJ (1984) Classification and regression trees. Wadsworth, Monterey.

Breiman L (1996) Bagging predictors. Mach Learn (24)2:123–140

Breiman L (2001) Statistical modeling: the two cultures. Stat Sci 16:199–215.

Brezger A, Kneib T, Lang S (2008) BayesX: software for Bayesian inference in structured additive regression models.
Buckland ST, Anderson DR, Burnham KP, Laake JL, Borchers DL, Thomas L (eds) (2001) Advanced distance sampling. Oxford University Press, Oxford.
Burnham KP, Anderson DR (2002) Model selection and multimodel inference: a practical information-theoretic approach, 2nd edition. Springer, New York.
Buskirk SW, Millspaugh JJ (2006) Metrics for studies of resource selection. J Wildl Manag 70:358–366.
Cade BS, Noon BR (2003) A gentle introduction to quantile regression for ecologists. Front Ecol Environ 1:412–420.
Cade BS, Noon BR, Flather CH (2005) Quantile regression reveals hidden bias and uncertainty in habitat models. Ecol 86:786–800.
Calenge C (2006) the package adehabitat for the R software: a tool for the analysis of space and habitat use by animals. Ecol Model 197:516–519.
Cameron AC, Trivedi PK (1998) Regression analysis of count data. Cambridge University Press, Cambridge.
Casella G, Edward IG (1992) Explaining the Gibbs sampler. Am Stat 46:167–174.
Chawla NV, Lazarevic A, Hall LO, Bowyer KW (2003) SMOTEboost: Improving prediction of the minority class in boosting. In: 7th European Conference on Principles and Practice of Knowledge Discovery in Databases, pp. 107–119.
Chase JM, Leibold MA (2003) Ecological niches. University of Chicago Press, Chicago, IL.
Chatterjee S, Hadi AS (2006) Regression analysis by example, 4th edition. Wiley, New York.
Chen C, Liaw A, Breiman L (2004) Using random forest to learn imbalanced data. http://oz.berkeley.edu/users/chenchao/666.pdf
Chrisman N (1989) Error in categorical maps: Testing versus simulation. In 9th Int. Symp Comput Assis Cart, ASPRS/ACSM, pp. 521–529.
Clark JS (2005) Why environmental scientists are becoming Bayesians. Ecol Lett 8:2–14.
Cohen J (1960) A coefficient of agreement for nominal scales. Educ Psych Meas 20:37–46,
Cohen J (1968) Weighted kappa: nominal scale agreement with provision for scaled disagreement or partial credit. Psych Bull 70:213–20.
Collins SL, Knapp AK, Briggs JM, et al (1998) Modulation of diversity by grazing and mowing in native tallgrass prairie. Science 280:745–747.
Cooch E, White G (2007) Program MARK: a gentle introduction, 6th edition. www.phidot.org.
Cooper AB, Millspaugh JJ (1999) The application of discrete choice models to wildlife resource selection studies. Ecol 80:566–575.
Cooper WS (1913) The climax forest of Isle Royale, Lake Superior, and its development. Botanical Gaz 55:1–235.
Craig, E., and F. Huettmann. (2008). Using "blackbox" algorithms such as TreeNet and Random Forests for data-mining and for finding meaningful patterns, relationships and outliers in complex ecological data: an overview, an example using golden eagle satellite data and an outlook for a promising future. Chapter IV in Intelligent Data Analysis: Developing New Methodologies through Pattern Discovery and Recovery (Hsiao-fan Wang, Ed.). IGI Global, Hershey, PA, USA.
Craiu RV, Duchesne T, Fortin D (2008) Inference methods for the conditional logistic regression model with longitudinal data. Biom J 50:97–109.
Cushman SA, McGarigal K (2003) Hierarchical, multiscale decomposition of species-environment relationships. Landsc Ecol 17:637–646.
Cushman SA, McKenzie D, Peterson DL, et al (2007) Research agenda for integrated landscape modeling. Gen. Tech. Rep. RMRSGTR-194. US Department of Agriculture, Forest Service, Rocky Mountain Research Station. Fort Collins.
De'ath G, Fabricius KE (2000) Classification and regression trees: a powerful yet simple technique for the analysis of complex ecological data. Ecol 81:3178–3192.
Dimitriadou E, Hornik K, Leisch F, Meyer D, Weingessel A (2009) e1071: Misc functions of the Department of Statistics (e1071), TU Wien. R package version 1.5–19.

Donovan TM, Hines J (2007) Exercises in occupancy modelling and estimation. www.uvm.edu/envnr/vtcfwru/spreadsheets/occupancy/occupancy.htm.
Dormann CF (2007) Effects of incorporating spatial autocorrelation into the analysis of species distribution data. Glob Ecol Biogeogr 16:129–138.
Drake JM, Randin C, Guisan A (2006) Modelling ecological niches with support vector machines. J Appl Ecol 43:424–432.
Elder JF (2003) The Generalization Paradox of Ensembles. J Computational Graph Stat 12: 853–864.
Elith J, Leathwick J (2007) Predicting species distributions from museum and herbarium records using multiresponse models fitted with multivariate adaptive regression splines. Divers Dist 13:265–275.
Elith J, Graham CH, Anderson RP, et al (2006) Novel methods improve prediction of species' distributions from occurrence data. Ecogr 29:129–151.
Elith J, Ferrier S, Huettmann F, Leathwick J (2005) The evaluation strip: a new and robust method for plotting predicted responses from species distribution models. Ecol Model 186:280–289.
Ellison, AM (2004) Bayesian inference in ecology. Ecol Lett 7:509–520.
Evans JS, Cushman SA (2009) Gradient modeling of conifer species using random forests. Landsc Ecol 24:678–683.
Fawcett, T (2006) An introduction to ROC analysis. Pattern Recognit Lett 27:861–874.
Ferrier S (2002) Mapping spatial pattern in biodiversity for regional conservation planning: where to go from here? Syst Biol 51:331–363.
Ferrier S, Drielsma M, Manion G, Watson G (2002) Extended statistical approaches to modelling spatial pattern in biodiversity in northeast New South Wales. II. Community-level modelling. Biodivers Cons 11:2309–2338.
Ferrier S, Manion G, Elith J, Richardson K (2007) Using generalized dissimilarity modelling to analyze and predict patterns of beta diversity in regional biodiversity assessment. Divers Dist 13:252–264.
Fielding AH, Bell JF (1997) A review of methods for the assessment of prediction errors in conservation presence/absence models. Env Cons 24:38–49.
Fletcher D, MacKenzie DI, Villouta E (2005) Modelling skewed data with many zeros: a simple approach combining ordinary and logistic regression. Env Ecol Stat 12:45–54.
Friedman JH (1991) Multivariate adaptive regression splines (with discussion). Ann Stat 19:1–141.
Gelfand AE, Latimer A, Wu S, Silander JA Jr (2006) Building statistical models to analyze species distributions. In: Clark JS, Gelfand AE (eds) Hierarchical modelling for the environmental sciences: statistical methods and applications, pp. 77–97. Oxford, New York.
Gelman A, Hill J (2007) Data analysis using regression and multilevel/hierarchical models. Cambridge University Press, Cambridge.
Gelman A, Carlin JB, Stern HS, Rubin DB (2004) Bayesian data analysis, 2nd edition. Chapman & Hall/CRC, New York.
Getz WM, Wilmers CC (2004) A local nearest-neighbour convex-hull construction of home ranges and utilization distributions. Ecography 27:489–505.
Getz WM, Fortmann-Roe S, Cross PC, Lyons AJ, et al (2007) LoCoH: nonparametric kernel methods for constructing home ranges and utilization distributions. PLOS Biol 2:e207. doi:10.1371/journal.pone.0000207
Gillies CS, Hebblewhite M, Nielsen SN, et al (2006) Application of random effects to the study of resource selection by animals. J Anim Ecol 75:887–898.
Glenn SM, Collins SL (1992) Effects of scale and disturbance on rates of immigration and extinction of species in prairies. Oikos 63:273–280.
Graham CH, Elith J, Hijmans RJ, et al (2008) The influence of spatial errors in species occurrence data used in distribution models. J Appl Ecol 45:239–247.
Graham CH, Ferrier S, Huettmann F, et al (2004) New development in museum-based informatics and applications in biodiversity analysis. Trends Ecol Evol 19:497–503.

Gu W, Swihart RK (2004) Absent or undetected? Effects of non-detection of species occurrence on wildlife-habitat models. Biol Cons 116:195–203.
Guisan A, Edwards TC Jr, Hastie T (2002) Generalized linear and generalized additive models in studies of species distributions: setting the stage. Ecol Model 157:89–100.
Guisan A, Graham CH, Elith J, et al (2007) Sensitivity of predictive species distribution models to change in grain size. Divers Dist 13:332–340.
Guisan A, Thuiller W (2005) Predicting species distribution: offering more than simple habitat models. Ecol Lett 8:993–1009.
Guisan A, Zimmermann NE (2000) Predictive habitat distribution models in ecology. Ecol Model 135:147–186.
Guo Q, Kelly M, Graham CH (2005) Support vector machines for predicting distribution of sudden oak death in California. Ecol Model 182:75–90.
Haegeman B, Loreau M (2008) Limitations of entropy maximization in ecology. Oikos 117:1700–1710
Hall LS, Krausman PR, Morrison ML (1997) The habitat concept and a plea for common terminology. Wildl Soc Bull 25:173–182.
Hand DJ, Till RJ (2001) A simple generalization of the area under the ROC curve to multiple class classification problems. Mach Learn 45:171–186
Harte J, Zillio T, Conlisk E, Smith AB (2008) Maximum entropy and the state-variable approach to macroecology. Ecol 89:2700–2711.
Hastie T, Tibshirani RJ (1990) Generalized additive models. Chapman & Hall, London.
Hastie T, Tibshirani RJ (1996) Discriminant analysis by Gaussian mixtures. J R Stat Soc Ser B 58:155–176.
He F, Gaston KJ (2000) Estimating species abundance from occurrence. Am Nat 156:553–559.
Hepinstall JA, Marzluff JM, Handcock MS, Hurvitz P (2004) Incorporating resource utilization distributions into the study of resource selection: dealing with spatial autocorrelation. In: Huzurbazur SV (ed) Resource selection methods and applications, pp. 12–19. Omnipress, Madison.
Hirzel AH, Guisan A (2002) Which is the optimal sampling strategy for habitat suitability modelling. Ecol Model 157:331–341.
Hirzel AH, Hausser J, Chessel D, Perrin N (2002) Ecological niche factor analysis: how to compute habitat-suitability maps without absence data? Ecol 83:2027–2036.
Hirzel AH, Posse B, Oggier P.-A., et al (2004) Ecological requirements of reintroduced species and the implications for release policy: the case of the bearded vulture. J Appl Ecol 41:1103–1116.
Hirzel AH, Le Lay G, Helfer V, Randin C, Guisan A (2006) Evaluating the ability of habitat suitability models to predict species presences. Ecol Model 199:142–152.
Hobbs NT, Hanley TA (1990) Habitat evaluation: do use/availability data reflect carrying capacity? J Wildl Manag 54:515–522.
Hochachka WM, Caruana R, Fink D, et al (2007) Data-mining discovery of pattern and process in ecological systems. J Wildl Manag 71:2427–2437.
Hosmer DW, Lemeshow S (2000) Applied logistic regression, 2nd edition. Wiley, New York.
Hsu C-W, Chang C-C, Lin C-J (2009) A practical guide to support vector classification. http://www.csie.ntu.edu.tw/~cjlin/papers/guide/guide.pdf
Huettmann F (2007). Constraints, suggested solutions and an outlook towards a new digital culture for the oceans and beyond: experiences from five predictive GIS models that contribute to global management, conservation and study of marine wildlife and habitat, in: Vanden Berghe, E. et al. (Ed.) Proceedings of 'Ocean Biodiversity Informatics': an international conference on marine biodiversity data management Hamburg, Germany, 29 November – 1 December, 2004. IOC Workshop Report, 202, VLIZ Special Publication 37: pp. 49–61. http://www.vliz.be/vmdcdata/imis2/imis.php?module=ref&refid=107201
Huettmann F. and A.W. Diamond (2006). Large-Scale Effects on the Spatial Distribution of Seabirds in the Northwest Atlantic. Landscape Ecology 21: 1089–1108.
Hutchinson GE (1957) Concluding remarks. Cold Spring Harb Symp Quant Biol 22:415–427.

Huzurbazar SV (ed) (2003) Resource selection methods and applications. Omnipress, Madison, WI.
Jachowski DS (2007) Resource selection by black-footed ferrets in relation to the spatial distribution of prairie dogs. Thesis, University of Missouri-Columbia, Columbia.
Johnson CJ, Seip DR (2008) Relationship between resource selection, distribution, and abundance: a test with implications to theory and conservation. Pop Ecol 50:145–157.
Johnson CJ, Nielsen SE, Merrill EH, et al (2006) Resource selection functions based on use-availability data: theoretical motivation and evaluation methods. J Wildl Manag 70:347–357.
Johnson CJ, Seip DR, Boyce MS (2004) A quantitative approach to conservation planning: using resource selection functions to map the distribution of mountain caribou at multiple scales. J Appl Ecol 41:238–251.
Johnson DH (1980) The comparison of usage and availability measurements for evaluating resource preference. Ecol 61:65–71.
Kearney M (2006) Habitat, environment and niche: what are we modelling? Oikos 115:186–191.
Keating KA, Cherry S (2004) Use and interpretation of logistic regression in habitat-selection studies. J Wildl Manag 68:774–789.
Kecman V (2005) Support vector machines: an introduction. In: Wang L (ed) Support vector machines: theory and applications, pp. 1–47. Springer, New York.
Kernohan BJ, Gitzen RA, Millspaugh JJ (2001) Analysis of animal space use and movements. In Millspaugh JJ, Marzluff JM (eds) Radio tracking and animal populations, pp 125–166. Academic, San Diego, CA.
Kothari R, Dong M (2001) Decision Trees for Classification: A Review and Some New Results. In: Pal SR, Pal NR (eds) Lecture Notes in Pattern Recognition. World Scientific, Singapore.
Koenker R (2005) Quantile regression. Cambridge University Press, Cambridge.
Kohonen T (1995) Self-Organizing Maps. Series in Information Sciences, Vol. 30. 2nd edition. Springer, Heidelberg.
Kynn, M (2005) Eliciting expert knowledge for Bayesian logistic regression in species habitat modelling. Dissertation. Queensland University of Technology, Brisbane.
La Morgia V, Bona F, Badino G (2008) Bayesian modelling procedures for the evaluation of changes in habitat suitability: a case study of roe deer in the Italian Alps. J Appl Ecol. doi: 10.1111/j.1365-2664.2008.01452.x.
Latimer AM, Wu S, Gelfand AE, Silander JA Jr (2006) Building statistical models to analyze species distributions. Ecol Appl 16:33–50.
Leathwick JR, Elith J, Hastie T (2006) Comparative performance of generalized additive models and multivariate adaptive regression splines for statistical modelling of species distributions. Ecol Model 199:188–196.
Leathwick JR, Rowe D, Richardson J, et al (2005) Using multivariate adaptive regression splines to predict the distributions of New Zealand's freshwater diadromous fish. Freshw Biol 50:2034–2052.
Lele SR, Keim JL (2006) Weighted distributions and estimation of resource selection probability functions. Ecol 87:3021–3028.
Leopold A (1933) Game management. Charles Scribner, New York.
Levin SA (1992) The problem of pattern and scale in ecology. Ecol 73:1943–1967.
Lilburne L, Gatelli D, Tarantola S (2006) Sensitivity analysis on spatial models: a new approach. In: Caetano M and Painho M (eds) Seventh International Symposium on Spatial Accur Assess Nat Res Env Sci, pp 329–338.
Lobo JM, Jimenez-Valverde A, Real R (2007) AUC: a misleading measure of the performance of predictive distribution models. Global Ecol Biogeogr 17:145–151.
Long JS (1997) Regression models for categorical and limited dependent variables. Sage, Thousand Oaks.
Lunn DJ, Thomas A, Best N, Spiegelhalter D (2000) WinBUGS-a Bayesian modelling framework: concepts, structure and extensibility. Stat Comp 10:325–337.
MacKenzie DI (2005) What are the issues with presence-absence data for managers? J Wildl Manag 69:849–860.

MacKenzie DI (2006) Modeling the probability of resource use: the effect of, and dealing with, detecting a species imperfectly. J Wildl Manag 70:367–374.
MacKenzie DI, Nichols JD, Lachman GB, et al (2002) Estimating site occupancy rates when detection probabilities are less than one. Ecol 83:2248–2255.
MacKenzie DI, Nichols JD, Royle JA, et al (2005) Occupancy estimation and modelling: inferring patterns and dynamics of species occurrence. Elsevier, San Diego, CA.
MacKenzie, DI, Royle JA (2005) Designing occupancy studies: general advice and allocating survey effort. J Appl Ecol 42:1105–1114.
Maclure M, Willett WC (1987) Misinterpretation and misuse of the kappa statistic. Am J Epidemiol 126:161–169.
Magness, D.R., F. Huettmann, and J.M. Morton. (2008). Using Random Forests to provide predicted species distribution maps as a metric for ecological inventory & monitoring programs. Pages 209–229 in T.G. Smolinski, M.G. Milanova & A-E. Hassanien (eds.). Applications of Computational Intelligence in Biology: Current Trends and Open Problems. Studies in Computational Intelligence, Vol. 122, Springer, Berlin. 428 pp.
Manel S, William HC, Ormerod SJ (2001) Evaluating presence-absence models in ecology: the need to account for prevalence. J Appl Ecol 38:921–931
Manly BFJ, McDonald LL, Thomas D, et al (2002) Resource selection by animals: statistical design and analysis for field studies, 2nd edition. Kluwer, Boston, MA.
Martin TG, Wintle BA, Rhodes JR, et al (2005) Zero tolerance ecology: improving ecological inference by modelling the source of zeros. Ecol Lett 8:1235–1246.
Marzluff JM, Millspaugh JJ, Hurvitz P, Handcock MA (2004) Relating resources to a probabilistic measure of space use: forest fragments and Steller's jays. Ecol 85:1411–1427.
McCarthy MA (2007) Bayesian methods for ecology. Cambridge University Press, Cambrige.
McCracken ML, Manly BFJ, Vander Heyden M (1998) The use of discrete-choice models for evaluating resource selection. J Agric Biol Environ Stat 3:268–279.
McCullagh P, Nelder JA (1989) Generalized linear models, 2nd edition. Chapman & Hall, London.
McDonald TL, Manly BFJ, Nielson RM, Diller LV (2006) Discrete-choice modelling in wildlife studies exemplified by northern spotted owl nighttime habitat selection. J Wildl Manag 70:375–383.
McLoughlin PD, Boyce MS, Coulson T, Clutton-Brock T (2006) Lifetime reproductive success and density-dependent, multi-variable resource selection. Proc R Soc B 273:1449–1454.
McLoughlin PD, Dunford JS, Boutin S (2005) Relating predation mortality to broad-scale habitat selection. J Anim Ecol 74:701–707.
Millspaugh JJ, Nielson RM, McDonald L, et al (2006) Analysis of resource selection using utilization distributions. J Wildl Manag 70:384–395.
Mitchell SC (2005) How useful is the concept of habitat? – A critique. Oikos 110:634–638.
Mobæk R, Mysterud A, Loe LE, Holand Ø, Austrheim G (2009) Density dependent and temporal variability in habitat selection by a large herbivore; an experimental approach. Oikos 118:209–218.
Monserud R A, Leemans R (1992) Comparing global vegetation maps with the *kappa* statistic. Ecol Model 62:275–293.
Mooney HA, Godron M (eds) (1983) Disturbance and ecosystems. Springer, New York.
Murphy MA, Evans JS, Storfer AS (Accepted) Quantifying ecological process at multiple spatial scales using landscape genetics: *Bufo boreas* connectivity in Yellowstone National Park. Ecol.
Nielsen SE, Johnson CJ, Heard DC, Boyce MS (2005) Can models of presence-absence be used to scale abundances? Two case studies considering extremes in life history. Ecography 28:197–208.
Nielsen SN, Boyce MS, Stenhouse GB, Munro RHM (2002) Modeling grizzly bear habitats in the Yellowhead ecosystem of Alberta: taking autocorrelation seriously. Ursus 13:45–56.
O'Connor R, Jones MT (1997) Hierarchical models to index the ecological health of the nation. In: Transactions of the 62nd North American Wildlife and Natural Resources Conference, pp. 501–608.

Özesmi SL, Özesmi U (1999) An artificial neural network approach to spatial habitat modelling with interspecific interaction. Ecol Model 116: 15–31.
Olden JD, Lawler JJ, Poff NL (2008) Machine learning methods without tears: a primer for ecologists. Q Rev Biol 83:171–193.
Popp JD, Neubauer D, Paciulli L, Huettmann F (2007). Using TreeNet for Identifying Management Thresholds of Mantled Howling Monkeys' Habitat Preferences on Ometepe Island, Nicaragua, on a Tree and Home Range Scale J Medical Biolog Sciences 1(2): 1–14. http://www.scientificjournals.org/journals2007/articles/1096.pdf
Pearce J, Boyce MS (2006) Modelling distribution and abundance with presence-only data. J Appl Ecol 43:405–412.
Pearce J, Ferrier S (2000) Evaluating the predictive performance of habitat models developed using logistic regression. Ecol Model 133:225–245.
Pearce J, Ferrier S (2001) The practical value of modelling relative abundance of species for regional conservation planning: a case study. Biol Cons 98:33–43.
Peterson AT, Papes M, Soberon J (2008) Rethinking receiver operating characteristic analysis applications in ecological modelling. Ecol Model 213:63–72
Peterson AT (2003) Predicting the geography of species' invasions via ecological niche modelling. Q Rev Biol 78:419–433.
Peterson AT (2006) Uses and requirements of ecological niche models and related distributional models. Biodiver Inform 3:59–72.
Peterson AT, Papeş M, Eaton M (2007) Transferability and model evaluation in ecological niche modelling: a comparison of GARP and Maxent. Ecography 30:550–560.
Phillips SJ, Dudík M, Schapire RE (2004) A maximum entropy approach to species distribution modelling. Proc 21st Conf Mach Learn 472–486.
Phillips SJ, Anderson RP, Schapire RE (2006) Maximum entropy modelling of species geographic distributions. Ecol Model 190:231–259.
Phillips SJ, Dudík M (2008) Modeling of species distributions with Maxent: new extensions and a comprehensive evaluation. Ecography 31:161–175.
Pinheiro JC, Bates DM (2000) Mixed-effects models in S and S-Plus. Springer, New York.
Potts JM, Elith J (2006) Comparing species abundance models. Ecol Model 199:153–163.
Prasad AM, Iverson LR, Liaw A (2006) Random forests for modeling the distribution of tree abundances. Ecosyst 9:181–199.
Pulliam HR (2000) On the relationship between niche and distribution. Ecol Lett 3:349–361.
Quinlan J R (1993) C4.5: Programs for Machine Learning. Morgan Kaufmann, San Fransisco, CA.
Rehfeldt GE, Crookston NL, Warwell MV, Evans JS (2006) Empirical analyses of plant-climate relationships for the western United States. Int J Plant Sci 167:1123–1150.
Reiners WA, Lang GE (1979) Vegetational patterns and processes in the balsam fir zone, White Mountains, New Hampshire. Ecol 60:403–417.
Rittenhouse CD, Millspaugh JJ, Cooper AB, et al (2008) Modeling resource selection using polytomous logistic regression and kernel density estimates. Environ Ecol Stat 15:39–47.
Salford Systems (2001) MARS 2.0 user's guide. Salford, San Diego, CA.
Sattler T, Bontadina F, Hirzle AH, Arlettaz R (2007) Ecological niche modelling of two cryptic bat species calls for a reassessment of their conservation status. J Appl Ecol 44:1188–1199.
Sawyer H, Nielson RM, Lindzey F, McDonald LL (2006) Winter habitat selection of mule deer before and during development of a natural gas field. J Wildl Manag 70:396–403.
Schneider DC (1994) Quantitative ecology: spatial and temporal scaling. Academic, San Diego, CA.
Scott JM, Heglund PJ, Samson F, et al (eds) (2002) Predicting species occurrences: issues of accuracy and scale. Island, Covelo.
Seaman DE, Powell, RA (1996) An evaluation of the accuracy of kernel density estimators for home range analysis. Ecology 77:2075–2085.
Shan Y, Paull D, McKay RI (2006) Machine learning of poorly predictable ecological data. Ecol Model 195:129–138.
Skrondal A, Rabe-Hesketh S (2004) Generalized latent variable modeling: multilevel, longitudinal, and structural equation models. Chapman & Hall, Boca Raton, FL.

Soberón J (2007) Grinnelian and Eltonian niches and geographic distribution of species. Ecol Lett 10:1115–1123.
Soberón J, Peterson AT (2005) Interpretation of models of fundamental ecological niches and species' distributional areas. Biodiver Inform 2:1–10.
Sousa WP (1984) The role of disturbance in natural communities. Ann Rev Ecol Systematics 15:353–391.
Stockwell D, Peters D (1999) The GARP modelling system: problems and solutions to automated spatial prediction. Int J Geog Inf Sci 13:143–158.
Strickland MD, McDonald LL (2006) Introduction to the special section on resource selection. J Wildl Manag 70:321–323.
Sutton T, de Giovanni R, de Siqueira MF (2007) Introducing openModeller: a fundamental niche modelling framework. OSGeo J 1:1–6.
Ter Braak CJF (1986) Canonical correspondence analysis: a new eigenvector technique for multivariate direct gradient analysis. Ecology 67:1167–1179.
Thomas DL, Taylor EJ (1990) Study designs and tests for comparing resource use and availability. J Wildl Manag 54:322–330.
Thomas DL, Taylor EJ (2006) Study designs and tests for comparing resource use and availability II. J Wildl Manag 70:324–336.
Thomas DL, Johnson D, Griffith B (2006) A Bayesian random effects discrete-choice model for resource selection: population-level selection inference. J Wildl Manag 70:404–412.
Turner MG, Gardner RH, O'Neill RV (2003) Landscape ecology in theory and practice. Springer, New York.
Tyre AJ, Tenhumberg H, Field SA, et al (2003) Improving precision and reducing bias in biological surveys: estimating false-negative error rates. Ecol Appl 13:1790–1801.
Van Horne B (1983) Density as a misleading indicator of habitat quality. J Wildl Manag 47:893–901.
Vaz S, Martin CS, Eastwood PD, et al (2008) Modelling species distributions using regression quantiles. J Appl Ecol 45:204–217.
Venables WN, Ripley BD (2002) Modern applied statistics with S, 4th edition. Springer, New York.
Watt AS (1947) Pattern and process in the plant community. J Ecol 35:1–22.
Welsh AH, Cunningham RB, Donnelly CF, Lindenmayer DB (1996) Modelling the abundance of rare species: statistical models for counts with extra zeros. Ecol Model 88:297–308.
White PS (1979) Pattern, process and natural disturbance in vegetation. Botanical Rev 45:229–299.
Whittaker RH (1975) Communities and Ecosystems. Macmillan, New York.
Wiens JA (1989) Spatial scaling in ecology. Funct Ecol 3:385–397.
Wiens TS, Dale BC, Boyce MS, Kershaw PG (2008) Three way *k*-fold cross-validation of resource selection functions. Ecol Model 212:244–255.
Wintle BA, Bardos DC (2006) Modeling species-habitat relationships with spatially autocorrelated observational data. Ecol Appl 16:1945–1958.
Wisz MS, Hijmans RJ, Li J, et al (2008) Effects of sample size on the performance of species distribution models. Divers Dist 14:763–773.
Wood SN (2006) Generalized additive models: an introduction with R. CRC/Chapman & Hall, London.
Wood SN, Augustin NH (2002) GAMs with integrated model selection using penalized regression splines and applications to environmental modelling. Ecol Model 157:157–177.
Worton BJ (1989) Kernel methods for estimating the utilization distribution in home-range studies. Ecol 70:164–168.
Yen P, Huettmann F, Cooke F (2004). Modelling abundance and distribution of Marbled Murrelets (*Brachyramphus marmoratus*) using GIS, marine data and advanced multivariate statistics. Ecol Mod 171: 395–413.

Appendix – Freely Available Software Resources

R Resources

R – The R Project for Statistical Computing (http://cran.r-project.org) – comprehensive free software package
Relevant R packages available for download from the CRAN website

- 'adehabitat' (ecological niche factor analysis)
- 'quantreg' (quantile regression)
- 'gbm', 'rpart', 'randomForest', 'party', 'tree', 'ipred' (tree-based models)
- 'gdm' (generalized dissimilarity modeling; http://www.biomaps.net.au/gdm)
- 'e1071' (support vector machines)
- 'caret' (tuning methods for a variety of machine learning models)
- 'earth', 'mda' (multivariate adaptive regression splines)
- 'GAMBoost', 'gam', 'mgcv', 'mda' (generalized additive models)
- 'caret', 'nnet' (neural networks)
- 'lme4', 'mgcv' (mixed models)
- 'R2WinBugs' (Bayesian models)
- R Task Views
- Bayesian (http://cran.r-project.org/web/views/Bayesian.html)
- Cluster (http://cran.r-project.org/web/views/Cluster.html)
- Environmetrics (http://cran.r-project.org/web/views/Environmetrics.html)
- Machine Learning (http://cran.r-project.org/web/views/MachineLearning.html)
- Spatial (http://cran.r-project.org/web/views/Spatial.html)

Model-Specific Software Resources

- Biomapper (http://www2.unil.ch/biomapper) – ecological niche factor analysis
- Maxent (http://www.cs.princeton.edu/~schapire/maxent) –entropy maximization
- DesktopGARP (http://www.nhm.ku.edu/desktopgarp) – genetic algorithm
- GeoSVM (http://www.unm.edu/~wyzuo/GEO.htm) – support vector machines
- LibSVM (http://www.csie.ntu.edu.tw/~cjlin/libsvm) – support vector machines
- OpenModeller (http://openmodeller.sourceforge.net/) – ensemble niche modeling
- Program MARK (http://www.phidot.org/software/mark/) – occupancy models
- Presence (http://www.mbr-pwrc.usgs.gov/software/presence.html) – occupancy models
- BLOSSOM (www.fort.usgs.gov/Products/Software/Blossom) – quantile regression
- Salford Systems Modeling Suite (http://www.salford-systems.com/) – CART, MARS, RandomForest and TreeNet

- Microsoft Excel Tools
- Neural Network Models in Excel (http://www.geocities.com/adotsaha/NNinExcel.html)
- Classification Tree in Excel (C4.5 algorithm) (http://www.geocities.com/adotsaha/CTree/CtreeinExcel.html)
- Self Organizing Map (SOM) in Excel (http://www.geocities.com/adotsaha/NN/SOMinExcel.html)
- BugsXLA, Bayesian Analysis in Excel (http://www.axrf86.dsl.pipex.com/)
- Bayesian Programs
- WinBUGS (http://www.mrc-bsu.cam.ac.uk/bugs/)
- OpenBUGS (http://mathstat.helsinki.fi/openbugs/)
- BayesX (http://www.stat.uni-muenchen.de/ bayesx/bayesx.html)
- Elicitor – Bayesian expert opinion models (http://silmaril.math.sci.qut.edu.au/~whateley)

Miscellaneous Spatial Software Resources

- Open Geospatial Consortium (http://en.wikipedia.org/wiki/Open_Geospatial_Consortium)
- CrimeStat (http://www.icpsr.umich.edu/CRIMESTAT/)
- Passage (http://www.passagesoftware.net/) Pattern analysis, spatial statistics, and geographic exegesis
- SADA (http://www.tiem.utk.edu/~sada/index.shtml) – Spatial Analysis and Decision Assistance
- SAM (http://www.ecoevol.ufg.br/sam/) – Spatial Analysis in Macroecology
- GeoDA (http://geodacenter.asu.edu/) – Exploratory spatial data analysis
- 'ade4' (http://pbil.univ-lyon1.fr/ADE-4/home.php?lang=eng) – R package for analytical functions to analyze ecological and environmental data in the framework of Euclidean exploratory methods
- STARS (http://regionalanalysislab.org/?n=STARS) –Space-Time Analysis of Regional Systems
- SatScan (http://www.satscan.org/) – Software for spatial, temporal, and space-time models
- Dr Pierre Legendre (http://www.bio.umontreal.ca/legendre/indexEn.html) – miscellaneous programs and scripts
- Weka (http://www.cs.waikato.ac.nz/ml/weka/) – Machine learning tools
- Diva-GIS (http://www.diva-gis.org

Chapter 17
Landscape Genetics

Kevin S. McKelvey, Samuel A. Cushman, and Michael K. Schwartz

In reading this book, you have observed that the spatial data used in landscape ecology come from many sources and in many forms. For many organisms, these data take the form of presence or absence at a location, or numbers of individuals at that same location. For species such as trees, where huge size differences exist between individuals, indices such as basal area, metric tons per hectare, or canopy cover are more useful than counts. For any measured species that is handled (or sampled noninvasively; Taberlet et al. 1999; Kendall and McKelvey 2008; Schwartz and Monfort 2008), an additional data source is available: the genetic data stored in the organism's tissue. If genetic samples are taken, then these data become another type of spatial data associated with the location where the organism was sampled. As such, genetic data can be analyzed with many of the same approaches used to analyze data of other types that vary spatially.

Genetic data, however, also have certain valuable properties that allow different sorts of analyses than are possible using many other data sources. Most data are representative of the present: where an organism is at the time of sampling, and what attributes are correlated with that presence. Genetic data allow us to look into the animal's past and ask where it was born, and where its parents were born. They also allow us to determine the individual's likely response to stressors such as heat, drought, or disease.

Landscape genetics has been viewed as a hybrid between population genetics and landscape ecology (Manel et al. 2003). Another way to view landscape genetics is as landscape ecology that uses spatial genetic patterns (e.g. genetic variation within or between individuals or groups). This is because landscape genetics treats the genetic patterns themselves as multivariate spatial data and seeks to infer ecological understandings by evaluating these patterns either in isolation or in conjunction with other spatial data. Because genetic data are very different from other data sources, we need to briefly discuss genetic data before we can explore their uses.

S.A. Cushman (✉), K.S. Mckelvey and M.K. Schwartz
US Forest Service, Rocky Mountain Research Station, 800 E Beckwith, Missoula, MT 59801, USA
e-mail: scushman@fs.fed.us

S.A. Cushman and F. Huettmann (eds.), *Spatial Complexity, Informatics, and Wildlife Conservation*
DOI 10.1007/978-4-431-87771-4_17, © Springer 2010

17.1 A Brief Primer on Genetics as they Relate to Landscape Analysis

The first thing to note about genetic data is that they represent an enormously rich data source. Even a primitive bacterium will generally have over one million base pairs in its genome. For example, the *Escherichia coli* K-12 genome contains 4,639,221-base pairs (Blattner et al. 1997). The human genome contains about 3 billion base pairs (Venter et al. 2001). Understanding the different patterns in the sequence of these base pairs is the business of molecular biology. How these patterns are arranged in geographic space is the business of landscape genetics or, more broadly, molecular ecology (Beebee and Rowe 2007).

Geneticists refer to genes, loci, and alleles. Genes are areas of genetic code associated with some function, loci are simply locations of interest within the code (although they can be genes), and alleles are different base pair patterns at a specific locus. Genetic variation is generated by random mutations at the level of the individual base pair of nucleotides (e.g. from adenine [A] to thymine [T], cytosine [C], or guanine [G]). Mutations lead to new alleles at loci which may or may not lead to different behavior of genes. If a mutation occurs in an area that codes for a protein (dubbed an "exon") and causes a different amino acid to be produced, most likely this mutation will prove detrimental, will therefore not persist, and will not be observed in samples from live individuals in a population. The removal of these mutations by natural selection conserves the original sequence, and therefore all individuals in a population will frequently have the same sequence (no variation at that locus). However, once in awhile a mutation in an exon will be beneficial. These mutations may persist transiently, with the new mutation being superior and the original allele will vanish or at least diminish in frequency (called directional selection). When either selection (or a process called genetic drift and discussed below) causes one allele to persist, this is called fixation. In some cases, both states will be maintained by selection (called diversifying selection or balanced polymorphisms). For example, it has been shown that the Human Leukocyte Antigen locus that recognizes antigens and thus allows the immune system to respond is under balancing selection, where there is an advantage for an individual to be heterozygous, in that this allows greater protection from different pathogen species or strains. Mutations that that are either removed or maintained by selection are not "neutral" but rather influence fitness and are therefore "adaptive".

On the other extreme are areas of the genome which, as far as we can tell, do nothing at all, dubbed "introns" (within genes) or intergenic DNA (between genes). Because mutations in these areas generally have no known effect on the organism's fitness, these areas are not conserved and therefore have the highest rates of observed mutations. In non-coding, or neutral areas, mutations build up at a predictable rate and therefore the number of mutations is related to time. In between these two extremes are non-coding sequences which are closely associated (physically linked by close proximity on a chromosome) with coding areas. Thus, there is likely some conservation within introns but they are variable enough to produce useful patterns. Further, their physical linkage to exons provides the potential to use intron patterns to track adaptive change.

While mutation is the ultimate source for genetic variation, the buildup of these mutations is a slow process. The movement of alleles across space, however, can be a rapid process. It occurs through reproduction and migration. Thus, spatial patterns of neutral genetic variation tend to be dominated by gene flow and drift and be little influenced by mutation. Gene flow is defined as exchange of genetic information (alleles) between populations through migration (Allendorf and Luikart 2007). Gene flow across landscapes makes populations or geographic areas more similar genetically. On the contrary, genetic drift causes populations or geographic areas across landscapes to become genetically differentiated. Genetic drift is defined as random changes in allele frequencies in populations between generations due to binomial sampling of genes during meiosis (gamete production). Genetic drift is more pronounced in small populations. For example, imagine a population of two individuals with two different alleles; by pure chance we may lose alleles not due to selection but by drift if one allele happens to not get passed down in a single generation.

For landscape genetics, patterns in neutral DNA, introns, and exons, each have different meanings, and can be used to ask different questions. Because changes in an organism's neutral DNA has no biological implications, these patterns are free to drift due to stochastic processes, and are free to flow due to organism movements. Patterns in neutral DNA, being free from natural selection, can be more easily interpreted, and a great deal of population and landscape genetics is based on analyzing these regions (Manel et al. 2003; Avise 2004; Allendorf and Luikart 2007). However, the evolution and movement of adaptive traits may be more pertinent for evaluating population status (e.g. long-tem adaptive potential or current extinction risk), with the observed neutral patterns providing a potential index of adaptive differentiation between populations. Populations that have been separated for long periods of time, and therefore have many differences in neutral DNA, will also tend to develop local adaptations (e.g., Storfer et al. 1999). While, in a broad sense, large inter-population differences in neutral markers indicate potential for large differences in adaptive traits, the relationship is by no means constant. For example, Bromham and Woolfit (2004) found no consistent patterns in rates of molecular evolution based on neutral markers and rates of observed evolution of phenotypes. Using the degree of differentiation in neutral markers to infer the degree of adaptive change or divergence is therefore somewhat speculative.

In the future we will almost certainly evaluate both adaptive and neutral patterns (Luikart et al. 2003) with large amounts of information contained not only in the patterns themselves, but in the covariance patterns as well. This approach is the focus of the Landscape Genomics Chapter (Chapter 9, this volume; see also Gebremedhin et al. 2009).

17.1.1 Maternal and Paternal Inheritance

Almost all of the DNA in higher eukaryotes is nuclear, and therefore inherited from both parents. However, there are sequences that are inherited exclusively either maternally or paternally. Exclusive maternally inherited DNA is associated with organelles (in animals mitochondria and in plants both mitochondria and chloroplasts) that contain their own DNA. They are largely not recombined with paternal DNA. In animals, the Y chromosome provides pure paternal inheritance in males; in plants, pollen contains extra-nuclear DNA providing richer data for tracking paternal lineages.

Genes inherited exclusively from maternal or paternal sources produce useful patterns because, lacking recombination (which shuffles DNA sequences), these patterns are easier to analyze, because differences in sex-related behavior can be inferred, and because there is only one "type" per individual (as opposed to nuclear DNA where there are two copies of each locus – one from each of the paired autosomal chromosomes). For example, in a famous case, analyses of mitochondrial DNA (mtDNA) for bears on the Admiralty, Baranof, and Chichagof (ABC) islands off the coast of Alaska indicated that the bears were highly divergent from mainland bears. However, nuclear DNA indicated that they were similar (Paetkau et al. 1998). Male bears crossed the passage to the mainland frequently, but apparently females never did. By comparing the patterns associated with the mixing of both sexes (nuclear) with those exclusively associated with females (mitochondrial), the movement patterns and relative isolation of the islands was elucidated.

The case of the ABC bears is an application of traditional population genetics, but the issues, movement and isolation, are commonly addressed through landscape genetics (especially when working at fine spatial scales). Indeed, the boundary between population and landscape genetics is blurry and in many cases a problem can be investigated using either approach. For the remainder of the chapter we will discuss the branch of landscape genetics that evaluates differences between individuals (e.g., Cushman et al. 2006) and does not seek to assign individuals to known populations (Holderegger and Wagner 2008 and Balkenhol et al. 2009). These approaches treat genetic differences as spatial gradients rather than as discrete clusters with relatively little internal pattern. In this simplified view, the difference between population and landscape genetics is analogous to the difference between ANOVA and regression. As with these statistical methods, the choice of approach should be dictated by the questions asked and the nature of the data; neither is uniformly superior. Similarly, just as regression using continuous variables and ANOVA represent two ends of a spectrum of models involving mixtures of categorical and continuous variables, landscape genetics can contain mixtures of gradients and groups.

17.2 The Nature of Genetic Data

As indicated above, genetics in general and landscape genetics in particular are huge fields; to avoid writing an entire book on it, we need to narrow down. For the remainder of the chapter, we will therefore discuss analyses of neutral nuclear DNA. For over a decade, the most common analyses of neutral DNA patterns have utilized "microsatellites", thus we will only consider this type of genetic data

(Luikart and England 1999). Microsatellites are composed of simple sequence repeats of 1–6 base pairs. For example cytosine-adenine (CA) is a common ("dinucleotide") repeat sequence found in genomes. In microsatellites, the same sequence of base pairs commonly will repeat 10–100 times (CA,CA,CA…CA) and mutations lead to variations in the number (n) of repeats (CA_n). Therefore genetic variation in microsatellites is scored in terms of the number of base pairs in the repeat (e.g. CA_{10}).

Microsatellite mutation occurs more frequently than single base pair mutations in sequences, and therefore microsatellite loci are often extremely variable. It is not uncommon to have ten or more lengths (alleles) of a particular microsatellite (locus) in a population (although some fish species have more than 45 alleles at a locus; Van Oppen et al. [1997]). They are also, as far as we can tell, truly neutral and different loci are often considered independent, although this can be tested with recent genome maps that are being produced. Because they are nuclear (and thus two alleles are contained in each individual at autosomal chromosomes), each locus can be either heterozygous (two different alleles) or homozygous (the same allele twice) increasing the between-individual variability. Because of the high number of potential combinations at any locus, we can often differentiate individuals (called DNA fingerprinting) in wild populations using 6–7 loci (Paetkau and Strobeck 1994).

17.2.1 Genetic Distance

Genetic distance is simply the degree to which genetic data between 2 individuals or populations vary. For individuals, it can be treated simply as the proportion of alleles different (e.g. Bray-Curtis, used in Cushman et al. 2006) or can be based on deviation from genetic expectations (e.g. a_r, Rousset 2000). Various distance metrics tend to be highly correlated; Cushman et al. (2006) found that the correlation between distance measured using the Bray–Curtis percentage dissimilarity measure (Legendre and Legendre 1998) and a_r (Rousset 2000) was 0.97 for their bear data.

The expected genetic distance between individuals in a well mixed population will related to the frequency within the population of alleles at each sampled locus. However, if individuals within a geographic area are related, the frequency of genotypes and alleles will differ from their global expectations. For example, if we were to sample from a family where all offspring had the same parents, no more than four alleles would exist at each locus across the entire family group regardless of the global allelic frequencies (this could also be as little as one allele if both parents were homozygous for the same allele at this locus). Similarly, this family would have, overall, more alleles in common than random expectations. Because this group is closer genetically than expectation, by definition it has to be further apart genetically from other family groups than random expectation. Similarly, if two populations are separated for a long time, allele frequencies will diverge due to the random processes of drift and mutation, and samples from individuals between the two populations will be genetically more dissimilar than would be expected if they were freely mixing. In landscape genetics, patterns of genetic distance are the spatial information of interest: mating behavior and movement patterns can be inferred by relating these patterns to the landscapes in which they occur.

A common spatial pattern of genetic distance in terrestrial organisms is that organisms that are spatially proximal will be more genetically similar than organisms that are distant. This is because, for animals, an organism is more likely to find a proximal organism than one that is distant, and the resulting offspring are more likely to settle nearby. For plants, pollen is more likely to be transferred to proximal plants (in some cases to other parts of the same plant), and seed densities likely are highest close to the parent plant. These types of patterns are referred to as "isolation by distance" (Wright 1943) and for many organisms can be considered the expectation distribution: on a completely featureless landscape, we would expect patterns to be associated with Euclidian distance. When population processes lead to isolation by distance, genetic distances will form gradients, and landscape genetic approaches are therefore appropriate. We can next consider what happens when landscape features interact with levels of gene flow and enhance or retard genetic exchange across the landscape.

17.3 Common Approaches in Landscape Genetics

As might be expected, early landscape genetic approaches sought to use spatial genetic patterns to define populations. Once genetic discontinuities (barriers) and populations were defined, standard population genetic approaches as well as individual based approaches could be applied (Manel et al. 2003 and references therein). The fundamental idea was that genetic differences formed a gradient surface with areas of rapid change and "flat" areas. Steep slopes would be related to barriers to gene flow, and flat areas with freely mixed or "panmitic populations. Rather than assuming prior knowledge of where the steep areas were, the genetic patterns would allow determination of population boundaries (Barbujani et al. 1989; Barbujani and Sokal 1990; Stenico et al. 1998; Simoni et al. 1999). While an excellent idea in theory, in practice genetic data are very noisy when used in this manner and are likely highly influenced by sampling scheme (Novembre and Stephens 2008; Schwartz and McKelvey 2009; Frantz et al. 2009). Thus early methods such as the Monmonier algorithm (Monmonier 1973) tended to get stuck in local minima and failed to reliably reproduce simulated populations. Of the various approaches to determining minima simulated annealing is the most robust and Dupanloup et al. (2002) were able to improve results using these methods. However, this approach has not become widespread. Instead, a posteriori grouping of genetic data into populations is largely done using non spatial (e.g. STRUCTURE; Prichard et al. (2000)) or semi-spatial (e. g. Geneland; Guillot et al. [2005]) clustering algorithms.

Coulon et al. (2004) correlated pair-wise genetic distance between samples with landscape features viewed as having different resistances to gene flow. Least cost resistance paths were created between all individuals and correlations between path resistances and genetic distance higher than those between Euclidian and genetic distance were viewed as supportive of the putative resistance model. More recently McRae (2006) conceived landscape resistance as a series of parallel circuits connected by resistors. Gene flow was viewed as analogous to electron flow, and, as this approach had been used in engineering to determine electron flow through metal plates, numerical approaches were available. The advantage of the circuit analogy is that gene flow would be altered not only by the minimum resistance between points, but also by the width and therefore number of alternative routes available. Recently, Garroway et al. (2008) utilized graph theoretic approaches similar to those that have been used to analyze social networks to evaluate genetic data. This approach allows the macro-scale characteristics of the network to be determined, potentially allowing deeper insights into the biology driving genetic interchange than are possible through evaluations of resistance to travel.

Once factors controlling (or at least correlating with) gene flow have been identified, the model can be applied to a landscape to determine likely areas for both corridors and barriers to flow. For many animals, the advent of GPS tracking should, in turn, allow the direct confirmation of the genetically derived connectivity rules, although genetics may reflect more historical gene flow integrated across many individuals, whereas GPS tracking could reflect more recent gene flow for a few select individuals as well as movements not related to mating, reproductive success, and gene flow.

17.4 Examples of Connectivity Modeling

In the remainder of this chapter we illustrate some of the potentials associated with landscape genetics with examples taken from our own work. The organisms involved are black bears (*Ursus americanus*) and wolverines (*Gulo gulo*) and in both cases our research group has used resistance modeling to infer corridors and barriers, and subsequently utilized these models to develop putative corridor structures. In the case of the wolverine, we had a coherent a priori habitat use model test. In the case of the black bear, we explored multiple models, but used "causal modeling" (Legendre and Troussellier 1988; Legendre 1993) to separate spurious correlations and identify biologically relevant factors.

17.4.1 Wolverine Example

Wolverine populations, at their southern extent are limited to high elevation areas. Not only do they need to den in snow (Magoun and Copeland 1998), but year-around habitat use is limited to high elevation areas, possibly due to aversion to summer heat (Hornocker and Hash 1981). Wolverines have large home ranges, on the order of 100 km^2 for females (Copeland et al. 2007). Hence, the mountainous areas where they live contain relatively few individuals (Squires et al. 2007) and continued presence therefore requires movement between the mountainous islands.

Aubry et al. (2007) suggest that wolverine range is limited by areas covered by persistent snow through the end of the denning period in mid May. In 8 radio-telemetry studies in the conterminous US and in British Columbia, Canada, areas characterized by persistent spring snow through the denning period (hereafter "snow layer") contained all recorded dens and 89% of all telemetry relocations (J. Copeland pers. Comm). Therefore, from a practical standpoint the snow layer defined the areas where wolverines both reproduced and utilized on a daily basis.

Based on their size, these islands are, from the standpoint of population stability, extremely small. Many likely contain a single family consisting of a mature male, several mature females and their progeny. For population stability, likely the single most important attribute is the ability to provide demographic and genetic rescue, and to rapidly recolonize areas when local extinctions occur (see Squires et al. 2007). We therefore needed to determine the dispersal rules and likely travel routes used by wolverines.

A spatially distributed sample of tissues from 210 wolverines existed for Montana, Idaho, and northern Wyoming. These data were derived from various studies and from trapping in Montana (Schwartz et al. in press). These samples have been analyzed in several papers (Kyle and Strobeck 2002; Cegelski et al. 2003, 2006; Guillot et al. 2005) and clustering algorithms have been used to define sub-populations. Additionally, autocorrelation analysis (Smouse and Peakall 1999) indicates a positive gradient of correlation extending approximately 180 km (Schwartz et al. in press). While both the clustering analyses and autocorrelation demonstrate that genetic data are spatially structured, the presence of strong gradients coupled with irregular sampling procedures likely lead to artificial clusters not indicative of actual movement and isolation patterns (Schwartz and McKelvey 2009). Gradient approaches, however, are appropriate for these types of genetic patterns and samples.

Using the approach of comparing pair-wise genetic distance with putative landscape resistance maps (Coulon et al. 2004, 2006; Cushman et al. 2006), Schwartz et al. tested the hypothesis that wolverines dispersed by keeping as much as possible within the mountainous areas where they both were born and to which they limited their daily movements. Because of the concordance between the snow layer, denning and daily movement, Schwartz et al. constructed resistance surfaces in which movements exterior to the areas of snow had higher resistance than movements staying within snow areas. Following Coulon et al. (2004, 2006) and Cushman et al. (2006) Schwartz et al. used least cost paths to construct the minimum resistance paths between each pair of individuals and used Mantel tests to measure the correlation between least cost path resistances and genetic distance. These correlations were tested against a Mantel test comparing genetic distance to Euclidian distance. Lastly, partial Mantel tests were used to separate the effects of the snow layer from the effects of Euclidian distance, and determine which was likely the primary cause of the observed correlations.

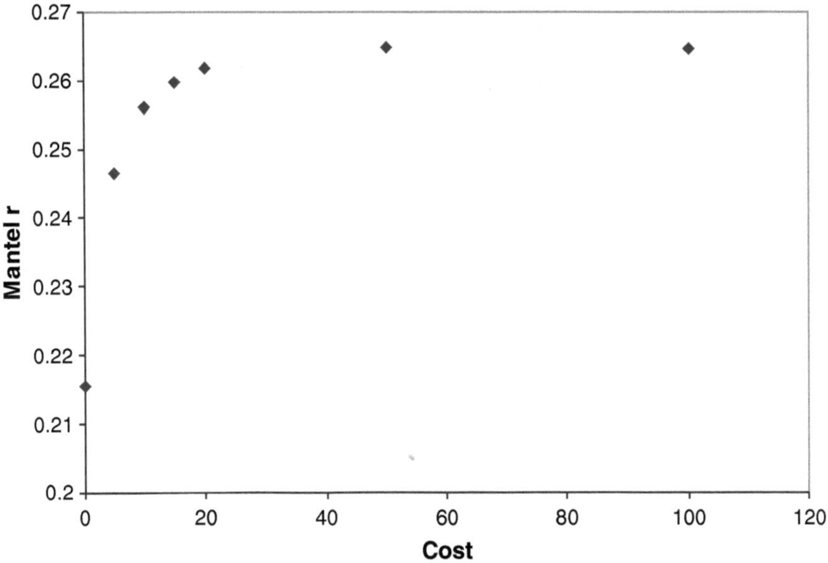

Fig. 17.1 Correlations against cost for various models testing whether areas characterized by persistent spring snow facilitated wolverine dispersal. Costs are the relative cost of moving through non-snow rasters (snow rasters always have Cost = 1)

For wolverines, rather extreme models, in which least cost resistance paths were maximally within the snow layer performed best, and all models with less resistance through snow areas were significantly better than the Euclidian distance model (Fig. 17.1).

The major goal of the analysis was to quantify the degree of population connectivity across a large portion of the Northern Rocky Mountains and identify important potential corridors or barriers to movement. As the identified resistance map only provides point-wise information and connectivity is mediated through movement, Schwartz et al. utilized the source–destination least path approach introduced by Cushman et al. (2009) to map broad scale landscape connectivity. Because wolverines were well distributed within the mountain islands and because all dens were within these areas, Schwartz et al. placed theoretical start locations in all snow areas large enough to hold a female wolverine home range (and hence a den). Using the best resistance model identified above, they built pairwise least cost paths between all theoretical start locations to determine the optimal routes from any area where a wolverine might be born to any area where a wolverine might breed and leave its genes behind. Kernels were draped across each path and the kernels added to produce a visualization of theoretical path densities across the landscape (Fig. 17.2).

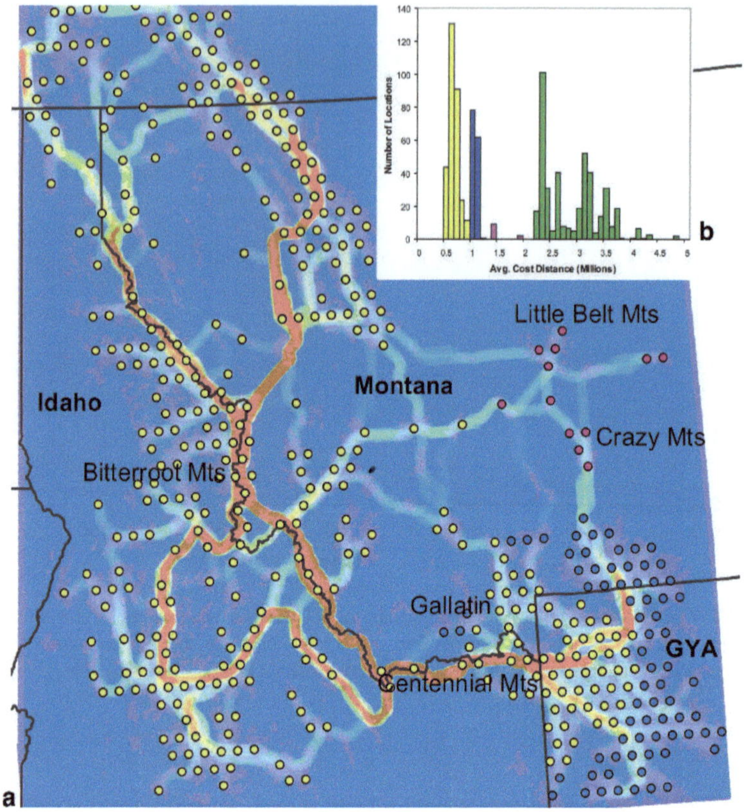

Fig. 17.2 (**a**) Map of the northern US Rockies with cumulative least-cost paths between systematically placed locations (circles) in spring snow cover cells. Areas in red are predicted to be used more often than those in cooler colors. The color of the circle corresponds to the average cost distance between that location and all other locations, based on modeling. (**b**) Cumulative plot of the average cost distance (in millions of cost units) between each systematically placed location and all other locations. The graph was divided into four modes (three within the northern US Rockies, and 1 between the Greater Yellowstone Area and Colorado). The yellow mode has the lowest average cost distances (within the Northern US Rockies), the blue bars the next lowest, the pink bars (Crazy and Little Belt Mountains) have the greatest average cost distances in the Northern US Rockies, and the green bars show the distances between all points from Colorado to the Greater Yellowstone Area

17.4.2 Bear Example

The second example is an effort to predict landscape resistance as experienced by American black bear as a function of elevation, slope, roads, land-use and forest cover. The study area consisted of an approximately 3,000-km^2 area of the Selkirk and Purcell mountain ranges in the extreme northern tip of Idaho State, USA (Cushman et al. 2006) and to apply this model to predict movement corridors and identify potential barriers for gene flow between Yellowstone National Park and the Canadian Border in the Montana, USA. (Cushman et al. 2009).

The topography in the Cushman et al. (2006) study area is mountainous, with elevation ranging from approximately 700–2,400 m. The Kootenai River trench bisects the study area, separating the Selkirk Mountains on the west from the Purcell Mountains on the east with a 5–8-km-wide unforested, agricultural valley and a broad, deep river. Mountains outside the valley are heavily forested, with *Abies lasiocarpa* (subalpine fir) and *Picea engelmannii* (Engelmann spruce) codominant above 1,300 m and a diverse mixed conifer forest dominating below 1,300 m.

We used noninvasive hair snaring to obtain genetic samples, following the protocols of Proctor et al. (2005). Sampling stations were set at 266 plots distributed on a 1.6-km grid. We detected no deviations from Hardy–Weinberg proportions, found no evidence of gametic disequilibrium, and expected heterozygosity in the Purcell and Selkirk mountain ranges were 0.78 and 0.80, respectively, while observed heterozygosity was 0.76 and 0.80 for each range, respectively (Schwartz et al. 2006).

A priori, we identified three potential drivers of genetic structure in this black bear population, including isolation by the Kootenai River valley as a barrier, isolation by geographic distance, and isolation by landscape-resistance gradients. Our goal was to determine the relative support for isolation by distance and barriers in comparison to isolation by gradients of landscape resistance.

In addition, there are many possible alternative models for landscape effects on population connectivity of this species. A priori, we proposed elevation, roads, topographical slope, land use and land-cover as factors hypothesized to affect gene flow in this species. In addition, each of these factors could influence bear gene flow in a number of potential ways. Therefore, we constructed a factorial of multiple levels of each of these factors, totaling 108 alternative landscape resistance hypotheses (Cushman et al. 2006).

We used the Bray–Curtis percentage dissimilarity measure to calculate genetic distance among individuals (Legendre and Legendre 1998), producing a matrix containing the genetic distances among all pairs of sampled bears. We used the ArcGIS to produce cost distance matrices that correspond to each of the 110 alternative models (108 landscape models, isolation by distance, and isolation by the Kootenai River barrier). The cost distance matrices corresponding to each landscape resistance hypotheses were created using the COSTDISTANCE function (ESRI 2003). This entailed creating cost matrices reflecting the least-cost distance from the location at which each bear's DNA sample was recorded to every other bear's location across each of the 108 resistance surfaces.

We next identified which of the seven models had the strongest support. All 110 resistance hypotheses were significantly related to the pattern of genetic difference among bears. The barrier hypothesis was ranked 102 out of 110 in terms of the magnitude of the Mantel correlation coefficient. The isolation-by-distance hypothesis was more strongly supported and was ranked 35 out of the 110 total models. The most supported model predicts that gene flow in this population is influenced predominantly by landscape-resistance gradients with no significant independent relationships with the Kootenai River barrier or geographical distance.

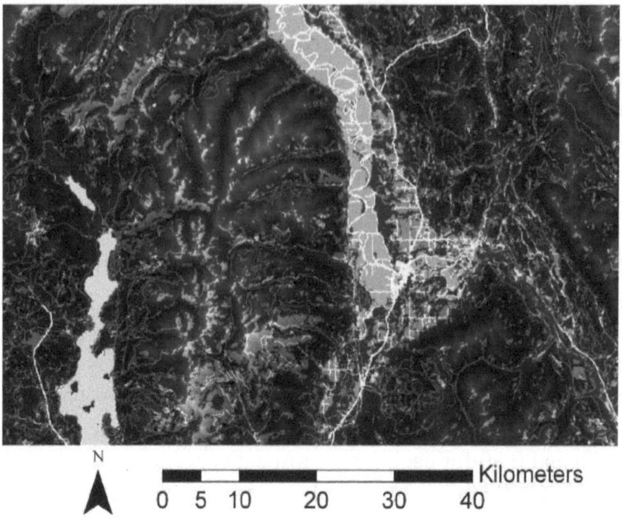

Fig. 17.3 Predicted landscape resistance to black bear gene flow in Idaho Study Area, inferred from causal modeling, Cushman et al. (2006). Color ramp ranges from low resistance (relative resistance 1) in black to high resistance (relative resistance maximum 62) in white

After the effects of distance are partialed out, ranking the ten significant partial models by the Monte Carlo permuted p-value value provided a means to determine which hypotheses have the greatest support and to identify the combination of landscape factors most related to the genetic structure of this population. The ten significant partial models were all concentrated in one small area of a four-dimensional factorial space (roads, elevation, forest cover, slope), indicating a unimodal peak of support. Within this small area of concentration, the best-supported models were associated with minimum resistance to movement at middle elevations, high resistance of nonforested habitat to movement, no relationship with slope, and equivocal support for the different levels of road resistance. A map extrapolating the understanding from the most highly supported model across the entire study area is shown in Fig. 17.3 .

Similar to the approach used to model wolverine corridors (above), we modeled movement from source locations distributed at 2-km intervals along the forested portion of the Canadian border between the Idaho–Washington state border and the eastern edge of Glacier National Park (160 locations) to 160 destination locations along the northern boundary of Yellowstone National Park (Cushman et al. 2009). We then computed the least-cost paths across the best-supported resistance surface developed by Cushman et al. (2006) for all combinations of the 160 northern and 160 southern points (25,600 paths).

Despite the large number of combinations, all least-cost paths followed only three major routes (Fig. 17.4). The most important route, in terms of the number of source–destination paths it contained, followed the eastern edge of the mountainous terrain along the Bridger and Belt mountains and crossed into the Bob Marshall Wilderness

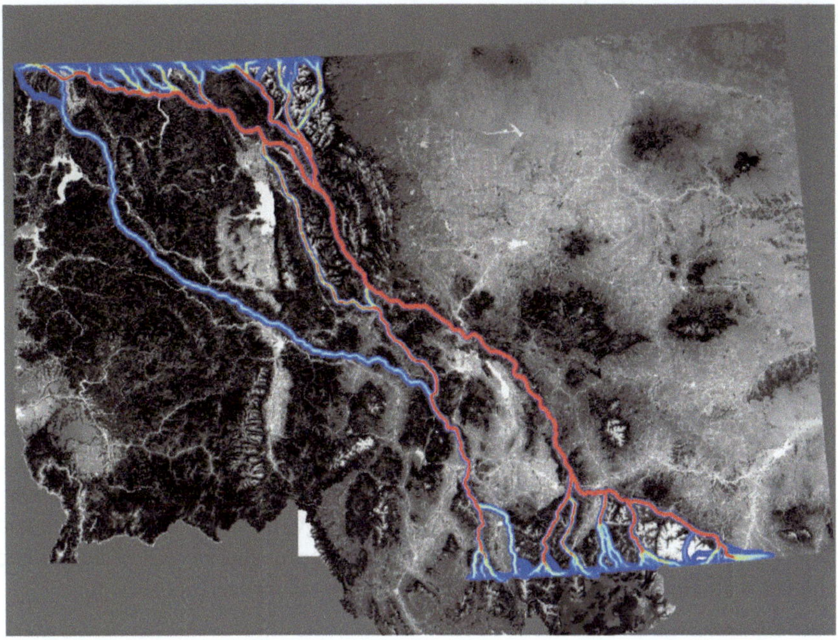

Fig. 17.4 Map of expected importance of least cost routes from Yellowstone National Park to the Canadian border based on the best resistance map identified in Cushman et al. 2006. The relative strength of the predicted route is shown in a color ramp from red (very highly important) to blue (relatively less important) based on the number of source–destination paths crossing that location

and then into Glacier National Park. The second major route proceeded north through the Gravelly and Tobacco Root ranges and then proceeded north roughly following the foothills along the east side of the Swan and North Fork Flathead rivers. The final route was much smaller, in terms of the number of source–destination paths it contained, but was important for connectivity between the northwestern part of the study area and the GYA. This route followed the Clark Fork River and continued northward though the Cabinet and Selkirk mountains to Canada. The majority of the length of the major paths fell within federal ownership (US Forest Service [USFS] and National Park Service [NPS]). Nevertheless, only 24% of the major paths length fell within designated wilderness or national park, and less than half were in wilderness, park, or designated roadless areas (Cushman et al. 2009).

17.5 Conclusions

Landscape genetics is still a nascent science. While we continue to be amazed both by the power and flexibility of these data, we are just beginning to scratch the surface. The strength and consistency of the results associated with the wolverine and bear examples in this chapter greatly exceeded our expectations. As new approaches emerge and, with the cost of genetic analyses halving every 2 years (Hauser and Seeb 2008), these approaches will likely become pervasive in the next decade (see Chapter 9, this volume). There are already strong movements to augment ongoing traditional monitoring efforts with genetic sampling. In many cases, the additional cost of collecting genetic material is trivial. This trend will lead to spatial genetic datasets that are much more regular and extensive than are currently available. In short, we expect one of the most commonly analyzed types of spatial datasets in the near future will be genetic datasets, as they provide powerful questions to long standing ecological and conservation questions. However, this rapid advance in availability of large genomic data sets will present major computational and database challenges. The size of genomic datasets is vast, which poses a major challenge for storage, organization and file access. In addition, analyzing these large data sets to extract meaningful information in the form of neutral markers, introns and exons is a massive datamining challenge. For the potential of landscape genetics to be realized there must be simultaneous advances in ecological informatics (Chapter 8, this volume), database architecture (Chapter 12, this volume), and major improvements in statistical data mining approaches to associate genetic patterns with landscape and environmental features (Chapter 16, this volume).

References

Allendorf FW, Luikart G (2007) Conservation and the genetics of populations. Blackwell, Hoboken, NJ

Aubry KB, McKelvey KS, Copeland JP (2007) Distribution and broadscale habitat relations of the wolverine in the contiguous United States. J Wildl Manag 71:2147–2158

Avise JC (2004) The hope, hype, and reality of genetic engineering. Oxford University Press, Oxford

Balkenhol N, Gugerli F, Cushman SA, Waits L, Coulon A, Arntzen J, Holderegger R, Wagner H (2007) Identifying future research needs in landscape genetics: where to from here? Landscape Ecol 24:455–463

Barbujani G, Oden NL, Sokal RR (1989) Detecting regions of abrupt change in maps of biological variables. System Zool 38:376–389

Barbujani G, Sokal RR (1990) Zones of sharp genetic change in Europe are also linguistic boundaries. Proc Nat Acad Sci 87:1816–1819

Blattner FR, Plunkett III G, Bloch CA, et al (1997) The complete genome sequence of *Escherichia coli* K-12. Science 277:1453–1462

Beebee, Rowe (2007) An introduction to molecular ecology. 2nd edition. Wiley, New York.

Bromham L, Woolfit M (2004) Explosive radiations and the reliability of molecular clocks: island endemic radiations as a test case. System Biol 53:758–766

Cegelski C, Waits L, Anderson N (2003) Assessing population substructure and gene flow in Montana wolverines (Gulo gulo) using assignment-based approaches. Mol Ecol 12:2907–2918

Cegelski CC, Waits LP, Anderson NJ, Flagstad O, Strobeck C, Kyle CJ (2006) Genetic diversity and population structure of wolverine (*Gulo gulo*) populations at the southern edge of their current distribution in North America with implications for genetic viability. Conservat Genet 7:197–211

Coulon A, Cosson J-F, Angibault JM, et al (2004) Landscape connectivity influences gene flow in a roe deer population inhabiting a fragmented landscape: an individual-based approach. Mol Ecol 13:2841–2850

Coulon A, Guillot G, Cosson J, Angibault J, Aulagnier S, et al (2006) Genetics structure is influenced by lansdcape features. Empirical evidence from a roe deer population. Mol Ecol 15:1669–1679

Copeland JP, Peek JM, Groves CR, et al (2007) Seasonal habitat associations of the wolverine in central Idaho. J Wildl Manag 71:2201–2212

Copeland JP, McKelvey KS, Aubry KB, et al (In review) The bioclimatic envelope of the wolverine: Do environmental constraints limit their geographic distribution?

Cushman SA, McKelvey KS, Hayden J, Schwartz MK (2006) Gene-flow in complex landscapes: testing multiple hypotheses with causal modeling. Am Nat 168:486–499

Cushman SA, McKelvey KS, Schwartz MK (2009) Identifying corridors for the movement of black bear (*Ursus* americanus) between Yellowstone National Park and the Canadian Border. Conservat Biol 23:368–376

Dupanloup I, Schneider S, Excoffier L (2002) A simulated annealing approach to define the genetic structure of populations. Mol Ecol 11:2571–2581

ESRI (2003) ArcGIS. Environmental systems research incorporated, Redlands, CA

Frantz AC, Cellina S, Krier A, Schley L, Burke T (2009) Using spatial Bayesian methods to determine the genetic structure of a continuously distributed population: clusters or isolation by distance? J App Ecol. doi: 10.1111/j.1365–2664.2008.01606.x

Garroway CJ, Bowman J, Carr D, Wilson PJ (2008) Applications of graph theory to landscape genetics. Evolut Appl 1:620–630

Gebremedhin B, Ficetola GF, Naderi S, Rezaei H.-R, Maudet C, Rioux D, Luikart G, Flagstad Ø, Thuiller W, Taberlet P (2009) Frontiers in identifying conservation units: from neutral markers to adaptive genetic variation. Anim Conserv 12:107–109

Guillot G, Mortier F, Estoup A (2005) Geneland: a computer package for landscape genetics. Mol Ecol Notes 5:712–715

Hauser L, Seeb JE (2008) Advances in molecular technology and their impact on fisheries genetics. Fish Fisheries 9:473–486

Hornocker MG, Hash HS (1981) Ecology of the wolverine in northwestern Montana. Canad J Zool 59:286–1301

Holderegger R, Wagner HH (2008) Landscape genetics. Biosci 58:199–207

Kendall, KC, McKelvey KS (2008) Hair Collection. In R. Long, P. MacKay, J. Ray, and W. Zielinski, editors. Noninvasive survey methods for North American Carnivores. Island. Washington, DC

Kyle CJ, Strobeck C (2002) Connectivity of peripheral and core populations of North American wolverines. J Mammal 83:1141–1150

Legendre P (1993) Spatial autocorrelation: trouble or new paradigm? Ecology 74:1659–1673

Legendre P, Legendre L (1998) Numerical ecology, 2nd edition. Elsevier, Amsterdam

Legendre P, Troussellier M (1988) Aquatic heterotrophic bacteria: modeling in the presence of spatial autocorrelation. Limnol Oceanography 33:1055–1067

Luikart G, England PR, Tallmon D, Jordan S, Taberlet P. (2003) The power and promise of population genomics: from genotyping to genome typing. Nat Rev Genet 4:981–994.

Luikart G, England PR (1999) Statistical analysis of microsatellite DNA data. Trends Ecol Evol 14:253–256.

Magoun AJ, Copeland JP (1998) Characteristics of wolverine reproductive den sites. J Wildl Manag 62:1313–1320

Manel S, Schwartz MK, Luikart G, Taberlet P (2003) Landscape genetics: the combination of landscape ecology and population genetics. Trends Ecol Evol 18:189–197

McRae BH (2006) Isolation by resistance. Evolution 60:1551–1561

Monmonier MS (1973) Maximum-difference barriers: an alternative numerical regionalization method. Geograph Anal 3:245–261

Novembre J, Stephens M (2008) Interpreting principal component analyses of spatial population genetic variation. Nat Genet 40:646–649

Paetkau D, Shields GF, Strobeck C (1998) Gene flow between insular, coastal and interior populations of brown bears in Alaska. Mol Ecol 7:1283–1292

Paetkau D, Strobeck C (1994) Microsatellite analysis of genetic variation in black bear populations. Mol Ecol 3:489–495

Prichard JK, Stephens M, Donnelly P (2000) Inference of population structure using multilocus genotype data. Genetics 155:945–959

Proctor MF, McLellan BN, Strobeck C, Barclay RMR (2005) Genetic analysis reveals demographic fragmentation of grizzly bears yielding vulnerability by small populations. Proc Roy Soc B: Biol Sci 272:2409–2416

Rousset F (2000) Genetic differentiation between individuals. J Evol Biol 13:58–62

Schwartz MK, Copeland JP, Anderson NJ, et al (In Press) Wolverine gene flow across a narrow climatic niche. Ecology

Schwartz MK, Cushman SA, McKelvey KS, Hayden J, Engkjer C (2006) Detecting genotyping errors and describing black bear movement in northern Idaho. Ursus 17:138–148

Schwartz MK, McKelvey KS (2009) Why sampling scheme matters: the effect of sampling scheme on landscape genetic results. Conserv Genet 10:441–452

Schwartz MK, Monfort SL (2008) Genetic and endocrine tools for carnivore surveys. R. Long, P. MacKay, J. Ray, and W. Zielinski (eds.) Noninvasive survey methods for North American Carnivores. Island, Washington, DC

Simoni L, Gueresi P, Pettener D, Barbujani G (1999) Patterns of gene flow inferred from genetic distances in the Mediterranean region. Human Biol 71:399–415

Smouse PE, Peakall R (1999) Spatial autocorrelation analysis of individual multiallele and multilocus genetic structure. Heredity 82:561–573

Squires JR, Copeland JP, Ulizio TJ, Schwartz MK, Ruggiero LF (2007) Sources and patterns of wolverine mortality in western Montana. J Wildl Manag71:2213–2220

Stenico M, Nigro L, Barbujani G (1998) Mitochondrial lineages in Ladin-speaking communities of the eastern Alps. Proc Roy Soc London, Ser B, Biol Sci 265:555–561

Storfer A, Cross J, Rush V, Caruso J (1999) Adaptive coloration and gene flow as a constraint to local adaptation in the streamside salamander, Ambystoma barbouri. Evolution 53:889–898.

Taberlet P, Waits LP, Luikart G (1999) Non-invasive sampling: look before you leap. Trends Ecol Evol 14:323–327.

Van Oppen MJH, Rico C, Deutsch JC, Turner GF, Hewitt GM (1997) Isolation and characterization of miicrosatellite loci in the cichlid fish Pseudotropheus zebra. Mol Ecol 6:387–388

Venter JC, Adams MD, Myers EW, et al (2001) The sequence of the human genome. Science 291:1304–1351

Wright S (1943) Isolation by distance. Genetics 28:114–138

Part III
Examples

Chapter 18
Using Geographical Mapping and Occupancy Modeling to Study the Distribution of the Critically Endangered Leopard (*Panthera pardus*) Population in Armenia

Igor G. Khorozyan, Alexander G. Malkhasyan, Shushanik G. Asmaryan, and Alexei V. Abramov

18.1 Introduction

Space limitations arising from human activities affect demographic structure and performance of mammalian populations and thus reduce their viability. This is especially true for wide-ranging wild cats (family Felidae) which generally lead solitary lives and require large tracts of good-quality habitats for survival (Sunquist and Sunquist 2001). As human activities leave more and more mosaics of modified lands behind, felid populations become fragmented and further impaired by the small and often unviable size of patches necessitating more complicated dispersal of individuals between patches (Reed 2004).

Prey abundance is the key factor determining the structure of female home ranges, whereas availability of females is most important for male home ranges (Stander et al. 1997). Thus, prey abundance determines felid requirements in space. However, space itself is also an important factor since solitary life and generally

I.G. Khorozyan (✉)
Zoological Institute, Russian Academy of Sciences, Universitetskaya nab. 1,
St. Petersburg 199034, Russia; WWF Armenia, Sarmeni Str. 96, Yerevan 0019, Armenia
e-mail: leopard_am@yahoo.com

A.G. Malkhasyan
WWF Armenia, Sarmeni Str. 96, Yerevan 0019, Armenia
e-mail: vipradde@yahoo.com

S.G. Asmaryan
Centre for Ecological Studies, National Academy of Sciences, Abovian Str. 68,
Yerevan 375025, Armenia
e-mail: ashuk@list.ru

A.V. Abramov
Zoological Institute, Russian Academy of Sciences, Universitetskaya nab. 1,
St. Petersburg 199034, Russia
a.abramov@mail.ru

S.A. Cushman and F. Huettmann (eds.), *Spatial Complexity, Informatics, and Wildlife Conservation*
DOI 10.1007/978-4-431-87771-4_18, © Springer 2010

exclusive home ranges of the same sex individuals in most felids force their populations to occupy large tracts of good habitats above some threshold to maintain viability. For example, leopard (*Panthera pardus*) population needs to number at least 31 individuals and the area at least 412 km^2 to remain viable (Smallwood 1999). That is why small and densely populated countries experience problems in maintaining viable leopard populations, even though prey densities can be high in some hotspots (Khorozyan et al. 2008).

Geographic range size is the principal factor of viability of carnivores, even though some exceptions, e.g. African wild dog (*Lycaon pictus*), do not always guarantee that wide range means conservation security for a population (Cardillo et al. 2004). The decline and extirpation of top carnivores from fragmented ecosystems may generate trophic cascades that alter the structure of ecological communities, so the persistence of these keystone species can indicate the levels of ecosystem health (Crooks 2002). These aspects are particularly relevant to big cats subsisting almost exclusively on wild ungulates which are vulnerable to human pressures, whereas the smaller felid species may even benefit from landscape fragmentation by feeding on rodents and other small prey thriving in human-dominated landscapes (Tigas et al. 2002).

Survival of fragmented felid populations relies heavily on the ability of sub-adults to successfully disperse, establish their own home ranges and then mate with non-relatives. Dispersal is an energetically costly process which forces sub-adult individuals to move long distances through the lower-quality habitats or hostile man-dominated inter-patch landscapes to reach the destination areas (Stander et al. 1997). Mortality of dispersing sub-adults accounts for a significant portion of overall population mortality (Stander et al. 1997; Haines et al. 2006). Hence, the maintenance and preservation of habitat integrity through the network of natural movement corridors used by cats has been among the most viable solutions to avert their extinction (Beier 1993).

The patchy pattern of populations is common among today's wild cats which tend to respond to anthropogenic pressures by retreating to inaccessible and less optimal habitats and to protected areas (Weber and Rabinowitz 1996). However, most countries cannot allocate sufficiently large tracts of undisturbed land for protected areas so cats, especially larger representatives, occasionally move outside zones of safety and die from human persecution (Woodroffe 2001). Generally, wild cats and other carnivores are intrinsically more prone to extinction than other mammals because of their position at top trophic levels which require large hunting areas and determine low densities and because of longer gestation lengths which underlie low capacities for recovery (Cardillo et al. 2004).

Hence, research on the spatial issues of felid ecology becomes an essential conservation tool as it enables to designate the priority conservation areas and corridors for the species of interest. At the global meta-population level, priority is given to preservation of sufficiently large and viable resident populations where prey resources, good habitats and ample lands suffice and human impact is minimal or none (Sanderson et al. 2002). At the national population level, the priority conservation areas are those which contain sufficiently large populations or sub-populations in the most pristine environments and/or those based on (in decreasing order of importance) breeding female home ranges, breeding male home ranges, temporary land tenures of dispersing sub-adults, landscape linkages and buffers against human disturbance (Ferreras 2001). The priority conservation areas designated for large mammals, such as big cats, can be used to identify and preserve the most representative biodiversity-rich areas and corridors between them (Allen et al. 2001). Also, the build-up of knowledge on relationships between felid distribution and environmental factors enables to predict the wide-range patterns of species distribution across the unstudied and less studied areas (Edwards et al. 1996).

In this paper, we apply scat counts to study the spatial issues of distribution and detection of the rare Caucasian leopard (*P. p. ciscaucasica*), synonym Persian leopard (*P. p. saxicolor*), in Armenia by using GIS mapping, occupancy modeling and analysis of multiple-season detection probability. Guesstimates show that no more than 10–15 leopards survive today in southern and south-western Armenia (Khorozyan et al. 2005). In the 2007 IUCN Red List of Threatened Species this cat is classified as globally "endangered", but in Armenia and elsewhere in the Caucasus it should be listed as "critically endangered" (IUCN 2003, 2007). We discuss distribution of the 16 habitat variables over the 16 study areas, their correlation and difference between the areas where leopard scats were found and not, leopard detection and habitat selectivity patterns across the range. Ultimately, we use this information to identify the Priority Leopard Conservation Areas (PLECAs) and propose the improved presence–absence survey design for this vanishing big cat.

18.2 Study Area

This study was carried out across the leopard range in south-western and southern Armenia within the Ararat, Vayk and Zangezur physico-geographical regions (Fig. 18.1). The Ararat region comprises 4 study areas: Kakavaberd (40°03′N/44°53′E), central and eastern Khosrov Reserve (39°58′N/44°57′E), Kharaba (39°55′N/44°59′E) and the Urts Ridge (39°49′N/44°49′E). The Vayk region holds 3 study areas: Elpin (39°48′N/45°06′E), Noravank (39°39′N/45°18′E) and Artavan (39°35′N/45°30′E). The Zangezur region includes the remaining 10 study areas: Salvard (39°28′N/45°55′E), Dastakert (39°20′N/46°01′E), Sisian (39°23′N/46°07′E), Ajubaj (39°15′N/46°02′E), Darmanadzor (39°15′N/46°10′E), Kapan (39°15′N/46°19′E), Khustup (39°08′N/46°19′E), Zangezur Ridge in Meghri district (38°53′N/46°09′E), central and western Meghri Ridge (38°57′N/46°19′E) and Nuvadi (38°57′N/46°26′E).

Vegetation zones within the leopard range are distributed as follows: (a) arid grassland: phrygana, tragacanths and tomillares at elevations 390–1,800 m above sea level; (b) xerophilous sparse forest: junipers (*Juniperus* spp.), almond (*Amygdalus fenzeliana*) and other trees with dense thorny scrubs at 800–2,240 m; (c) mesophilous broad-leaved forest: oaks (*Quercus* spp.), European ash (*Fraxinus excelsior*), Caucasian hornbeam (*Carpinus caucasica*) and shrubs at 800–2,400 m; (d) mountain grassland and subalpine meadow: cereals, dicotyledons, honey plants and other herbs at 1,000–2,800 m; and (e) alpine meadow: herbaceous vegetation at 2,800–3,100 m. Climate is continental, mean air temperature ranges from −10–13°C to 0.9°C in

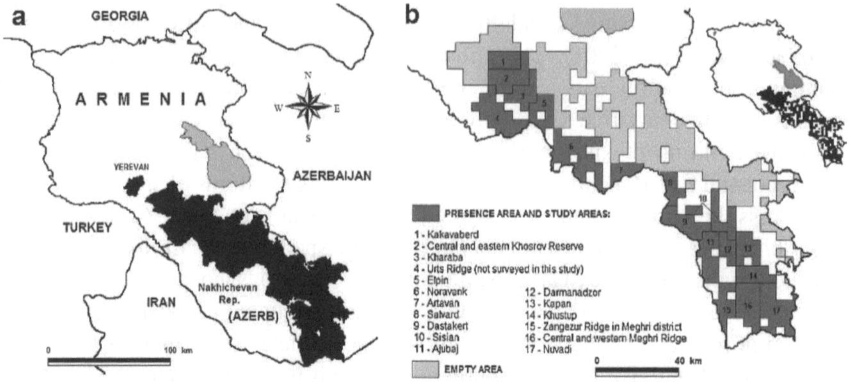

Fig. 18.1 The basic map of the leopard (*Panthera pardus*) range (**a**) and location of study areas (**b**) in Armenia

January and from 12.8°C to over 25°C is July depending on landscapes. Annual precipitation varied from 250–400 mm/year in arid grassland to 600–900 mm/year in alpine meadow (Aivazyan 2006).

We do not consider semi-deserts, nival and harsh nival belts which fall beyond the leopard range.

18.3 Material and Methods

The 1:200,000 georeferenced topographic map, recommended elsewhere for big cat studies (Stith and Kumar 2002), was used as the basis for our GIS map. We produced a GIS map of south-western and southern Armenia which included the landscape belts, dirt roads impassable for vehicles and main roads passable for vehicles, settlements (villages and towns), isohypses and slope aspects. We employed the software ArcView 3.2 and its extensions 3D Analyst and Spatial Analyst and then upgraded it to ArcGIS 9.2 (ESRI Inc., USA). The range boundary was delineated along the boundaries of semi-desert, nival and harsh nival landscapes (see above) and the national borders. Thus, we have produced the basic leopard range map of Armenia (Fig. 18.1a). Then, a specific map of the leopard range was produced by overlaying the basic range map with the cartographic layer of grid of 4 × 4 km cells and removing those grid cells which contained inhabited settlements as they are spatially exclusive with leopard distribution (Khorozyan 2003).

Using our field experience and information from local people, we created two areas: empty area – which is not used by leopards, but occasional penetrations during dispersals and displacements are possible; and presence area – which is inhabited constantly. The presence area was divided into 17 study areas according to their topographic and geographic distinctiveness, of which 16 were surveyed during this study (Fig. 18.1b). The Urts Ridge was not surveyed as its ownership status was unclear.

We have measured the 16 habitat variables across the grid cells and then extrapolated them for study areas (Tables 18.1 and 18.2). The sizes of study areas, areas of landscapes, road lengths, distances to the nearest village and the areas of southern and northern slopes were measured using the ArcView or ArcGIS measuring tool. Landscape diversity (*ldiv*) was calculated as follows:

$$ldiv = \sum_{i=1}^{n} Pi \times \ln Pi \tag{18.1}$$

where *Pi* is the proportion of the area of the *i*-th landscape to the area of all landscapes (Khorozyan et al. 2005). Terrain ruggedness index (*rugg*) was calculated as follows:

$$rugg = \frac{TNC \times TNF}{TNC + TNF} \tag{18.2}$$

where *TNC* is the total number of topographic contours (isohypses) intersecting the selected transect (top-right corner to down-left corner diagonal of the grid cell) and

Table 18.1 Summarized statistical information on the habitat variables across the 16 study areas within the leopard (*Panthera pardus*) presence area in Armenia. SE – standard error, Min – minimum value, Max – maximum value, *p* – significance level of variable difference between the study areas over the mean estimated by the χ^2-test, ns – the result is not significant at $p > 0.05$

Variable	Code	Total	Mean	SE	Min	Max	χ^2	p
Size of study area, km^2	size	2641.0	165.1	19.6	79.5	375.9	231.1	*
Total route length, km	leng	1201.9	75.1	19.1	15.6	331.3	46.9	**
Route density, km/km^2	rout		0.5	0.1	0.1	1.6	1.2	ns
Mean landscape diversity index	ldiv		0.5	0.1	0.2	0.8	0.5	ns
Area of arid grassland, km^2	agra	239.0	14.9	5.9	0.0	70.9	210.0	*
Area of sparse forest, km^2	spar	595.5	37.2	13.7	0.0	203.9	445.5	*
Area of mountain grassland, km^2	moun	647.6	40.5	15.4	0.0	233.0	503.0	*
Area of subalpine meadow, km^2	suba	719.8	45.0	4.8	8.0	72.0	80.4	*
Area of alpine meadow, km^2	alpi	439.1	27.5	5.6	0.0	81.6	120.2	*
Length of dirt roads, km	dirt	1405.6	87.9	10.2	44.5	177.8	127.4	*
Length of main roads, km	main	166.5	10.4	2.5	0.0	36.0	69.5	*
Mean terrain ruggedness index	rugg		9.0	0.5	4.7	12.5	3.9	ns
Mean distance to the nearest village, km	vill		6.5	0.7	3.6	14.1	7.0	ns
Mean wild fire index	fire		1.9	0.1	0.5	2.9	1.4	ns
Area of southern slopes, km^2	sout	903.7	56.5	7.9	21.9	129.6	117.4	*
Area of northern slopes, km^2	nort	772.3	48.3	6.8	19.6	133.5	86.0	*

* the result is significant at $p < 0.01$
** the result is significant at $p < 0.05$

Table 18.2 Distribution of 16 habitat variables over the 16 study areas in this study. The areas where we found leopard scats are marked by bold. Variable codes are the same as in Table 18.1. CE Khosrov – central and eastern Khosrov Reserve, Z Meghri – Zangezur Ridge in Meghri district, CW Meghri – central and western Meghri Ridge

	Habitat variables															
Area	size	leng	rout	ldiv	agra	spar	moun	suba	alpi	dirt	main	rugg	vill	fire	sout	nort
Kakavaberd	116.7	66.9	0.6	0.8	25.3	38.9	14.3	23.7	14.5	53.0	7.8	9.6	14.1	2.2	46.3	27.4
CE Khosrov	207.9	94.6	0.5	0.5	46.2	57.3	81.7	8.0	14.7	90.7	13.1	10.1	11.5	2.3	91.0	40.4
Kharaba	160.0	79.4	0.5	0.5	4.3	0.0	81.3	62.6	11.8	118.4	5.1	10.5	8.3	2.0	55.3	45.8
Elpin	155.1	27.9	0.2	0.4	1.6	0.0	86.9	50.5	16.1	59.5	21.2	8.8	4.9	2.0	63.4	29.5
Noravank	375.9	53.8	0.1	0.3	70.9	0.0	233.0	72.0	0.0	156.0	36.0	7.8	4.2	2.1	93.6	133.5
Artavan	103.8	27.5	0.3	0.5	0.0	0.0	15.5	55.1	33.2	44.5	1.0	6.9	6.2	1.5	21.9	45.1
Salvard	108.8	23.3	0.2	0.4	0.0	0.0	40.0	45.5	23.3	57.3	14.7	6.0	7.0	1.6	26.1	37.0
Dastakert	162.9	54.2	0.3	0.7	0.0	0.0	21.1	60.2	81.6	88.6	0.0	8.9	5.6	1.3	42.7	62.2
Sisian	168.0	123.3	0.7	0.6	1.0	13.0	73.8	61.8	18.4	77.4	3.6	10.5	3.6	1.7	33.4	71.7
Ajubaj	79.5	126.9	1.6	0.3	0.0	0.0	0.0	9.8	69.7	48.7	0.0	4.7	5.8	0.5	31.8	26.5
Darmanadzor	81.5	22.0	0.3	0.7	0.0	16.9	0.0	35.2	29.4	60.9	0.0	11.4	4.5	1.6	31.4	19.6
Kapan	151.1	33.1	0.2	0.8	13.4	59.6	0.0	54.9	23.2	88.8	14.4	12.5	4.8	2.2	52.5	44.4
Khustup	138.0	15.6	0.1	0.7	2.6	68.6	0.0	49.5	17.3	92.6	8.1	10.4	5.4	2.2	29.5	56.6
Z Meghri	125.2	50.9	0.4	0.6	0.3	33.4	0.0	39.8	51.7	51.2	6.2	7.6	5.2	1.6	53.4	32.0
CW Meghri	209.7	71.2	0.3	0.6	11.9	103.9	0.0	63.6	30.3	140.2	14.2	10.3	6.7	2.1	101.8	35.4
Nuvadi	296.9	331.3	1.1	0.2	61.5	203.9	0.0	27.6	3.9	177.8	21.1	7.6	5.6	2.9	129.6	65.2
Total	2641.0	1201.9			239.0	595.5	647.6	719.8	439.1	1405.6	166.5				903.7	772.3

TNF is the total number of changes in topographic aspect along the same transect (Khorozyan et al. 2005). Wild fire index (*fire*) was calculated as:

$$fire = \sum_{i=1}^{n} Pi \times Fi \tag{18.3}$$

where *Pi* is defined above and *Fi* is the score of landscape propensity to burning on the basis of precipitation and wind patterns, maximum air temperatures in summer and dominating vegetation (score 3 for arid grassland and sparse forest, 2 for mountain grassland and subalpine meadow and 1 for alpine meadow – Khorozyan and Abramov 2005). The southern, south-western and south-eastern slopes were merged into the southern slopes and the northern, north-western and north-eastern slopes into the northern slopes. The mean landscape diversity, mean terrain ruggedness, mean distance to the nearest village and mean wild fire indices of study areas were calculated as the arithmetic means of the respected values over the constituent grid cells. The χ^2-test was employed to estimate the significance of variables over their means across the study areas (Quinn and Keough 2002).

The 16 study areas were surveyed on foot during at least two survey periods each with an interval of several months in the snow-free seasons from April 2004 to November 2006 (Fig. 18.1b). We walked one route per day along the wildlife trails and visually identified the origin of scats on the basis of their appearance, smell, deposit place (mostly on the ridgetops) and/or accompanying presence signs (scrapes or tracks). Earlier we have shown that our ability to visually recognize leopard scats is reliable as supported by fecal bile acid thin-layer chromatography and that relative abundances of leopards estimated from visually and chromatographically identified scats were statistically similar (Khorozyan et al. 2007).

The routes crossed all local landscapes and they were selected arbitrarily on a basis of their use by wildlife. All scat samples were collected to avoid their repeated counts during the subsequent surveys. The locations and elevations of scat sites, as well as the lengths of daily routes walked were recorded by the handheld GPS device for their plotting on the GIS map. The route density was calculated as the ratio of total route length in the study area (km) to the size of that area (km^2). Forty-four routes (36.4%) were walked in spring, 16 (13.2%) in summer, 48 (39.7%) in autumn and 13 (10.7%) in winter. In winter, the surveys were conducted only in the southernmost Nuvadi area which is the warmest subtropical area of Armenia.

To compare sampling efforts across the study areas and find possible bias (oversampling of smaller areas and under-sampling of larger ones), we studied distribution of seasons, total route lengths and route densities in study areas and used the χ^2-test. For this analysis, seasons were assigned the dummy variables, from 1 for spring to 4 for winter.

Multiple discriminant analysis was used to find differences between the areas where we found the leopard scats and where we did not (Clevenger et al. 2002).

Habitat selectivity by the leopard was estimated by Jacobs' preference index (*D*):

$$D = \frac{R - P}{R + P - 2RP} \tag{18.4}$$

where R is the ratio of the scat number found in a specific landscape to the total number of scats and P is the ratio of the area of a specific landscape to the total size of study areas (Jacobs 1974). D changes from −1 (always ignored) through 0 (indifference) to +1 (restricted to that landscape). The landscapes having the highest value of D were identified as the critical habitats (Khorozyan 2003). The values of P were calculated from total areas of landscapes presented in Tables 18.1 and 18.2.

Relative abundance of leopards was estimated as the number of scats found per 10km of survey (Wilson and Delahay 2001). To find correlation between relative abundance of leopards and 16 habitat variables, we constructed the Pearson's correlation matrix and considered as correlated those variables whose correlation coefficient (r_p) was higher than 0.5 or lower than −0.5 (Quinn and Keough 2002).

The multi-season subprogramme of programme PRESENCE 2.0 (<www.mbr-pwrc.usgs.gov/software>) was employed to calculate the occupancy and detection probabilities of leopards in Armenia within the multiple-year frame. Occupancy (ψ) is the probability that an area is occupied by the species or, alternatively, is the proportion of an area occupied by the species. Detection probability (p) is the probability of detecting the species, given presence, in each survey within the period (T) (MacKenzie et al. 2006). As the multi-season subprogramme of PRESENCE offers the year as a reasonable unit of T for long-living animals, we used three T's (first – year 2004, second – 2005 and third – 2006).

As the model covariates, we used 6 uncorrelated habitat variables from the Pearson's correlation matrix: size of study area, mean landscape diversity index, area of sparse forest, area of subalpine meadow, mean terrain ruggedness index and mean distance to the nearest village. In the input spreadsheet, we inserted 1 if scats were found in a given survey, 0 if they were not found and – if no survey was conducted and incorporated information about the site and sampling covariates. We manipulated with occupancy and detection probability to make them change over years, depend on the selected habitat variables or stay constant. We used 10,000 bootstraps. Total statistical summary of 192 occupancy models was ranked in an order of decreasing Akaike's Information Criterion (AIC) weights, thus indicating the most important to the least important models. The sum of AIC weights is 1 and the lower the AIC value of a model, the better that model.

The number of surveys in a study area (m) required to reach the desired probability of successfully obtaining one or more detections (power of area surveys or Pa) under a given detection probability (p) was calculated as (Reed 1996; Stauffer et al. 2002):

$$m = \frac{\log(1 - Pa)}{\log(1 - p)} \tag{18.5}$$

The number of study areas within the range (n) to be surveyed to reach the desired probability of successfully obtaining one or more detections during the surveys in the entire range (power of range surveys or Pr) under a given probability of occupancy (ψ) was calculated as (Stauffer et al. 2002):

$$n = \frac{\log(1 - Pr)}{\log(1 - \psi \times Pa)} \tag{18.6}$$

Processing of statistical information throughout this study was done in SPSS 13.0 and MS Excel 2003 software.

18.4 Results

In total, 121 daily routes (surveys) of total length 1201.9 km were walked and 31 leopard scats were found. Mean daily route was 9.9 ± 0.4 (range 3.0–35.0) km which did not differ between the study areas ($\chi^2 = 0.4$, $p > 0.5$). Route density and season also were similar and unbiased across the study areas, thus indicating spatial uniformity of sampling effort (Table 18.1). The study areas significantly differed in their size and, correspondingly, in total route lengths (Tables 18.1 and 18.2).

The Nuvadi area held most of leopard scats (67.7%), followed by central and eastern Khosrov Reserve (19.4), whereas contribution of the central and western Meghri Ridge (6.5), Sisian and Ajubaj areas (3.2 each) was low. Relative abundance of leopards was the highest in the Nuvadi area and the central and eastern Khosrov Reserve (0.63 scats/10 km in each), followed by the central and western Meghri Ridge (0.28), Sisian and Ajubaj (0.08) areas (Fig. 18.1).

Relative abundance of leopards positively correlated with the areas of arid grassland ($r_P = 0.60$, $p < 0.05$) and sparse forest ($r_P = 0.74$, $p < 0.001$), lengths of dirt roads ($r_P = 0.52$, $p < 0.05$) and the areas of southern slopes ($r_P = 0.75$, $p < 0.001$). However, the lengths of dirt roads and areas of southern slopes in their turn correlated with each other and with the areas of arid grassland and sparse forest (r_P varied from 0.63 and 0.83, $p < 0.001$). The areas of arid grassland and sparse forest are uncorrelated and can be considered as principal predictors of leopard occurrence. No negative correlation was found between relative abundance and habitat variables.

The eleven areas where we did not find leopard scats and the five ones where we found them significantly differed over the three habitat variables: the area of sparse forest, mean wild fire index and mean terrain ruggedness index. The statistical results are: 100% of variance, eigenvalue = 65.3, canonical correlation = 0.99, Wilk's lambda = 0.02, chi-square value = 31.4, standardized discriminant coefficient (SDC) of sparse forest = 26.0, SDC of mean wild fire index = 21.9, SDC of mean terrain ruggedness index = 14.4, significance level $p = 0.003$. The mean wild fire index is intrinsically correlated with the area of sparse forest (see Material and Methods; $r_P = 0.69$, $p < 0.005$), but the area of sparse forest and

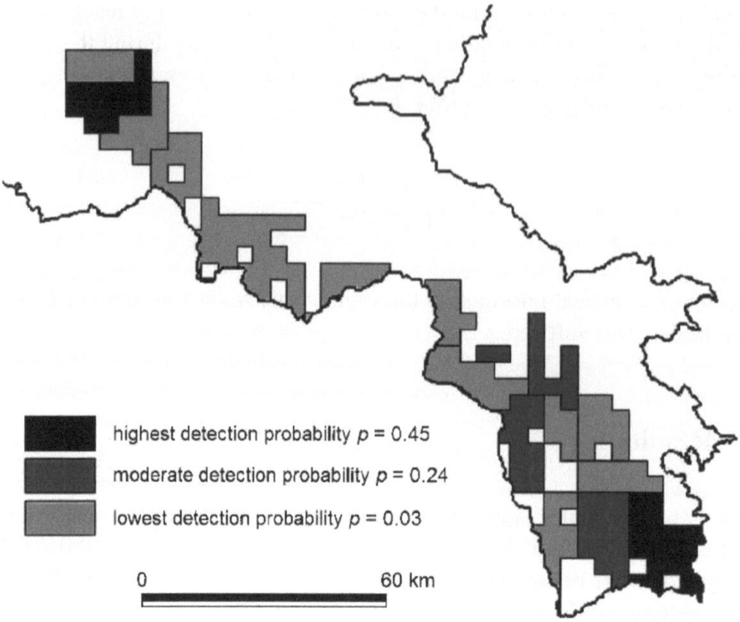

Fig. 18.2 Distribution of detection probability (p) across the study areas

mean terrain ruggedness index do not correlate ($r_p = 0.16$, $p > 0.05$). So, mean terrain ruggedness index can be considered as an additional predictor of leopard occurrence uncorrelated with the other two predictors (area of sparse forest and area of arid grassland).

Out of 31 leopard scats found in this study, 18 were collected in sparse forest, 7 in arid grassland, 4 in mountain grassland and 2 in subalpine meadow. So, the leopard is highly selective for sparse forest ($D = 0.64$) and arid grassland ($D = 0.50$) which are the critical habitats, avoiding mountain grassland ($D = -0.38$) and subalpine meadow ($D = -0.71$) and ignoring alpine meadow ($D = -1$).

Mean elevation of the scat sites was 1537.6 ± 109.8 m above sea level (range 747–2767 m, $n = 31$). Distribution of the scat site elevations (y) against the record months (x, from 1 for January to 12 for December) was statistically strong and curvilinear ($y = -11.30x^3 + 165.65x^2 - 395.29x + 1091.19$, $R^2 = 0.79$, ANOVA: $F_{3,27} = 34.60$, $p < 0.001$). The highest elevations were used by leopards from late spring to late autumn (1,863–2,507 m) and the lowest – from early winter to mid-spring (747–1,450 m in the Nuvadi area and 1,566–2,267 m in the central and eastern Khosrov Reserve).

Detection probability of leopard scats was year-dependent and stable across the models within the survey years, but sharply declined from 2004 ($p = 0.45$) to 2006 ($p = 0.03$) (Fig. 18.2). Meanwhile, leopard occupancy remained high and stable at 0.85 or 85% of presence area.

To calculate the number of surveys in a study area (m), we used the sequence of desired Pa (0; 0.1; 0.2; 0.3; 0.4; 0.5; 0.6; 0.7; 0.8; 0.9; 0.95) and three empirical levels of p (0.45; 0.24; 0.03) for each number of the sequence. For n, we used empirical $\psi = 0.85$, p (0.45; 0.24; 0.03), sequences of desired Pr (0.8; 0.9; 0.95) and m (5; 10; 20; 30; 40; 50). The resulting graphs are illustrated in Fig. 18.3.

The number of surveys to be undertaken in a study area to reach the desired power of area surveys Pa depends on detection probability p, especially when it is the lowest. At the 95% Pa, five surveys should be carried out when $p = 0.45$, 11 when $p = 0.24$ and 130 when $p = 0.03$ (Fig. 18.3). Similar pattern is observed in the relationship between the numbers of study areas to be surveyed and the power of range surveys Pr. At the 95% Pr, one to two study areas are sufficient when $p = 0.45$ regardless of m. When $p = 0.24$, 1–2 study areas are sufficient when m varies from 10 to 50 surveys and increase to three study areas when $m = 5$ surveys. When $p = 0.03$, the number of study areas strongly depends on m. At the 95% Pr, n changes from three study areas when $m = 40–50$ surveys to 22 study areas when only five surveys are conducted (Fig. 18.3).

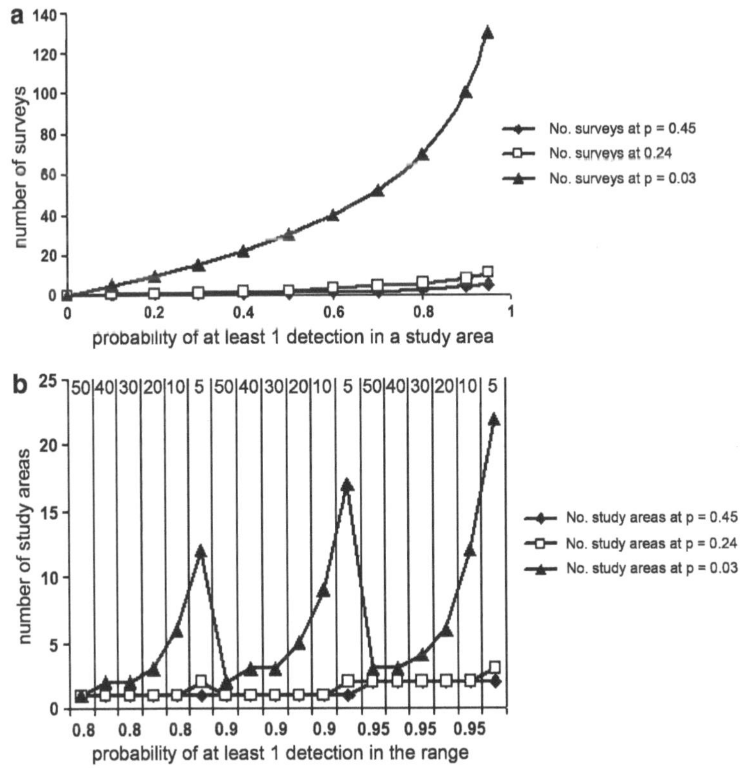

Fig. 18.3 Distribution of the numbers of surveys required to reach the desired probability of ≥1 detection in a study area (power of area surveys) at different levels of detection probability p (**a**) and distribution of the numbers of study areas to be surveyed to reach the desired probability of ≥1 detection in the entire range (power of range surveys) at the empirical occupancy $\psi = 0.85$ (**b**). The numbers of surveys per study area are indicated on the top of picture (**b**)

18.5 Discussion

The leopard is the rarest and most elusive mammal of Armenia's fauna distributed in the southern and south-western portions of the country over the presence area of 2,856.8 km^2 (2,641.0 km^2 of the 16 surveyed study areas and 215.8 km^2 of the unstudied Urts Ridge) (Fig. 18.1). So, it is essential to know the distribution of this predator and how this is related to the factors of ambient environment.

The leopard occurrence strongly correlates with arid grassland, xerophilous sparse forest and rugged terrain. Arid grassland and sparse forest hold sufficient prey base and prove to be the critical habitats for local leopards. Recent presence–absence occupancy modeling has shown that biomass of ungulate prey, including the bezoar goat (*Capra aegagrus*), wild boar (*Sus scrofa*) and roe deer (*Capreolus capreolus*), in the Nuvadi area is high (720.37 ± 142.72 kg/km^2) and capable of supporting many more leopards than actually live there (Khorozyan et al. 2008). Preference of these habitats, especially the sparse forest, by leopards agrees with our earlier data obtained from Khosrov Reserve and its vicinities (Khorozyan 2003).

Availability of precipitous rocky terrain and screes is an essential requirement for the leopard existence since they provide plenty of secluded nooks for shelters, dens and ambush sites, harbour the leopard's staple prey (bezoar goat) and limit access by humans and livestock.

On the other hand, preference of arid grassland and sparse forest can be caused by that these landscapes are distributed over the southern slopes and the leopard, being a species of tropical origin, will prefer southern exposure where snow accumulation is minimal. Gavashelishvili and Lukarevskiy (2008) have also found that snow cover is an important factor limiting leopard distribution in the Middle East, particularly in the Caucasus. Also, these landscapes are easier to be used by leopards and other wildlife as they contain most of dirt roads. Leopards live on higher elevations during the snow-free seasons, moving mainly along the narrow ridgetops, and descend to the foothills when snowfalls come. The statement by Gavashelishvili and Lukarevskiy (2008) that in the Middle East leopards avoid deserts and human settlements is indirectly confirmed in our study, as we *a priori* knew the pattern of such avoidance and excluded semi-desert (Armenia has no deserts) and inhabited settlements from the leopard distribution map.

Avoidance of mountain grassland and subalpine meadow and ignorance of alpine meadow, which are situated on plain mountaintop plateaus, are caused by prey scarcity resulting from intensive livestock breeding, deficiency of permanent water sources, shelter and adequate cover and accumulation of deep snow in the autumn–spring period (Khorozyan et al. 2005).

We failed to obtain a statistically robust logistic relationship between the leopard presence–absence data and the habitat variables, a result which hinted at a possible significant role of non-detections, also called false negatives (i.e. species is present but goes undetected) (Gu and Swihart 2004). The presence–absence models dealing with rare or elusive species, but ignoring their detection probability, suffer from overestimated absence and underestimated presence. In this case, the naïve assumption that if a species is present it would be definitely detected, i.e. its detection probability is 1, has been violated (Reed 1996; Moilanen 2002; Tyre et al. 2003; Gu and Swihart 2004; Wintle et al. 2005; MacKenzie et al. 2006).

This is a case for leopard in Armenia. Even in the best study areas, the central and eastern Khosrov Reserve and the Nuvadi area, which were surveyed in 2004 detection probability was 0.45. Then, in 2005 it dropped to 0.24 as we moved to the less optimal study areas but continued to survey the Nuvadi area and to 0.03 in 2006 when we discontinued the surveys in Nuvadi and have concentrated only on the worst areas (Fig. 18.2). The occupancy of the predator was kept high in all study years, at 85% of presence area. So, the leopard is a widely occurring, but seldom detectable predator. Despite detection probability of leopard in this study is definitely low, we discriminate three levels of it: high (0.45), moderate (0.24) and low (0.03).

Knowing detection probability and occupancy of the species, it is possible to estimate the number of surveys per study area and the number of study areas to be surveyed so that to obtain the desired probability of one or more detections of the species (i.e., power of area and range surveys, otherwise known as confidence level) or, alternatively, be sure that the species is extinct (Reed 1996; Stauffer et al. 2002). In our case, at the 95% power of area surveys the number of surveys to be undertaken is the lowest when detection probability is high, but increases moderately at the medium detection probability and sharply at the lowest level of this probability (Fig. 18.3). At the 95% power of range surveys, one to two study areas are sufficient to be surveyed at the medium and high detection probabilities regardless of the numbers of surveys in each, but their number increases to three at the medium detection probability and the least number of surveys. When detection probability is the lowest, much more study areas should be surveyed when the number of surveys is limited (Fig. 18.3). It is more efficient to increase the numbers of study areas and survey them less intensively than vice versa, especially at the lowest detection probability (Fig. 18.3; see also Stauffer et al. 2002; Wintle et al. 2005).

Detection probability, particularly in relation to scat counts, is often affected by non-random, or predetermined, bias caused by detection-favouring habitats, seasons, fecal decay and/or observers (Reed 1996; Wilson and Delahay 2001; Gu and Swihart 2004; Wintle et al. 2005). In our study, no such biases were observed as scats remain identifiable much longer in Armenia's arid mountains than other presence signs (tracks and scrapes), survey seasons did not vary between the areas, no particular habitats favoured detection of leopard scats against the others, and the observer bias was absent as the same researchers (IGK and AGM) were involved in all surveys.

We suggest the optimized leopard presence–absence survey design in Armenia. To attain the 95% confidence level, 5–10 surveys per study area and 1–3 study areas are sufficient at the medium to high levels of detection probability. When detection probability is the lowest, 12–22 study areas should be studied by conducting 5–10 surveys in each to gain the same confidence level. The standardized range-wide survey could look as nine surveys per study area conducted as three surveys per survey period over three survey periods, one period per snow-free season (spring to autumn). The larger the study area, the longer routes are to be walked to keep sampling effort unbiased. The interval between subsequent survey periods should be 3–4 months to allow leopards living at low density to visit an area and leave scats there, but prevent disappearance or loss of identifiability of scats. Under this scheme, at least one detection means presence and zero detection means true absence, i.e. extinction (Reed 1996; Stauffer et al. 2002). This design is not fixed and can be reasonably manipulated to comply with constraints of survey budget.

Our results agree with those of other authors that at least three surveys should be conducted in a study area to obtain the usable estimates of detection probability and the numbers of surveys, trails and study areas should be maximized whenever possible, even at the expense of route lengths (Van Sickle and Lindzey 1991; Kendall et al. 1992; Stander 1998; Tyre et al. 2003; MacKenzie and Royle 2005). As the leopard is wide-ranging, more frequent surveys would be more expedient in its research than extension of survey period which works well for animals with small home ranges (Wintle et al. 2005).

In Armenia, relative abundance of leopards is maximal in the Nuvadi area and the central and eastern Khosrov Reserve, whereas in the other three areas (central and western Meghri Ridge, Sisian and Ajubaj areas) it is much lower. As fresh scats were recorded during all survey periods in the Nuvadi and Khosrov Reserve, we suppose these areas are constantly inhabited by leopards and propose them as the Priority Leopard Conservation Areas (PLECAs) where this predator must be protected and studied first.

In central and western Meghri Ridge, Sisian and Ajubaj areas leopard scats were found irregularly at medium detection probability, so these areas are possibly used by cats occasionally as true corridors. It is unclear whether the eleven areas where we did not find scats at all are used as corridors as we certainly surveyed them insufficiently in light of the above-mentioned survey design (potential corridors) (Fig. 18.2). Most of the true and potential corridors are situated in the Zangezur Ridge which is stretched along the state border of Armenia and Nakhichevan Republic, an enclave of Azerbaijan. This area was devastated during the Armenian-Azerbaijani war over Nagorno-Karabakh in 1989–1994 and, since 1995 when the cease-fire regime was proclaimed, it suffers from recommencement of human activities: military training and testing grounds, border posts, agriculture, mining and re-settling of previously abandoned villages. Thus, the status of the Zangezur Ridge as a suite of movement linkages for leopards and other wildlife is in jeopardy. The habitats dominating in the Zangezur Ridge and its branches are mountain grasslands, subalpine and alpine meadows which are avoided or ignored by leopards, but can potentially be used as movement conduits during the snow-free seasons.

No one of the study areas surveyed by us, as well as the Urts Ridge, are large enough to maintain the viable leopard populations since each of them is smaller than the minimum area likely to support a leopard population, 412 km^2, known as threshold area (Smallwood 1999). Even the largest protected areas of southern and south-western Armenia, such as reserves (Khosrov Reserve, 239 km^2), sanctuaries (Jermuk Hydrological Sanctuary, 180 km^2) and forest management territories (Kapan Forestry, 393.4 km^2) are below the threshold area (Aivazyan 2006; A. Aghasyan, personal communication, 2007). Meantime, all together the study areas provide an ample space capable of retaining the viable leopard population if the efficient conservation and enforcement measures would have been taken. As Armenia is small and cannot afford to set aside the large tracts of lands for protected areas or even sufficiently enlarge the existing reserves, priority must be given to development of the community-based conservation schemes. Much attention should be paid to the leopard presence–absence surveys in the true and potential corridors and the Urts Ridge. Conservation of the designated corridors should be simultaneous and large-scale to prevent insularization of the PLECAs.

Current population of leopard in Armenia is too small to be viable even in a short run. The principal threats are poaching, human disturbance and wild fire which, if continuing at current rates, can make this species disappear from the national fauna in a few years (Khorozyan et al. 2005). Historically and especially now, its survival has been ensured by immigration of individuals from Iran which is a leopard stronghold in the Middle East. The powerful borderline Arax River contains many suitable fords for immigration and the barbed-wire border infrastructure does not hinder leopard crossings (I. Khorozyan, personal observation). Maintenance of this gene flow must become a priority for development of transboundary conservation projects in Armenia and Iran.

From the most skeptical view, Armenia can be considered as the peripheral and sink part of the largest Iranian pool so conservation of its leopard population would look impractical (Peterson 2001). We argue this opinion from two viewpoints. First, despite the leopard range in Iran being vast (885,300 km^2), its guessed abundance and crude density are low (550–850 individuals or 0.06–0.1 individuals/100 km^2) (Kiabi et al. 2002), so efforts in Armenia are worth taking anyway. Second, the perception that carrying capacities of small countries like Armenia for wide-ranging carnivores are low must not detract the researchers and conservationists from taking essential actions to avert local extinction. Even small areas and countries are capable to retain good leopard populations if prey is sufficient and human pressure is minimal.

Acknowledgments This paper is a part of the first author's PhD dissertation at the Zoological Institute, Russian Academy of Sciences, St. Petersburg, Russia. We sincerely thank A. Aghasyan, S. Shaboyan and S. Hovhannisyan for administrative support and local villagers, especially M. Boyajyan, M. Arzumanyan, A. Petrosyan, S. Martirosyan, V. Martirosyan and the rangers of Khosrov Reserve for hospitality and assistance. Special thanks go to D.I. MacKenzie and J.E. Hines for useful advices on using PRESENCE 2.0. We are also grateful to M. Hayward, D.H. Reed and an anonymous reviewer for valuable comments on the earlier version of the draft.

Financial support was provided by People's Trust for Endangered Species (UK) and Doue la Fontaine Zoo (France).

References

Aivazyan HM (2006) Nature of Armenia. Series "Family Encyclopedia", book 3. Haikakan Hanragitaran, Yerevan

Allen CR, Pearlstine LG, Kitchens WM (2001) Modeling viable mammal populations in gap analyses. Biol Conserv 99:135–144

Beier P (1993) Determining minimum habitat areas and habitat corridors for cougars. Conserv Biol 7:94–108

Cardillo M, Purvis A, Sechrest W, Gittleman JL, Bielby J, Mace GM (2004) Human population density and extinction risk in the world's carnivores. PLoS Biol 2:909–914

Clevenger AP, Wierzchowski J, Chruszcz B, Gunson K (2002) GIS-generated, expert-based models for identifying wildlife habitat linkages and planning mitigation passages. Conserv Biol 16:503–514

Crooks KR (2002) Relative sensitivities of mammalian carnivores to habitat fragmentation. Conserv Biol 16:488–502

Edwards TC, Deshler E, Foster D, Moisen GG (1996) Adequacy of wildlife habitat relation models for estimating spatial distributions of terrestrial vertebrates. Conserv Biol 10:263–270

Ferreras P (2001) Landscape structure and asymmetrical inter-patch connectivity in a metapopulation of the endangered Iberian lynx. Biol Conserv 100:125–136

Gavashelishvili A, Lukarevskiy V (2008) Modelling the habitat requirements of leopard *Panthera pardus* in west and central Asia. J Appl Ecol 45:579–588

Gu W, Swihart RK (2004) Absent or undetected? Effects of non-detection of species occurrence on wildlife-habitat models. Biol Conserv 116:195–203

Haines AM, Tewes ME, Laack LL, Horne JS, Young JH (2006) A habitat-based population viability analysis for ocelots (*Leopardus pardalis*) in the United States. Biol Conserv 132:424–436

IUCN (2003) Guidelines for application of IUCN Red List criteria at regional levels: version 3.0. IUCN Species Survival Commission, Gland and Cambridge. www.iucnredlist.org Accessed 20 February 2008

IUCN (2007) The 2007 IUCN Red List of Threatened Species. www.iucnredlist.org Accessed 20 February 2008

Jacobs J (1974) Quantitative measurement of food selection – a modification of the forage ratio and Ivlev's electivity index. Oecologia 14:413–417

Kendall KC, Metzgar LH, Patterson DA, Steele BM (1992) Power of sign surveys to monitor population trends. Ecol Appl 2:422–430

Khorozyan I (2003) Habitat preferences by the endangered Persian leopard (*Panthera pardus saxicolor* Pocock, 1927) in Armenia. Zool Middle East 30:25–36

Khorozyan IG, Abramov AV (2005) The spatial structure and conservation of the Persian leopard population in Armenia. In: Rozhnov VV, Tembotova FA (eds) Mlekopitayuschie gornykh territorii. KMK Scientific, Moscow

Khorozyan IG, Cazon A, Malkhasyan AG, Abramov AV (2007) Using thin-layer chromatography of fecal bile acids to study the leopard (*Panthera pardus ciscaucasica*) population. Biol Bull 34:361–366

Khorozyan IG, Malkhasyan AG, Abramov AV (2008) Presence-absence surveys of prey and their use in predicting leopard (*Panthera pardus*) densities: a case study from Armenia. Integr Zool 3:322–332

Khorozyan I, Malkhasyan A, Asmaryan S (2005) The Persian leopard prowls its way to survival. Endang Spec Update 22:51–60

Kiabi BH, Dareshouri BF, Ghaemi RA, Jahanshahi M (2002) Population status of the Persian leopard (*Panthera pardus saxicolor* Pocock, 1927) in Iran. Zool Middle East 26:41–47

MacKenzie DI, Nichols JD, Royle JA, Pollock KH, Bailey LL, Hines JE (2006) Occupancy estimation and modeling: inferring patterns and dynamics of species occurrence. Academic, Boston, MA

MacKenzie DI, Royle JA (2005) Designing occupancy studies: general advice and allocating survey effort. J Appl Ecol 42:1105–1114

Moilanen A (2002) Implications of empirical data quality to metapopulation model parameter estimation and application. Oikos 96:516–530

Peterson AT (2001) Endangered species and peripheral populations: cause for reflection. Endang Spec Update 18:30–31

Quinn GP, Keough MJ (2002) Experimental design and data analysis for biologists. Cambridge University Press, Cambridge

Reed DH (2004) Extinction risk in fragmented habitats. Anim Conserv 7:181–191

Reed JM (1996) Using statistical probability to increase confidence of inferring species extinction. Conserv Biol 10:1283–1285

Sanderson EW, Redford KH, Chetkiewicz CLB, Medellin RA, Rabinowitz AR, Robinson JG, Taber AB (2002) Planning to save a species: the jaguar as a model. Conserv Biol 16:58–72

Smallwood KS (1999) Scale domains of abundance amongst species of mammalian Carnivora. Env Conserv 26:102–111

Stander PE (1998) Spoor counts as indices of large carnivore populations: the relationship between spoor frequency, sampling effort and true density. J Appl Ecol 35:378–385

Stander PE, Haden PJ, Kaqece, Ghau (1997) The ecology of asociality in Namibian leopards. J Zool 242:343–364

Stauffer HB, Ralph CJ, Miller SL (2002) Incorporating detection uncertainty into presence–absence surveys for marbled murrelet. In: Scott JM, Heglund PJ, Morrison ML, Haufler JB, Raphael MG, Wall WA, Samson FB (eds) Predicting species occurrences: issues of accuracy and scale. Island, Washington, DC

Stith BM, Kumar NS (2002) Spatial distributions of tigers and prey: mapping and the use of GIS. In: Karanth KU, Nichols JD (eds) Monitoring tigers and their prey. Centre of Wildlife Studies, Bangalore

Sunquist ME, Sunquist F (2001) Changing landscapes: consequences for carnivores. In: Gittleman JL, Funk SM, MacDonald D, Wayne RK (eds) Carnivore conservation. Cambridge University Press, Cambridge

Tigas LA, Van Vuren DH, Sauvajot RM (2002) Behavioral responses of bobcats and coyotes to habitat fragmentation and corridors in an urban environment. Biol Conserv 108:299–306

Tyre AJ, Tenhumberg B, Field SA, Niejalke D, Parris K, Possingham HP (2003) Improving precision and reducing bias in biological surveys: estimating false-negative error rates. Ecol Appl 13:1790–1801

Van Sickle WD, Lindzey FG (1991) Evaluation of a cougar population estimator based on probability sampling. J Wildl Manag 55:738–743

Weber W, Rabinowitz A (1996) A global perspective on large carnivore conservation. Conserv Biol 10:1046–1054

Wilson GJ, Delahay RJ (2001) A review of methods to estimate the abundance of terrestrial carnivores using field signs and observation. Wildl Res 28:151–164

Wintle BA, Kavanagh RP, McCarthy MA, Burgman MA (2005) Estimating and dealing with detectability in occupancy surveys for forest owls and arboreal marsupials. J Wildl Manag 69:905–917

Woodroffe R (2001) Strategies for carnivore conservation: lessons from contemporary extinctions. In: Gittleman JL, Funk SM, MacDonald D, Wayne RK (eds) Carnivore conservation. Cambridge University Press, Cambridge

Chapter 19
Mapping Landscape Resistance to Identify Corridors and Barriers for Elephant Movement in Southern Africa

Samuel A. Cushman, Michael Chase, and Curtice Griffin

19.1 Introduction

One of Africa's greatest conservation successes is the recovery of elephant (*Loxodonta africana*) populations within protected areas (e.g. Aleper and Moe 2006), such as those in northern Botswana. This recovery poses several challenges, however. First, habitat within protected areas is becoming degraded from high intensity elephant browsing. Second, the increasing elephant and human populations in the region have led to large increases in human–elephant conflict along the periphery of protected areas (Sitati et al. 2005; Lee and Graham 2006). Management options include facilitating natural dispersal, active relocation, and culling. Relocation is prohibitively expensive as a population-level solution given the high per capita cost. Culling is politically unpopular given Botswana's booming wildlife tourist industry. Simultaneously, large areas of the neighboring countries of Namibia, Zambia and Angola have low elephant densities. Some of these governments desire to increase elephant populations within their protected areas to promote the growth of wildlife tourism. Thus, facilitated dispersal of elephants from high density areas of northern Botswana to protected areas in other countries with low elephant densities is an attractive potential solution.

S.A. Cushman (✉)
US Forest Service, Rocky Mountain Research Station, 800 E Beckwith, Missoula, MT 59801, USA
e-mail: scushman@fs.fed.us

M. Chase
Elephants Without Borders, Kasane, Botswana
e-mail: er@info.bw

C. Griffin
Department of Natural Resources Conservation, University of Massachusetts, Amherst, MA 01003, USA
e-mail: cgriffin@nrc.umass.edu

S.A. Cushman and F. Huettmann (eds.), *Spatial Complexity, Informatics, and Wildlife Conservation*
DOI 10.1007/978-4-431-87771-4_19, © Springer 2010

Recently, several approaches have been shown to be effective in developing rigorous, species specific landscape resistance maps, which represent the resistance to organism movement as functions of multiple variables from a variety of spatial scales. Such resistance maps, if reflective of the factors and scales at which organisms respond to environmental conditions in their movement behaviour, provide a key foundation for a variety of applied analyses of landscape connectivity, including identification of factors that influence landscape connectivity and mapping of movement corridors. Cushman et al. (2006) developed landscape resistance maps using molecular genetic data and least cost path modelling. Such landscape genetic approaches hold tremendous promise for evaluating the factors that affect gene flow over time scales of several to many generations. However, many of the issues of most conservation importance are incipient in time such that they have not yet left a genetic signature in the population. In addition, for some conservation questions it is the movement of organisms rather than of genes that is the critical parameter.

To address the factors that affect organism movement directly on time scales of less than the life span of individual organisms, landscape resistance modelling with telemetry data perhaps holds the greatest promise (e.g. Osborn and Parker 2003). GPS telemetry data can provide spatially precise records of the movement paths of individual animals at a temporal sampling rate that allows direct assessment of the influences of landscape features on movement path selection. This enables the development of species-specific landscape resistance models in which the resistance of any location, or pixel, in a landscape is a function of multiple landscape features measured at one or several scales.

Cushman et al. (2005) investigated the pattern of temporal autocorrelation of elephant movements monitored thorough satellite telemetry in Botswana. They found that autocorrelation of elephant movements is long-term, complex and seasonally related. During much of the year, locations as much as 30 days apart were significantly autocorrelated. This autocorrelation presents a problem for traditional analyses that treat individual locations as statistically independent replicates for statistical analysis. Some researchers have advocated subsampling these autocorrelated movement data streams to achieve statistical independence. However, this approach does not in fact remove spatial dependence (Fortin and Dale 2005) and results in unacceptable information loss (Cushman et al. 2005). In addition, the spatial patterns of movement that create autocorrelation are the biological signal that should be investigated. Thus, alternative methods that do not depend on statistically independent individual locations are essential.

In this study, we use a path-level spatial randomization method to assess the effects of multiple landscape features on elephant movement in the transboundary region of Botswana, Namibia and Zambia. The first goal of this study was to evaluate the influences of water sources, roads, wildlife fences and human settlements on elephant movements, and use this information to produce a map of landscape resistance to elephant movements.

Movement resistance models are essential foundations for applied analyses of population connectivity. However, resistance maps are not in themselves sufficient to answer many questions of greatest concern. For example, pixel level resistance to movement does not in itself provide sufficient information to evaluate the existence, strength and location of barriers and corridors. Where resistance maps are point specific, connectivity is route specific. Connectivity must be evaluated as the path and cost of moving across a landscape resistance model from a source to a destination. The resistance model is the foundation for these analyses, but it is explicit consideration of movement paths across the resistance surface that provides the key information for conservation and management.

By adding source-destination least cost path analysis to species-specific landscape resistance mapping, it is possible to comprehensively analyze the effects of landscape structure on animal movement such that the strength and location of movement corridors and barriers can be rigorously evaluated (Cushman et al. 2009). The second goal of this paper is to map potential movement corridors and barriers between northern Botswana and Sioma National Park in Zambia, and evaluate the relative impact of water sources, wildlife fences and human settlements on elephant movement routes and degree of habitat isolation.

19.2 Materials and Methods

19.2.1 Movement Data

This study used GPS location data from four elephant herds and landscape maps of rivers, roads, fences and settlements to identify corridors, barriers and to produce a map of landscape resistance to elephant movement. The GPS data consist of fixes acquired at 8 hour intervals (Fig. 19.1). Elephant 1 was monitored from July 14, 2005 to September 18, 2006, elephant 2 from August 9, 2003 to December 3, 2004, elephant 3 from June 26, 2005 to September 18, 2006, and elephant 4 from August 19, 2003 to April 30, 2005.

19.2.2 Path Randomization and Landscape Resistance Hypotheses

Our analysis tests alternative hypotheses of landscape resistance against the movement paths selected by individual elephants. Elephant movements may be influenced by landscape features (Sitati et al. 2003, Murwira and Skidmore 2005). A priori, we proposed six landscape features that we believe may influence elephant movements, including distance to water (Chamaille-Jammes 2006), roads, wildlife fences, towns, villages and subsistence huts (Lee and Graham 2006). These features can be combined to create many alternative models of possible landscape resistance to movement.

To determine relative support among the many possible alternative models of landscape resistance, we compared utilized paths to available paths in a two-step

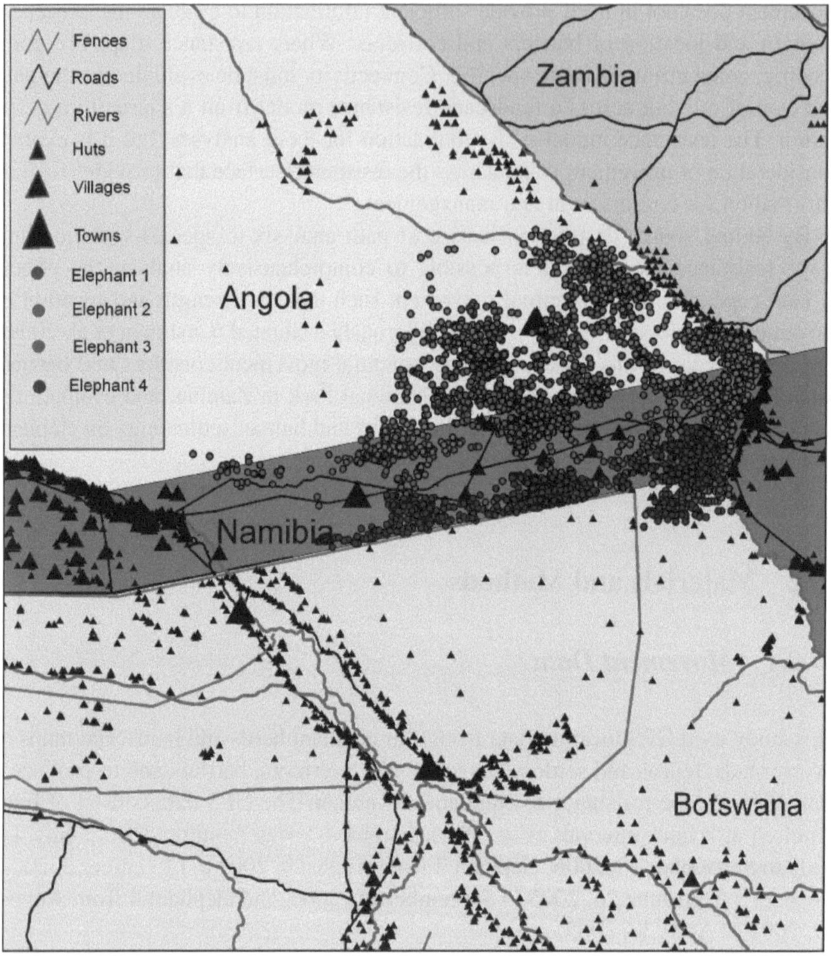

Fig. 19.1 Map showing the study area, landscape features used in the resistance hypotheses (fences, roads, rivers and settlements) and the locational data for each of the four elephants included in the analyses

process. First, the utilized path was created by converting the series of sequential point locations for each elephant into a line in ArcInfo workstation (ESRI 2005). Second, 199 available paths with identical topology were created by randomly shifting and rotating this utilized path. The available paths were randomly shifted a distance between 0 and 30 km in x and y, and randomly rotated between 0° and 360°. This resulted in a population of 199 available paths with identical spatial topology for each utilized path (Fig. 19.2). The statistical support for the resistance model is determined by calculating the number of standard errors the cumulative cost for the utilized path is from the distribution of costs of the randomized sample of 199 available paths.

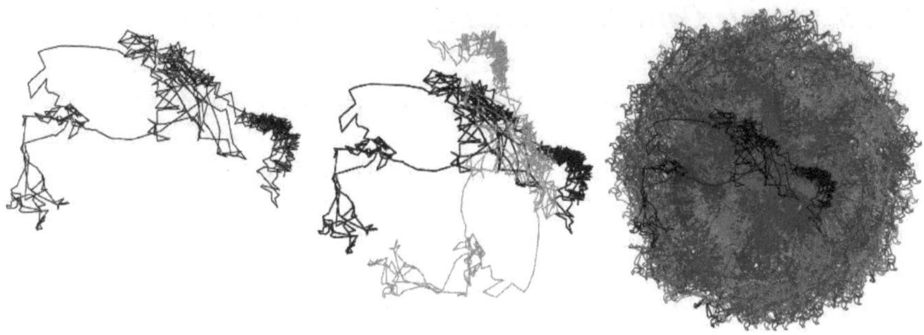

Fig. 19.2 Utilized path and random paths for elephant 1. The utilized path is shown on the left, a single random path overlaid in the middle, and the full set of 199 random paths on the right

19.2.3 Scaling of Landscape Resistance Factors

It is important to determine the spatial scale at which each landscape feature maximally influences elephant movement (Wiens 1989). We conducted a scaling analysis for each landscape feature independently by computing the standard errors of the utilized path from the distribution of available paths for each factor for multiple scales. We investigated scaling relationships of towns, villages and huts at five scales. These scales were created by placing a unimodal resistant kernel (e.g. Compton et al. 2007) of varying width over each town, village or hut and summing the kernels into a resistance surface. The five scales we compared corresponded to kernel widths of 1, 5, 10, 20, and 40. Similarly, we conducted a scaling analysis for distance to water. We compared resistance scaled as linear distance, square root of distance and the square of the distance to water. Roads and fences were analyzed at a single scale. We treated these as potential barriers, such that resistance only accumulated in crossing these features, and not as functions of their proximity.

19.2.4 Factorial Weighting Analysis

In addition to scaling, we also conducted factorial weighting analyses to determine the most supported combination of weights among the resistance factors for each individual elephant. A priori, we specified three levels of relative weight, 1, 5 and 10. A factor given a weight of five has five times the maximum resistance value of a factor given weight 1, and ½ the maximum resistance value of a factor given weight 10. The analysis then proceeds by computing combined resistance values for each cell in the landscape based on the sum of resistances due to the different landscape features. This was conducted across a factorial combination of the three relative weights given to each landscape feature.

19.2.5 Monte Carlo Randomization to Assess Support

We used a Monte Carlo permutation procedure to test for global support. When hypotheses are constructed across a quantitative range of values for a parameter, it is possible to evaluate the degree to which the analysis indicates a unimodal peak of support for a global best model. This is done by computing the differences in support (in our case standard errors from mean of the distribution of available paths) among all neighbouring cells in the hypothesis cube and comparing the sum of those differences to the distribution of the sum of differences from a large number of randomizations of the hypothesis cube.

In our case, we have two 3 × 3 × 3 hypothesis cubes for each elephant, totalling 27 cells per cube. We computed the sum of the differences in the standard errors from the mean of the available path distribution between each pair of neighbouring hypotheses in the 27 cell hypothesis cube. We then randomized the locations of each value of the standard error within the 27 cell cube 99,999 times, recalculating the sum of the differences in standard errors each time. The test evaluates the significance of a unimodal peak of support for the best model in the hypothesis cube.

19.2.6 Model Averaging

The scaling and weighting analyses identified a best resistance model for each elephant. We produced a global model across elephants by averaging the four individual resistance models on a cell-by-cell basis. This produced a single, average resistance model which was used for all subsequent analyses.

19.2.7 Least Cost Path Analysis

We mapped movement corridors and identified potential barriers by computing the density of least cost paths across the resistance map between 1,183 points in northern Botswana and the geographical center of Sioma National Park. The 1,183 points were selected systematically to provide source points at 5 km spacing throughout northern Botswana, to give a comprehensive view of movement routes from all of northern Botswana to Sioma national Park.

First, we computed the cost distance from Sioma National Park to all points in the study area, using the COSTDISTANCE function in ArcInfo Workstation (ESRI 2005). This provided a measure of isolation of each location in northern Botswana from Sioma National Park, based on the resistance map. Next, we calculated the difference in cost distance between the current landscape and the predicted resistance for the same landscape in the absence of fences, roads and human settlements, to provide a measure of the effects of human development on population isolation across the study area.

Then, we computed least cost paths between each of the 1,183 source pixels and the destination pixel using the COSTPATH function in ArcInfo Workstation (ESRI 2005). These least cost paths are single pixel in width, and record the route of a least cost path from the source to the destination pixel. We smoothed these lease cost paths by applying a parabolic kernel with a 3000 m radius, on the belief that actual paths taken will imperfectly follow least cost routes due to stochastic behavioural choices of individual animals. The kernel smoothed least cost paths were then summed to provide maps of the density of least cost routes from northern Botswana to Sioma. We computed these summed least cost path maps for two resistance models: (1) the full landscape resistance model, (2) landscape resistance predicted in the absence of settlements, fences and roads. The comparison of these two enables us to identify both the areas of highest importance for connectivity in the current landscape, and to evaluate the effects of anthropogenic barriers in blocking historical movement corridors.

19.3 Results

19.3.1 Scaling Analyses

There was high consistency among elephants in the scales at which each factor was most strongly related to movement path selection. All four elephants showed strong avoidance of towns, with three of the four showing strongest avoidance at a kernel width of 5 km (Table 19.1). Similarly, all four elephants showed significant avoidance of villages and huts, with three of four showing strongest avoidance at a kernel width of 1 km, in both cases (Tables 19.2 and 19.3). These results show that the movements of these four elephants are negatively related to the presence of human settlements on the landscape, with strong avoidance of towns at distances of up to 5 km, and avoidance of villages and huts at a finer spatial scale of up to 1 km.

Interestingly, there is an apparent positive relationship between elephant movements and the presence of huts and villages at scales of over 20 km and with towns at scales of over 40 km. This is a result of spatial covariation between the

Table 19.1 Scaling results showing avoidance of towns by elephants across scales from 1 to 40 km. Numbers in the table refer to the number of standard errors from the mean of the distribution of available paths that the utilized path fell in terms of cumulative resistance due to towns, at each of the five spatial scales. The table indicates that all four elephants significantly avoided towns, with three of the four most strongly avoiding towns at a scale of 5 km

	1 km	5 km	10 km	20 km	40 km
Elephant 1	−26.42	−38.71	−34.16	−15.28	67.13
Elephant 2	−51.02	−35.63	−13.46	27.34	30.63
Elephant 3	−4.8	−6.69	−4.68	0.76	13.91
Elephant 4	−51.02	−52.41	−52.41	−52.07	−10.5

Table 19.2 Scaling results for villages across scales from 1 to 40 km. The table indicates that all four elephants significantly avoided villages, with three of the four showing strongest avoidance at a scale of 1 km

	1 km	5 km	10 km	20 km	40 km
Elephant 1	−2.8	−3.63	2.2	11.67	9.08
Elephant 2	−48.2	−41	−23.8	0.47	12.63
Elephant 3	−6.35	−5.72	4.86	15.48	18.85
Elephant 4	−51.02	−29.45	−24.28	−12.8	−7.91

Table 19.3 Scaling results for huts across scales from 1 to 40 km. The table indicates that all four elephants significantly avoided huts, with three of the four showing strongest avoidance at a scale of 1 km

	1 km	5 km	10 km	20 km	40 km
Elephant 1	−68.9	−50.65	−2.57	51.28	76.67
Elephant 2	−33.52	−34.62	−19.09	10.05	30.54
Elephant 3	−86.36	−64.77	3.55	64.1	57
Elephant 4	−41.75	−38.82	−17.79	60.4	103.05

Table 19.4 Scaling results for distance to water. Sqrt D – Square root distance to water, D – Euclidean distance to water, Dsq – Square of the distance to water. The table indicates that all three elephants significantly selected movement paths near water. Two of the four elephants had highest support for a model where resistance increases as the square root of distance to water. Elephant 4 had statistically equal support for SqrtD and D, while elephant 2 had statistically equal support for D and Dsq

	SqrtD	D	Dsq
Elephant 1	−15.24	−12.41	−5.04
Elephant 2	−14.44	−15.8	−15.9
Elephant 3	−11.34	−5.87	−2.18
Elephant 4	−17.61	−17.83	−10.93

location of human settlements and water (Fig. 19.1). Human settlements tend to be located near permanent rivers. As shown below, elephants select areas near rivers preferentially for movement. This results in an apparent positive relationship between settlements and elephant movements at large spatial scales, but is an artefact of elephants selecting routes near water but that avoid coming into close proximity to human settlements.

Movements of all four elephants were strongly related to distance to water, with strong selection for movement paths relatively close to permanent rivers (Table 19.4). Two of the four elephants showed significantly stronger relationships with the square root of distance to water than to Euclidean distance or distance squared. Elephant 4 showed marginally more support for Euclidean distance, but it was not significantly more supported than the square root of distance to water. Elephant 2, however, had statistically equal support for selection of movement paths based on proximity to water as measured by Euclidean distance or the square of Euclidean distance.

Table 19.5 Avoidance of crossing fences and roads. All four elephants showed significant avoidance of crossing wildlife fences, with elephant 4 showing very strong avoidance. In contrast, only elephant 1 showed significant avoidance of crossing roads. This suggests that fences are a much stronger barrier to elephant movements in the study area than are roads

	Fence	Road
Elephant 1	−6.8	−10.65
Elephant 2	−25.4	−0.236
Elephant 3	−15.1	0.512
Elephant 4	−95.1	−0.621

All four elephants showed significant avoidance of crossing fences (Table 19.5), with elephant 4 showing much stronger avoidance than the others. In contrast, only elephant one showed a significant avoidance of crossing roads (Table 19.5).

19.3.2 Weighting Analyses

19.3.2.1 Towns–Villages–Huts

We conducted a weighting analysis for resistance due to settlements across a full factorial of three levels of relative weighting (1, 5, 10). The factorial of three levels of resistance for each of towns, villages and huts is represented as a $3 \times 3 \times 3$ hypothesis cube. Elephants 1 and 3 had identical models receiving highest support. For these elephants, the relative influence of huts appears to be ten times that of villages or towns. In contrast, the most supported weighting hypothesis for elephant 2 suggests that the relative influence of villages is twice that of towns and ten times that of huts. Finally, the most supported weighting hypothesis for elephant 4 suggests that villages have twice the influence of huts and ten times the influence of towns.

19.3.2.2 Water–Fences–Settlements

We conducted a factorial weighting analysis to determine the relative importance of settlements, water and fences to elephant movement path selection for each of the four elephants, incorporating the optimal scaling for each elephant from the scaling analysis across a full factorial of three levels of relative weighting (1, 5, 10). The factorial of three levels of resistance for each factor is represented as a $3 \times 3 \times 3$ hypothesis cube. Elephants 1, 2 and 3 had identical models receiving highest support. For these elephants, the relative influence of settlements and fences are equal, and ten times that of distance to rivers. Finally, the most supported weighting hypothesis for elephant 4 suggests that the relative influence of fences is five times that of settlements or water.

19.3.2.3 Roads

Only one elephant (Elephant 1) showed significant relationships with roads. We combined the optimal water–fences–settlements hypothesis for elephant 1 with the three possible levels of roads (1, 5, 10). The most supported combined model for Elephant 1 indicated that maximum road and water effects are approximately equal and much weaker than the effects of settlements or fences.

19.3.3 Monte Carlo Randomization Tests

19.3.3.1 Towns–Villages–Huts Factorial

We compared the actual sum of differences between adjacent cells in the Towns–Villages–Huts hypothesis cubes for each elephant with the distribution of summed differences from 99,999 random permutations of the hypothesis cubes (Table 19.6). For each elephant, the actual sum of differences of adjacent cells in the hypothesis cube is lower than any of those obtained in permuting the adjacencies randomly. For all elephants, there is a very strong and highly significant unimodal peak of support. This suggests a unimodal peak of support within the tested model space.

19.3.3.2 Settlements–Water–Fences Factorial

The comparison of the differences between adjacent cells in the Settlements–Water–Fences hypothesis cubes with the distribution of summed differences from 99,999 random permutations suggests that for all elephants the tested models form a highly unimodal pattern of support (Table 19.7). For each of elephants 1–3, the actual sum of differences of adjacent cells in the hypothesis cube is lower

Table 19.6 Comparison of actual sum of differences between adjacent cells in the towns–villages–huts hypothesis cube with the distribution of summed differences from 99,999 random permutations. For each elephant, the actual sum of differences of adjacent cells in the hypothesis cube is lower than any of those obtained in permuting the adjacencies randomly. For all elephants there is a very strong and highly significant unimodal peak of support, indicating that the most supported cell in the hypothesis cube is a peak of support

Elephant	Minimum sum of differences of adjacencies across 99,999 permutations	Actual sum of differences of adjacencies	Probability of no difference
1	74.7	68.01	<0.00001
2	165.5	154.9	<0.00001
3	88.1	85.8	<0.00001
4	138.2	117.8	<0.00001

Table 19.7 Comparison of actual sum of differences between adjacent cells in the Settlements–Water–Fences hypothesis with the distribution of summed differences from 99,999 random permutations of the hypothesis cubes. For elephants 1–3, the actual sum of differences of adjacent cells in the hypothesis cube is lower than any of those obtained in permuting the adjacencies randomly. In the case of elephant 4, the actual sum of differences of adjacencies was ranked 29th from the least of 99,999. For all elephants there is a very strong and highly significant unimodal peak of support. The most supported cell in the hypothesis cube is a peak of support

Elephant	Minimum sum of differences of adjacencies across 99,999 permutations	Actual sum of differences of adjacencies	Probability of no difference
1	73.3	60.6	<0.00001
2	264.4	213.8	<0.00001
3	117.1	93.8	<0.00001
4	228.4	245.8	0.00029

than any of those obtained in permuting the adjacencies randomly. In the case of elephant 4, the actual sum of differences of adjacencies was ranked 29th from the bottom of 100,000. For all elephants there is a strong and highly significant unimodal peak of support.

19.3.4 Mapping the Average Model

The scaling and weighting analysis identified a best model for each elephant. In all cases, this best model was the top of a unimodal peak of support in the parameter space. We combined these four best models into a global model by averaging the resistance surfaces predicted by these models across the four elephants (Fig. 19.3). The value of each pixel in this map is the expected resistance of that location to elephant movement, as predicted by the combined model. Resistance in the map ranges from a minimum of 1, for example along rivers far from human settlements, to a maximum resistance of 10.

19.3.5 Cost Distance Mapping

Cost distance increases away from the destination pixel in Sioma National Park as a function of the least cumulative cost across the resistance map (Fig. 19.4). Figure 19.4 illustrates that fences seem to exert a dominant effect on isolation of parts of northern Botswana from Sioma, with human settlements also contributing substantially to isolation in the north eastern part of the study area. Figure 19.5 shows the relative change in least cost distance from each pixel in the study area to the Sioma destination cell between the current landscape, including fences, settlements and roads, and a hypothetical historic resistance landscape without human development.

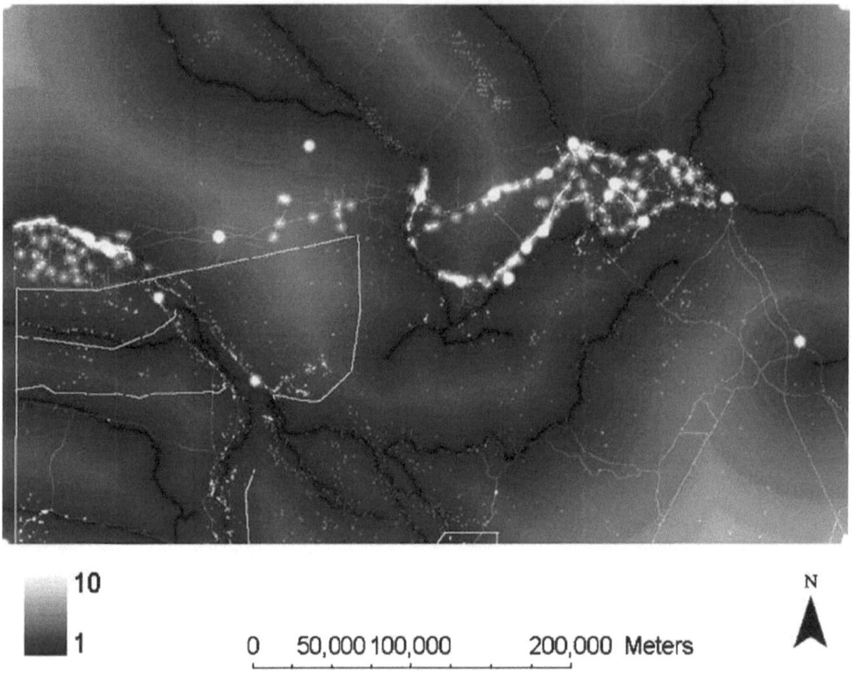

Fig. 19.3 Best resistance model, created by averaging the maps produced by the scaled and weighted resistance models for each of the four elephants. Lighter shades indicate higher resistence

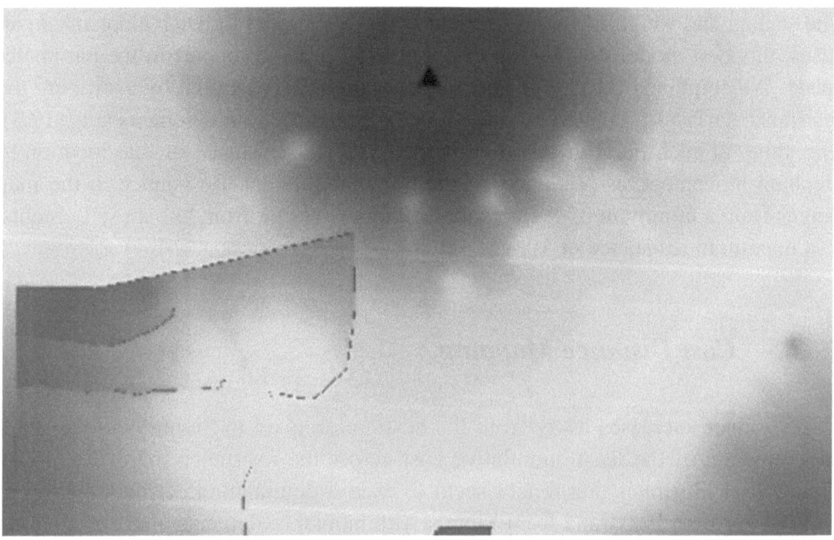

Fig. 19.4 Map of cost distance from every cell in the study area to the destination cell in Sioma National Park. Veterinary fences in the southwest corner of the study area have a dominant effect on cost distance, with settlements in the northwest part of the study area also having a substantial influence on cost distance to Sioma National Park

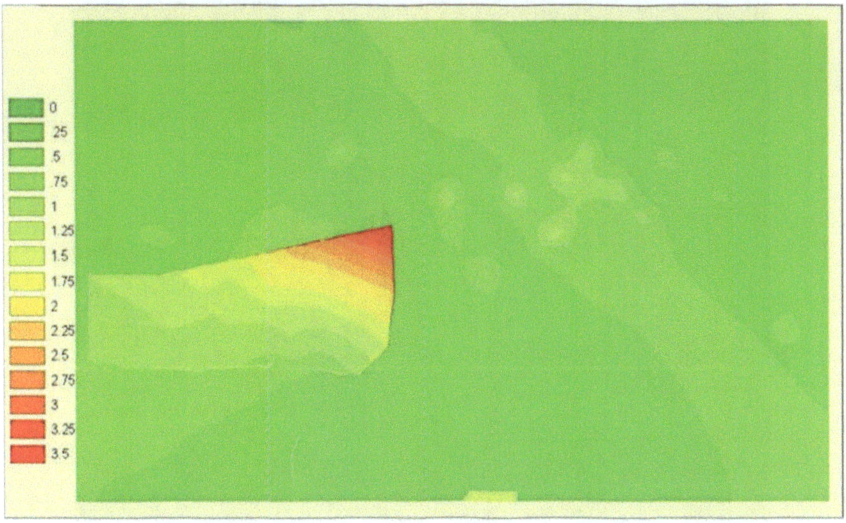

Fig. 19.5 Relative change in cost distance between the current landscape and a hypothetical landscape without human settlements, fences or roads. Areas within the perimeter of the Border Cordon and Northern Buffalo fence are predicted to have an increase of between 100 and 400% in cost distance to Sioma National Park. The dense human settlements in the Caprivi Strip result in much less increase in cost distance in the northeast portion of the study area

Areas in dark green are those for which there is little or no change in cost distance to Sioma National Park. Areas in light green are predicted to have between 100 and 175% increase in cost distance in the current landscape compared to historic. Areas in yellow and orange are predicted to have an increase of between 175 and 300%, and red over 300% in cost distance to Sioma National Park.

19.3.6 Cost Path Corridor Mapping

Figure 19.6 shows the density of least cost paths to Sioma National Park in a historic landscape without human development (Fig. 19.6a), and the current landscape (Fig. 19.6b). The corridor analysis for the historic landscape indicates that the least cost route of elephants to Sioma will be approximately straight lines, except when the path moves into proximity to rivers, in which cases the paths are altered to preferentially follow the river courses.

Figure 19.6 shows several major corridors, most notably a large central corridor flowing along the Kwando and Botetti Rivers, which collects the paths from most of the central portion of the study area. Three other notable corridors exist also. First, a corridor is predicted from the upper Okavango panhandle across the dry uplands of Caprivi and south east Angola. Second, a substantial corridor is predicted from the Chobe/Linyanti region in the east-central portion of the study area and

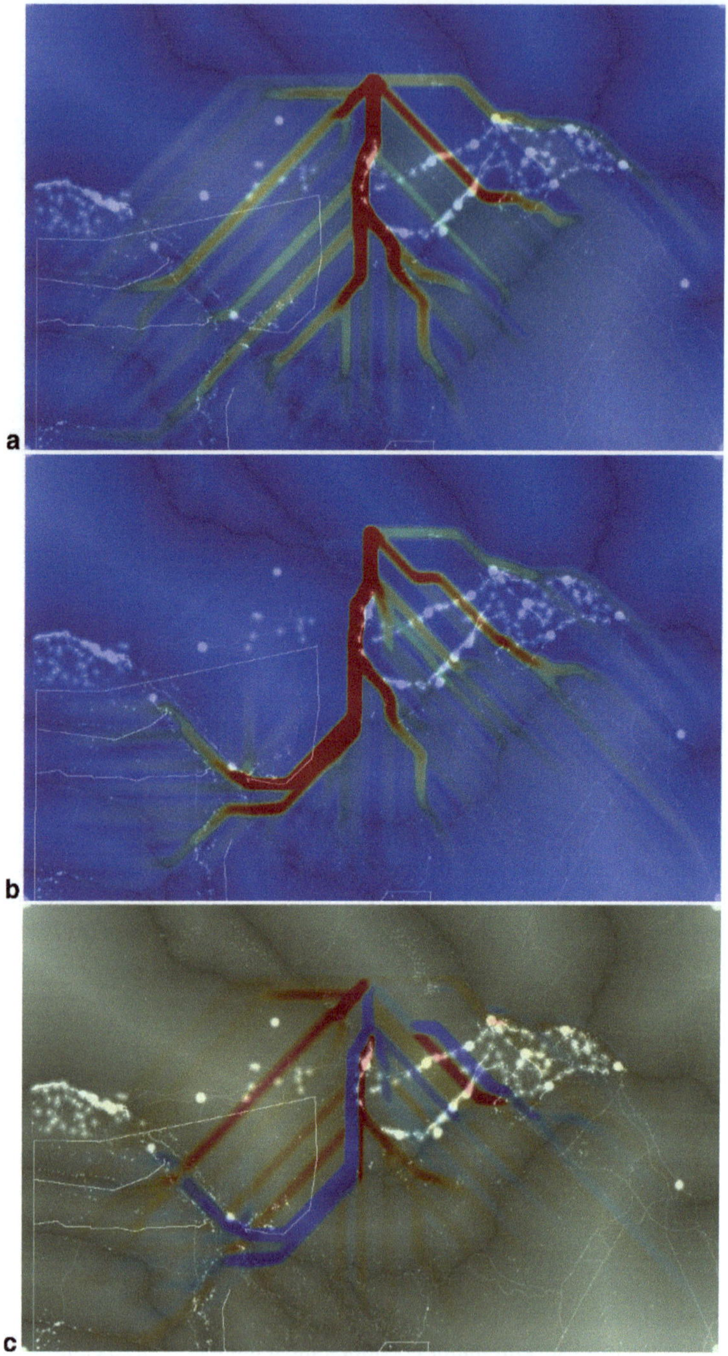

Fig. 19.6 Map of least cost path corridors from 1183 points uniformly distributed across the study area to Sioma National Park for (**a**) the landscape in the absence of human settlements, roads or fences, (**b**) the current landscape including human settlements, roads and fences, and (**c**) the proportional difference between current and historic corridors. Maps 5a and 5b are scaled from blue, reflecting very low density of least cost paths, to red, reflecting very high density of least

across Caprivi and south west Zambia to Sioma. Finally, a relatively minor corridor is predicted to Sioma along the Zambeizi River corridor.

Figure 19.6b shows the expected corridor density in the current landscape and 6c shows the difference between the two corridor maps, scaled as a proportion change of the maximum of Fig. 19.6a. The most notable difference is the elimination of the movement corridors from the Okavango Delta in the west central part of the study area to Sioma, and their rerouting south and around to the east to connect with the central Kwando–Botetti corridor due to the barrier effects of the Border Cordon and Northern Buffalo veterinary fences. Another notable change is the rerouting of much of the south east branch of the Kwando–Botetti corridor in the current landscape north between gaps between towns and villages in the Caprivi strip due to human settlement along the Kwando River along the Angola–Zambia border. A third notable change is the slight rerouting of the Chobe-Linyanti corridor to the northeast to pass through gaps between towns in the Caprivi strip.

19.4 Discussion

19.4.1 Resistance of Utilized Pathway Compared to Available Paths

We used a multi-factorial approach to assess the influences of multiple landscape factors on the selection of elephant movement paths. Focusing on the entire movement path as an observational unit, rather than individual relocation points, resolves several challenges, including spatial autocorrelation among locations, pseudoreplication of observations and most importantly allowed us to powerfully assess the cumulative cost of elephant movement paths. The path randomization procedure produces a large number of available paths of identical spatial topology with which to compare to the utilized path for each individual elephant. This provides a strong means to evaluate use versus availability based on cumulative resistance of movement paths, while holding the length and shape of the paths constant, which is necessary for meaningful comparison among paths.

Formal scaling analyses are critical to identify the spatial scale at which each landscape feature had the strongest relationship with the selection of elephant movement paths. Given the strong differences observed in the apparent relationships

Fig. 19.6 (continued) cost paths. Areas in yellow to red indicate major predicted movement corridors from the study area to Sioma National Park. Map 5c shows the difference between current and historically available corridors, scaled as proportion of maximum of 5a. In 5c areas in grey are predicted to have very little change from historic to current in the density of least cost paths. Areas in blue are areas that were predicted to be corridors in the historic landscape that are not longer available due to human settlements. Areas in yellow to red in 5c are areas in the current landscape that are predicted to be corridors that were not corridors, or were weaker corridors, in the historic landscape between different kinds of human settlements and elephant movement paths, it is clear that careful consideration of scaling relationships between landscape features and animal movement path selection is critical to avoid spurious results.

Factorial weighting analysis is useful to assess the relative influence of each factor and identify a combined model that was maximally supported. The large differences in the degree of support across the range of weighting combinations for each elephant illustrate the importance of proper weighting of resistance factors. Failure to conduct this weighting analysis would at best leave the analysis as a single weighting without evaluation of the relative predictive power of alternative variable weights. At worst, it could result in dramatically incorrect conclusions.

19.4.2 Evaluating Unimodality of Support

The factorial weighting analysis also enabled us to evaluate the unimodality of support across a multidimensional cube of alternative hypotheses. In this paper we presented a new approach to assess unimodal peaks of support among multiple hypotheses using a permutation procedure. We can use the level of homogeneity or unimodality across a quantitative hypothesis cube to assess global significance. If there is a single optimum in the parameter space at which the significance of the chosen statistical test is highest, and significance decreases monotonically away from that peak in all dimensions of the space, then any permutation of this space would result in lower values of the test statistic. The factorial permutation procedure provides a statistical test of the significance of a unimodal peak of support for a globally best model in cases where the tested hypotheses comprise a quantitative cube of parameter combinations.

19.4.3 Consistency Among Elephants in Scaling and Weighting

The analysis indicated that all four elephants strongly avoided towns, villages and huts, and that towns had a larger distance effect (5 km in ¾ of tested elephants) than either villages or huts (1 km in ¾ of tested elephants). Similarly, the analysis showed that the movement paths of all tested elephants were significantly related to distance to water, and that in ¾ of the tested elephants the square root of distance to water was statistically the best or tied for the best scaling of effects of water on movement. These scaling results show both strong effects of these landscape features and high consistency among individual elephants in the scales at which they are most important. This ability to identify the correct scale in pattern–process relationships is a central challenge in ecology (Wiens 1989; Levin 1992) that has been largely neglected in studies of animal movement.

Similarly, the factorial weighting analyses showed high levels of consistency among elephants. Three of the four elephants showed identical patterns of support across the hypothesis cube for Settlements–Water–Fences. This analysis suggests that the maximum resistance due to settlements is equal to that of fences, and 10 times that of distance to water, at the pixel level. This does not imply that water effects are globally subordinate because water effects extend synoptically across the landscape. In contrast, fence effects only accrue when an elephant encounters a fence pixel and settlement effects only accrue within the specified kernel distance of a town, village or hut. Thus, water effects actually dominate path selection at the broadest spatial scales, but are highly subordinate to settlement and fence effects at fine spatial scales. The three elephants that uniformly avoided human settlements at fine scales, did not cross wildlife fences, and selected movement paths preferentially based on the square root of distance to water. In contrast, fence effects were greatest for elephant 4, with five times the effects of either settlements or water. The reason for this difference is evident from this elephant's elongated east–west movement path bounded on the south by the Caprivi Border Fence along the Botswana border (Fig. 19.1). This fence is a double, electrified, high tensile fence that creates is an effective barrier to elephants and other wildlife.

19.4.4 Landscape Resistance, Barriers and Corridors: Implications for Conservation

Combining multiscale analysis of landscape resistance (Fig. 19.3) with cost distance (Fig. 19.4) and least cost path mapping (Fig. 19.6) provides a comprehensive picture of both the factors driving connectivity and the functional effects of landscape structure in creating movement corridors and barriers. This analysis identified several major historical movement corridors between northern Botswana and Sioma National Park. The location and strength of these historic corridors may be useful to guide managers in identifying priority areas for conservation or mitigation to maximally facilitate elephant movement. In addition, comparing the historical to current corridors provides managers with explicit information about the effects of fences and human settlement on historical elephant movement corridors (e.g. Osborn and Parker 2003).

Our analysis also indicated that veterinary fences in north eastern Botswana have a dominant effect of landscape connectivity for elephants. The Border Cordon and Northern Buffalo Fence are predicted to cut off several major movement corridors, most notably between the Okavango panhandle and Sioma. The fence also largely separates the panhandle from the rest of the Okavango Delta. The Okavango is an area of extremely high ecological importance, which supports a very dense elephant population. The veterinary fences result in an increase in cost distance of between 200 and 400% between the northern parts of the Okavango Delta and Sioma, which probably effectively isolates the northern Okavango Delta from much of the rest of the study area. Given the apparent dominant effects of the fence system on elephant population connectivity, it is important for managers to be aware of their influences and consider ways to reduce their negative effects on migration and dispersal, while also preserving the substantial protections the fences provide in places to wildlife from encroaching livestock and human populations

The analysis indicates that the relatively high density of human development in the Caprivi strip and along the Kwando River may act as a partial barrier to elephant movement. However, our analysis suggests that this barrier is highly porous and that it acts to reroute and filter elephant movements, but does not, at existing development levels, block potential dispersal routes to Sioma. The analysis identified three key corridors through this area of relatively high human development (Fig. 19.6b, c). These should be the focus of conservation and mitigation efforts designed to maintain the integrity of the corridor. Similar to Sitati et al. (2003), our results suggest that human settlement density is a major factor affecting elephant movement. The most effective way, therefore, to maintain the integrity of these corridors will likely be to limit future human development within them. Assuming governmental will and ability to direct patterns of future development, an effective strategy may involve limiting development in the corridors we identified and directing it to areas predicted to be less important for habitat connectivity (Osborn and Parker 2003). Of course, habitat connectivity for elephants is only one of many environmental and economic concerns, and decisions about future development must consider other factors, such as protecting critical habitat for other species and economic costs and benefits (Sitati et al. 2003; Lee and Graham 2006; Chamaille-Jammes et al. 2007).

19.5 Conclusion

In the trans-frontier region of northern Botswana, Namibia, Angola and Zambia, effective management of a growing elephant population will depend in part on managers' ability to facilitate dispersal from Botswana to neighboring countries. Understanding the factors that affect elephant movements between and within these nations is essential, as is the application of this knowledge to identify critical movement corridors and barriers. The combination of empirically-derived landscape resistance mapping and least cost path analysis provides a powerful analytical framework for assessing habitat isolation and identifying corridors and barriers to organism movement. In this study we evaluated the degree of isolation of Sioma National Park in Zambia from a large area of northern Botswana and mapped corridors connecting northern Botswana to Sioma National Park. We identified several major movement routes and found that human development has likely substantially altered regional population connectivity for elephants, with veterinary fences and human settlements both increasing isolation of portions of the study area and changing the routes of least cost movement corridors. This information on how human development has affected regional population connectivity and detailed predictions of the location of specific corridors and barriers will be valuable in ongoing efforts to conserve the spectacular elephant population in northern Botswana.

Acknowledgments This project was made possible with funding from the African Elephant Conservation Fund of the U.S. Fish and Wildlife Service, Conservation International, Elephants Without Borders, Player Crosby, Gale Family Foundation, Swiss Agency for Development and Cooperation, Shikar Safari Club, the United States Forest Service and the University of Massachusetts. T. Nkala, K. Landen, L. Patterson, and P. Pearlstein provided critical logistical support for collaring elephants. We wish to thank the Botswana Department of Wildlife and National Parks and the Namibia Ministry of Environment and Tourism for allowing us to undertake this study. We are grateful for the support and encouragement of J. Hanks, O. Langrand, L. Braack, P. Lindeque, H. Kolberg, R. Mojaphoko, D. Mughogho, C. Taolo, T. Mangwe, S. Sekute, G. Gwapela, M. Vandewalle and K. Alexander.

References

Aleper D, Moe SR (2006) The African savannah elephant population in Kidepo Valley National Park, Uganda: changes in size and structure from 1967 to 2000. Afr J Ecol 44:157–164

Chamaille-Jammes S, Valiex M, Fritz H (2007) Managing heterogeneity in elephant distribution: interactions between elephant population density and surface-water availability. J Appl Ecol 44:625–633

Chamaille-Jammes S (2006) Under the African sky: from climatic fluctuations to the regulation of populations, the role of surface-water in the spatio-temporal dynamics of a large elephant population. Ph.D. Thesis, University Pierre et Marie Curie, Paris

Compton BW, McGarigal K, Cushman S, Gamble L (2007) A resistant-kernel model of connectivity for amphibians that breed in vernal pools. Conservat Biol 21:788–799

Cushman SA, Chase M, Griffin C (2005) Elephants in space and time. Oikos 109:331–341

Cushman SA, McKelvey KS, Hayden J, Schwartz MK (2006) Gene flow in complex landscapes: testing multiple hypotheses with causal modelling. Am Nat 168:486–499

Cushman SA, McKelvey KS, Schwartz MK (2009) Identifying corridors for the movement of black bear (*Ursus* americanus) between Yellowstone National Park and the Canadian Border. Conservat Biol 23:368–376

ESRI (2005) ArcInfo Workstation. ESRI, Redlands, California

Fortin M-J, Dale M (2005) Spatial analysis: a guide for ecologists. Cambridge University Press, Cambridge

Lee PC, Graham MD (2006) African elephants *Loxodonta Africana* and human-elephant interactions: implications for conservation. Int Zoo Yearbk 40:9–19

Levin SA (1992) The problem of pattern and scale in ecology. Ecology 73:1943–1967

Murwira A, Skidmore AK (2005) The response of elephants to the spatial heterogeneity of vegetation in a Southern African agricultural landscape. Landsc Ecol 20:217

Osborn FV, Parker GE (2003) Linking two elephant refuges with a corridor in the communal lands of Zimbabwe. Afr J Ecol 41:68–74

Siati NW, Walpole MJ, Leader-Williams N (2005) Factors affecting susceptibility of farms to crop raiding by African elephants: using a predictive model to mitigate conflict. J Appl Ecol 42:1175–1182

Sitati NW, Walpole MJ, Smith RJ, Leader-Williams N (2003) Predicting spatial aspects of human-elephant conflict. J Appl Ecol 40:157–164

Wiens JA (1989) Spatial scaling in ecology. Funct Ecol 3:385–397

Chapter 20
Habitat Fragmentation Effects Depend on Complex Interactions Between Population Size and Dispersal Ability: Modeling Influences of Roads, Agriculture and Residential Development Across a Range of Life-History Characteristics

Samuel A. Cushman, Bradley W. Compton, and Kevin McGarigal

20.1 Introduction

Habitat loss and fragmentation are widely believed to be the most important drivers of extinction (Leakey and Lewin 1995). The habitats in which organisms live are spatially structured at a number of scales, and these patterns interact with organism perception and behavior to drive population dynamics and community structure (Johnson et al. 1992). Anthropogenic habitat loss and fragmentation disrupts these patterns and is expected to have large, negative effects on biodiversity (Flather and Bevers 2002; Haila 2002; Fahrig 2003). The majority of theoretical studies suggest that the effect of habitat fragmentation is weak relative to the effect of habitat loss (Fahrig 1997; Henein et al. 1998; Collingham and Huntley 2000; Flather and Bevers 2002; Fahrig 2003), although some studies have predicted larger fragmentation effects (Boswell et al. 1998; Burkey 1999; Hill and Caswell 1999; Urban and Keitt 2001). In addition, some theoretical studies suggest that the effects of fragmentation per se should become apparent only at low levels of habitat amount, for example below approximately 20–30% of the landscape (Fahrig 1998; Flather

S.A. Cushman (✉)
US Forest Service, Rocky Mountain Research Station, 800 E Beckwith, Missoula, MT 59801, USA
e-mail: scushman@fs.fed.us

B.W. Compton and K. McGarigal
Department of Natural Resources Conservation, University of Massachusetts, Amherst, MA 01003, USA

S.A. Cushman and F. Huettmann (eds.), *Spatial Complexity, Informatics, and Wildlife Conservation*
DOI 10.1007/978-4-431-87771-4_20, © Springer 2010

and Bevers 2002), although there is little empirical evidence available to test this prediction (Fahrig 2003).

The results of empirical studies of habitat fragmentation are often difficult to interpret because many studies fail to address habitat fragmentation at the landscape-level, and most do not distinguish between habitat loss and habitat fragmentation (McGarigal and Cushman 2002; Fahrig 2003). Fragmentation is a landscape-level process and its effects cannot be resolved through fragmentation it is necessary to experimentally or statistically control for the effects of habitat loss (McGarigal and McComb 1995; McGarigal and Cushman 2002). In order to understand the population and species level implications of these relationships it is necessary to move from site-specific inferences to assessments of how the influences of multiple factors interact across large spatial extents in influencing population size and population connectivity (Ruggiero et al. 1994; Bowne and Bowers 2004; Cushman 2006). Non-spatial studies conducted at local scales do not provide a basis for inferences at the landscape or regional level (McGarigal and Cushman 2002). There is often a gross mismatch between the scale of ecological research and population-level responses (Kareiva and Anderson 1988; Ruggiero et al. 1994). Landscape-level studies that explicitly include the spatial patterns of the environment in a representation relevant to the organisms of question, and that address species-specific movement and abundance parameters are essential to extend fine-scale species environment relationships to the level of regional populations (Cushman 2006).

In this paper we present an analysis of habitat fragmentation by roads and residential and agricultural land uses on a broad range of hypothetical vernal pool breeding animals in western Massachusetts. Our analysis models the distribution and expected densities of dispersing organisms in the terrestrial environment based on a factorial implementation of least-cost dispersal models. Our major goal is to quantify the relative influences and interaction of roads and land cover on the area and configuration of occupied terrestrial habitat for 90 different hypothetical organisms representing a factorial combination of population sizes and dispersal abilities. We test five specific hypotheses:

1. Habitat connectivity, as measured by correlation length (McGarigal et al. 2002), will increase with both population size and dispersal ability.
2. Thresholds will exist where habitat connectivity drops dramatically at low population sizes and low dispersal abilities.
3. Population size and dispersal ability will interact such that thresholds of habitat connectivity will be exacerbated at when population size and dispersal ability are both low.
4. The effects of habitat fragmentation by roads will have relatively smaller effects than habitat loss due to land cover change.
5. The effects of habitat loss and fragmentation will be disproportionately high for species with large dispersal abilities.

20.2 Methods

20.2.1 Input Data

20.2.1.1 Vernal Pool Data

We used the distribution of vernal pools in Western Massachusetts as breeding sites in the models. Vernal pools provide important habitat for a variety of species, including some amphibians that breed exclusively in vernal pools. Locations of vernal pools were obtained from the NHESP Potential Vernal Pools coverage photo-interpreted by the Commonwealth of Massachusetts Natural Heritage & Endangered Species Program (Compton et al. 2007). This data layer identifies the locations of more than 29,000 potential vernal pool habitats. These pools provide a spatially realistic pattern of source populations in a real landscape, which adds an important degree of realism to our simulations.

20.2.1.2 Ecoregional Data

The US Environmental Protection Agency has delineated thirteen ecoregions in Massachusetts, based on geology, hydrology, climate, and the distribution of species. In this study we consider the ten ecoregions that comprise Western Massachusetts. We combined portions of the *Taconic Mountains* and *Western New England Marble Valleys/Berkshire Valley/Houstonic and Hoosic Valleys* ecoregions due to disjunctions, small size and irregular shape. This served to reduce edge effects, and make the areas comparable to the other ecoregions, which is important for comparability of model output. Also, given the relative ecological similarity of these two ecoregions we felt this union did not introduce appreciable subjectivity into the analysis. Thus, we implemented the analysis for eight ecoregional units in Massachusetts (Fig. 20.1).

20.2.1.3 Road and Landcover Data

The roads data were derived from USGS 1:100,000 Roads Digital Line Graphs (DLGs) with additional linework from the Massachusetts Highway Department (MHD). The road classes in this layer are listed in Table 20.1. The landcover map used in this analysis was provided by MassGIS, and contains 37 land use categories interpreted from 1:25,000 aerial photography taken in 1999. We converted the coverage to a raster grid, with a 15 m cell size. In addition, we reclassified the map into seven classes which we deemed to be most relevant to the ecologies of vernal pool breeding amphibians (Table 20.1).

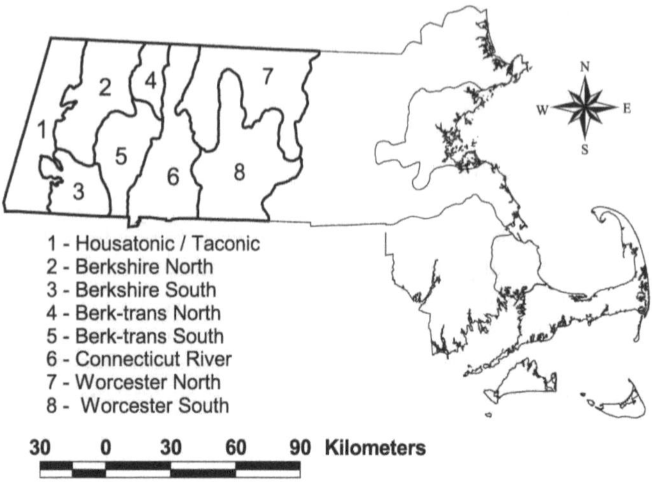

Fig. 20.1 Map of the area covered in this study. The study area consists of the eight ecoregions shown here, which cover approximately 15,000 km^2 of Western Massachusetts. These ecoregions were defined by the United States Environmental Protection Agency

Table 20.1 Road and landcover classes used in the Road and Road + Land Use scenarios. The resistance values reflect the relative cost of traveling through a cell of that class, in comparison to the minimum value of 1

Road classes	Resistance value	Landcover classes	Resistance value
Minor street	20	Forest	1
Major arterial	30	Grass/open land	10
Two-lane highway	50	Low density residential	10
Multi-lane highway	75	Row crop	20
Limited access multi-lane highway	100	High-density residential	30
		Urban	50
		Water	50

20.2.2 *Model of Terrestrial Phase Amphibian Density*

Our model is based on least-cost dispersal from point sources. The sources in our case are the locations of individual pools. The model calculates the probability of an organism being present in each pixel around the source, given the number of individuals which originated in the source, the dispersal ability of that species, the nature of the dispersal function, and the resistance of the landscape (Compton et al. 2007). In our analysis we use a normal probability density function as the basis of the dispersal model. In a homogeneous environment the expected density of dispersing organisms in the environment surrounding the source will be Gaussian, with the peak at the source, and the density dropping off according to the normal distribution with a standard deviation set according to the dispersal ability of the species. The standard deviation in this dispersal function represents the dispersal ability of the modeled organism. We wished to bracket the range of dispersal abilities of animal species breeding in Massachusetts vernal pools. Accordingly, we ran the models over nine levels of dispersal ability (D), corresponding to standard deviations of the normal dispersal function of 100, 200, 300, 400, 500, 600, 700, 800, and 900m. In addition, habitat connectivity may be influenced by the population size originating at the breeding sites. Accordingly, we simulated ten levels of populations size (P), ranging from 100 to 1,000 individuals per pool, by units of 100.

The model works by clipping an area of a given number of standard deviations around each pool and calculating the expected density within that. We calculated expected densities within 2.5 standard deviations of dispersal ability, which will include over 99% of the dispersing individuals. Within the clipped area the model calculates the expected density according to the normal dispersal function, accounting for the resistance of the surrounding landscape. A resistance value is given to each class in the land-cover and roads maps. The resistance value is in the form of the cost of crossing that cover type relative to the least cost cover type. These costs are used as weights in the dispersal function, such that the expected density in a pixel is down-weighted by the cumulative cost from the source, following the least cost route (Compton et al. 2007). Once the expected density around each pool is calculated, the values for all pools at all locations are summed to give the total expected density at each pixel. The resistance values for each land-cover class and road class are shown in Table 20.1. These resistance values are hypothetical and are not intended to optimally represent the responses of any particular species, although they are consistent with empirical data for several vernal pool breeding species in Massachusetts (Gamble unpublished data). The results of the model are surfaces of expected density of dispersing organisms in the upland environment (Fig. 20.2).

20.2.3 Modeling Scenarios

We conducted a four-way factorial modeling experiment. The four factors are (1) scenario, (2) dispersal ability, (3) abundance level, and (4) ecoregion. There were three levels of the factor scenario. These are null, roads, and roads plus land use. In the null scenario the expected density of amphibians is modeled across a null landscape where every cell has a resistance of 1. This provides a baseline prediction of distribution expected in the absence of any differential resistance in the landscape. In the roads scenario, roads were given resistance according to the definitions provided above, but the remainder of the landscape retained a resistance of 1. This scenario models the influences of roads in the absence of any differential resistance due to landcover, and provides a measure of the relative influence of roads. The third scenario includes the resistance of both roads and land use. The differences between the second and third scenario provide a measure of the relative influence of land use, and the third scenario itself gives a measure of the combined influences of

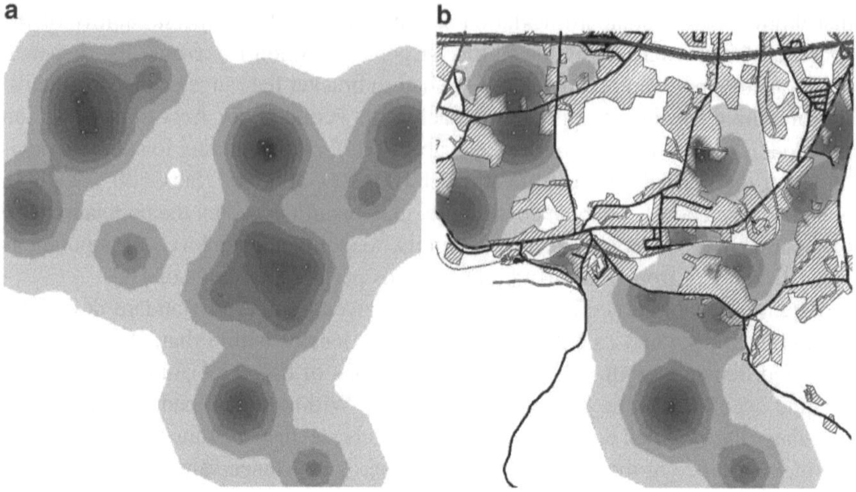

Fig. 20.2 Example of changes in population connectivity predicted for a small area of one ecoregion between the Null (**a**) and Roads + Land Use scenario (**b**). The relative darkness of the filled area indicates the expected density of dispersing juveniles in the uplands based on the least-cost kernel dispersal model. In the Roads + Land Use scenario roads are represented by solid lines and residential areas by cross-hatched polygons. These features reduce both the predicted area of dispersal habitat and the expected density of amphibians within it

roads and land use, and provides a measure of the full effect of human development on habitat connectivity (Fig. 20.3).

The factor dispersal ability had ten levels, abundance nine levels and ecoregion eight levels, as discussed above. We ran the model over the 1920 combinations of these four factors. For each combination we created an output grid showing the expected density of the organisms in each cell of the landscape. These grids have a 15 m cell size and the cell value is equal to the expected density in that cell.

20.2.4 Fragstats Analysis of Resulting Maps

The 1920 output grids were reclassified into binary maps for analysis in FRAGSTATS. We recoded the output grids into binary maps showing the areas predicted to have 5 or more adults per hectare We used FRAGSTATS (McGarigal et al. 2002) to calculate the correlation length of predicted habitat for each cover map. The correlation length is the area weighted mean radius of gyration, and provides a measure of the average distance an organism can move within a patch before encountering the patch boundary from a random starting point. When aggregated at the class or landscape level, the correlation length represents the average traversability of the landscape for an organism that is confined to remain within a single

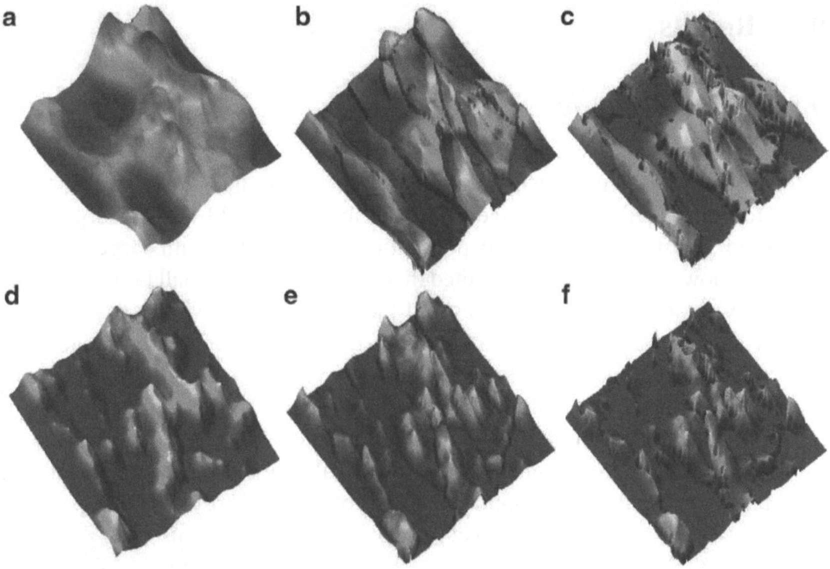

Fig. 20.3 Visual depiction of predicted density of organisms in the terrestrial environment in one small section of one ecoregion, across two levels of dispersal ability and three scenarios: (**a–c**) 400 dispersing individuals; (**d–f**) 800 dispersing individuals; (**b** and **d**) Null scenario; (**b** and **e**) Roads Scenario; (**c** and **f**) Roads + Land Use scenario. The height of the surface is proportional to expected density of organisms in the landscape

patch. It gives a global measure of the habitat connectivity in the landscape and is a more relevant functional measure of habitat availability for vernal pool breeding amphibians than more basic measures such as patch size, nearest neighbor distance and percentage of the landscape in occupied habitat (McGarigal et al. 2002).

20.2.5 Analysis of Habitat Correlation Length Across Life-History Space

To visualize the relationships between dispersal ability, population size and habitat connectivity, we formed matrices of the correlation-lengths of dispersal habitat predicted by the model across the factorial of dispersal ability and population size, for each of the three scenarios. We computed the average response surface for each scenario across the eight ecoregions. Then we computed the proportion of decrease in predicted correlation length of dispersal habitat from the null scenario to the roads and Roads + Land Use scenarios. These difference surfaces measure the relative impact of roads and land use on the correlation length of occupied terrestrial habitat across the life-history space.

20.3 Results

20.3.1 Null Scenario

Correlation length of predicted occupied habitat in the null model was strongly related to both population size and dispersal ability (Fig. 20.4). Not surprisingly, correlation was highest for large populations of animals with large dispersal ability, and lowest for those with limited dispersal ability and small population size. There were strongly non-linear, threshold-like relationships with both population size and dispersal ability. There are three distinct zones which are of interest in the correlation length surface across life-history space surface. The first is a zone of highly interconnected occupied habitat. This roughly corresponds to a zone of Fig. 20.4 above 10,000 m of correlation length. Species occupying this zone are able to occupy large interconnected portions of the landscape. The second zone of interest corresponds to the opposite set of conditions, those where populations occupy relatively small and disjunct patches. This corresponds roughly to the zone below 3,000 m in Fig. 20.4. In this zone, unoccupied habitat is the matrix,

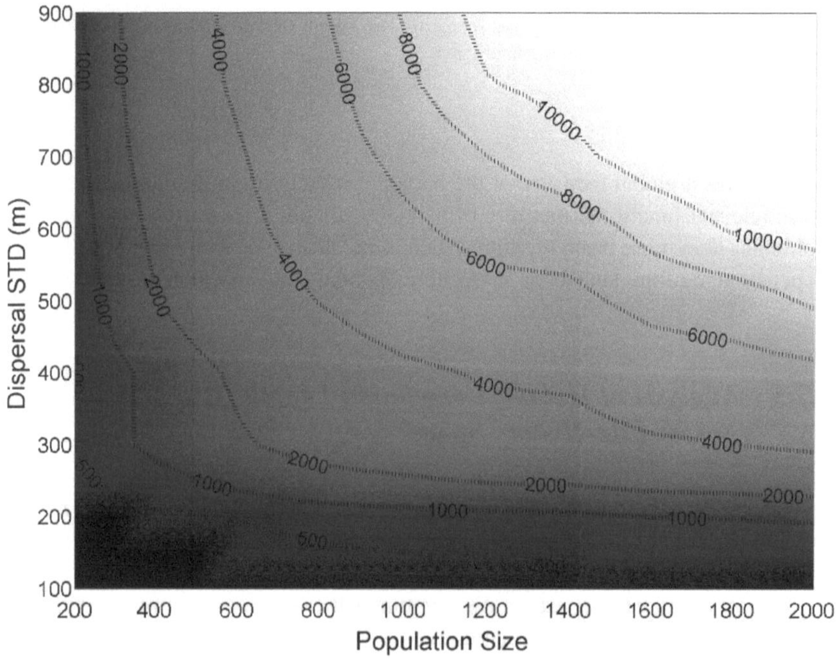

Fig. 20.4 Correlation length of habitat occupied by a minimum of five individuals per hectare across a factorial combination of dispersal ability and population size for the Null scenario. The dashed contours and indicate correlation length of occupied habitat. The surface shows strong non-linear interactions between population size and dispersal ability

with occupied habitat accounting for less than 30% of the landscape in disjunct patches. The third zone corresponds to where correlation length rapidly increases, and where small, disjunct patches of occupied habitat rapidly coalesce into large interconnected blocks. This zone roughly corresponds to the area between 4,000 and 8,000 m in Fig. 20.4. The proportion of the landscape covered by occupied habitat in this zone is between 35 and 55%.

20.3.2 Roads Scenario

The average correlation length among the eight ecoregions decreased dramatically when the resistance of roads was included in the model (Fig. 20.5). This large reduction occurred across nearly all of the life-history space. Well over half of the life-history space experienced at least a 75% reduction in expected correlation length of occupied terrestrial habitat (Fig. 20.5). The largest reductions occurred

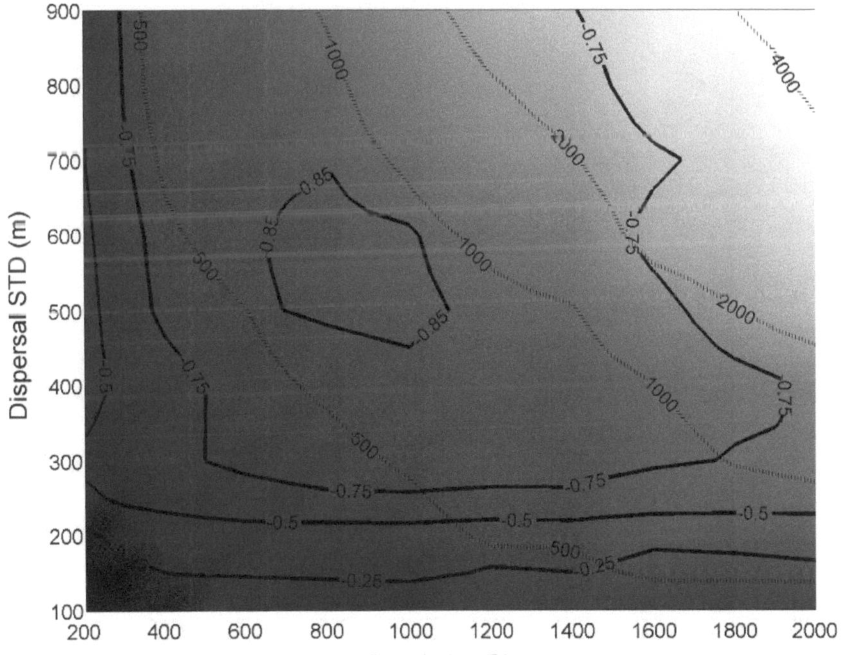

Fig. 20.5 Correlation length of habitat occupied by a minimum of five individuals per hectare for the Roads scenario. The dashed contours and indicate correlation length of dispersal habitat. The solid contours indicate the proportional decrease in correlation length in the Roads scenario from that of the Null scenario. Over 60% of the life-history space is predicted to experience at least a 75% reduction of occupied habitat correlation length due to roads in comparison with the null scenario, and approximately 10% of the life-history space experiences over 85% reduction in habitat connectivity

in the center of the life-history space. Organisms with between 400 and 600 m dispersal standard deviations and population sizes between 700 and 1,200 individuals were expected to experience decreases in correlation length of dispersal habitat of over 85%. Because of its position near this threshold, the center of the life-history space appears to be particularly vulnerable to habitat fragmentation due to roads.

20.3.3 Roads and Land Use Scenario

The average correlation length among the eight ecoregions decreased substantially when the resistance of land use categories was added to that of roads (Fig. 20.6). Interestingly, this reduction was somewhat smaller than that due to road effects. Approximately 20% of the life-history space experienced decreases in correlation

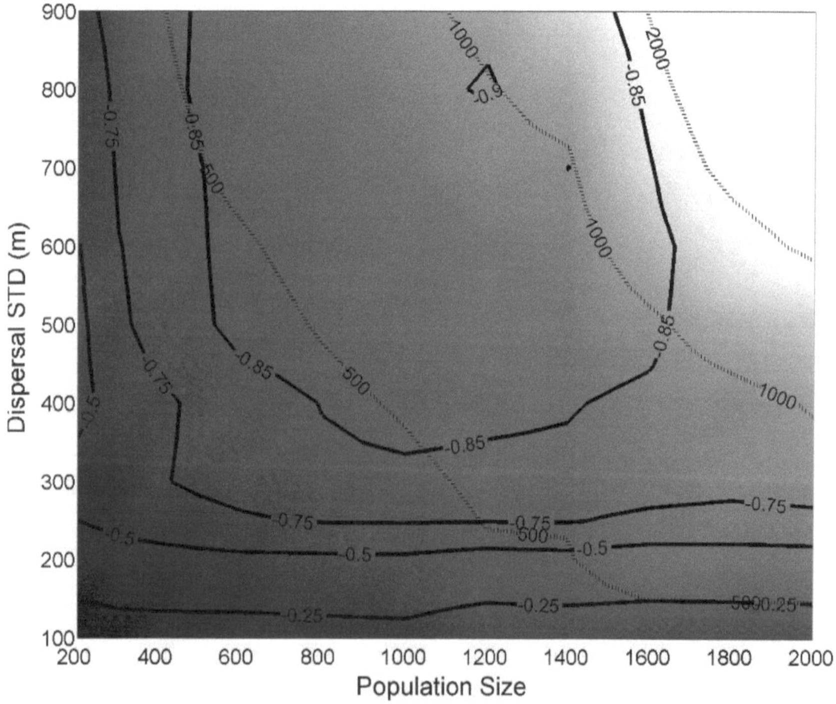

Fig. 20.6 Correlation length of habitat occupied by a minimum of five individuals per hectare for the Roads + Land Use scenario. The dashed contours indicate correlation length of occupied habitat. The solid contours indicate the proportional decrease in correlation length in the Roads + Land Use scenario from that of the Null scenario. Over 40% of the life-history space is predicted to experience at least an 85% reduction of dispersal habitat correlation length due to roads in comparison with the null scenario

length of over 40% beyond those of the roads scenario. With the combined effects of roads and residential/urban development, the proportion of the life-history space predicted to experience over 85% reduction in habitat connectivity increased from less than 10% to nearly 50% (Fig. 20.6).

20.4 Discussion

20.4.1 Hypothesis 1: Habitat Connectivity Will Increase with Both Population Size and Dispersal Ability

Holding dispersal ability constant, increasing population size will result in increased height of the surface at all points in the landscape, resulting in larger areas that have expected densities above the lower limit of one animal per hectare. Holding population size constant and increasing dispersal ability will result in the surface spreading out. Peaks in local abundance will lower, but the area occupied will increase.

As expected, the simulations suggest that there is a strong interaction between population size and dispersal ability in influencing habitat connectivity. Animals with large population sizes and high dispersal abilities are predicted to have high levels of population connectivity (Fig. 20.4). Conversely, those with low population sizes and low dispersal abilities are predicted to have low levels of population connectivity. The predicted correlation length of occupied habitat spanned over an order of magnitude in the null model, from less than 500 to over 10,000 m.

The structure of the correlation length surface implies potential life-history tradeoffs between animals with high dispersal and those with high population sizes. Specifically, assuming that there are fitness costs to increasing either dispersal ability or population size through adaptation, an isocline of correlation length could reflect an evolutionary stable strategy (Martin 1995; Charnov 1997). If actual organisms were distributed along the isloclines in Fig. 20.4, this would be consistent with life-history tradeoffs. However, this is presently untestable given lack of knowledge of the population sizes and dispersal abilities of actual organisms.

20.4.2 Hypothesis 2: Thresholds will Exist where Habitat Connectivity Drops Dramatically at Low Population Sizes and Low Dispersal Abilities

A critical threshold is an abrupt, nonlinear change that occurs in an organism's response across a small range of habitat loss and fragmentation (With and King 1999). Both empirical data (Carlson and Stenberg 1995; Doncaster et al. 1996; Jansson and Angelstam 1999) and theoretical models predict critical thresholds of habitat where ecological relationships change abruptly (O'Neill et al. 1988; Turner and Gardner 1991; With and Crist 1995; Bascompte and Sole 1996; Flather and Bevers 2002). For example, some theoretical and empirical work suggests that major population declines will occur when habitat area drops below 10–30% (With and Crist 1995; Hill and Caswell 1999; Jansson and Angelstam 1999; Fahrig 2001; Flather and Bevers 2002).

As anticipated by our hypothesis, habitat connectivity changes non-linearly with dispersal ability and population size. Correlation length of occupied habitat was predicted to be uniformly low at either low population sizes or low dispersal abilities. Of these two factors, dispersal ability seems to have a larger influence, with correlation length decreasing sharply below dispersal abilities of 300 m (Fig. 20.4). Below a dispersal distance standard deviation of 300–400 m, increasing population size cannot compensate for decreased dispersal ability. Organisms with less than 400 m dispersal standard deviation are predicted to exist in naturally fragmented populations regardless of their population sizes. This has potentially important ecological implications, as most invertebrate species inhabiting vernal pools have dispersal abilities at or below this level. Similarly, when populations are reduced below 600–400 individuals, increasing dispersal ability cannot compensate for decreasing population size. Populations less than approximately 400 individuals are predicted to exist in disconnected fragments of habitat regardless of dispersal ability.

20.4.3 Hypothesis 3: Population Size and Dispersal Ability will Interact such that Thresholds of Habitat Connectivity will be Exacerbated when Population Size and Dispersal Ability are both Low

Several simulation studies have suggested that habitat fragmentation effects become important to population viability when habitat areas are small (Fahrig 1997; Flather and Bevers 2002). Our hypothesis, in contrast, centers on relationships between habitat connectivity, population size and dispersal ability. The hypothesis that population size and dispersal ability will interact non-additively to reduce habitat connectivity at low population sizes and dispersal abilities is not supported. At low levels population size and dispersal ability act largely independently over large ranges of each factor (Fig. 20.4). While there are thresholds along dispersal ability and along population size where habitat connectivity decreases dramatically, these parameters do not appear to interact negatively at low levels. Thus there is no evidence of habitat fragmentation due to low dispersal ability being exacerbated by low population size. In contrast, in the roads and Roads + Land Use scenarios, the correlation length isoclines angle from upper left to lower right throughout the life-history space (Figs. 20.5 and 20.6). This indicates that they interact throughout the life history space, with no evidence of decreasing interaction when population size and dispersal abilities become small or become large. Thus, our third hypothesis is not supported.

While dispersal ability and population size interact in the Roads and Roads + Land Use scenarios, this interaction does not increase as population size and dispersal ability decrease.

20.4.4 Hypothesis 4: The Effects of Habitat Fragmentation by Roads will Have Relatively Smaller Effects Than Habitat Loss Due to Land Cover Change

A number of theoretical (Fahrig 1997, 2003) and empirical (Villard et al. 1999; Trzcinski et al. 1999; Cushman and McGarigal 2004) studies suggest that the effects of habitat loss are generally greater than those of habitat fragmentation (Fahrig 2003). Also, effects of habitat fragmentation are predicted to increase below some level of habitat loss. Some theoretical studies suggest that the effects of fragmentation per se should become apparent only when habitat area drops below approximately 20–30% of the landscape (Fahrig 1998, 2003; Flather and Bevers 2002). Threshold studies considering only a single factor at a single scale are limited in their applicability to real-world systems, because habitat loss and fragmentation in real landscapes typically involve multiple factors operating at several scales.

Our fourth hypothesis addresses these issues by evaluating the relative impacts of fragmentation by roads and habitat-loss due to land use changes. We expect that habitat fragmentation due to roads will have relatively smaller effects on habitat connectivity than habitat loss and fragmentation due to agricultural, residential and urban development. Contrary to our expectation, roads had a substantially greater impact on habitat connectivity than did Land Use (Figs. 20.5 and 20.6). These results show that the road network in western Massachusetts may result in substantial reductions in habitat connectivity for species that have difficulty or avoid crossing roads. Roads may subdivide habitat for these species into patches and attenuate migration, reducing the area of occupied habitat. The western Massachusetts landscape is very heavily roaded but is forest dominated. Thus the predominance of road impacts probably reflects the nature of this study area, which is highly dissected by roads but only moderately perforated by agricultural, residential and urban development. While land use effects are quantitatively less than road effects in our study area, their impacts are synergistic. Habitat fragmentation due to residential and urban development are proportionally greatest in the zones of life history space that were least severely impacted by roads alone. Thus, the relative importance of roads vs. land use may be different than in other landscapes that differ in the density of roads and the degree of habitat loss due to agricultural, residential or other land uses.

The observed importance of road effects is consistent with results reported for pond breeding amphibians (Cushman 2006). Roads can have substantial negative effects on species that avoid or experience high mortality risk when crossing roads (Vos and Chardon 1998, Carr and Fahrig 2001). Habitat fragmentation by roads and other barriers decreases dispersal (Gibbs 1998, deMaynadier and Hunter 2000), increases mortality (Fahrig et al. 1995; Carr and Fahrig 2001) and reduces genetic diversity (Reh and Seitz 1990). Habitat fragmentation can increase extinction risk by reducing demographic and genetic input from immigrants and reducing the chance of recolonization after extinction (Lande 1988; Sjögren-Gulve 1994). The relative importance of habitat loss due to land use and fragmentation by roads will depend on interactions between the habitat requirements and dispersal ability of the organisms and the pattern of roads and land uses in the subject landscape.

20.4.5 Hypothesis 5: The effects of Habitat Loss and Fragmentation will Be Disproportionately High for Species with Large Dispersal Abilities

A number of recent empirical studies have found that the negative effects of habitat fragmentation increase with increasing dispersal ability (Carr and Fahrig 2001; Cushman 2006). Carr et al. (2002) suggest that highly vagile organisms may be at a disadvantage in landscapes with roads because of increased likelihood of mortality. In a fragmented landscape individuals of species with large dispersal abilities will generally encounter roads and other anthropogenic barriers at higher rates than less vagile species. This will tend to increase mortality risk and decrease habitat connectivity.

Our model results largely match these predictions. Consistent with the expectation of our hypothesis, as population sizes and dispersal abilities increase in fragmented landscapes, the relative impact of habitat fragmentation increases (Figs. 20.5 and 20.6). In the Roads scenario, the proportional reduction in correlation length from the null scenario increased with increasing dispersal ability, to a maximum decrease at dispersal abilities between approximately 500 and 600 m (Fig. 20.5). Similarly, in the Roads + land Use scenario the largest decrease in correlation length from the null scenario occurred at a dispersal distance standard deviation of approximately 800 m. In both cases, there is a clear pattern of increasing impacts of fragmentation on habitat connectivity with increasing dispersal abilities, particularly as dispersal distance standard deviation increases from 200 to 500 m. Our results suggest that in western Massachusetts current patterns of roads, agriculture and residential development have a disproportionate impact on species with mid-sized populations and relatively large dispersal abilities.

20.4.6 Validating and Extending Results

There are several ways that one could empirically test the predictions of these models for particular species. First, one could conduct large-scale, mark-recapture studies (e.g. Gamble et al. submitted) of dispersing juveniles of several species, and quantify the similarity of actual movement rates and patterns to those predicted by the model. The challenge in these studies is one of cost and sample sizes. Large-scale mark-recapture metapopulation studies are exceptionally expensive to implement, take a number of years to produce reliable results, and generally do not provide large landscape-level sample sizes due to financial and logistical constraints. An alternative to mark-recapture movement studies is to use molecular genetic methods to empirically derive rates of gene flow among ponds and effective population sizes (Schwartz et al. 1998; Manel et al. 2003; Funk et al. 2005; Cushman et al. 2006). Molecular genetic methods offer a particularly attractive approach to quantifying gene flow across heterogeneous landscapes, as the logistical and financial costs of extensive mark-recapture study grids are obviated, and the genetic characteristics of subpopulations at each sampled pond can provide both information on its effective population size and the degree to which it differs genetically from other ponds (Funk et al. 2005). Such methods allow one to quantify rates of gene flow between ponds, and quantify resistance to movement due to gradients of landscape conditions (Cushman et al. 2006).

References

Bascompte J, Sole RV (1996) Habitat fragmentation and extinction thresholds in spatially explicit models. J Anim Ecol 65:465–473

Boswell GP, Britton NF, Franks NR (1998) Habitat fragmentation, percolation theory and the conservation of a keystone species. Proc R Soc London Ser B 265:1921–1925

Bowne DR, Bowers MA (2004) Interpatch movements in spatially structured populations: a literature review. Landsc Ecol 19:1–20

Burkey TV (1999) Extinction in fragmented habitats predicted from stochastic birth-death processes with density dependence. J Theor Biol 199:395–406

Carlson A, Stenberg I (1995) Vitryggig hackspett (Dendrocopos leucotos). Biotopval och sårbarhetsanalys. In: Swedish Rapport 27 Swedish University of agricultural sciences, Department of wildlife ecology.

Carr LW, Fahrig L (2001) Effect of road traffic on two amphibian species of different vagility. Conservat Biol 15:1071–1078

Carr LW, Pope SE, Fahrig L (2002) Impacts of landscape transformation by roads. In: Gutzwiller KJ (ed.) Concepts and applications of landscape ecology in biological conservation. Springer, New York

Charnov EL (1997) Trade-off invariant rules for evolutionary stable life-histories. Nature 387:393–394

Collingham YC, Huntley B (2000) Impacts of habitat fragmentation and patch size upon migration rates. Ecol Appl 10:131–144

Compton BK, McGarigal SA, Cushman SA, Gamble L (2007) A resistant kernel model of connectivity for vernal pool breeding amphibians. Biol Conservat 21:788–799

Cushman SA (2006) Effects of habitat loss and fragmentation on amphibians: a review and prospectus. Biol Conservat 128:231–240

Cushman SA, McGarigal K (2004) Hierarchical analysis of forest bird species-environment relationships in the Oregon Coast Range. Ecol Appl 14:1090–1105

Cushman SA, McKelvey KS, Hayden J, Schwartz MK (2006) Gene flow in complex landscapes: confronting models with data. Am Nat 168:486–499

deMaynadier PG, Hunter ML Jr (2000) Road effects on amphibian movements in a forested landscape. Nat Areas J 20:56–65

Doncaster CP, Micol T, Jensen SP (1996) Determining minimum habitat requirements in theory and practice. Oikos 75:335–339

Fahrig L (1997) Relative effects of habitat loss and fragmentation on population extinction. J Wildl Manage 61:603–610

Fahrig L (1998) When does fragmentation of breeding habitat affect population survival? Ecol Model 105:273–292

Fahrig L (2001) How much habitat is enough? Biol Conservat 100:65–74

Fahrig L (2003) Effects of habitat fragmentation on biodiversity. Annu Rev Ecol Systemat 34:487–515

Fahrig L, Pedlar JH, Pope SE, Taylor PD, Wegner JF (1995) Effect of road traffic on amphibian density. Biol Conservat 73:177–182

Flather CH, Bevers M (2002) Patchy reaction-diffusion and population abundance: the relative importance of habitat amount and arrangement. Am Nat 159:40–56

Funk WC, Blouin MC, Cron PS, Stephen, Maxell BA, Pilliod DS, Amish S, Allendorf FW (2005) Population structure of Columbia spotted frogs *(Rana luteiventris)* is strongly affected by the landscape. Mol Ecol 14:483–496

Gibbs JP (1998) Amphibian movements in response to forest edges, roads, and streambeds in southern New England. J Wildlife Management 62:584–589

Haila Y (2002) A conceptual genealogy of fragmentation research: from island biogeography to landscape ecology. Ecol Appl 12:321–334

Harrison S (1991) Local extinction in a metapopulation context: an empirical evaluation. In: Gilpin ME, Hanski I (eds) Metapopulation dynamics: empirical and theoretical investigations. Academic, London

Henein K, Wegner J, Merriam G (1998) Population effects of landscape model manipulation on two behaviourally different woodland small mammals. Oikos 81:168–186

Hill MF, Caswell H (1999) Habitat fragmentation and extinction thresholds on fractal landscapes. Ecol Lett 2:121–127

Jansson G, Angelstam P (1999) Threshold levels of habitat composition for the presence of the long-tailed tit (*Aegithalos caudatus*) in a boreal landscape. Landsc Ecol 14:283–290

Johnson AR, Milne BT, Wiens JA, Crist TO (1992) Animal movements and population dynamics in heterogeneous landscapes. Landsc Ecol 7:63–75

Kareiva P, Anderson M (1988) Spatial aspects of species interactions: the wedding of models and experiments. In: Hastings A (ed) Community ecology. Springer, New York

Lande R (1988) Genetics and demography in biological conservation. Science 241:1455–1460

Leakey R, Roger L. The Sixth Extinction: Patterns of Life and the Future of Mankind (1995) Morell, Virginia

Manel S, Schwartz MK, Luikart G, Taberlet P (2003) Landscape genetics: combining landscape ecology and population genetics. Trends Ecol Evol 18:189–197

Martin TE (1995) Avian life history evolution in relation to nest sites, nest predation and food. Ecol Monogr 65:101–127

McGarigal K, Cushman SA (2002) Comparative evaluation of experimental approaches to the study of habitat fragmentation effects. Ecol Appl 12:335–345

McGarigal K, McComb WC (1995) Relationship between landscape structure and breeding birds in the Oregon Coast Range. Ecol Monogr 65:235–260

McGarigal K, Cushman SA, Neel MC, Ene E (2002) FRAGSTATS: spatial pattern analysis program for categorical maps. Computer software program produced by the authors at the University of Massachusetts, Amherst. www.umass.edu/landeco/research/fragstats/fragstats.html.

O'Neill RV, Krummel JR, Gardner RH, Sugihara G, Jackson B, DeAngelis DL, Milne BT, Turner MG, Zygmunt B, Christensen SW, Dale VH, Graham RL (1988) Indices of landscape configuration. Landsc Ecol 1:153–162

Reh W, Seitz A (1990) The influence of Land Useon the genetic structure of populations of the common frog *(Rana temporaria)*. Biol Conservat 54:239–249

Ruggiero LF, Hayward GD, Squires JR (1994) Viability analysis in biological evaluations: concepts of population viability analysis, biological population, and ecological scale. Conservat Biol 8:364–372
Schwartz MK, Tallmon DA, Luikart G (1998) Review of DNA-based census and effective population size estimators. Anim Conservat 1:293–299
Sjögren-Gulve P (1994) Distribution and extinction patterns within a northern metapopulation of the pool frog, *Rana lessonae*. Ecology 75:1357–1367
Trzcinski MK, Fahrig L, Merriam G (1999) Independent effects of forest cover and fragmentation on the distribution of forest breeding birds. Ecol Appl 9:586–593
Turner MG, Gardner RH (eds) (1991) Quantitative methods in landscape ecology. Springer, New York
Urban D, Keitt T (2001) Landscape connectivity: a graph-theoretic perspective. Ecology 82:1205–1218
Villard M-A, Trzcinski MK, Merriam G (1999) Fragmentation effects on forest birds: relative influence of woodland cover and configuration on landscape occupancy. Conservat Biol 13:774–783
Vos CC, Chardon JP (1998) Effects of habitat fragmentation and road density on the distribution pattern of the moor frog Rana arvalis. J Appl Ecol 35:44–56
With KA, Crist TO (1995) Critical thresholds in species' responses to landscape structure. Ecology 76:2446–2459
With KA, King AW (1999) Extinction thresholds for species in fractal landscapes. Conservat Biol 13:314–326

Chapter 21
Linking Cetaceans to Their Environment: Spatial Data Acquisition, Digital Processing and Predictive Modeling for Marine Spatial Planning in the Northwest Atlantic

Simon J. Pittman and Bryan Costa

21.1 Introduction

Cetaceans are large bodied, long-lived and highly mobile marine animals that exhibit extensive migrations, as well as, high site fidelity in areas where they aggregate for feeding, socializing, mating or calving. The marine environment in which they live is characterized by complex spatial and temporal heterogeneity. Cetaceans respond to this dynamic spatial structure at a range of scales, as denoted by their space-use patterns (Kenney et al. 2001; Baumgartner & Mate 2005). Space-use patterns provide important information about distributions of cetaceans and resource managers need these patterns to develop targeted conservation policies and resource management strategies. Despite this urgent need, adequate, spatially-explicit datasets do not exist for many regions of the world. Often resource managers that are charged with protecting endangered or threatened cetaceans have to rely on datasets that are sparse in both space and time. In order to address these knowledge gaps, resource managers urgently require quantitative, spatially explicit data on cetacean species distributions and species–environment relationships at ecologically and operationally relevant scales.

This pressing management need presents a major analytical challenge made considerably more difficult by a lack of ecological data and by deficiencies in our knowledge about the spatial ecology of the majority of cetacean species. Despite the magnitude of these challenges, spatial predictive modeling techniques have been developed which effectively address these deficiencies. Spatial predictive models

S.J. Pittman (✉)
Biogeography Branch, Center for Coastal Monitoring & Assessment, National Oceanic and Atmospheric Administration, Silver Spring, MD 20910, USA; Marine Science Center, University of the Virgin Islands, St Thomas, VI 00802, USA
e-mail: Simon.Pittman@noaa.gov

B. Costa
Biogeography Branch, Center for Coastal Monitoring & Assessment, National Oceanic and Atmospheric Administration, Silver Spring, MD 20910, USA

S.A. Cushman and F. Huettmann (eds.), *Spatial Complexity, Informatics, and Wildlife Conservation*
DOI 10.1007/978-4-431-87771-4_21, © Springer 2010

have demonstrated significant utility in providing continuous data on spatial distributions, even for complex species–environment relationships and in locations where the best available data was patchy in both time and space (Manly et al. 2002; Guisan and Thuiller 2005). They have done so by quantitatively linking the spatial occurrence, abundance and richness of species to the surrounding environmental conditions (e.g., prey abundance, water temperature, water depth, etc.) across a range of scales. The maps of species distributions, abundance and biodiversity hotspots developed from predictive models can then be used to help mitigate potentially harmful human activities and to support the design and management of MPAs (Marine Protected Areas), including the delineation of ecologically meaningful boundaries. In addition, quantitative data on species–environment relationships provide valuable information on the environmental characteristics of high-use areas and, when non-linear techniques are used, can identify species-specific threshold values (Hamazaki 2002; Redfern et al. 2006; Leathwick et al. 2006).

Linking highly mobile marine animals to a complex, dynamic, multi-dimensional environment presents several spatial data management and analysis problems. First of all, sufficient spatially referenced (and ideally effort corrected) sightings data should be acquired to provide reliable spatial estimates of abundance or occurrence. Second, appropriately broad scale, accurate and ecologically meaningful environmental data should be acquired at relevant spatial and temporal resolutions to be useful as predictors. If possible, these datasets should include prey distribution data, as food availability is often a key driver of space-use. In some cases, spatially explicit prey data may be replaced by an adequate, collinear environmental variable often referred to as a "surrogate" or "proxy" variable. Third, in order to determine key ecological linkages and develop accurate spatial predictions, techniques are required that are capable of linking animal census data to environmental data in a mathematically and ecologically meaningful way. Such techniques need to allow for multiple interactions between predictor variables and need to be capable of handling complex non-linear relationships (including threshold effects) that exist between fauna and their environment (Moisen and Frescino 2002; Leathwick et al. 2006).

In this chapter, we describe a predictive modeling approach that involves the integration of historical cetacean sightings data from multiple survey programs for the southern Gulf of Maine (USA). These datasets were spatially and temporally referenced, effort corrected and then statistically linked to 29 spatially continuous environmental variables including data on geographical setting, prey distributions, surficial sea conditions, bathymetric features, water depth and water stratification. Data were either archived on internet servers or available by written request to the data providers. A Geographic Information System (GIS) was used to visualize and process the data on cetacean abundance patterns and environmental variables. Predictive models were developed using two complementary nonlinear modeling techniques: CART (Classification and Regression Trees), a recursive partitioning technique, and MARS (Multivariate Adaptive Regression Splines), a regression technique that is capable of incorporating multiple interactions between predictors.

This project was undertaken as part of a broader ecological characterization of the Stellwagen Bank National Marine Sanctuary (SBNMS) and surrounding region (http://ccma.nos.noaa.gov/products/biogeography/stellwagen/), in support of the marine sanctuary management review process and their implementation of an ecosystem-based approach to management (NOAA 2006). Ecosystem-based approaches to management require managers to fully and judiciously consider, incorporate and represent the ecosystem characteristics and connections between the environment, both within and beyond the boundaries of the managed area (Holling 1978; Walters 1986; Sherman et al. 2005). The importance of applying an ecosystem-based management approach was emphasized by the Pew Oceans Commission in 2003 and the U.S. Commission on Ocean Policy in 2004. Government agencies including the U.S. National Oceanic and Atmospheric Administration (NOAA) are currently developing and implementing Ecosystem Approaches to Management (EAM) as an adaptive and holistic management strategy.

21.2 Methods

21.2.1 Study Area

The Gulf of Maine is a 93,239 km^2 semi-enclosed sea in the Northwest Atlantic Ocean. It is considered a distinct biogeographical region based on unique oceanographic and ecological characteristics, such as water masses, patterns of primary production and geographic patterns of species assemblages (Cook and Auster 2006). The Gulf of Maine region, which includes Georges Bank, is regularly utilized by at least 13 species of cetacean (Hain and Waring 1994) primarily for feeding, although these cetaceans have also been observed mating and nursing their young in the area. The southern Gulf of Maine, particularly the region from the Great South Channel to Stellwagen Bank and Jeffreys Ledge, is considered to support the highest densities of cetaceans on the northeast U.S. continental shelf (Kenney and Winn 1986). Consistent aggregations of baleen whales occur in Cape Cod Bay and Massachusetts Bay in late winter and early spring; in the Great South Channel in late spring; and in the lower Bay of Fundy, Scotian Shelf and Jeffreys Ledge in the summer and fall (Reeves and Kenney 2003). The Stellwagen Bank NMS is considered an important feeding and transiting area for baleen whales, including humpback whales (*Megaptera novaengliae*) and the critically endangered North Atlantic right whale (*Eubalaena glacialis*). The southern Gulf of Maine is also an important food, recreation and transportation resource for humans, with high levels of ship traffic (Fig. 21.1) due to commercial shipping vessels, fishing vessels, whale-watching vessels, gas facilities and dredge spoil dumps. This overlap of high-use areas has inevitably resulted in a large number of human-cetacean interactions, some of which have resulted in the injury and mortality of cetaceans.

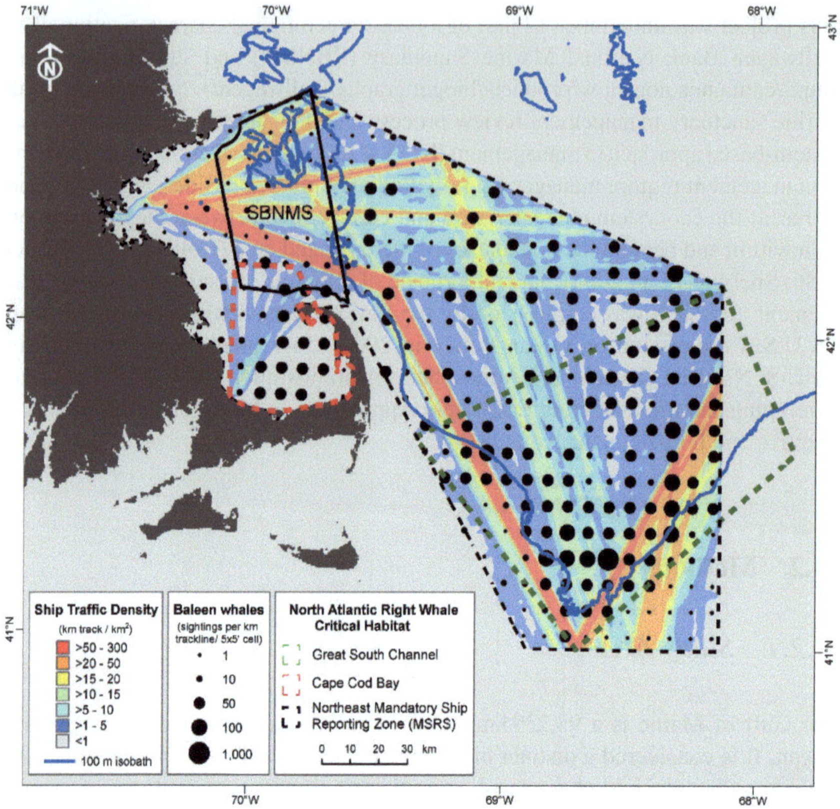

Fig. 21.1 Relative ship traffic density (kilometers of ship track per square kilometer) representing data from the first 3 years (1999–2002) of the northeast Mandatory Ship Reporting System. Several managed areas are show, including: the Federally designated Cape Cod Bay and Great South Channel right whale critical habitats, and Stellwagen Bank National Marine Sanctuary (SBNMS). The northeast mandatory ship reporting boundary is also shown (map adapted from Ward-Geiger et al. 2005)

21.2.2 Spatial Data Acquisition, Data Management and Integration of Cetacean Data

Data on cetacean sightings were acquired from two sources: (1) the North Atlantic Right Whale Consortium (NARWC – http://www.rightwhaleweb.org/) sightings database held at the University of Rhode Island (Kenney 2001); and (2) the Manomet Bird Observatory (MBO) database. The MBO database (MBO 1980) was compiled for the Cetacean and Seabird Assessment Program commissioned between 1980 and 1988 by the Northeast Fisheries Science Center (NEFSC – http://www.nefsc.noaa.gov/) of NOAA's National Marine Fisheries Service (NMFS).

NARWC offered the most comprehensive compilation of cetacean sightings data for the Gulf of Maine, however, some of the data incorporated within NARWC was also available online as individual cruises via the Ocean Biogeographic Information System (OBIS; http://www.iobis.org/). In contrast, the MBO data archived by NEFSC, was not available online and had not been used by the scientific community since the late 1980's.

21.2.2.1 Data Selection Criteria for NARWC Cetacean Data

The NARWC data contained survey effort and sightings data from ship and aerial surveys and opportunistic sources between 1970 and 2005. In order to minimize error due to environmental conditions (e.g., low visibility, rough seas) only records from dedicated aerial surveys and platforms-of-opportunity (Kenney and Winn 1986; Shoop and Kenney 1992) that met the data selection criteria were used to calculate an index of relative abundance. Data records were selected if they contained sufficient information to construct survey tracks and if they were conducted in relatively calm seas (Beaufort Sea State ≤3 for small cetaceans and ≤ 4 for large cetaceans), with good visibility (≥2 nautical miles). Furthermore, only aerial surveys conducted at altitudes less than 366 m above sea level were included.

A total of 653,725 km of survey track and 34,589 cetacean observations were provisionally selected for analysis. In order to minimize bias from the uneven allocation of survey effort in both time and space, we calculated the relative abundance of cetacean species as sightings-per-unit-effort (SPUE) (Equation 21.1; Kenney and Winn 1986; Shoop and Kenney 1992). SPUE allowed comparisons between discrete spatial units and temporal subsets within the study area. The spatial extent of the study area and spatial resolution of the grid (i.e., 5x5 minute cells) were selected a priori through negotiation with the NARWC data providers.

$$\text{SPUE} = 1{,}000 \times (n/e) \tag{21.1}$$

Where: n = number of individual cetaceans; e = km of surveyed trackline. The factor of 1,000 was included to avoid very small decimal SPUE values. Attributes of the SPUE dataset included the following: (1) *Effort* – calculated as km of valid track line surveyed; (2) *Animals* – calculated as the number of individual cetaceans sighted; and (3) SPUE – calculated as the number of cetaceans per 1,000 km of trackline surveyed. Few surveys provided enough information (i.e., distance from platform) to calculate probability of detection functions for individual species. We therefore assumed equal detectability for each species across survey platforms. It is likely that some of this bias in animal detection was minimized by our data selection criteria, although relative abundance may still have been underestimated because of animals diving at the time of the census. In addition, many sightings were recorded without information on behavior such as feeding, vocalizing or speed of movement, so that data were pooled regardless of individual or group behavior.

The MBO data and other regional sightings data had never before been integrated. In order to prepare the MBO dataset for integration with NARWC's SPUE data, a series of data selection and geoprocessing steps were required (Fig. 21.2).

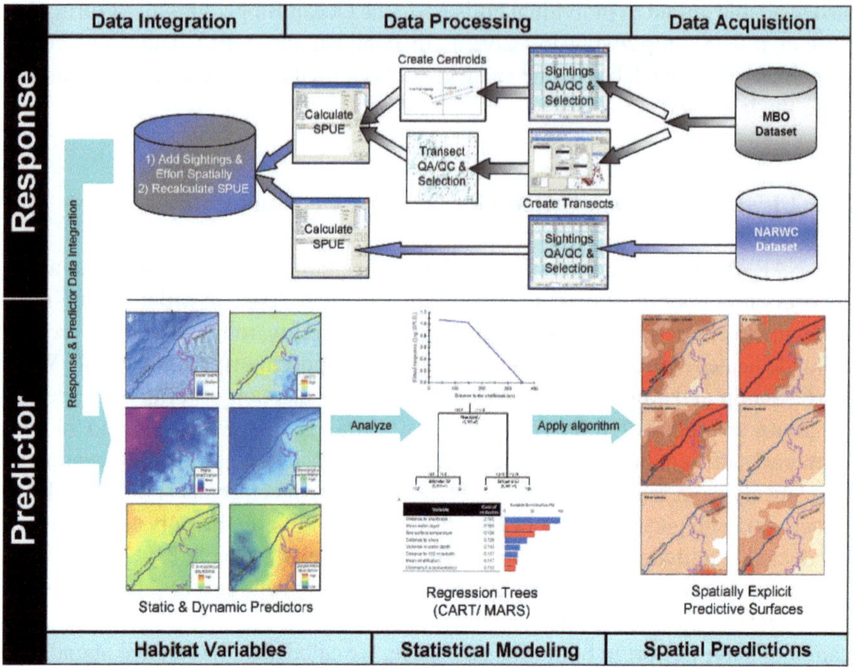

Fig. 21.2 The sequence of the major analytical processes applied to ecological data in the linking of cetaceans to their environment

This process included extracting cetacean sighting records; updating database field names to match the NARWC database; creating geometry to represent survey tracklines and applying the same data selection criteria as was used for the NARWC dataset (except altitude and visibility). The NARWC altitude criterion did not apply to the MBO dataset because all of the surveys were conducted from ships. Visibility data were not provided in the MBO dataset, therefore we assumed acceptable visibility (i.e., visibility of at least two nautical miles) for all transects that were recorded as "on effort". These selection criteria resulted in exclusion of 145 sightings. Furthermore, the original MBO sightings data were assigned to a single coordinate representing only the beginning of the transect. This presented a problem, particularly where transects crossed two or more grid cells, since the sighting may have occurred anywhere along a transect. In order to explore the extent of the problem, we calculated the proportion (%) of the total transect length that fell within each grid cell it intersected (Equation 21.2).

$$\% D = (Di / D) \times 100 \tag{21.2}$$

Where: $\%D$ = proportion of transect length (%) Di = intersected transect length D = total transect length.

Next, we decided on an appropriate error threshold by calculating the effect of excluding transects with less than 50%, 70% and 90% of their total length falling within an individual 5 × 5 min cell. In order to maximize the number of transects in the analyses, but minimize uncertainty in our spatial assignment of cetacean data, only transects that had ≥70% of their length in a single 5 × 5 min cell were selected and exported for use in the biogeographic assessment. This selection criterion removed 596 out of 2,162 transects and 341 sightings out of 1,431 sightings in our area of interest. Each poly-line transect was converted to a centroid using XTools Pro 3.0 "convert shapes to centroids" function and cetacean sightings attributes were transferred to the corresponding centroid for each transect.

The sightings data for both NARWC & MBO datasets were aggregated in 5 × 5 min grid cells by month (1997–2004) and season (1978–2005) and by cetacean species and group (dolphin/porpoise, baleen). Next, the number of animals and amount of effort were combined using unique grid cell IDs to link the two datasets geographically. The sightings were corrected for effort using the SPUE Index Equation (21.1). Cells with less than 2.5 km of effort were considered to have "no effort" and were removed from the analysis. Combined valid survey effort for the southern Gulf of Maine region was 567,955 km of survey track for small-bodied cetaceans and 658,935 km for large-bodied cetaceans.

21.2.3 Environmental Data

In order to develop cetacean–environment models to explain patterns of relative abundance, we processed spatial data for a wide range of environmental variables some of which were considered a priori to be ecologically meaningful to cetaceans (e.g., water depth, actual prey species) and others that were of unknown importance (e.g., distance to shore, benthic habitat richness, potential prey species, proxy prey species). We included predictor variables that were of unknown importance because little was known about the spatial ecology of some of the species of interest. Twenty nine environmental variables representing means and standard deviations were calculated for each cell of a 5 × 5 min grid using a Visual Basic (VB) script to interface with Spatial Analyst's Zonal Stats in ArcGIS v9.1 (Table 21.1). Environmental data types included: prey abundance (fish and zooplankton), water depth, bathymetric slope, water stratification, chlorophyll *a* concentration, sea surface temperature, turbidity, habitat richness and area of substratum type, distance to shore, and distance to the 200- and 100-m isobaths (Table 21.2). Slope was calculated in degrees using the slope function found in ArcGIS's Spatial Analyst. Distances to the shore and to the 200- and 100-m isobaths were calculated using a VB script called "Minimum Distance 2 Layers" (Chasen 2005) (Table 21.1). Seasonal water stratification was calculated by interpolating water density values at 0 and 50 m, and subtracting the resulting raster surfaces. Water stratification was calculated for the study region using a 98 year dataset (1912–2004) and a ten year dataset (1994–2004), although both were highly collinear. Pixel values from SeaWIFS (chlorophyll *a* and turbidity)

Table 21.1 A selection of GIS based spatial tools used by this project to manage spatial data

GIS tools	Version	Author	Functionality	Source
Xtools Pro v3.0	ArcGIS 9.X	Data East, LLC	Multiple geoprocessing functions	http://www.xtoolspro.com/
Hawths Tools	ArcGIS 9.X	Hawthorne Beyer	Multiple geoprocessing functions	http://www.spatialecology.com/htools
Minimum Distance2 Layers	ArcGIS 9.X – VB	Rob Chasan	Takes two feature layers and finds the minimum distance between the feature selected in the source layer and the features selected in the target layer.	http:// arcscripts.esri.com/details.asp? dbid=12829
Calculate Zonal Stats & Join/Export Summary Tables	ArcGIS 9.1 – VB	Eric Finnen	(1) Masks legend and no data values; (2) calculates empirical chlorophyll, turbidity and sea surface temperature values; (3) calculates zonal statitics for rasters; (4) joins summary tables to shapefile and exports shapefile; (5) intersects 2 polygon layers and calculates percent area based on individual bins	N/A

Table 21.2 Environmental variables used to explain spatial patterns in cetacean densities in the southern Gulf of Maine study area

	Environmental (predictor) datasets	Data type	Temporal resolution	Spatial resolution	Original source
Benthic	Water depth (m)	Raster Dataset	N/A	360 m	USGS – Marine GIS library for Massachusetts Bay
	Mean & SD water depth (m)	Polygon (GIS Derivative)	N/A	5 minute	NOAA CCMA - Ecological Characterisation for SBNMS
	Mean & SD slope (degree)	Polygon (GIS Derivative)	N/A	5 minute	NOAA CCMA - Ecological Characterisation for SBNMS
	Sediment types	Polygon shapefile	N/A	N/A	USGS – Marine GIS library for Massachusetts Bay
	Amount of each sediment type (%)	Polygon (GIS Derivative)	N/A	5 minute	NOAA CCMA - Ecological Characterisation for SBNMS
	Number of sediment types	Polygon (GIS Derivative)	N/A	5 minute	NOAA CCMA - Ecological Characterisation for SBNMS
	Minimum distance to shore (km) from 5×5 min cell centroid	Polygon shapefile	N/A	5 minute	NOAA CCMA - Ecological Characterisation for SBNMS
	Minimum distance to 100 m isobath (km) from 5×5 min cell centroid	Polygon shapefile	N/A	5 minute	NOAA CCMA - Ecological Characterisation for SBNMS
	Minimum distance to shelfbreak (km) from 5×5 min cell centroid	Polygon shapefile	N/A	5 minute	NOAA CCMA - Ecological Characterisation for SBNMS
Oceanographic	Water salinity, temperature, density	Tabular Dataset	1912–2004	N/A	Canadian Fisheries and Oceans - OES database
	Water density and stratification	Raster (GIS Derivative)	1912–2004	2.4 & 1.1 km	NOAA CCMA - Ecological Characterisation for SBNMS
	Mean & SD water stratification (Sigma t)	Polygon (GIS Derivative)	1912–2004	5 minute	NOAA CCMA - Ecological Characterisation for SBNMS
	Mean & SD water stratification (Sigma t)	Polygon (GIS Derivative)	1994–2004	5 minute	NOAA CCMA - Ecological Characterisation for SBNMS
	SST (°C)	Raster Dataset	1985–2004	4.0 & 1.1 km	NASA – AVHRR
	Mean & SD SST (°C)	Polygon (GIS Derivative)	1985–2004	5 minute	NOAA CCMA – Ecological Characterisation for SBNMS
	Chlorophyll a (ug/L)	Raster Dataset	1998–2005	4.0 & 1.1 km	NASA – SeaWiFS
	Mean & SD chlorophyll a (ug/L)	Polygon (GIS Derivative)	1998–2005	5 minute	NOAA CCMA – Ecological Characterisation for SBNMS

(continued)

Table 21.2 (continued)

	Environmental (predictor) datasets	Data type	Temporal resolution	Spatial resolution	Original source
Prey	Fish Trawls	Tabular Dataset	1975–2005	N/A	NOAA NEFSC - MARMAP program
	Mean fish density (all species)	Polygon (GIS Derivative)	1975–2005	5 minute	NOAA CCMA - Ecological Characterisation for SBNMS
	Interpolated fish density (all species)	Polygon (GIS Derivative)	1975–2005	5 minute	NOAA CCMA - Ecological Characterisation for SBNMS
	Mean sandlance density	Polygon (GIS Derivative)	1975–2005	5 minute	NOAA CCMA - Ecological Characterisation for SBNMS
	Interpolated sandlance density	Polygon (GIS Derivative)	1975–2005	5 minute	NOAA CCMA - Ecological Characterisation for SBNMS
	Mean herring density	Polygon (GIS Derivative)	1975–2005	5 minute	NOAA CCMA - Ecological Characterisation for SBNMS
	Interpolated herring density	Polygon (GIS Derivative)	1975–2005	5 minute	NOAA CCMA - Ecological Characterisation for SBNMS
	Mean mackerel density	Polygon (GIS Derivative)	1975–2005	5 minute	NOAA CCMA - Ecological Characterisation for SBNMS
	Mean hake (silver, red & white) density	Polygon (GIS Derivative)	1975–2005	5 minute	NOAA CCMA - Ecological Characterisation for SBNMS
	Mean sandlance, herring, hake combined density	Polygon (GIS Derivative)	1975–2005	5 minute	NOAA CCMA - Ecological Characterisation for SBNMS
	Zooplankton Bottom Trawls	Tabular Dataset	1977–1988	N/A	NOAA NEFSC - MARMAP program
	Mean zooplankton density	GIS Derivative	1980–1988	5 minute	NOAA CCMA - Ecological Characterisation for SBNMS
	Mean interpolated zooplankton density	GIS Derivative	1980–1988	5 minute	NOAA CCMA - Ecological Characterisation for SBNMS
	Mean log C. finmarchicus density (log)	GIS Derivative	1980–1988	5 minute	NOAA CCMA - Ecological Characterisation for SBNMS

and AVHRR (SST) satellite derived imagery were converted to standard units, and the land and no data values were re-classed as null values using a VB script. The temporal overlap between predictors and response variables varied slightly, with zooplankton prey data exhibiting the least overlap with SPUE (the best available zooplankton data was only for 1980–1988) (Table 21.2).

21.2.4 Spatial Data Storage, Visualization and Predictive Modeling

All spatial data layers and GIS analytical tools and routines were integrated using an ArcGIS personal geodatabase with a common spatial framework (i.e., coordinate system) and supported by FGDC compliant metadata. A personal geodatabase (PGDB) is an object-oriented database that provides services for managing geographic data and is stored inside relational database management systems (RDBMS). This architecture provides greater flexibility in interacting with datasets, as it allows users to access the embedded functionality of the RDBMS, in addition to the data management services provided through ArcCatalog. These enhanced services and functionality allow the user to perform a variety of tasks. In particular, the user may: (1) assign management requests, perform queries and develop summary reports using MS Access; (2) create topological validation rules to ensure data quality and uniformity; (3) nest datasets hierarchically to facilitate indexing; and (4) document relevant metadata making the data easy to catalog, query and identify. For this project, the PGDB was implemented as the data format of choice, given these advantages over the use of individual rasters or shapfiles. Additionally, the PGDB was chosen over enterprise geodatabases because of the small size of the usergroup (i.e., <4 people) and as well as the moderate storage size needed for the data (i.e., <30 GB). While the PGDB format was optimal for use in the Stellwagen Bank NMS project, ESRI has introduced a new type of geodatabase, called a file geodatabase (FGDB), with its release of ArcGIS 9.2. This new geodatabase format offers unique advantages over the PGDB, namely storage space, and should also be evaluated when choosing among non-enterprise storage formats (Table 21.3). In addition to file geodatabases, the newly released Arc Marine data model should also be evaluated when choosing a data format. This data model facilitates the integration, sharing and exchange of data, and may also prove valuable in predictive modeling studies using complex multi-dimensional data (Wright et al. 2007).

There are several challenges associated with developing spatial databases. The most notable of these challenges include: (1) gaining an understanding of the multiplicity of data; (2) identifying the errors in each dataset; (3) referencing data to the same coordinate system; (4) logically resampling data to a uniform spatial and temporal resolutions; and (5) indexing data so that they could easily be found. It is important that decisions regarding spatial data architecture and management be made at the outset of a project, so as to avoid issues related to geographic projections, spatial and temporal

Table 21.3 Comparison of ESRI's ArcGIS personal geodatabases (PGDB) and file geodatabases (FGDB). Adapted from ESRI, 2007

Key characteristics	File geodatabase	Personal geodatabase
Description	A collection of various types of GIS datasets held in a file system folder	Original data format for ArcGIS geodatabases stored and managed in Microsoft Access data files
	This is the recommended native data format for ArcGIS stored and managed in a file system folder.	This is limited in size and tied to the Windows operating system.
Number of Users	Single user and small workgroups	Single user and small workgroups with smaller datasets
	Some readers and one writer per feature dataset, standalone feature class or table.	Some readers and one writer.
	Concurrent use of any specific file eventually degrades for large numbers of readers.	Concurrent use eventually degrades for large numbers of readers.
Storage Format	Each dataset is a separate file on disk A file geodatabase is a file folder that holds its dataset files.	All the contents in each personal geodatabase are held in a single Microsoft Access file.
Size Limits	One terrabyte for each dataset. Each file geodatabase can hold many datasets	Two gigabytes per MS Access database
	Each feature class can scale up to hundreds of millions of vector features per dataset.	Effective limit before performance degrades is typically between 250 and 500 MB per Access database file.
Versioning Support	Not supported	Not supported
Platforms	Cross-platform	Windows only
Security and Permissions	Operating file system security	Windows file system security
Database Administration Tools	File system management	Windows file system management
Notes	Allows you to optionally store data in a read-only compressed format to reduce storage requirements	Often used as an attribute table manager (via Microsoft Access). Users like the string handling for text attributes.

misalignments, versioning and size or record limitations. Projection and misalignment issues occur because some geoprocessing functions can not reproject on the fly. Versioning issues occur when there are many users, but there is no way to reconcile the individual edits. Dataset size or record limitation issues occur if the dataset(s) grow to be too large to be reliably handled by the implemented data format.

For predictive modeling, we developed a relatively generalized and single-scale approach for analyzing the linkages between cetaceans and their environment. We linked spatial variability in environmental variables to cetacean abundance by season across a 5 × 5 minute spatial grid. By scaling up our temporal resolution from month to season we expected to minimize some of the error that may have occurred, while preserving the strong patterns that are useful for resource management decision making. Discussions with resource managers and exploratory analyses of temporal variability within and between seasons indicated that this approach was likely to be both ecologically meaningful and strategically useful (NOAA 2006). The intention of model development was to: (1) identify the most influential environmental variables that determine seasonal patterns of cetacean distribution and relative abundance; (2) characterize the attributes of the environment that support highest mean abundance; and (3) identify thresholds in the statistical representation of the cetacean–environment relationship.

To achieve these objectives, we first used regression trees (CART™ by Breiman *et al.,* 1998 and Salford Systems Inc. - http://www.salford-systems.com/) in order to derive a set of breakpoints that quantitatively described the multivariate environmental conditions associated with varying levels of cetacean abundance. Second, we used multivariate adaptive regression splines (MARS™ by Friedman, 1990 and Salford Systems Inc.) to fit the cetacean data more precisely to the environmental data, accounting for interactions and to determine the form of nonlinear relationships. MARS is a novel and flexible multivariate nonparametric regression technique that combines the most useful aspects of recursive partitioning and spline fitting to perform piecewise linear regressions. MARS also has the additional flexibility to model relationships that are nearly additive or involve multiple interactions between predictors (Friedman and Roosen 1995; Moisen and Frescino 2002; Leathwick et al. 2006). The technique is capable of ignoring redundant variables and reliably tracking the complex data structures that are often ensconced in high-dimensional ecological datasets. MARS estimates variable importance by comparing the estimated cost of omission of each variable with the highest cost of omission of all variables. The cost of omission is estimated by calculating the model goodness of fit after the variable has been excluded. SPUE data from 1997 to 2005 were used for the predictive modeling.

21.3 Final Data Products, Results and Discussion

Some geographic areas consistently exhibited high densities for a number of species and were identified as high-use areas for special management concern. However, individual species also showed distinctive spatio-temporal patterns across seasons, and interpolations of SPUE were used to identify and visualize high-use patches and high-use corridors connecting patches. Overall, the 100 m isobath was identified as a cetacean superhighway that was utilized by many species, particularly large baleen species, to enter, exit and navigate through the southern Gulf of Maine. Visual examination of environmental patterns revealed that the 100 m isobath functioned as an edge or ecotone environment, with elevated primary and secondary productivity in close proximity. Although several baleen whale species used the 100 m isobath intensively, our study showed that species utilized adjacent areas differently (i.e., centered over the oceanic versus coastal side of the 100 m isobath). Little is known about the behavior associated with this key bathymetric feature or the cues that result in its popularity for cetaceans. Overall, there was significant non-linearity in the relationships between cetaceans and their environment. Comparisons between model performance using linear regression and non-linear regression revealed that an average of 16% more variation in cetacean abundance was explained using non-linear models in the spring and 14% more in the summer (Table 21.4).

Here we present as an example, an integrated ecological interpretation of the modeling results from North Atlantic right whale SPUE in spring and summer. To do so, we use the numerical outputs from the MARS and CART models to determine the statistical relationship between right whale SPUE and a set of environmental predictors. We also examine ranked variable importance to determine the best predictors of SPUE, as well as visually interpretated maps of right whale SPUE overlayed on maps of the density of right whale's preferred prey, *Calanus finmarchicus*, which were interpolated from historical plankton trawls.

In the Gulf of Maine, *Calanus* early life stages coincide with the spring phytoplankton blooms on which they feed before beginning a gradual move to deeper water. In midsummer they enter diapause (fourth and fifth copepodite stages) and spend the remainder of the year at depths of 50–300 m (Bigelow, 1926). Examination of right whale SPUE and *Calanus* abundances across the spring and summer seasons revealed a clear geographic shift in whale abundance. This shift broadly tracked seasonal shifts in *Calanus* abundance hotspots. In spring, *Calanus* and right whale hotspots were located along the northern slope of Georges Bank, the Great South Channel, Cape Cod Bay, western SBNMS and some of the deep basins in the central southern Gulf of Maine. These high abundance areas were

Table 21.4 Summary of MARS model results showing model performance (r^2) and the most influential environmental variables (which were determined by calculating the change in model performance when each variable is excluded from the model). Only predictors that contributed more than 50% to models are listed

Species/group	Season	r^2	Best model predictors (>50% contribution)
Humpback whale	Spring	0.63	Dist. 100 m isobath, Dist. shelf, Dist. shore
	Summer	0.60	Dist. 100 m isobath
Fin whale	Spring	0.47	Dist. 100 m isobath
	Summer	0.43	Dist. 100 m isobath, SST
Minke whale	Spring	0.36	Dist. 100 m isobath, Water depth, Dist. shore, Zoopl. abund.
	Summer	0.37	Dist. 100 m isobath
North Atlantic right whale	Spring	0.61	Dist. shelf, water depth, SST
	Summer	0.51	Chlorophyll a, Dist. shelf, Water depth
Sei whale	Spring	0.41	Dist. shelf, Water depth
	Summer	0.42	Water depth, Dist. shelf, Dist shore
Atlantic white-sided dolphin	Spring	0.46	Water depth, Dist. shelf, Dist shore, All fish abund., SST
	Summer	0.47	Zooplankton abund., SST, Dist. shelf, Water depth

generally located along the inside edge of the 100-m isobath. The 100-m isobath denotes approximately the boundary between the deeper waters of the Gulf and the shoals or banks. The relatively high abundance of *Calanus* along the 100-m isobath in spring is thought to be related to the advection of a low-salinity surface plume, which moves south along the 100-m isobath from the northern Gulf of Maine (Wishner et al. 1988; Beardsley et al. 1996). In summer, *Calanus* hotspots shifted offshore towards the central, southern Gulf of Maine. A corresponding distributional shift was also evident for North Atlantic right whale, with fewer individuals using nearshore regions such as Cape Cod Bay and the northeastern slopes of Nantucket Shoals in the summer. Although a clear shift was evident, the strength of the spatial association was possibly weakened by the fact that zooplankton trawls may not have adequately sampled the horizontal or vertical distribution of zooplankton, thereby underestimating the concentration of prey available to the whales in some areas. Perhaps for the same reason, prey abundance patterns (all zooplankton and *Calanus finmarchicus*) were not the strongest predictors in statistical models. They did, however, still make a significant contribution to the summer predictive model and demonstrated quantitatively that a complex spatio-temporal linkage exists between the distribution of right whales and their zooplankton prey (Fig. 21.3).

Statistical modeling techniques provided some numerical substantiation for these broad scale seasonal distribution patterns. The MARS model ($r^2 = 0.61$) for spring right whale SPUE determined that distance to shelfbreak (mostly negative direction) and mean water depth (mostly positive direction) were the strongest predictors. Plots of SPUE and water depth showed twin peaks with high SPUE in shallow waters (<50 m) and high SPUE over deeper waters (100–200 m). These two peaks highlighted the importance of known feeding areas in shallow nearshore waters of Cape Cod and deeper waters adjacent the 100-m isobath, including the Great South Channel. CART models showed that highest use areas were characterized by waters warmer than 6.3°C, mean water depth less than 85.3 m and where the substratum was more than 18.5% sand and gravelly sand. These environmental conditions were characteristic of sites inside Cape Code Bay. Another high use area was characterized by slightly colder water, with a mean stratification index of between −1.1 and −0.2 (weak stratification to mixed waters) and with relatively high (>2.4) phytoplankton (chlorophyll *a* concentration) and high *Calanus* abundance (>3.9 log density). These environmental conditions were characteristic of sites in the Great South Channel and along the 100-m isobath.

For the summer model, chlorophyll *a* concentration and distance to shelf break contributed most to the final MARS model ($r^2 = 0.51$). The plot of SPUE versus distance from the 100-m isobath showed a more gradual decline in SPUE than was shown in spring (Fig. 21.3), indicative of the movement away from the edge towards the deeper water basins of the Gulf. The relationship with water stratification was also pronounced showing a strong increase in SPUE where sigma-*t* was greater than approximately −2.8 (well-stratified). Similarly, the CART model indicated that highest SPUE values were associated with well-stratified waters (≥2.8 sigma-*t*) and zooplankton abundance that exceeded the regional average (Fig. 21.3). Abundance of zooplankton (all species combined) and calanoid cope-

a

Variable	Cost of omission
Distance to shelfbreak	0.192
Mean water depth	0.165
Sea surface temperature	0.136
Distance to shore	0.126
Variance in water depth	0.119
Distance to 100 m isobath	0.117
Mean stratification	0.117
Chlorophyll *a* concentration	0.116

b

Variable	Cost of omission
Chlorophyll *a* concentration	0.092
Distance to shelfbreak	0.082
Mean water depth	0.081
Sea surface temperature	0.08
Zooplankton abundance	0.08
C. finmarchicus abundance	0.078
Clay	0.077

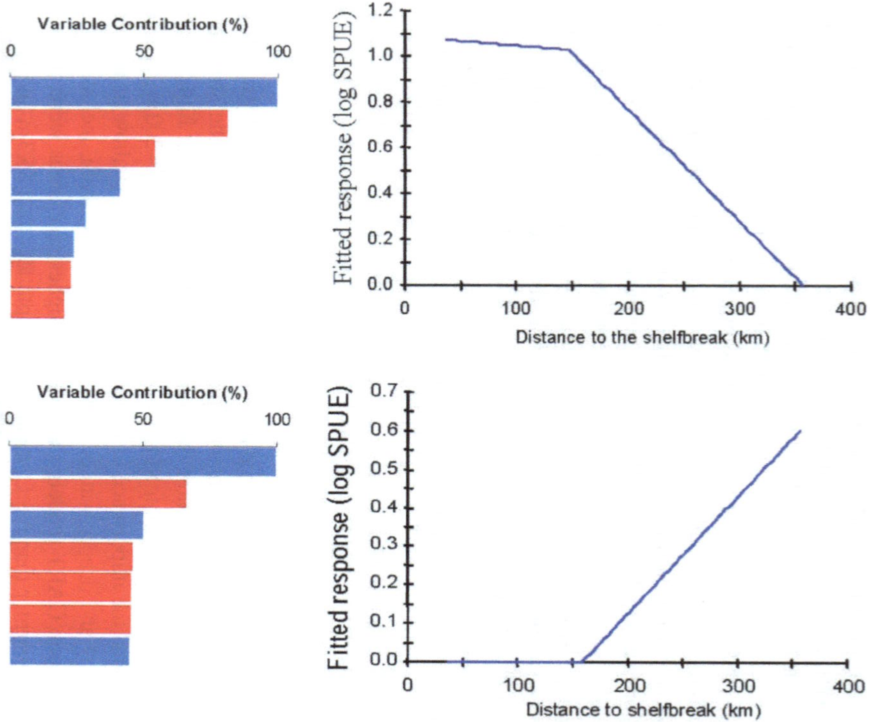

Fig. 21.3 (*Left*) List of the most influential variables determining spatial patterns in North Atlantic right whale abundance in the southern Gulf of Maine for (**a**) spring and (**b**) summer based on the cost of omission calculated by MARS™ software. Only variables with more than 20% contribution are listed. Contribution was based on non-linear models. Directionality (blue = negative; red = positive) was derived from the linear relationship. (*Right*) Plots showing the MARS™ model fit between North Atlantic right whale and the top environmental predictors for (**a**) spring (**b**) summer. To improve visual interpretation these plots were generated from additive models, where no interactions between predictors were allowed

pod (*Calanus finmarchicus* – all life stages combined) contributed 46 and 45% to the MARS model, respectively. These apparent habitat preferences and threshold values are of ecological interest since in the summer the right whale's primary prey are distributed vertically in discrete layers (Winn et al. 1995; Woodley and Gaskin 1996; Baumgartner and Mate 2003). Preference for deeper, well-stratified waters in summer may reflect the proximity to colder, deeper waters supporting a higher abundance of overwintering *Calanus*. Baumgartner and Mate (2003) tracked right whales in the Gulf of Maine and found that they repeatedly dove between 80 and 175 m, where the average dive depth was strongly positively correlated with the average depth of peak *Calanus* C5 abundance.

In summary, our analyses of spatial and temporal patterns of North Atlantic right whales abundance confirmed that the arrival and departure of whales to and from the Gulf of Maine and their space-use patterns within the Gulf of Maine are synchronized with the spatio-temporal distribution of the *Calanus* and its life-cycle movements. Although it is well known that the movements of this species in the Gulf of Maine are closely linked to the life-history and spatial distribution of its main prey, few studies have actually examined and quantified the statistical and spatial relationship between predator and prey abundance.

21.4 Relevance to Sustainable Resource Management

Information on the specific environmental characteristics of high-use areas provides useful ecological insights and essential baseline information to support individual species management, as well as ecosystem-based management. In addition, these spatial and temporal cetacean–environment relationships may be used to help future studies develop targeted hypotheses to examine specific ecological mechanisms. Spatially explicit modeling techniques used in this study provided numerical outputs in a format that can be easily incorporated into a GIS to develop spatial predictions (Guisan and Thullier 2005; Leathwick et al. 2006b; Pittman et al. 2007a, b). For example, MARS provided a series of basis functions equivalent to beta weights in a linear regression, and CART provided a set of binary splits on variables (i.e., numerical decision rules) in the form of a simple logical query (i.e., SQL – Structured Query Language). Overall, CART and MARS share the ability to extract high-order interactions and can deal with a large number of variables, nonlinear relationships and multi-collinearity among variables.

The combination of environmental variables included in our model appeared to have useful predictive power. In addition, the robustness of the resulting relationships were enhanced by the use of a spatially extensive, effort-corrected data set, aggregated by season to minimize variability in the response and to obviate much of the concern about bias in the sampling design. However, smaller sample sizes and lower abundance in fall and winter resulted in lower model performance. In addition, the use of aggregated data in both time and space is a potential limitation in the ability to accurately and precisely capture the ecologically important variability or heterogeneity in both animal distributions and environmental conditions. As such, our analyses focused on large and persistent differences between cetacean space-use patterns and dominant characteristics of their environment. Our conclusion that static features explained more of the variability than more dynamic features of the environment suggests that additional modeling efforts may gain improved model performance by using all years of data (1970–2005 instead of 1997–2005) regardless of the interannual variability. Further studies may also usefully explore the interdecadal linkages between spatio-temporal patterns of cetacean abundance, their prey and the North Atlantic Oscillation (Greene et al. 2003).

One major limitation in our ecological interpretation is that we currently do not know which high-use areas function for feeding, corridors of movement, social aggregations, or any combination of these. Also, we have little information on the movement patterns of individual animals, including diving behavior, which further limits our ability to evaluate the importance of environmental features, such as the slope waters above the 100-m isobath. Future survey and telemetry studies will allow us to map individual space-use patterns and provide information necessary to evaluate the ecological significance of areas that we have found to be highly utilized (based on the patterns of cetacean sightings). Individual movement data will also provide further insights on how animals respond to structural features in their environment, such as prey patchiness, bathymetric morphology and water column structures (including thermoclines, haloclines and pycnoclines). Recent studies have indicated that marine animals respond to structure in their environment at multiple spatial scales (Kenney et al. 2001; Schneider 2001; Pittman and McAlpine 2003).

In addition, our selected spatial resolution subsumed much of the within-cell heterogeneity existing in the original data sets, thereby potentially smoothing out important fine-scale anomalies and coarsening spatio-temporal gradients. This may have influenced the performance of the models and possibly mitigated the importance of prey abundance. Furthermore, future animal–environment studies should be carried out at a range of spatial scales using data with a range of spatial resolutions (Guisan et al. 2007; Pittman et al. 2007a, b). Such a scalar approach is needed because often little is known about the scales at which environmental heterogeneity is important for a species or a group of species. The analytical framework we developed here can be easily adapted to conduct exploratory analyses at multiple spatial scales by aggregating grid cells or selecting different temporal resolutions (i.e., weeks, months, years, decades). From a temporal perspective, we were primarily interested in detecting seasonal patterns and found that available environmental information was more complete when aggregated by season. However, analyses by month may have more closely aligned spatial heterogeneity in the response variables with spatial heterogeneity in the predictor variables. In addition, monthly data may reveal more details on the trajectories of the habitat shifts that were evident at the seasonal level of data aggregation. For example, Hamazaki (2002) modeled monthly cetacean sightings (from June to August), which showed a gradual northward shift in the distribution of many cetacean species on the northeastern continental shelf.

Ultimately, model performance was based on a statistical relationship between cetacean sightings and predictor environmental variables. The accuracy and precision of the models consequently will depend on the quality of predictor variables and sightings data, spatial and temporal resolution of the data, the relationship between response and predictor, and the algorithm used to develop the model. Measurement errors that occur during in situ environmental sampling and during the acquisition of remotely sensed data can often be difficult to quantify. Additional errors are introduced when interpolating point data to provide continuous surfaces. We recognize these issues and have only just begun to examine the complex relationships using the best available data. We suggest that results from these analyses be used to guide future research in spatial ecology and generation of hypotheses.

21.5 Conclusions

The integration and analyses of available digital spatial ecological data revealed key species–environment relationships. The broad-scale and spatially-explicit predictive modeling approach developed here linked (for the first time) a diverse and spatially comprehensive suite of environmental predictors with cetacean sightings data to explore spatial patterns in a relatively well surveyed region of the earth. This approach provided a powerful and flexible analytical framework for exploring cetacean–environment linkages, integrating best available data at multiple resolutions and generating hypotheses for future studies. In general, we found that relatively static features, most notably the 100-m isobath and the continental shelf edge, were better predictors of cetacean SPUE distribution patterns than were relatively dynamic features, such as SST and chlorophyll *a* concentrations. One important characteristic of high relief areas is the generation of heterogeneous boundary conditions that facilitate elevated secondary productivity that is typically persistent in time (Hyrenbach et al. 2000). Such bathymetric edge habitats appear to provide high quality resources for many species. This indicates that for some cetacean species, useful predictions can be achieved using information on bathymetry alone, since prey distributions themselves are strongly linked with bathymetric features. Bathymetric data (and analytical derivatives of bathymetry such as slope and rugosity) are widely available for many regions of the earth and should become prime candidates for use as predictors in studies attempting to map the distribution, abundance and diversity of cetaceans. As noted by Hyrenbach (2000), such information will be of great value in establishing networks of marine protected areas.

Prey abundance patterns were important components of most models, but made only relatively minor contributions to explaining spatial patterns in cetacean SPUE. The absence of a strong relationship may be due to high spatial and temporal variability in prey abundance, sampling error or scale mismatch between response and predictor. The challenge now is to develop improved techniques to map prey abundance.

This study indicates that cetacean–environment studies must not ignore variable interactions when developing predictive models, as interactions between variables were important in explaining variability. Notably, some of the interactions between bathymetry, hydrodynamic processes, temperature regimes, phytoplankton, zooplankton and planktivorous fish need further investigation in order to better explain spatial patterns and key mechanisms controlling the patterns of cetaceans abundance, and to predict the effects of change in the southern Gulf of Maine.

We demonstrated here that considerable amounts of ecological information can be acquired for many species across broad expanses of the ocean from freely available online data providers. The internet is a valuable resource for the dissemination of data that (when integrated in a spatial framework) provides hosts of potentially valuable and previously untapped ecological information. To facilitate these studies, greater efforts are needed to develop and distribute adequate metadata with existing datasets, so as to make cetacean sightings data readily available to researchers and resource managers without restrictions on resolution or data types. The distribution of raw data instead of summary data may prevent potentially important information being inadvertently excluded. Furthermore, the effective use and ecologically meaningful interpretation of existing data can be greatly enhanced through closer collaboration with the original data collectors and analysts possessing intimate and often undocumented knowledge of data limitations and caveats.

21.6 Epilogue

Immediately after the release of this study, NOAA's Office of Protected Species used the seasonal spatial data on cetacean SPUE to evaluate and mitigate the potential impacts of a proposed offshore gas facility in Massachusetts Bay on cetaceans, including the highly endangered North Atlantic right whale. The cetacean distribution patterns have also been used to evaluate proposals to relocate the Boston Harbor shipping lane to minimize vessel strikes to cetaceans and for ocean zoning as part of the Massachusetts Ocean Plan.

Acknowledgements We thank all the contributors of cetacean sightings data archived in both NARWC and MBO datasets. We also thank the National Marine Sanctuary Program for initiating this project including the staff of the Stellwagen Bank National Marine Sanctuary; Robert Kenney of the University of Rhode Island for contributing data and metadata and the Remote Sensing Team of NOAA's Center for Coastal Monitoring and Assessment for processing the satellite data.

References

Baumgartner MF, Mate BR (2003) Summertime foraging ecology of North Atlantic right whales. Mar Ecol Prog Ser 264:123–135

Baumgartner MF, Mate BR (2005) Summer and fall habitat of North Atlantic right whales (*Eubalaena glacialis*) inferred from satellite telemetry. Can J Fish Aquatic Sci 62:527–543

Beardsley RC, Epstein AW, Chen C, Wishner KF, Macaulay MC, Kenney RD (1996) Spatial variability in zooplankton abundance near feeding right whales in the Great South Channel. Deep-Sea Res. II 43:601–625

Bigelow HB (1926) Plankton of the offshore waters of the Gulf of Maine. Bull Bureau Fisheries 40: 1–509

Breiman L, Friedman JH, Olshen RA, Stone CJ (1998) Classification and regression trees. Wadsworth, Belmont, CA

Cook R, Auster P (2006) A bioregional classification for the continental shelf of the northeastern North America for conservation analysis and planning based on represaentation. Marine Sanctuaries Conservation Series NMSP-06–09. National Oceanic and Atmospheric Administration, National Marine Sanctuary Program, Silver Spring

ESRI (Environmental Systems Research Institute) (2007) An Overview of the Geodatabase: Types of Geodatabases. ArcGIS 9.2 Desktop Help Online (http://webhelp.esri.com)

Friedman JH (1990) Multivariate adaptive regression splines. Department of Statistics Technical Report 102, August 1990, Stanford University, California

Friedman JH, Roosen CB (1995) An introduction to multivariate adaptive regression splines. Statl Meth Med Res 4:197–217

Greene CH, Pershing AJ, Kenney RD, Jossi JW (2003) Impact of climate variability on the recovery of endangered North Atlantic right whales. Oceanography 16:96-101

Guisan A, Thuiller W (2005) Predicting species distributions: offering more than simple habitat models. Ecol Lett 8:993–1009

Guisan A, Graham C, Elith J, Huettman F, the NCEAS species distribution modeling Group (2007) Sensitivity of predictive species distribution models to change in grain size: insights from a multi-models experiment across five continents. Diversity Distribution 13:332–340

Hain JHW, Waring GT (1994) Marine mammals in New England shelf waters. In: Langton RW, Pearce JB, Gibson JA (eds.) Selected living resources, habitat conditions and human perturbations of the Gulf of Maine: environmental and ecological considerations for fishery management. NOAA Technical Memorandum NMFS-NE-106. Northeast Fisheries Science Center, Woods Hole, MA

Hamazaki T (2002) Spatiotemporal prediction models of cetacean habitats in the mid-western North Atlantic Ocean (from Cape Hatteras, North Carolina, USA to Nova Scotia, Canada). Mar Mammal Sci 18:920–939

Holling CS (1978) Adaptive environmental assessment and management. Wiley, London

Hyrenbach DK, Forney KA, Dayton PK (2000) Marine protected areas and ocean basin management. Aquat Cons: Mar Freshw Ecosyst 10:437–458

Kenney RD (2001) The North Atlantic right whale consortium databases. Maritimes Fall 01:3–5

Kenney RD, Winn HE (1986) Cetacean high-use habitats of the northeast United States continental shelf. Fish Bulletin 84:345–357

Kenney RD, Mayo CA, Winn HE (2001) Migration and foraging strategies at varying spatial scales in western North Atlantic right whales. J Cet Res Manag Special Issue 2:251–260

Leathwick JR, Elith J, Hastie T (2006a) Comparative performance of generalized additive models and multivariate adaptive regression splines for statistical modelling of species distributions. Ecol Model 199:188–196

Leathwick JR, Elith J, Francis MP, Hastie T, Taylor P (2006b) Variation in demersal fish species richness in the oceans surrounding New Zealand: an analysis using boosted regression trees. Ecol Model 199:288–242

Manly BFJ, McDonald LL, Thomas DL, McDonald RL, Erickson WP (2002) Resource selection by animals: statistical design and analysis for field studies, 2nd edn. Kluwer, Dordrecht

Manomet Bird Observatory (1980) Marine Observer Program cruise reports. Department of Energy Contract No.DEAC0278EV04706A002 report

Meise CJ, OReilly JE (1996) Spatial and seasonal patterns in abundance and age-composition of *Calanus finmarchicus* in the Gulf of Maine and on Georges Bank: 1977–1987. Deep-Sea Research 43:1473–1501

Moisen GG, Frescino TS (2002) Comparing five modelling techniques for predicting forest characteristics. Ecol Model 157:209–225
NOAA National Centers for Coastal Ocean Science (NCCOS) (2006) An ecological characterization of the Stellwagen Bank National Marine Sanctuary Region: Oceanographic, Biogeographic and Contaminants Assessment. Prepared by NCCOS's Biogeography Team in cooperation with the National Marine Sanctuary Program. Silver Spring, MD. NOAA Technical Memorandum NOS NCCOS 45
Pittman SJ, McAlpine CA (2003) Movement of marine fish and decapod crustaceans: process, theory and application. Advan Mar Biol 44:205–294
Pittman SJ, Christensen JD, Caldow C, Menza C, Monaco ME (2007a) Predictive mapping of fish species richness across shallow-water seascapes in the Caribbean. Ecol Model 204:9–21.
Pittman SJ, Caldow C, Hile SD, Monaco ME (2007b) Using seascape types to explain the spatial patterns of fish in the mangroves of SW Puerto Rico. Mar Ecol Prog Series 348:273–284
Redfern JV, Ferguson MC, Becker EA, Hyrenbach KD, Good C, Barlow J, Kaschner K, Baumgartner MF, Forney KA, Balance LT, Fauchald P, Halpin P, Hamazaki T, Pershing AJ, Qian SS, Read A, Reilly SB, Torres L, Werner F (2006) Techniques for cetacean-habitat modeling: a review. Mar Ecol Prog Series 310:271–295
Reeves RR, Kenney RD (2003) Baleen whales: Eubalaena spp., and allies. In: Feldhamer GA, Thompson BC, Chapman JA (eds) Wild mammals of North America: biology, management, and economics, 2nd edn. Johns Hopkins University Press, Baltimore, MD
Schneider DC (2001) The rise of the concept of scale in ecology. Bioscience 51:545–554
Sherman K, Sissenwine M, Christensen V, Duda A, Hempel G, Ibe C, Levin S, Lluch-Belda D, Matishov G, McGlade J, O'Toole M, Seitzinger S, Serra R, Skjoldal HR, Tang Q, Thulin J, Vandeweerd V, Zwanenburg K (2005) A global movement toward an ecosystem approach to management of marine resources. Mar Ecol Prog Series 300:275–279
Shoop CR, Kenney RD (1992) Seasonal distributions and abundances of sea turtles off the northeastern United States. Herpetological Monogr 6:43–67
Walters CJ (1986) Adaptive management of renewable resources. Macmillan, New York
Ward-Geiger LI, Silber GK, Baumstark RD, Pulfer DL (2005) Characterization of Ship Traffic in Right Whale Critical Habitat. Coastal Management 33:263–278
Winn HE, Goodyear JD, Kenney RD, Petricig RO (1995) Dive patterns of tagged right whales in the Great South Channel. Cont Shelf Research 15:593–611
Wishner K, Durbin E, Macaulay M, Winn H, Kenney R (1988) Copepod patches and right whales in the Great South Channel off New England. Bull Mar Sci 43:825–844
Woodley TH, Gaskin DE (1996) Environmental characteristics of North Atlantic right and fin whale habitat in the lower Bay of Fundy, Canada. Can J Zool 74:75–84
Wright DJ, Blongewicz MJ, Halpin PN, Breman J (2007) Arc Marine: GIS for a blue planet. ESRI, Redlands, CA

Chapter 22
Multi-spectral Satellite-Airborne Management of Ice Form Marine Mammals and Their Habitat in the Presence of Climate Change Using a "Hot Spots" Approach

Vladimir V. Melentyev and Vladimir I. Chernook

22.1 Satellite-Airborne Survey as a Tool for Monitoring Sea Life Ecology: Populations, Methodological Aspects, Instruments and Data Base

Until the beginning of a world-wide harvest in the 1600s, there was a high abundance of marine mammals in the western Arctic. Especially strong decreases of whales, walruses and phocid species in the Russian Arctic took place during last quarter of the nineteenth and the beginning of the twentieth century (Nansen 1924). Large-scale precautionary measures in the 1960s helped to maintain some diminished populations and preserve them from further declines. Unfortunately, today the population number of virtually all representatives of ice-related mammals is lower than what it was at the first decade of the twentieth century. New population decreases and a worsening of the health status of ice-related marine mammals is reported from many field-based, ship-borne investigations; this situation is directly associated with anthropogenic pressures (Hansen et al. 1996; Ridgway and Harrison 1981, 1989; Zannutdin et al. 2006).

In recent years, climate change became the most relevant factor in this situation because of the shorter duration of the freeze-up period and due to the dramatic reduction coverage of sea ice in marginal Arctic seas and sub-Arctic inland water bodies (Hydromet 1991; Kondratyev and Donchenko 1999; Kondratyev et al. 1996).

V.V. Melentyev (✉)
International Foundation "Nansen International Environmental and Remote Sensing Center" (NIERSC), 14 Linia Vasil'evsky Ostrov, 7A, 199034 St. Petersburg, Russia;
State University of Aerospace Instrumentation (GUAP), B. Morskaya Str., 67, 190000 St. Petersburg, Russia
e-mail: vladimir.melentyev@niersc.spb.ru

V.I. Chernook
Research Institute of Marine Fishery Fleet Designing (GYPRO Rybflot), M. Morskaya Str., 18–20, 190000 St. Petersburg, Russia

S.A. Cushman and F. Huettmann (eds.), *Spatial Complexity, Informatics, and Wildlife Conservation*
DOI 10.1007/978-4-431-87771-4_22, © Springer 2010

Climate warming overburdens the already ongoing negative effects - thinning and moistening of ice that provoke mass mortality of newborn generations during reproduction time (Melentyev et al. 2004; Melentyev and Chernook 2008). According to reasonable assumptions, and if the softening of winter seasons in the Russian Arctic will continue, the disappearance of many aquatic species can come into place and will resemble the situation that occurred in the Holocene with Greenland seals in Baltic Sea (Melentyev and Chernook 2008).

From this situation stems the demand for designing a National decision-making system for ecomonitoring and for a satellite-airborne-based spatial management of sea life ecology. In Russia, and since the Soviet epoch, stock-taking of population numbers and regional control for animal welfare of various aquatic species is the mandate of different branches of industries, academic institutions, applied marine research centers and universities alike.

The White Sea (Fig. 22.1) for example represents the habitat of walruses, white whales ('beluhas') and harp seals as well the ringed seals and a unique population of grey seals (Hydromet 1991). This vast aquatic area totaling 95,000 km^2 is the mandated zone for several institutes and marine centers related to the Ministry of Fisheries Russian Federation. They are responsible for regular multi-disciplinary investigations through the provision of round-the-year monitoring of this water basin as habitat of the mentioned species by using ship-borne observations, in situ measurements, as well as aerial photography.

In the Russian Federation airborne hydrobiological investigations became an indispensable part of the fishing industry and spatial management of sea life ecology from 1920s onwards (Hydromet 1991; Melentyev et al. 1998). Subsequently, after the 1960s different 'flying laboratories' that belonged to research centers and equipped for specific applications got designed, such as the internationally known institutes Voeikov Main Geophysical Observatory (MGO), Polar Research Institute of Marine Oceanography and Fisheries (PINRO), Research Institute of Fishery Fleet Designing (GIPRO Rybflot) (Kondratyev et al. 1975, 1992; Kondratyev and Melentyev 1996; Levashov 2003; Melentyev et al. 2004). They were supported through research aircrafts such as Ilushin-18 MGO, Ilushin-18 "Pomor", Ilushin-18 DORR (long-range explorer for fishery surveys), Antonov-26 "Arktika" and L-410 "Nord" - Fig. 22.2.

All of these aircrafts are fully equipped with digital photo cameras, thermal IR scanners, lidars, spectrometers, microwave radiometers and SAR/SLR radar system, including various accessory devices that allow for attaching meteorological and flight-technical parameters (Melentyev et al. 2004; Melentyev and Chernook 2008).

A computer system provides for the immediate processing onboard the aircraft of spatial hydrobiological, ecological, hydrometeorological, as well as ice information. As an outcome, "the primary product" representing the results of automated multi-spectral airborne data can immediately be delivered via radio to any user in the operational regime and in the form of composite thematic charts accompanied with annotated legends. The "end-product", and based on the interpretation and comprehensive analysis of multi-level data, could then already be obtained locally with just a short delay right at the airport.

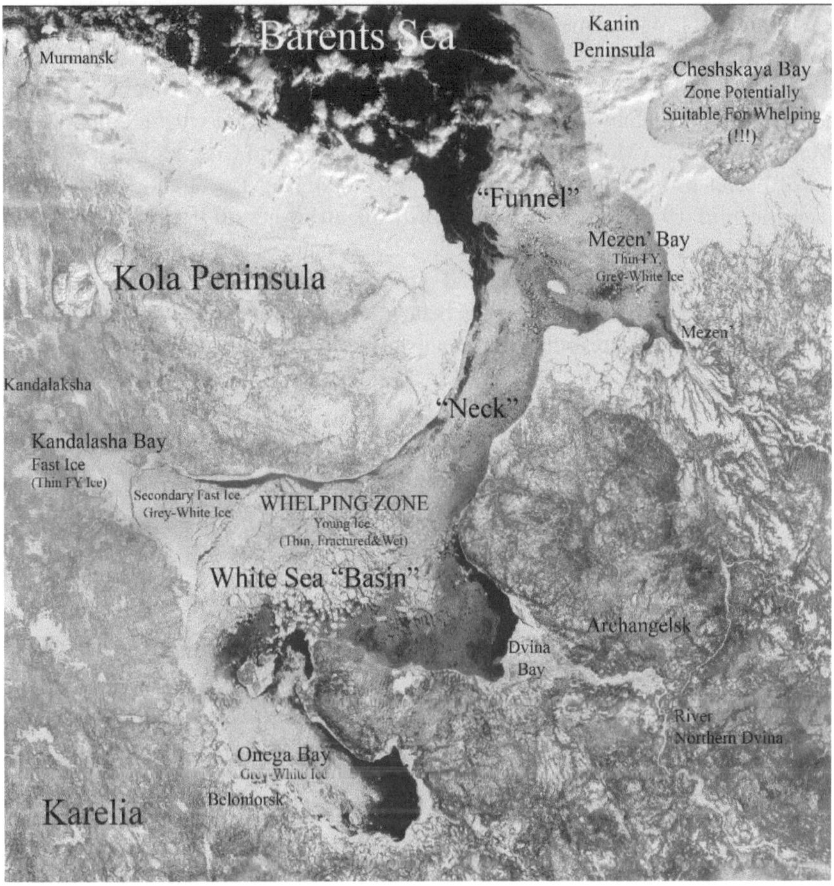

Fig. 22.1 Ice coverage of the White Sea and contiguous waters. Satellite "Terra", visual range, 18 March 2008, 08.09 GMT. Very mild 2007/08 winter season with domination new and young types of ice poor suitable or unfit for whelping of harp seals

Fig. 22.2 Research aircraft L-410 "Nord"

The aircraft L-410 of the Research Institute GIPRO Rybflot is used for our study since spring 2005. It was already applied in different parts of the Russian Arctic (the study zone of our activity extends from the White Sea and Barents region till the Bering and Chukchy seas) for multi-spectral monitoring of what is called system "Mammal-Media" (Melentyev et al. 2007; Melentyev and Chernook 2008).

We began to apply the satellite data for monitoring the aquatic environment since the mid-1970s; currently we fulfill all airborne studies in combination with SAR/ASAR satellite surveys - Fig. 22.3. A main objective of the present-day investigation is to monitor the Arctic and sub-Arctic icy waters, which are the ice habitat for marine mammals. Studying the behavior ecology and calculating the number of marine mammals are other important tasks that help in the understanding of

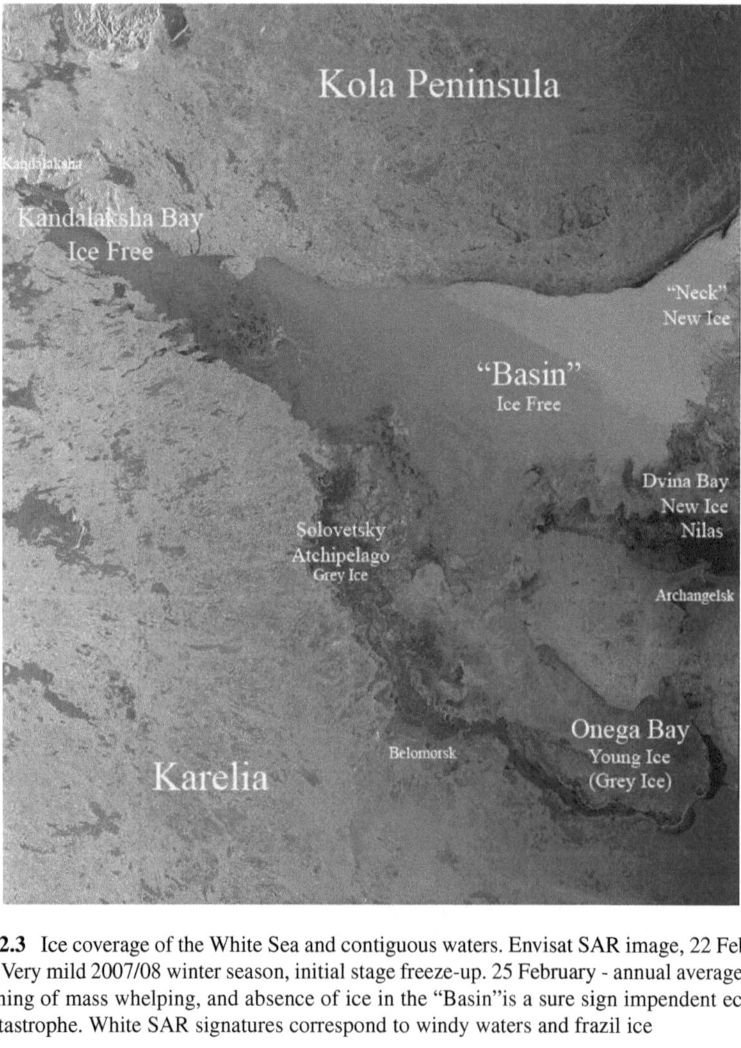

Fig. 22.3 Ice coverage of the White Sea and contiguous waters. Envisat SAR image, 22 February 2008. Very mild 2007/08 winter season, initial stage freeze-up. 25 February - annual averaged date beginning of mass whelping, and absence of ice in the "Basin"is a sure sign impendent ecological catastrophe. White SAR signatures correspond to windy waters and frazil ice

trends of the complex ecodynamics in Polar Regions. Originally designed as oceanographic studies, we have high hopes that behavior ecology of marine mammals can also be used to improve our knowledge of the ice regime and winter hydrology of marginal Arctic seas and inland water bodies - Fig. 22.4.

The explicit application of satellite data for the ecomonitoring of sea mammals has been attractive for marine biologists since the very beginning of the satellite era. But the use of satellite data for monitoring migration features of sea mammals so far was constrained by the coarse spatial resolution. Additional limits came with daytime illumination restriction and with the dominating clouds in Arctic ecosystems. This situation changed after launching satellites equipped with radar system ("Almaz" and "Okean" in USSR, ERS/Envisat/ Radarsat - EU and Canada). These side-looking radar (SLR) and synthetic aperture radars (SAR) provide us with

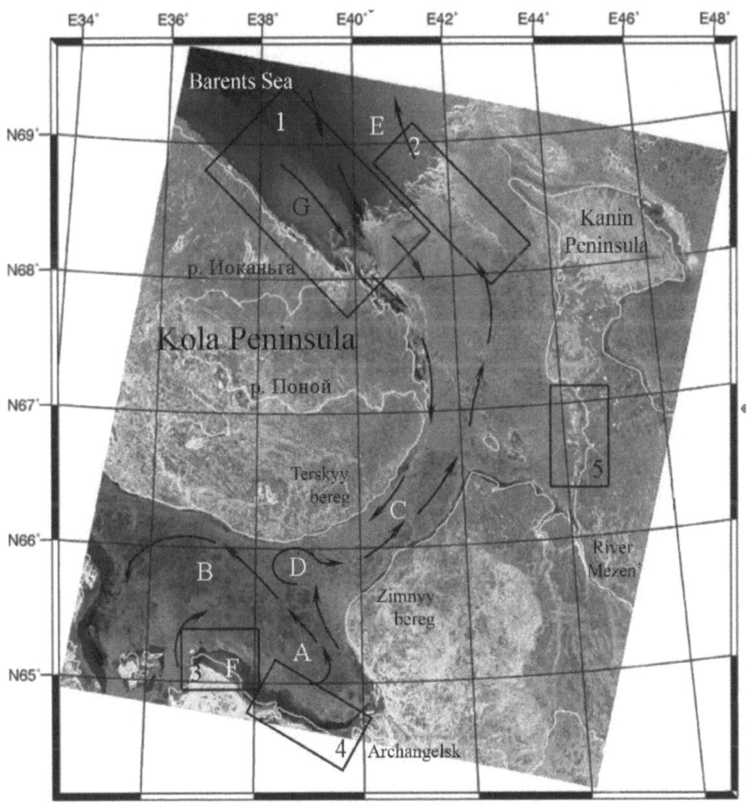

Fig. 22.4 Ice coverage of the White Sea and contiguous waters. Radarsat SAR image. 27 February 1998, 03.54 GMT. 1997/98 severe winter season with domination ice types suitable for whelping. Deep dark signatures correspond to solid thin fragmented FY ice; grey-white signatures - to young and new ice types (nilas, slush, shuga). A - zone unfit for whelping. B, C, D - zones suitable for whelping, F - shore polynya, G - frazil ice, E - windy waters, 1 - incoming stable current, 2 - stable flowing current, 3 - diffused ice edge, 4 - compacted ice edge, 5 - ice ridging zone

all-weather soundings of icy waters using wavelengths of 9.6 cm and 5.0 cm, and correspondingly with a resolution of 10–12 and 25–30 m. An additional advantage of SAR/SLR surveys is the possibility for obtaining sophisticated sub-surface ice information.

It was during February and March 1998 when the first experiment was carried out that focused on the study of ice as an abiotic factor for the ecology of Greenland/ harp seals (*Phoca groenlandica*) in the White Sea with using satellite SAR in frame sub-satellite experiments carried out onboard the Russian nuclear icebreaker "Taymir" (Melentyev et al. 1998). These allow for linking the variability of SAR signatures to different types of ice, and then can establish a connection between the location of whelping rookeries of seals with specific types of brackish-water ice, with the severity of winter, as well as with the patterns of ice and water exchange between the White Sea and its surrounding basins.

The feasibility for applying satellite SAR technology for mammological applications represented a new approach in marine biology: It allowed for the integration of spatial information in sea life ecology, and subsequently it also reveals so far unknown features of ice. Another significant result of these experimental studies was the clarification of reasons why Greenland seals fulfill such distinct mass migrations from the central part of the Arctic to the White Sea and other similar semi-closed water basins, and why they specifically select this marginal basin for whelping and molting (Melentyev et al. 1998).

The sub-surface investigation of ice with SAR (Fig. 22.4) helps us to disclose that the ice massif that seals use for whelping (panels B, C, D) has a contrasting radar signature, when compared to neighboring ice zones (panels A, F, G). Therefore, radio-physical and electrical parameters differ that are detected and selected for by these seals. Further, differences exist in the mechanical ice patterns.

Ice zones that harp seals selected for whelping (Fig. 22.5a) and for molting (Fig. 22.5b) represent ice breccia that distinguishes oneself by a particular durability and solidity because it has a fresh-water origin from the Northern Dvina river.

These ice floes consolidate themselves as one ice massif situated in the Basin - (zone "A") and that way they can resist the break-up (Fig. 22.6). It is this vast zone that leads to the mass accumulation of seals for whelping associated with an anti-cyclonic spiraling eddy A (monopole type) - Fig. 22.7. Another vastly elongated ice area and conveniently to be used for whelping (zone "B") is associated with a meandering front, called the "Neck" of the White Sea.

Both these ice zones can help to safeguard whelps and help them to avoid the deadly contacts with water. Only this specific type of ice can protect newborn generations from possible moistening and supercooling during the 2–3 weeks period of lactation and changing the uterine fur.

The subsequent hydro-biological studies on the behavior ecology of marine mammals and their foraging details and reproduction migration are presented in the framework of many national and international projects. And it is in this framework, that the habitat of different aquatic species (harp seals and grey seals, ringed seals inhabitant of different regions in the Arctic and sub-Arctic, whales, walruses, sea birds and fishes) located in marine and an inland water bodies was investigated.

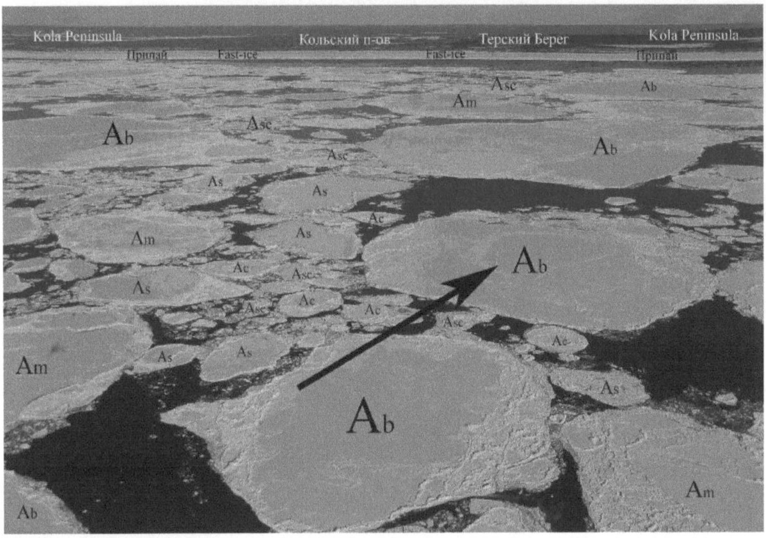

Fig. 22.5 Panoramic photo the whelping zone situated in the "Neck" of the White Sea, research aircraft L-410, 17.03.08 - 12.48.28 Msk. Thin first-year white ice, first stage development, H = 30–50 cm, concentration 4/10–6/10 tenths (open floating ice) and different forms (size of floes). Durable and solid ice floes originated from the water yield of river Northern Dvina have various range convenience for whelping (S - suitable, PS - poor suitable, U - unfit ice). Ab - big floes with diameter d = 0, 5 - 2 km (S), Am - medium floes, d – 100–500 m (S), As - small floes, d = 20–100 m (S), Ac - ice cake, d < 20 m (PS), Asc - small ice cake, d < 2 m (U). Black arrow - general direction of Zimneberegny stable current

22.2 Satellite–Airborne Integrated Technology for the Spatial Management of Aquatic Species and for Detecting Biological Active Zones (BAZ): The "hot spots" Approach

We perform our studies on ice forms and with marine mammals in different parts of the western Arctic, North Atlantic and North Pacific. Marine ecosystems of the White, Barents and Kara seas, of the Norwegian and Greenland seas as well as the Bering Sea and Sea of Okhotsk are investigated by using research aircrafts "Arktika" and L-410 "Nord" equipped with modern sensors. Further, specific features related to the reproduction of seals inhabiting peculiar inland (fresh) water bodies such as the Baltic and Caspian Seas as well the Lake Ladoga were also studied. SAR/ASAR/ALSAR surveys were used to monitor the polytypic icy waters and for the management of different phocid species: harp and hooded seals, which are pelagic, and ringed and bearded seals, which are coastal and strictly Arctic, and finally the grey and harbor seals, which are coastal and sub-Arctic (Melentyev et al. 2004; Melentyev and Chernook 2008).

As a result, an integrated technology for the spatial management of the aquatic environment and monitor biological active zones (BAZ), the so-called biological

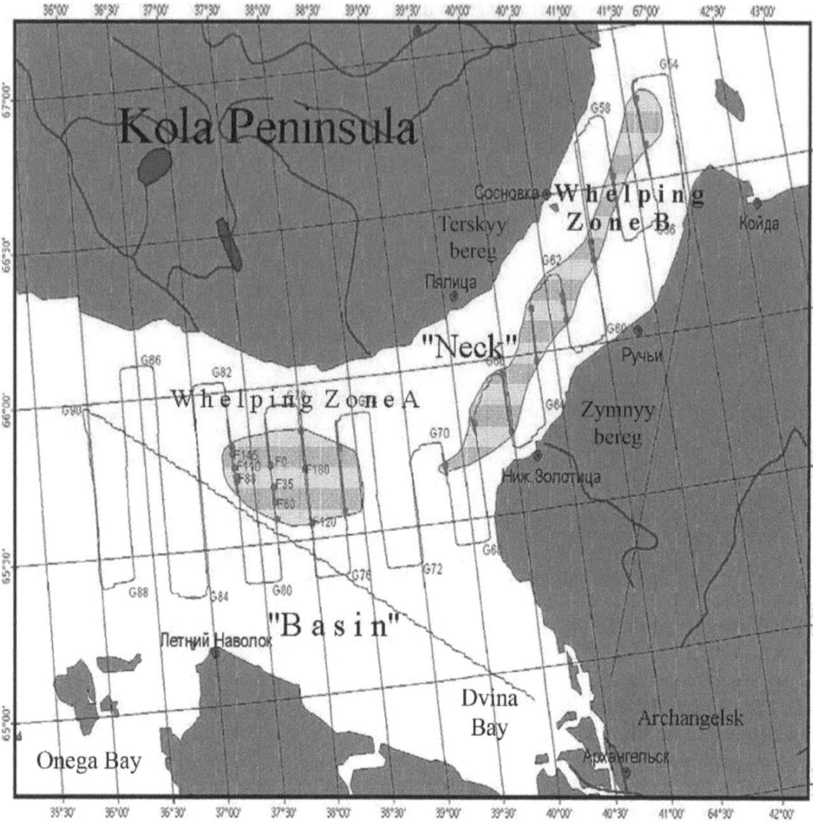

Fig. 22.6 Results airborne charting of whelping zones of Greenland seals in the White Sea. Research aircraft Antonov-26 "Arktika", H = 200 m, 13 March 2000. Zone of mass accumulation of seals in the "Basin" confined to stationary spiraling eddy (vortex); zone of whelping inside the "Neck" connects with meandering front of Zimneberegny stable current

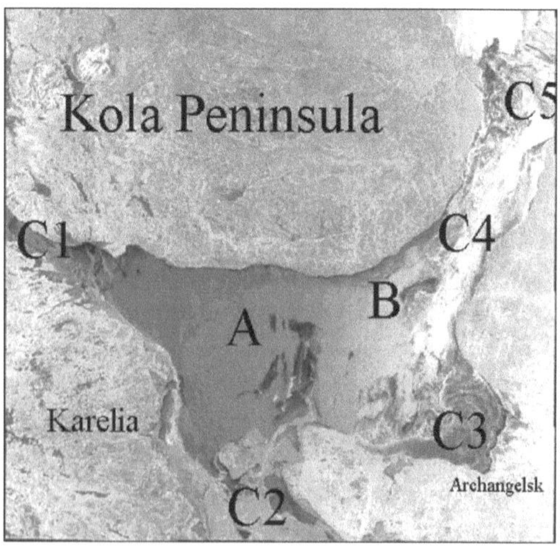

Fig. 22.7 Ice coverage of the White Sea. Envisat SAR image, 22 January 2004. A - anti-cyclonic spiraling eddy (monopole) in the "Basin", initial stage of ice formation, B - drift ice zone suitable for whelping (S), C1, C2, C3 - fast ice zones unfit for whelping, C4, C5 - ice zones inside the "Neck" and "Funnel" of little use for whelping (PS)

"hot spot" approach, was suggested by Melentyev et al. (2004). These biological hot spot zones represent specific water areas where biological activity is concentrated in the aquatic environment (Melentyev and Sjoberg 2004; Melentyev et al. 2004; Melentyev and Chernook 2008; Palacios et al. 2006; Zannutdin et al. 2006), and thus it receives specific study attention.

The specific objectives of our studies are to provide multi-level satellite - airborne - in situ experiments, thematic interpretation of multi-spectral data and the detection of biological "hot spots". They reveal and classify specific features of BAZ with reference to different aquatic species: white whales ('beluhas'), harp seals, Pacific walruses, sea birds and others. Further, we do multi-disciplinary investigations of BAZ as an indicator of the sustainable development of aquatic environment and the welfare status of marine mammals. We also work out detailed suggestions for designing decision-making procedures, based on systematic satellite-airborne management of the aquatic environment and the creation of the National Service of Sea Life Protection.

The origin of BAZ goes back to the finding that Arctic seas and large lakes can look spacious and vast, but biological activity and mass aggregations of marine mammals, sea birds and fishes are mainly concentrated in separate isolated marine areas - BAZs. The presence of marine mammals and sea birds at these places is a sure sign of environmental stability and safety.

According to our definition, BAZs could be determined as specific zones where biological activity in the aquatic environment is concentrated temporally or lasting longer. Two specific types of BAZs can be differentiated in spatial terms; and sometimes these got captured in our satellite-airborne studies: foraging and reproduction BAZs.

The causes of biological "hot spots" in Arctic seas and fresh-water basins are connected with different oceanological and hydrological processes. Those are frontal zones and meandering structures, convergence zones, mushroom structures, zones splitting and joined stable currents (warm and cold streams) as well zones of intensive vertical water exchange, i.e., eddies, upstreams, upwellings, etc.

Meteorological processes and atmospheric phenomena, and first of all, surface winds, can provoke formations specific to the BAZ, or they contribute to the intensification and strengthening of biological activity within existing BAZ.

It was discovered (Melentyev et al. 2004, 2007; Melentyev and Chernook 2008; Palacios et al. 2006) that the appearance of a BAZ is tightly connected with a spatial heterogeneous bathymetry, i.e., with the spatial variability of seafloor patterns. Orographically induced atmospheric processes and the heterogeneity of the shoreline can also contribute to the formation of BAZs in coastal zones and shallow waters.

Manifestations of BAZ, and uncertainties in their parameters. Manifestations of biological "hot spots" could be manifold; they tend to consist of a distinct anomaly of surface temperature or salinity with surrounding waters. Usually, the thermal contrast of a BAZ is a valuable identification feature, and it can be captured well through the use of airborne-satellite sensors. A BAZ could be described as a clearly distinguishable hydrological differential gradient from the surrounding waters. The ice regime and ice features inside the BAZ have sizeable differences with contiguous waters and different aquatic species including ecologically significant pelagic fishes, sea birds and ice forms of marine animals (whales, seals and walruses) that select the BAZ as temporal or continuous habitat. Mass aggregation of aquatic species in separate isolated water areas and when compared to a background is a good indicator for the presence of a BAZ.

Relative stability of the major underlying factors that create the BAZ is the reason why geographical position as well as hydro-biological patterns of "hot spots" are relatively stable and could be well described. Only these recurring certainties are utilized in our studies for designing integrated technology for the spatial management of the aquatic environment. In our studies we elaborated on climatic and seasonal inconsistencies of BAZ parameters - Fig. 22.8.

Foraging BAZs. The overwhelming majority of BAZs represent foraging "hot spot" zones - specific areas that present high primary production. The trophic "web" - the availability of high supply of food is a reason why different predators (e.g. fishes, birds and marine mammals) use these water areas for mass aggregation in order to forage.

The multi-disciplinary study on 'foraging BAZs' in Russian Arctic has a long history: airborne investigations started as early as the 1920s, and continued with satellites since the 1960s (Hydromet 1991; Filatov and Shilov 1996; Glazov et al. 2006). Figure 22.9 shows an example of a foraging BAZ that was charted in the central part of the Norwegian Sea. Interrelation of the locations of fishes shoals, zones of mass aggregation of dolphins, killer whales, medusas and sea birds (with densities of ten birds per square km) with a spatial variability of SST and a concentration of chlorophyll *a* is evident. Many years of flights of the aircraft "Arktika" with the aim of

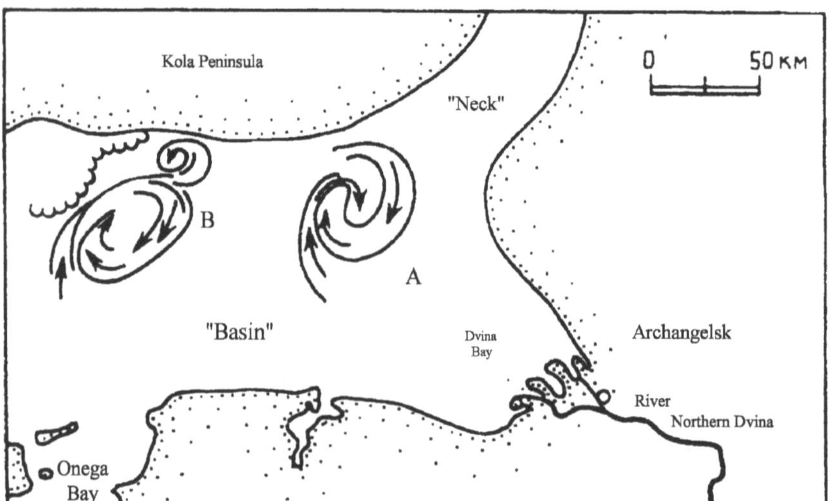

Fig. 22.8 Results satellite charting of the White Sea. A - stationary spiraling anti-cyclonic eddy (monopole) that will determine location of zone of mass accumulation of seals for whelping in the "Basin", B - modification of the stationary eddy, its spatial displacement and transformation in dipole resulted climate change

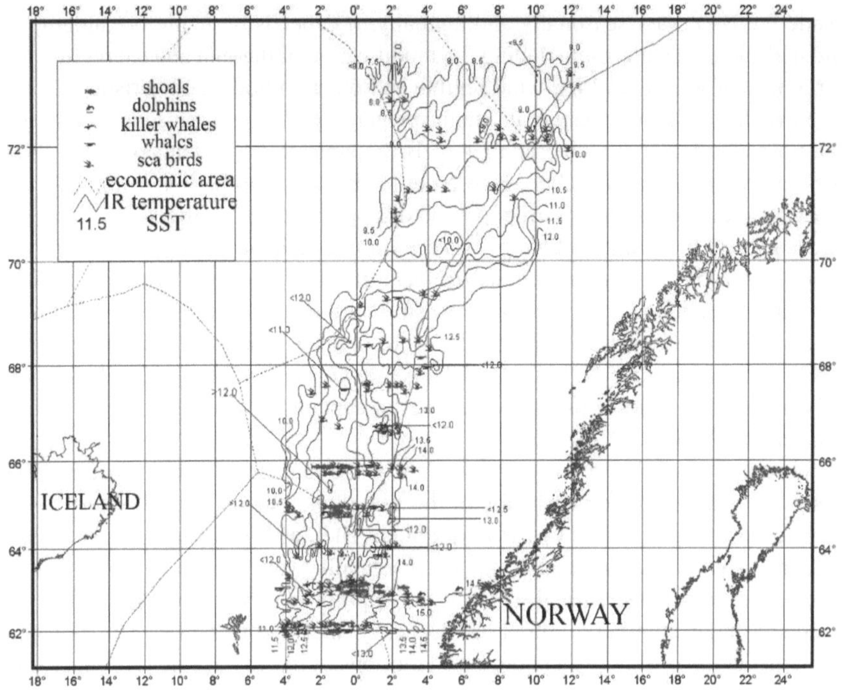

Fig. 22.9 Spatial arrangement of fishes, dolphins, killer whales and sea birds - indicator species of biologically active zone (BAZ). Research aircraft "Arktika", Norwegian Sea, 23–29. 07. 2003

"fish reconnaissance" at this highly bio-productive region helped us to assess these regularities in the distribution of marine mammals and sea birds - which represent known indicator species of sustained stability of the Norwegian Sea ecosystem. Airborne data for example allowed us to capture the delayed appearance of predators in the studied area from the third decade onwards (starting July 2003).

Reproduction type of BAZs were investigated applying multi-spectral airborne sensors and satellite tagging (Melentyev and Sjoberg 2004; Melentyev et al. 2007; Melentyev and Chernook 2008). Different aquatic biological communities use these areas for discharge and for essential functions of the life cycle that is not directly concerned with the trophic "web" (at this time, some species interrupt feeding; they 'fast'). They use these BAZs for securing different stages of reproduction: finding a mate and proper mating, finding a birth-place and for producing young, linked actions for protection of the new-born generation, and for avoiding predators. Fishes, sea birds and marine mammals undertake a long-range migration for searching zones suitable and secure for reproduction. In some cases, reproduction was found to be partly overlapping in space and time with feeding.

Both types of hot spots were found to be linked with explicit hydro-physical processes. This is the reason why BAZs are environmentally vulnerable: the time of their appearance as well as the spatial displacement and intensity processes inside the BAZ vary seasonally and annually. It depends on many oceanographic and hydrological patterns, including the modification of thermal and ice regime, the variability of winter severity, wind regime and other natural characteristics.

Nowadays, climate change and anthropogenic pressures represent a continual determinative factor for uncertainties of BAZ parameters (Melentyev et al. 2004; Melentyev and Chernook 2008; Hansen et al. 1996). A further extension of these processes could cause the destruction and disappearance of entire BAZs.

Shortcomings of the "sighting" survey approach requires the design of "hot spot" technology for the spatial management of BAZs. Systematic documentations of visual detection locations became possible in the nineteenth century when first field studies of the marine environment were organized onboard ships (Hydromet 1991). Up to that time the basic knowledge about seals, whales and walruses was only obtained from logbooks and narratives of captains and hunters of whale-boats and sealers (Nansen 1924).

Surveys were done as ship-borne counts of the numerous aquatic species. Two observers were placed on each side of the vessel and one additional observer was located on the mast. Surveys used equipment such as binoculars and fish finder systems. This survey method could be practically well applied in open waters, whereas icy waters were usually closed for ship-borne survey studies.

Small-range ship-borne observations carry many constraints. The practical implementation of survey technology is laborious; it has to be done for longer time periods due to the necessity to cross the often large regions of interest. And there is an important problem how to count migrating and moving mammals. It's this challenging context that provides the reason why survey work often only allows calculating population numbers poorly.

Nevertheless, sighting data gets stored in our databases for planning airborne and satellite survey and for filling information gaps. It helps to further capture the exact scale of mammal's distribution and also helps to understand its temporal and spatial variability.

Figure 22.10 shows the habitat of harp seals at the end of the nineteenth century (Nansen 1924). Unfortunately, the map based on sightings only gives us a general overview of the inhabited area. It shows, for example, the existing presence of harp seals in the entire White Sea but without relevant indication about the annual variability. According to sighting data, whelping zones hundred years ago were situated in the "Funnel" and Mezen Bay (Nansen 1924). But now, mammals select for reproduction zones located in "Basin" and "Neck" (Ridgway and Harrison 1981, 1989, 1990; Melentyev et al. 2004; Melentyev and Chernook 2008).

Why does this happen? Why do mammals change their reproduction zones? Regular sighting data couldn't explain this fact. But a clarification of these questions could be achieved with using new satellite and multi-spectral airborne sensors.

Integrated satellite-airborne technology for the spatial management of BAZ. In spite of some shortcomings we use this survey approach in concert with airborne observations, and tried initially to overlap the whole study area of the White Sea and contiguous icy waters. Following the relevant administrative divisions, the following parts of the sea should be controlled: "Funnel" –23,290; "Neck" –8,113 km^2;

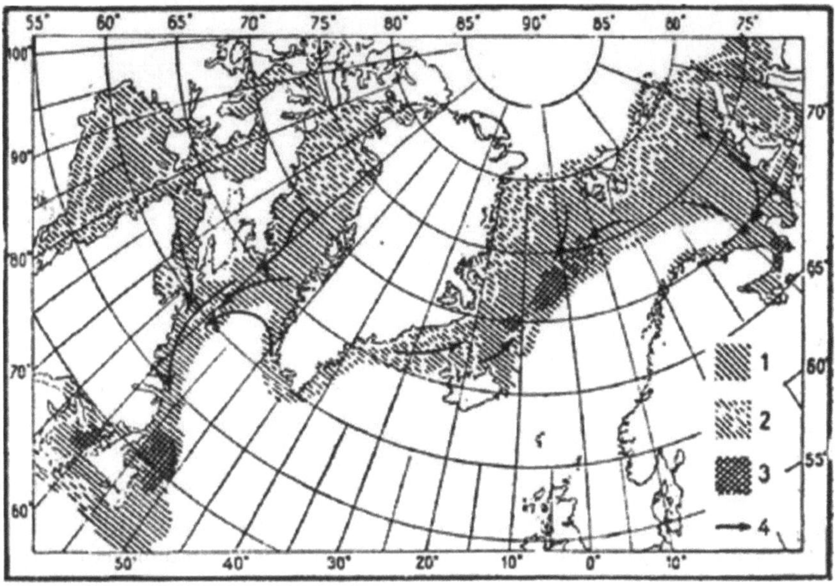

Fig. 22.10 Distribution harp seals in the Western Arctic and North Atlantic at the end of nineteenth century according to sighting point fixation (Nansen 1924). 1 - zones mass aggregation, 2 - rare presence zones, 3 - zones of reproduction, 4 - migration routes

Kandalakshskiy, Dvinskiy and Onegskiy bays –19,510, 12,426 and 13,146 km^2, correspondingly. The total protected area adds up to 90,185 km^2 (Hydromet 1991).

A crucial instrument for the airborne counting of ice-related mammals is the thermal IR scanner "Malakhit M". For the recognition of seals lying on the ice, as well as for white whales in the water, three photo cameras of the type Nikon D 70s got also applied (F = 50 and 200 mm). Usually, they are applied at small altitudes: H varies within 150–200 m till 500–1,000 m. IR-scanners ensure a resolution of 1, 3 m, and photo cameras of 0, 16 m and 0, 4 m - Fig. 22.11. In the case of H = 1000 m widths surveyed strips represent 100, 500 and 3,500 m, correspondingly.

Satellite SAR/ASAR surveys are used for the monitoring of biological "hot spots" as a tracer of ecodynamics of the studied region. Fig. 22.12a, b present an example of a satellite SAR survey with the aim for detecting and monitoring non-volatile solid ice (dark SAR signatures) and for revealing 1-day drifting features of ice (white arrows showing the direction and spread) in Dvinsky Bay during March 1998.

The airborne range is larger than what shipborne observations can cover. But the high cost of research flight does not allow monitoring big areas. For example, the coverage of the surveyed area during investigations of white whales in the White Sea in June–July 2,000 averaged to about 28.1%, in July 2002 - 9.2%, in July 2005 - 35.2%, and in July 2006 - 25.2% of the whole required zone. The total number of white whales including calves was estimated in 2,000 as 174 individuals, in 2002 it was 584 animals, and 2,245 animals for the year of 2005 (Glazov et al. 2006).

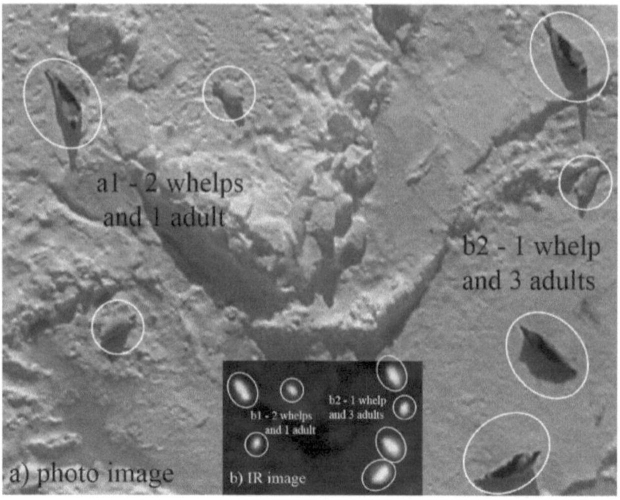

Fig. 22.11 Example superposed visual - infrared IR image that's applied for counts of harp seals. Research aircraft L-410, 28 April 2006, the "Neck", H = 400 m. Small bright IR targets corresponds to whelps, large IR targets - to adults; females together with whelps - zone a1 - b1, male group - zone a2 - b2

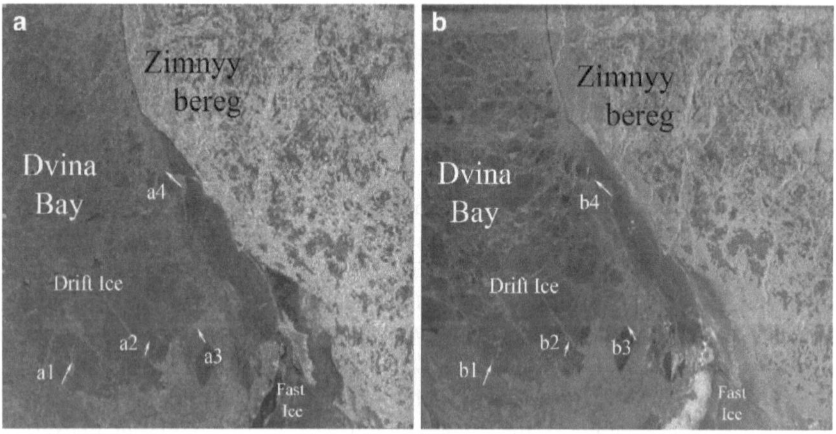

Fig. 22.12 a, b Results satellite SAR revealing ice drift parameters (white arrows). White Sea, Dvinsky Bay: (**a**) RADARSAT SAR image, 17 March 1998; (**b**) ERS SAR image, 18 March 1998

This is important as the number of detected whales is closely related to the location of reproduction "hot spots".

Figure 22.13 shows the scheme of full-length surveys that was achieved onboard L-410 for the study of the distribution of white whales in July 2006 when single whales and groups - females together with calves - were detected. Different diameters of circles represent a different number of mammals inside each group.

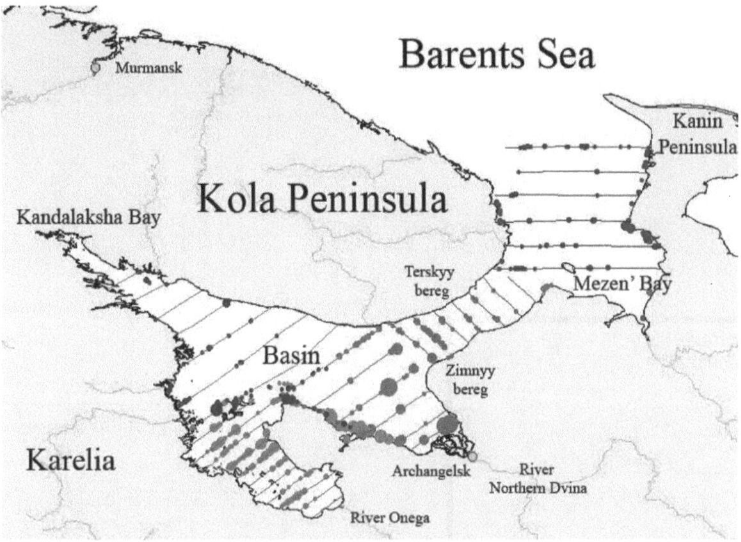

Fig. 22.13 Results airborne charting white whales in the White Sea using full-length routing scheme ("sighting approach"). Research aircraft L-410, July 2006

It's clear from Fig. 22.13 that the majority of beluha's aggregation was situated in coastal zones (Zimny, Konyushinsky Shores, and Solovetsky Archipelago). Another majority of this species is located in the central part of Dvinsky and Onegsky Bays. We think that the distribution of "hot spots" is determined by the water mass arrangement and influence of specific hydrological processes. In coastal zones it was for instance the tidal effect and upwelling. The frequent occurrence of beluhas in the central part of the Dvinsky Bay could be explained by the mixing zone of river waters coming from Northern Dvina with sea waters. Upwelling areas of highly productive 'mushroom' structures arise here in the frontal zone (Hydromet 1991).

In March–April of 2005 and 2006 two series of large-scale joint Russian–American satellite - airborne - ship-borne investigations of the Pacific population of walruses that inhabited the Bering Sea were organized in cooperation with scientists from Alaska, USA (Melentyev et al. 2007; Melentyev and Chernook 2008). Envisat/Radarsat SAR surveys allowed for revealing some regularities of the ice regime of the Anadyrsky Gulf that can influence the behavior ecology of ice forms of marine mammals. Satellite tagging of walruses was done from the icebreaker "Magadan".

Figure 22.14 shows results of the airborne investigation of the Pacific walruses (*Odobenus rosmarus*) using "hot spots" approach in April 2006. Taking into account our previous study of winter hydrology of the Bering Sea (Kondratyev et al. 1973, 1975, 1992, 1996) we did not carry out an airborne survey of the entire Russian part as we did in April 2005. Instead, we now concentrated our studies on charting the reproduction "hot spot" zones that are situated in the Gulfs of Anadyrsky and Mechigmensky.

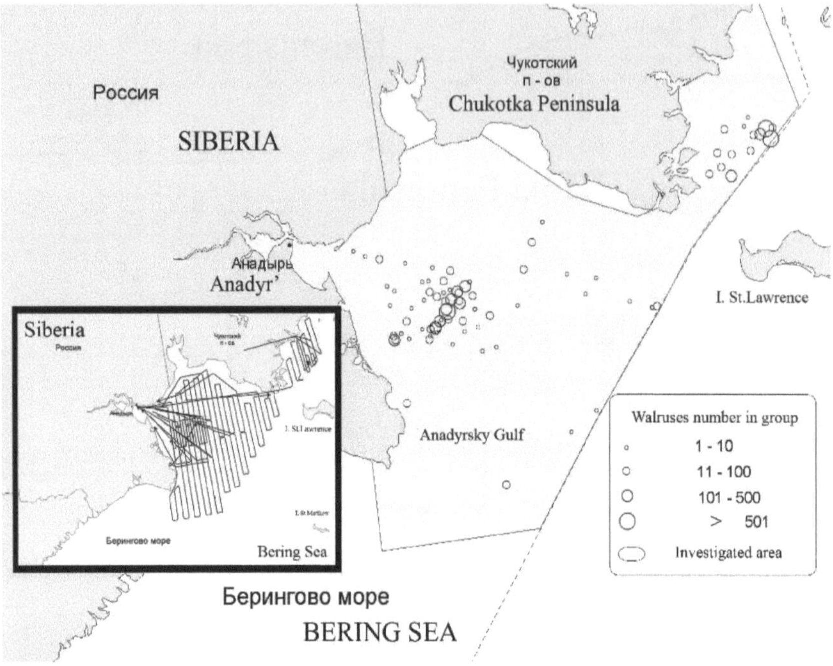

Fig. 22.14 Results airborne charting the Pacific walruses using "hot spots" approach. Russian economical zones in the Bering Sea. Research aircraft L-410, 4–24 April 2006

Haul-outs situated inside the ice massif were investigated with the use the IR scanner "Malakhit M" and satellite data.

Ship-borne observations help validating SAR satellite data and fixing prevalent types of ice inside of the reproduction "hot spot" zones. This work was performed onboard icebreaker "Magadan". We found mostly ice breccia - very close floating thin first-year ice with a concentration of 8/10–9/10 tenths and a thickness of 30–50 cm. The aggregations of walruses were situated close to narrow cracks and fractures (width <50 m), seemingly at recurring polynya covered with grey-white ice and 'nilas'. Non-homogeneous bathymetry was also assessed.

Airborne and shipborne studies show a distinct conservative behavioral feature of the Pacific walruses: Air temperature during airborne surveys was below minus 20–25°C when females and pups were located on the ice. However, all males (adults) were found in the water: this is to conserve haul-outs and to prevent a freeze-up of the polynya.

Recent L-410 flights were conducted in March of 2008 with goal to inventory the numbers of harp seals. Note that consequences of climate change were fixed as "poor ice year" patterns when satellite SAR survey shows ice free conditions in the White Sea till the third decade of February - Fig. 22.3. The detected situation can provoke the imminent ecological catastrophe because of the danger threatening seals population (Hydromet 1991).

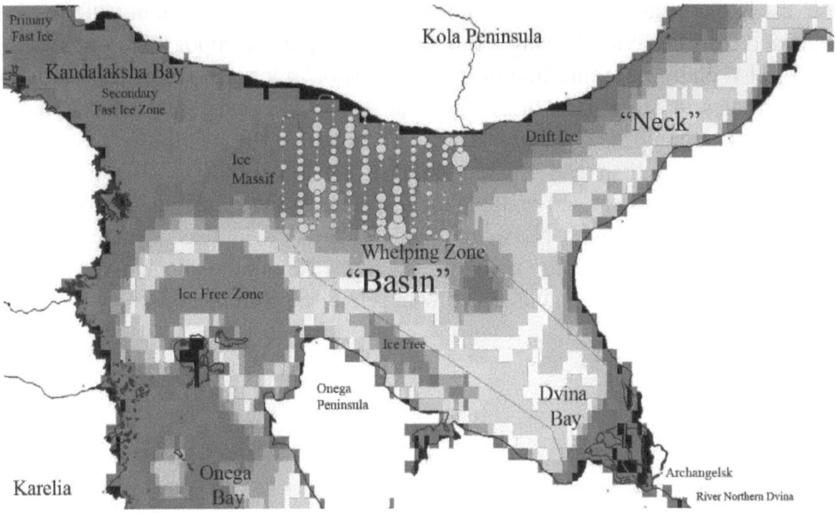

Fig. 22.15 Results airborne charting the Greenland seals in the White Sea using "hot spots" approach. Research aircraft L-410, 14 March 2008

We currently have limited financial means for providing only three flights (14, 15, and 17 March 2008), and the "hot spots" approach was used for counting of seals - Fig. 22.15. And airborne investigations confirmed the perception about the decrease of population numbers when zone of mass accumulation of seals was concentrated as narrow "ice belt" in the north-western part of the "Basin". And additional disaster was fixed, this whelping zone it was fragmented by cargo transportations from ships. According to the L-410 data, the number of whelps in half reduced by the middle of 1990s (Hydromet 1991; Melentyev and Chernook 2008).

In conclusion, one should take advantage of the economical and methodological efficiencies inherent in the presented "hot spot" survey method, and then apply a related spatial management that reduces the need for large-scale investigations of marine mammals. This helps to avoid studies of "biologically dead and absence zones" in the vast aquatic environments.

As a further and final argument we would like to remind the reader about the specific and extreme features of the weather conditions in the Polar region, and the restrictions of flying in the Arctic, whereas the "hot spot" approach looks especially effective and economically favorable.

We plan to provide further development for the described technologies regarding other aquatic species. Our more global idea is to design a decision-making system for a satellite-airborne-based management of the aquatic environment, and which is in support of the establishment of a National Service of Sea Life Protection.

As a first step of this idea, we currently develop a practical realization of satellite-airborne detection and monitoring of the "hot spot" zones in the Black and Azov seas, and study there the behavior ecology of dolphins. Another and related interest is connected with the continuation of comprehensive studies of Caspian seals and their behavioral changes due to climatic and anthropogenic pressures.

We also plan to use and fine-tune the "hot spot" approach further for studying white whales in other parts of the Russian Arctic and seals, grey whales in Sea of Okhotsk around the Sakhalin Island. We hope for the possibility to continue joint Russian–Swedish and Russian–Finnish satellite-airborne investigations of the aquatic environment and on the phocid species in the Baltic and Ladoga Lake. Finally, and together with Norwegian and Canadian scientists, we are ready to conduct a comparative inquiry on harp seals in the Newfoundland and West Ice, as well as on Pacific walruses and joined with scientists from the USA.

References

Filatov NN, Shilov IO (1996) The investigation peculiarities of changeability of the thermo hydrodynamics fields of the White Sea using the satellite information. Earth Obs. Rem Sens 3:89–98

Glazov DM, Chernook VI, Boltunov AN, et al. (2006) Aerial survey of white whales (*Delphinapterus leucas* Pall.) in the White Sea in 2000, 2002, 2005. "Marine,ammals of the holarctic". Proceedings of the 4th Int Conference, St. Petersburg, 10–14 October 2006. MMC Publ House 148–150

Hansen JR, Hansson R, Norris S (eds) (1996) The state of the European Arctic environment. The European Environment Agency Environmental Monograph No 3 Luxemburg

Hydromet (1991) Hydrometeorology and hydrochemistry of the seas of USSR. Vol 2 The White Sea, Hydromet Publ House 273 p (in Russian)

Kondratyev KYa, Donchenko VK (1999) Ecodynamics and geopolitics. Issue 1 Global problems, RCES RAS Publ House, 1039 p (in Russian).

Kondratyev KYa, Johannessen OM, Melentyev VV (1996) High latitude Ccimate and remote sensing. Wiley-Praxis Series in Remote Sensing, Chichester, UK 200 p

Kondratyev KYa, Melentyev VV, Rabinovich YuI (1973) The Soviet-American Experiment "Bering". Meteorology and Hydrology 11:3–10 (in Russian)

Kondratyev KYa, Melentyev VV, Nazarkin OM (1992) Remote Sensing of Water areas and Watersheds", Hydromet Publ House, 248 p (in Russian)

Kondratyev KYa, Melentyev VV (1996) Flying laboratory Il-18 of the Voeikov Main geophysical observatory (30 years on scientific program). Earth Obs Rem Sens 4:114–120

Kondratyev KYa., Nordberg W, Rabinovich YuI, Melentyev VV (1975) The USSR/USA Bering Sea Experiment "BESEX". Proceedings COSPAR Plenary Meeting, Varna, Bulgaria

Levashov DE (2003) Technique of the expeditionary investigations. VNIIRO Publ House 400 p (in Russian)

Melentyev VV, Chernook VI, Johannessen OM (1998) Analysis of ice dynamics of Arctic seas for research of the harp seals migration in the White Sea using satellite data. Earth Obs Rem Sens 5:76–93

Melentyev VV, Chernook VI, Tikhomirov AI, Vlasov VP, Chernook VI, Melentyev KV, Pettersson LH (2004) New ice and nilas types of lake ice: experience of SAR diagnostics and validation by using airborne SAR and SLR observations (co-authors)//Earth Observations and Remote Sensing 5:28–43

Melentyev VV, Chernook VI, Pettersson LH (2003) Integrated satellite-airborne technology for monitoring ice cover parameters and ice-associated seals in the Arctic and perspectives of sea mammal global management. In: Harmancioglu NG, Geerders P (eds) Integrated technologies for environmental monitoring and information production. NATO Science Series IY, Earth and Environmental Sciences 23:211–226 Kluwer, Dordrecht

Melentyev VV, Sjoberg M (2004) Post-breeding foraging migration of grey seals in the Baltic and it connection with the type of winter severity: results of studies by using satellite tagging and ice coverage cartography. Proceedings of the 3rd Int Conference "Marine Mammals of Holarctic", Koktebel, Crimea, 11–17 October 2004. MMC Publ House 372–375

Melentyev VV, Chernook VI, Sjoberg M, Sandven S (2004) ERS/RADARSAT/ENVISAT SAR satellite studies of the behavior ecology of ice form of seals affected by differences of winter severity, winter hydrology and climate change. Proceedings of Envisat Symposium, Salzburg, Austria, 10–15 September 2004

Melentyev VV, Chernook VI, Pettersson LH (2007) Integrated satellite-airborne technology for monitor of biological active zones ("hot spots") and diagnosis of aquatic environment sustained stability. Earth Obs Rem Sens 4:39–50

Melentyev VV, Chernook VI (2008) Integrated use of multispectral satellite and airborne sensors as a tool for detecting biological hot spots and monitor of aquatic environment security. In: Gonca Coscun H (ed) Integration of information for environmental security. NATO Science for Peace and Security Series, Environmental Security, Springer, Dordrecht

Nansen F (1924) Blant Sel og Bjorn Frilufts-Lif. Kristiania, Norge. 663 p. (In Norwegian).

Palacios DM, Bograd SJ, Foley DG, Schwing FB (2006) Oceanographic characteristics of biological hot spots in the North Pacific: a remote sensing perspective. Deep-Sea Research 53:250–269

Ridgway SH, Harrison RJ (1981) Handbook of marine mammals. The walrus, sea lions, fur seals and sea otter. Academic, 1:235

Ridgway SH, Harrison RJ (1989) Handbook of marine mammals. River dolphins and the larger toothed whales. Academic 4:442

Ridgway SH, Harrison RJ (1990) Handbook of marine mammals. Seals. Academic 2:359

Zannutdin M, Kiyofuji H, Saitoh K, Saitoh SI (2006) Using multi-sensor satellite remote sensing and catch data to detect ocean hot spots for albacore (*Thunnus alalunga*) in the northwestern North Pacific. Deep-Sea Research 53:419–431

Chapter 23
How Spatial Information Contributes to the Conservation and Management of Biodiversity

Dawn Robin Magness, John M. Morton, and Falk Huettmann

23.1 Introduction

Reliable ecological information is a necessary component of sustainable management practices (Walters 1986). Land managers need to understand the spatial distribution and population status of species and habitats in regional landscapes. The Millennium Assessment, a global assessment of human well-being, identified biodiversity as a crucial ecosystem service that increases the capacity of ecosystems to adapt to environmental change and maintain productivity (http://www.millenniumassessment.org/en/index.aspx). Biodiversity is widely defined as the variety of compositional, structural, and functional biological components available across multiple scales including landscapes, ecosystems, species, and genetics (Noss 2001). As biodiversity occurs at a multitude of scales, species conservation and sustainable management requires that planning also occur at these scales. Planning for biodiversity conservation is critical because regional landscapes are increasingly compromised by global anthropogenic influences (Vitousek et al. 1997). More than 75% of habitable, ice free land is already altered by human residence and land-use (Ellis and Ramankutty 2008; Usher et al. 2005; Vitousek et al. 1997).

Climate and other environmental change further increase the necessity for multi-scale, coordinated planning. Species, across multiple taxa, have already responded to climatic changes by shifting distributions northward and upward in elevation (Parmesan and Yohe 2003). However, individual species are responding

D.R. Magness (✉)
Kenai National Wildlife Refuge, U.S. Fish & Wildlife Service, Soldotna, AK 99669, USA;
EWHALE lab- Biology and Wildlife Department, Institute of Arctic Biology,
University of Alaska-Fairbanks, 419 IRVING I, Fairbanks, AK 99775-7000, USA
e-mail: Dawn_Magness@fws.gov

J.M. Morton
Kenai National Wildlife Refuge, U.S. Fish & Wildlife Service, Soldotna, AK 99669, USA

F. Huettmann
EWHALE lab- Biology and Wildlife Department, Institute of Arctic Biology,
University of Alaska-Fairbanks, 419 IRVING I, Fairbanks, AK 99775-7000, USA
e-mail: fffh@uaf.edu

S.A. Cushman and F. Huettmann (eds.), *Spatial Complexity, Informatics, and Wildlife Conservation*
DOI 10.1007/978-4-431-87771-4_23, © Springer 2010

independently and not as ecologically coherent units like ecosystems, resulting in novel species assemblages, new forced neighborhood relationships, new host–parasite relationships, and trophic "mismatch" (Parmesan 2006; Root and Schneider 2001). To adapt to climate change impacts as much as possible, managers need to maintain landscape connectivity, provide suitable policies, manage for diverse conditions, and integrate the possibility of surprise into management (Inkley et al. 2004).

Although climate change is a pressing management problem, most managers in the United States currently lack the baseline information necessary to identify climate change impacts or to project likely future conditions for planning (GAO 2007). Data-gaps in time and space that prohibit an understanding of baseline ecological conditions and the projection of future conditions are not unique to the United States though. For example, research groups in the Netherlands also identified the two most pressing areas of research for climate change programs to be: (1) how current and future climate influence the spatial and temporal distribution patterns and viability of species and ecosystems, and (2) the biological mechanisms underlying the responses of species and ecosystems to climate change (Kappelle et al. 1999).

The goals of developing indicators of current ecological conditions (spatially explicit baseline) and developing conceptual models of system structure that can be used to project future ecological response have been described as falling under two research frameworks with complementary, but different objectives (Bella et al. 1992). For documenting baseline conditions, spatially explicit models linking the distribution, abundance, or density of a species to environmental conditions can be used to produce an indicator (i.e., map, population estimate for landscape). The ecological indicator can then be used to detect future changes to baseline (Magness et al. 2008; Scott et al. 2002). Managers can use unexpected changes or trends in the baseline indicator as an early warning to direct attention to new and unforeseen problems that may arise with environmental change (Bella et al. 1992). Reliable indicator models must produce accurate predictions. In a spatial as well as aspatial context, data-mining approaches provide the most accurate indicator models, as measured by predictive accuracy, for the sampled areas and require little a priori knowledge of the ecological system (Breiman 2001; Cutler et al. 2007; Elith et al. 2006; Hochachka et al. 2007).

In contrast, and in order to project future condition, managers need vast ecological understanding to develop integrated, conceptual models of the processes affecting resources of interest. For projections, simulation modeling to understand system dynamics, hypothesis testing of model structure, and empirical data to validate model outputs are all needed. The two research frameworks are complementary because future projections cannot be useful without understanding the underlying ecological processes, but environmental changes (i.e., climate change) have the potential to restructure the underlying system unexpectedly as ecosystems adapt. Therefore, baseline indicators provide an alternative approach that is useful for detecting surprises and alerting managers of the need to evaluate assumptions about system response (Bella et al. 1992).

Although these two complementary research frameworks are conceptually different, spatially explicit data are still needed for both. Well-designed spatial monitoring programs can provide a cost-effective source of data that is robust, flexible and can be used for a wide variety of purposes. However, conservation planners and global climate change groups have prioritized the need for predictive modeling of species distributions (GAO 2007; Rodriguez et al. 2007). Species distribution models provide a powerful tool to fill conservation knowledge gaps regarding current ecological conditions. To deliver this type of conservation information effectively, a management framework is required to integrate monitoring programs with spatially explicit sampling frames, data management, and data delivery of spatial products like distribution maps. We begin by reviewing the history of general approaches used to extrapolate wildlife data across space. We then review two monitoring programs to provide examples of spatially explicit monitoring designs, and the Gap Analysis program to provide an example of information synthesis. Finally, we outline the components of a management framework for spatial data that includes approaches to synthesize data collection, data management and data delivery.

23.2 History of Spatial Information for Conservation and Management

Early on, resource managers recognized the utility of merging spatial data for planning (McHarg 1969), but overlaying maps manually was cumbersome. Spatial information became more accessible for management and conservation with the advent of Geographic Information Systems (GIS). GIS are computer-based systems used to store, manage, manipulate, and analyze geo-referenced data (Koeln et al. 1994). The use of GIS has expanded since the 1990s due to advancements in computer hardware technology, increased access to remotely sensed data, and the increased spatial documentation of wildlife sampling efforts due to the affordability of Global Positioning Systems (GPS) (Koeln et al. 1994; Lunetta 1998). Although museum collections include spatially referenced locations for species spanning several hundred years (Chapman and Busby 1994), the inclusion of spatial information in management and research has become commonplace with the increased availability of inexpensive and user-friendly GPS units (Graham et al. 2004). Access to observational data, along with increased availability of environmental layers, has lead to the development of modeling approaches, like species distribution models, that increase the biogeographic information available to managers.

GIS, GPS, and increased computing power have proven to be extremely useful for management and planning, but these technologies also increase the roles and responsibilities of resource managers. Spatial wildlife management requires that monitoring program design include spatial considerations and increases the importance of data management and archiving that is compatible across space and therefore institutions. Finally, managers need tools to visualize spatial data products that can be linked to planning and policy review.

23.2.1 Overview of Spatial Modeling Approaches for Evaluating Wildlife Habitat

With sparse or spatially-biased spatial data, expert knowledge and published literature were used to link wildlife to habitat. An early approach used by the U.S. Fish & Wildlife Service was the habitat suitability index (HSI) (Morrison et al. 1992). HSI analysis used expert opinion to choose environmental variables that were thought to be important to the species of interest. An index, ranging from 0 to 1, was developed to link the availability of the environmental variables to the overall habitat suitability. The HSI was mainly to be viewed as a hypothesis and not representing a causal relationship. However, they were touted as an easily documented assessment procedure that could be used to spatially compare how alternative management plans would affect carrying capacity (Morrison et al. 1992). HSI started the culture of modeling in wildlife management and challenged old-fashioned species distribution map making. Critiques of HSI centered around the lack of a link to population size. Habitat capability (HC) models were developed as an approach that was similar to HSI but used to describe the habitat conditions necessary to maintain specific population sizes (Berry 1986). Expert-based models, such as HSI and HC, may be less useful for understanding climate change because the habitat associations identified by experts may restructure in the future.

Next, Resource Selection Functions (RSF) provided one of the first empirical approaches to quantitatively link wildlife and habitat variables (Manly et al. 1993). RSFs build on decades of conceptual and real-world experience with habitat relationships in the wildlife and management community. RSFs can compare used (or alternatively unused habitat) versus available habitat, or used habitats versus unused habitats. RSFs represent habitat preference (as opposed to selection) because a researcher only knows that an animal is exploiting certain habitats in a manner that is disproportional to the availability of those habitats on the landscape. RSFs can also be used to estimate population size. If the population size of a small area is known, the population size of the entire landscape can be estimated by extrapolating the density of the known area to the entire study area as scaled by the RSF (Boyce and McDonald 1999). Alternatively, distribution models may be built using density estimates as the dependent variable, and total population size can then be estimated by summing the densities predicted across the entire study area (Yen et al. 2004). Researchers have also developed methods to estimate species distributions based solely on presence locations (see Busby 1991; Hirzel et al. 2002; Phillips et al. 2006; Stockwell and Peters 1999).

Initially, traditional statistical approaches like general linear models (GLMs) commonly represented the relationship between a species and the environment (Manly et al. 1993). Traditional statistical approaches require an a priori conceptual model (i.e., linear) be defined, and model evaluation centers on how well data fit the conceptual model (Breiman 2001). Often, model selection between competing conceptual models became the focus of an analysis (Burnham and Anderson 2002). Alternatively, data-mining approaches use predictive ability to evaluate model performance and do not require a priori knowledge of the system (Breiman 2001; Hochachka et al. 2007). When predictive ability is the modeling goal, data-mining approaches outperform traditional statistical approaches (Elith et al. 2006).

23.2.2 Examples of Spatially Explicit Monitoring Programs

Although the extent and resolution of sampling efforts are critical for spatial management, few large-scale, long-term monitoring efforts have incorporated spatially-explicit sampling frames into their design. Some notable exceptions include the Forest Inventory and Analysis Program (FIA) and the National Resources Inventory (NRI).

The U.S. Forest Service organized FIA, a statistically reviewed and rigorous monitoring program, to provide information about the extent, condition, and trends of forest resources as the global drivers of ecosystems change (Smith 2002). FIA uses a systematic, national grid as the sampling frame in order to collect consistent and compatible data across forested landscapes with a variety of management regimes and political boundaries. Additionally, the FIA sampling is hierarchal in resolution. At its coarsest level (Phase 1) FIA delineated forested lands using remote sensed data. In Phase 2, FIA established~125,000 permanent plots randomly assigned within tessalated 6000-acre hexagons that are systematically distributed over the forested landscape regardless of ownership. Each of these permanent plots, composed of four subplots to sample within-site variability, are sampled every 7–15 years for 300 landscape and vegetation variables. In Phase 3, permanent plots within every 1 of 16 hexagons are subsampled for additional data about the health and condition of the forest ecosystem. The FIA integrated framework ensures standard protocols and database management across the national grid. FIA data have already been used to model current distributions of 80 tree species along with future range shifts given climate change (Iverson et al. 2004).

The US Department of Agriculture's National Resources Conservation Service developed NRI in 1977 to assess the condition of soil, water, and other natural resources on non-federal lands in the United States (Nusser and Goebel 1997). NRI has evolved into a spatial-explicit sampling design using the political boundaries originally divided by the Public Land Survey. The political boundaries of Counties (36 × 36 miles), Townships (6 × 6 miles), and Sections (1 × 1 mile) delineate a hierarchical grid. In areas not organized by the Public Land Survey, NRI imposed an analogous grid system. NRI uses 40–640 acre grid cells, selected through a spatially constrained (by Township) randomization procedure, as the primary sampling unit. The primary sampling unit size and sampling intensity (2–6% of land area) are determined by the heterogeneity of the landscape with a smaller proportion of larger grid cells being sampled in homogenous regions. Within each selected primary sampling unit, sample points are selected using a restricted randomization procedure. Researchers collect data for multiple resources at the sample point locations, for the entire primary sampling unit, and for larger polygons delineated based on geographic features intersected by county boundaries. Generally, points are resampled every 5 years to provide longitudinal data, but annual data collection can be incorporated into the sampling design to meet specific needs or to monitor rapid change. The NRI sampling design is flexible: data collection methods can change as long as change effects are estimated, variables can be added, and primary sampling units can be added or dropped. The NRI framework produces a longitudinal database of agricultural and ecological variables along with associated software for database analysis and well-documented data collection protocols.

23.2.3 Gap Analysis: An Effort to Merge Information for Landscape Planning

Monitoring alone is not sufficient for conservation delivery; spatial products must be linked with planning (Nichols and Williams 2006). Gap Analysis (http://gapanalysis.nbii.gov/portal/server.pt) was among the first landscape-scale coordinated efforts to use spatial information and modeling for wildlife conservation and land management planning (Scott et al. 1993). Gap projects now exist in 49 states of the USA (Gotthardt, Payare and Huettmann, personal communication for Alaska). Gap Analysis aims to merge spatial information about land cover, species distributions, and land-management regimes to provide a course-filter framework for designing conservation networks. Land-cover information is obtained through computer-based, supervised training or visual interpretation of satellite imagery to create a classified layer of vegetation alliance based on the National Vegetation Classification System (Stoms 2007). Gap Analysis combines knowledge about range limits and habitat associations to generate maps of predicted species distributions (Csuti and Crist 2000). For each species, the species distribution analysis phase does result in a database of species occurrence locations, a literature review of habitat preferences, and a map consisting of a 635 km^2 hexagon grid covering the known range extent (Jennings 2000). Each hexagon grid cell can be assigned as a documented occurrence or a predicted occurrence (as based on habitat preferences). Finally, land-management regimes are represented in a layer of land-status based on ownership and classified as permanently protected, protected with some management activities, protected with extractive resource use, or not protected (Jennings 2000). Once overlaid, the land cover, species distributions, and land-management regime information can be used strategically to identify gaps within the network of conservation lands (Scott et al. 1993).

Gap Analysis has been useful for large-scale planning and identifying gaps where data should be collected. Linking the creation of occurrence datasets, remotely sensed land-cover layers, and management boundaries create a useful process for gathering and storing information. Gap Analysis has occurred in every state, and the creation and presentation of models tends to improve the local culture and technical skill within each state (i.e., species data, management and modeling as such). Gap Analysis data still have more potential to influence planning, for example in environmental impact studies and as a state infrastructure. However, Gap Analysis is based on historical relationships between species and habitat that may be irrelevant in a rapidly changing climate.

23.3 Components of Adaptive Spatial Management Programs

The future of spatial information management, with respect to wildlife, is the overwhelming consensus among conservation planners and global climate change groups of the need for predictive modeling of species distributions (GAO 2007; Rodriguez et al. 2007). Furthermore, predictive models need to be linked to planning and policy evaluation. The ecological changes expected with climatic changes, like shifting distributions, necessitate flexible management approaches that adapt to changing conditions. For example, stepping-stone refugia may have high conservation value during a time-period of transition, but become less important later. Flexible management approaches need to change temporally and spatially to match conservation needs (Chapin et al. 2004). This will require the development and distribution of data sets that currently do not exist because managers need higher resolution surveys for implementing management at local scales. A framework to integrate higher-resolution surveys into a larger informational network would be extremely useful for adaptive spatial management. Spatial management programs that are adaptive and relevant for planning will need to integrate across several implementation steps including data collection, data management, and data delivery. Here we suggest an efficient framework to collect spatial data for spatial wildlife management that would include remote sensing and grid-based sampling designs. Data management will require the integration of data across space and institutions, as well as data archiving that includes rigorous metadata standards and open access (Huettmann 2005). Finally, modeling can be used to produce crucial metrics and visualizations, both of which will be imperative for planning and policy review.

23.4 Data Collection: Grid-Based Designs and Remote Sensing

Wildlife data and environmental layers are needed to provide landscape-scale products via predictive modeling. All attempts to synthesize wildlife data into spatial products are limited by the sampling design (extent, resolution) of geo-referenced data (Scott et al. 2002). In this section, we review data available for spatial products that are currently collected by national monitoring initiatives.

We believe that monitoring design has been conventionally constrained by a focus on detecting temporal change in a metric (e.g., power analysis), and neither on detecting true, spatial change nor on model building. For future monitoring initiatives, we suggest grid-based monitoring designs in conjunction with remote sensing efforts. Along with other benefits, grid-based monitoring designs provide a simple underlying framework for organizing spatial reference that can provide cohesion between datasets.

23.4.1 Currently Available Wildlife Data

Many spatial products currently used for conservation were modeled with occurrence locations that were originally collected for other purposes. For example, the Breeding Bird Survey (http://www.pwrc.usgs.gov/BBS/) and North American Amphibian Monitoring Program (http://www.pwrc.usgs.gov/naamp/) are long-term monitoring programs designed to calculate aspatial, regional population trends using routes sampled by volunteers. Although routes are geo-referenced, routes are spatially biased based on road access and volunteer interest, not taking autocorrelation and similar problems into account. In addition, the exact spatial distribution of sampling routes is not considered explicitly in the survey design except in terms of the how the spatial variation influences the power to detect a population trend. Further, many county, state, and federal land managers develop aspatial species inventories; e.g., most National Wildlife Refuges maintain published checklists of bird species. Although inventories and checklists may be aspatial at the local scale, they become spatial at large scales (regional, continental, global). For example, a refuge checklist is aspatial at the refuge scale because checklists do not provide information about where to find an animal within the refuge boundary. However, the same checklist can be used spatially, albeit with limited precision, when used to indicate species presence within a 10 km^2 grid cell. Other examples of such data-sources include bird-band recovery programs and museum collections, which provide opportunistically collected locations with variable spatial precision. State Natural Heritage Programs also maintain databases of species observations obtained in field surveys and historical accounts which are maintained on the NatureServe website (http://www.natureserve.org). Locations from telemetry studies may also be used, but these locations can be biased by the individual preferences of the animals sampled and the locations include autocorrelation based on the temporal resolution of the data collection. Almost all of such data have not been made publicly available yet. Numerous other groups collect monitoring and scientific information based on local needs, but these data are usually collected for small spatial extents with limited data access.

New initiatives, such as the Avian Knowledge Network (AKN; http://www.avianknowledge.net/content/download) and National Biodiversity Information Infrastructure (NBII; http://www.nbii.gov/portal/server.pt), are beginning efforts to link small-scale efforts into coordinated databases. Similarly, coordinated databases such as E-bird (http://ebird.org/content/ebird/) consolidate volunteer field observations. Although volunteer field observations contain more variability in observer expertise and spatial precision, these data can be utilized for detecting population trends, changes in phenology, and species distributions (Droege et al. 1998; Hochachka et al. 2007).

23.4.2 *Grid-Based Sampling Design*

We suggest grid-based designs when new data must be collected. For ground plots, grid-based designs provide an efficient, underused but powerful sampling design for monitoring and inventory. Spatial reference, in terms of data location and resolution, can provide the underlying structure to unify monitoring and inventory efforts. A gridded system provides a simple but rich framework for organizing spatial reference. For example, the UTM system, with coordinates conveniently expressed in meters, could provide an international sampling frame that monitoring and inventory programs could easily scale and subsample based on program objectives.

Grid-based designs are flexible because users can draw representative samples based on boundaries (geographic or strata-based) delineated by the study purpose (Nusser and Goebel 1997; Smith 2002). Therefore, grid-based designs provide data that can be engaged to explore and test a wide variety of conceptual models about underlying ecological structure. For example, Prasad et al. (2006) have already used FIA data to explore if predicted tree species distributions were driven by climatic variables or edaphic variables.

Grid-based designs also provide a powerful sampling framework for generating baseline ecological indicators with predictive models. A grid-based design produced more accurate species distribution models for lichen species, as assessed by prediction accuracy on independent data, than a clustered sampling design (Edwards et al. 2006). In the lichen example, the models from grid-based designs provided a more representative sample of locations where species were both present and absent. Other research also found that sampling designs that do not provide a representative sample of environmental conditions within the study area result in biased distribution models (Austin and Heyligers 1989; Kodric-Brown and Brown 1998). Species occurrence data collected at grid-based locations can be used to develop species distribution models, but species detectability must be considered (MacKenzie et al. 2006). Species detectability refers to the probability that a surveyed area will be falsely recorded as an absence. Using false negatives can bias distribution models and reduce the predictive capacity (Gu and Swihart 2004; Lobo 2008). Finally, grid-based designs are particularly useful for generating species distributions because the landscape is sampled beyond the spatial extent of an identified target population and therefore shifts in the location and extent of the population can be identified (Nusser and Goebel 1997).

Monitoring designs, like grid-based designs, that sample multi-taxa at the same spatial location are particularly useful. Community information can be extracted from the associations of species and species diversity at known locations. This information is critical as new species assemblages can result from climate change and exotic invasions.

23.4.3 Remote Sensing

We need to consider program designs in which remote sensing is the primary monitoring approach because this approach is relatively inexpensive and can be implemented across large spatial scales. Remote sensing can also provide the environmental layers necessary for modeling species distributions. Remote sensing refers to techniques that utilize satellites or aircraft to conveniently gather data without directly touching the object (Koeln et al. 1994). Remote sensing requires technical expertise to prescreen data for quality, to mosaic multiple images, to geometrically correct images for topography, and to radiometrically correct images (Lunetta 1998). Monitoring programs can utilize remote sensing technology for image classification and for change detection. Image classification is the process of classifying clusters of similar remote sensing data, such as spectral signatures, into categories that are meaningful to the researcher (Lunetta 1998). For example, the Multi-Resolution Land Characteristics program (MRLC) classified Landsat TM data into vegetation categories for North America (http://www.epa.gov/mrlc/nled.html). Researchers can use classified images to generate inventory metrics representing the spatial distribution of categories of interest and to ensure categories are adequately represented in ground-based sampling designs.

Change detection compares satellite images from two timeframes in order to detect changes in land cover or land use. Two general approaches, post-classification and pre-classification can be used (Lunetta 1998). In post-classification, the researcher classifies both images into the same categories independently and then compares the classifications by category. Post-classification is advantageous because different types of data can be compared without normalization to reduce differences due to atmospheric effects and detector performance. Disadvantages include high cost due to processing time, a lack of consistency due to interpreter error, and the propagation of classification errors from the individual images into the change detection product. The pre-classification approach transforms the two images into one single or multi-band image that is analyzed to detect areas of change. In some cases, researchers can combine pre and post-classification techniques for a deeper analysis (Franklin 2001). For example, a study on the Alaskan Seward Peninsula used a pre-classification approach to quantify the area of change across three images and post-classification to document that most change was due to an increase in shrub cover (Stow et al. 2004).

Researchers should use periodic ground-based plots for both assessing change in monitoring variables and for validating/calibrating remote sensing data. Plot information can be used to modify and refine the categories assigned to spectral signatures. In addition, ground-based plots can be utilized to conduct an accuracy assessment of remote sensing products. For accuracy assessment, an error matrix is generally constructed to generate omission and commission errors for each category (Lunetta et al. 1998). Such concepts should belong in any monitoring program and be facilitated with online data.

23.5 Data Management: Database Integration, Archiving, and Open Access

All monitoring efforts require a database management to ensure the data is archived and accessible. Often, monitoring databases include data sampled for one or few taxonomic groups with sampling protocols that are specific to the monitoring program. Within the past years, the Convention of Biological Diversity identified a need to merge observational data collected by multiple institutions with multiple sampling methods into interoperable databases. Access to all scientific observations within a geographic area will greatly increase options for modeling to help achieve management goals. Data integration requires accessible standards and protocols for data collection and data exchange with open access (Canhos et al. 2004; see for instance Darwin Core). The integration of data collected for different purposes (i.e. museum collection, observational counts) and with different field protocols could be supported through software workbenches such as Kepler (http://kepler-project.org/). The Kepler Project helps scientists create and execute workflows of their analysis from data acquisition and integration to complex computations.

Although the Kepler Project may be beyond the scope of many studies, a (manual) data flow should be part of any project, and is to be defined regardless if data are to be handy for management decisions (Nichols and Willams 2006). For the integration of databases from multiple sources two main topics can be defined (1) inclusion of data into that environment, and (2) accessing the grid of databases. For the first section, DiGIR (http://digir.sourceforge.net/) plays a bigger role. It can provide the tools and protocol to get data out on the Internet and join the huge global network of biological data. Once this is achieved, DiGIR also allows queries with other sources, and which leads to part (2). The Science Environment for Ecological Knowledge (SEEK) initiative for instance is working to create a cyber-infrastucture to facilitate ecological research (http://seek.ecoinformatics.org/).

Differences in data collection protocols must be documented with high-quality spatial (http://www.fgdc.gov/) and biological (http://www.nbii.gov/) metadata such as FGDC NBII or compatible. When possible, standard data collection protocols would subsequently increase analysis power. An additional protocol library could provide managers with easily accessible, rigorous sampling methodologies; therefore, reducing planning effort for project implementation. Suggested data collection protocols should be flexible for analysis. For example, collecting distance information at survey sites is better than simply recording occurrence because distance information can be easily condensed into occurrence information or be used for density estimate. Database networks also need standards to ensure that information from different sources will truly integrate. A schema must be designed that includes the elements needed for data exchange and retrieval. For example, the mentioned Darwin Core and its updates provides an example of an XML schema that can be used for data from all taxonomic groups. Once integrated, databases need to be easily queried and with open access.

We suggest that a global data grid system could provide an underlying spatial structure to organize multiple sampling efforts. However, with this approach, georeferencing becomes a crucial aspect of data collection. Therefore, we recommend that managers stringently adhere to best practices for georeferencing. The BioGeomancer software project (http://www.biogeomancer.org/) for instance has provided a detailed manual for georeferencing best practices (Chapman and Wieczorek 2006). Among other issues, BioGeomancer best practices provide guidelines to document spatial accuracy and reliability. It's a key feature of BioGeomancer that a location estimate also carries uncertainty, which needs to be expressed as well. Ignoring this accuracy estimate renders data bases less useful.

For spatial data delivery, web publishing services should be considered. In order to publish spatial data, a platform to store the data must exist. PostgreSQL (http://en.wikipedia.org/wiki/PostgreSQL) is the open source solution for the database engine, and can include spatial data with the PostGIS extension. While a standard database can store numeric and character data, a spatial database can store geometry objects (e.g., points, lines and polygons). Spatial information is stored by PostGreSQL in a data type called *geometry* based on the Open Geospatial Consortiums (OGC; http://en.wikipedia.org/wiki/Open_Geospatial_Consortium) Simple Features data specifications. The added functionality of a spatial database includes the ability to run numerous spatial queries. For example, records can get selected based on geospatial proximity, calculated measurements (distance between point, lined, polygons), calculated areas, or on created new features such as buffers, and numerous others. Additionally, spatial databases can create spatial indexes to improve the timing for queries. Several open source programs are available to view the PostGis data layers, such as GRASS (http://grass.osgeo.org/), or uDig (http://udig.refractions.net/).

A widely used open source platform for publishing and displaying spatial data over the Internet is MapServer (http://mapserver.org/). Originally developed in the 1990s at the University of Minnesota MapServer is currently supported by developers worldwide, including the Army Corps of Engineering. MapServers enable developers to easily develop spatial web content, i.e., maps with their data displayed and to serve their data directly through a Web Mapping Services. Despite ongoing technical development efforts, well documented and transparent data products still rarely enter the management and policy arena.

23.6 Data Delivery: Modeling and Visualizations as Planning Tools

The final component of spatial management will be data delivery in the form of models and visualizations that are accessible to a wide variety of stakeholders. Models that extrapolate wildlife data across space and time, like species distribution models, have well developed methods (e.g., Scott et al. 2002). However, web-based applications and decision support tools need to be better developed to communicate modeling efforts (see OpenModeler at http://openmodeller. sourceforge.net/; GBIF http://www.gbif.org/ and OBIS http://www.iobis.org/). Although not linked to modeling per se, Google Earth (http://earth.google.com/) can provide such a platform to display spatial information on the Internet. For general visualization of complex online databases see for instance http://www.readwriteweb.com/archives/the_best_tools_for_visualization.php.

Decision support tools are defined as "computer-based systems that provide information by means of forecasting models and access to databases, for example geographical information systems or climatic databases, in order to support a decision-maker in complex and un-/semi-structured management issues"(Tremblay et al. 2004). Much development is currently underway on this topic, and it is hoped that these will feed directly into management and policy, and in a transparent fashion.

23.7 Examples of Spatial Management: Linking Information to Planning for Conservation Delivery

Ultimately, for spatial information to be useful it needs to be applied in a way that resource management is improved. One of the best examples of conservation delivery is the co-evolution of the permanent easement program in the Prairie Pothole Region of the U.S. with landscape-level monitoring and spatial modeling conducted by the U.S. Fish and Wildlife Service's Habitat and Population Evaluation Team (HAPET; Niemuth et al. 2008). Using five waterfowl species as indicators of priority wetlands for conservation, HAPET surveyed breeding pairs and relative surface area of water on ponds within 4-mile^2 blocks that were randomly-selected from townships within the Prairie Pothole Region. General linear models of the relationships between productivity and wetland size were developed for each of the five species. These models were then applied to the landscape using a moving-windows analysis on 40-acre pixels to develop maps that prioritize areas for land acquisition and conservation easements. Collectively, waterfowl production areas and easements protect >2.7 million acres of wetlands and grasslands in the Prairie Pothole Region.

Using a landscape-level approach more focused on biodiversity than single species, the Kenai National Wildlife Refuge has developed their Long Term Ecological Monitoring Program as a formally-designated adjunct inventory to the FIA program. Extending the FIA's Phase 2 sampling grid over both forested and non-forested habitats, occurrences of vascular and nonvascular plants, breeding landbirds, and selected arthropods were rapidly sampled on permanent points systemically arrayed at 5-km intervals across the 2-million acre refuge (Morton et al. 2009; see Figs. 1 and 2). To date, over 1000 species have been detected on the grid. The collection of occurrence data for multiple species at each grid point set the stage for statistically-robust, spatially-explicit modeling of the distributions of individual species (Magness et al. 2008) and species richness (Bowser and Morton 2009). Plans for this program include the incorporation of scaled-down estimates of future temperature and precipitation from global circulation models (http://www.snap.uaf.

edu/) to forecast re-distributions of species over the next century, and the development of management approaches to facilitate anticipatory adaptation.

The HAPET and Kenai National Wildlife Refuge monitoring program are good examples of using a spatial information framework to manage wildlife. These examples include a grid design for monitoring and modeling products linked to planning. However, both have yet to share data and products via open access. Therefore, data have not been incorporated into larger, biodiversity data management schema, yet. Data management, including archiving and ensuring integrity, is internal and may not be available for other purposes. These are next steps, and they will increase the capacity for science-based conservation of biodiversity.

Acknowledgements The University of Alaska's Integrative Graduate Education and Research Traineeship (IGERT) program provided fellowship support to Dawn Magness. Mark Laker, Kenai National Wildlife Refuge, provided helpful comments. The second author is fully appreciative of the cooperative nature of staff at the FIA's Pacific Northwest Research Station which allowed the development of LTEMP at Kenai National Wildlife Refuge.

References

Alidina HM, Fisher D, Stienback C, Ferdana Z, Lombana A, Huettmann F (2008) Assessing and managing data. In: Ardron J, Possingham H, Klein C (eds) Marxan good practices handbook. Vancouver, Canada http://www.pacmara.org/

Austin MP, Heyligers PC (1989) Vegetation survey design for conservation: gradsect sampling of forests in north-eastern New South Wales. Biol Conserv 50:13–32

Bella D, Li H, et al (1992) Ecological indicators of global climate change: proceeding of a US Fish and Wildlife Service global climate change workshop held at Oregon State University, Corvallis, Oregon 13–15 November 1990

Berry KH (1986) Introduction: development, testing, and application of wildlife-habitat models. In: Verner J, Morrison ML, Ralph CJ Wildlife 2000: modeling habiat relationships of terrestrial vertebrates. The University of Wisconsin Press, Madison, WI

Bowser ML, Morton JM (2009) Modeling terrestrial arthropod diversity on the Kenai National Wildlife Refuge. In: McWilliams W, Moisen G, Czaplewski R Forest inventory and analysis (FIA) Symposium, 21–23 October 2008, Park City, UT. Proc RMRS-P-56CD USDA Forest Service, Rocky Mountain Research Station

Boyce MS, McDonald LL (1999) Relating populations to habitats using resouce selection functions. Trends Ecol Evol 14:268–272

Breiman L (2001) Statistical modeling: the two cultures. Stat Sci 16:199–231

Burnham KP, Anderson DR (2002) Model selection and multimodel inference: a practical information-theoretic approach. Springer, New York

Busby JR (1991) BIOCLIM - a bioclimatic analysis and prediction system. In: Austin MP, Margules CR Conservation: cost effective biological surveys and data analysis. CISIRO, Melbourne, Australia

Canhos VP, Souza S, et al (2004) Global biodiversity informatics: setting the scene for a "new world" of ecological modeling. Biodiver Inform 1:1–13

Chapin FS III, Peterson G, et al (2004) Resilience and vulnerability of northern regions to social and environmental change. Ambio 33:344–349

Chapman AD, Busby JR (1994) Linking plant species information to continental bidoversity inventory, climate and environmental monitoring. In: Miller RI Mapping the diversity of nature. Chapman & Hall, London

Chapman AD, Wieczorek J (2006) Guide to best practices for georeferencing. Copenhagen: Global Biodiversity Information Facility

Csuti B, Crist P (2000) Methods for developing terrestrial vertebrate distribution maps for gap analysis. In: Scott JM, Jennings MD A handbook for gap analysis. Version 2.0 Idaho Cooperative Fish and Wildlife Research Unit, University of Idaho. Moscow

Cutler DR, Edwards TC Jr, et al (2007) Random forests for classification in ecology. Ecology 88:2783–2792

Droege S, Cyr A, et al (1998) Checklists: and under-used tool for the inventory and monitoring of plants and animals. Conserv Biol 12:1134–1138

Edwards TC Jr, Cutler DR, Zimmermann NE, Geiser L, Moisen GG (2006) Effects of sample survey design on the accuracy of classification tree models in species distribution models. Ecological Modelling 199:132–141

Elith J, Graham CH, et al (2006) Novel methods improve prediction of species' distributions from occurrence data. Ecography 29:129–151

Ellis EC, Ramankutty V (2008) Putting people in the map: anthropogenic biomes of the world. Frontiers in Ecol Environ 6:439–447

Franklin SE (2001) Remote sensing for sustainable forest management. Lewis, Boca Raton, FL

GAO (2007) Climate change: agencies should develop guidance for addressing the effects on federal land and water resources. US Government Accountability Office Report to Congressional Requesters

Graham CH, Ferrier S, Huettmann F, Moritz C, Peterson AT (2004) New developments in museum-based informatics and applications in biodiversity analysis. Trends Ecol Evol 19:497–503

Gu W, Swihart RK (2004) Absent or undetected? Effects of non-detection of species occurrence on wildlife-habitat models. Biol Conserv 116:195–203

Hirzel AH, Hausser J, et al (2002) Ecological-niche factor analysis: how to compute habitat-suitability maps without absence data? Ecology 83:2027–2036

Hochachka WM, Caruana R, et al (2007) Data-mining discovery of pattern and process in ecological systems. J Wildl Manag 71:2427–2437

Huettmann F (2005) Databases and science-based management in the context of wildlife and habitat: toward a certified ISO standard for objective decision-making for the global community by using the internet. The Journal of Wildlife Management 69 (2):466–472

Inkley DB, Anderson MG, et al (2004) Global climate change and wildlife in North America. Wildl Soc: 34

Iverson LR, Schwartz MW, et al (2004) How fast and far might tree species migrate in the eastern United States due to climate change? Global Ecol Biogeogr 13:209–219

Jennings MD (2000) Gap analysis: concepts, methods, and resent results. Landsc Ecol 15:5–20

Kappelle M, Vuuren MMIV, et al (1999) Effects of climate change on biodiversity: a review and identification of key research issues. Biodiver Conserv 8:1383–1397

Kodric-Brown A, Brown JH (1998) Incomplete datasets in community ecology and biogeography: a cautionary tale. Ecol Appl 3:736–742

Koeln GT, Cowardin LM, et al (1994) Geographic information systems. In: Bookhout TA Research and management techniques for wildlife and habitats. The Wildlife Society, Bethesda, MD

Lobo JM (2008) More complex distribution models or more representative data? Biodiver Inform 5:14–19

Lunetta RS (1998) Applications, project formulation, and analytical approach. In: Lunetta RS, Elvidge CD Remote sensing change detection: environmental monitoring methods and applications. Ann Arbor, Chelsea, Michigan

Lunetta RS, Lyon JG, et al (1998) North American Landscape Characterization: triplicate data sets and data fusion products. In: Lunetta RS, Elvidge CD Remote sensing change detection: environmental monitoring methods and applications. Ann Arbor, Chelsea, Michigan

MacKenzie DI, Nichols JD, et al (2006) Estimation and modeling: inferring patterns and dynamics of species occurrence. Elsevier, San Diego, CA

Magness DR, Huettmann F, et al (2008) Using random forests to provide predicted species distribution maps as a metric for ecological inventory & monitoring programs. In: Smolinski TG,

Milanova MG, Hassanien A-E Applications of computational intelligence in biology: current trends and open problems. Studies in computational intelligence, Springer, Berlin
Manly B, McDonald L, et al (1993) Resource selection by animals: statistical design for field studies. Chapman & Hall, London
McHarg IL (1969) Design with nature. Doubleday & Company, Garden City, NJ
Morrison ML, Marcot BG, et al (1992) Wildlife-habitat relationships: concepts and applications. The University of Wisconsin Press, Madison
Morton J, Bowser M, et al (2009) Long term ecological monitoring program on the Kenai National Wildlife Refuge: an FIA adjunct inventory. In: McWilliams W, Moisen G, Czaplewski R Forest inventory and analysis (FIA) symposium, 21–23 October 2008, Park City, UT, USDA Forest Service, Rocky Mountain Research Station RMRS-P-56CD
Nichols JD, Williams BR (2006) Monitoring for Conservation. Trends Ecol Evol 21:668–673
Niemuth ND, Reynolds RE, et al (2008) Landscape-level planning for conservation of wetland birds in the US Prairie Pothole Region. In: Millspaugh JJ, Thompson FR III Models for planning wildlife conservation in large landscapes. Elsevier, Amsterdam
Noss RF (2001) Beyond Kyoto: forest management in a time of rapid climate change. Conserv Biol 15:578–590
Nusser SM, Goebel JJ (1997) The national resources inventory: a long-term multi-resource monitoring programme. Environ EcolStat 4:181–204
Parmesan C (2006) Ecological and evolutionary responses to recent climate change. Ann Rev Ecol Evol Syst 37:637–669
Parmesan C, Yohe G (2003) A globally coherent fingerprint of climate change impacts across natural systems. Nature 421:37–42
Phillips SJ, Anderson RP, et al (2006) Maximum entropy modeling of species geographic distributions. Ecol Model 190:231–259
Prasad AM, Iverson LR, et al (2006) Newer classification and regression tree techniques: bagging and random forests for ecological prediction. Ecosystems 9:181–199
Rodriguez JP, Brotons L, et al (2007) The application of predictive modeling of species distribution to biodiversity conservation. Diversity Distributions 13:243–251
Root TL, Schneider SH (2001) Climate change: overview and implications for wildlife. In: Schneider SH, Root TL Wildlife responses to climate change: North American case studies. Island, Washington, DC
Scott JM, Davis F, et al (1993) Gap analysis: a geographic approach to the protection of biological diversity. Wildl Monogr 123
Scott JM, Heglund PJ, et al (2002) Predicting species occurrences: issues of accuracy and scale. Island, Washington, DC
Smith WB (2002) Forest inventory and analysis: a national inventory and monitoring program. Enviro Pollut 116:S233–S242
Stockwell D, Peters D (1999) The GARP modelling system: problems and solutions to automated spatial prediction. Internat J Geograph Info Sci 13:143–158
Stoms DM (2007. Actual vegetation layer. In: Scott JM, Jennings MD A handbook for gap analysis. Version 2.0 Idaho Cooperative Fish and Wildlife Research Unit, University of Idaho, Moscow
Stow DA, Hope A, et al (2004) Remote sensing of vegetation and land-cover change in Arctic tundra ecosystems. Remote Sensing of Environment 89:281–308
Tremblay J-P, Hester A, et al (2004) Choice and development of decision support tools for the sustainable management of deer-forest systems. For Ecol Manag 191:1–16
Vitousek PM, Lubchenco J, et al (1997) Human domination of Earth's ecosystems. Science 277:494–499
Walters C (1986) Adaptive management of renewable resources. Blackburn, Caldwell, NJ
Yen PPW, Huettmann F, et al (2004) A large-scale model for the at-sea distribution and abundance of Marbled Murrlets (Brachyramphus marmoratus) during the breeding season in coastal British Columbis, Canada. Ecol Model 171:395–413

Chapter 24
Future and Outlook: Where Are We, and Where Will the Spatial Information Management in Wildlife Ecology Be in 50 Years from Now?

Samuel A. Cushman and Falk Huettmann

24.1 Looking Back

In this final chapter we briefly look back over what we have attempted in this book, and then look toward the future to discuss the outlook for overcoming the challenges we face within our fields of ecological science and in the greater application of this knowledge to enhance the prospect for a sustainable future for the biosphere. Looking back, we have tried to present a view of ecological science at the cusp of major change. The change is all around us, driven by vast improvements in computational power, analytical tools, spatial databases, economy and related governance structures within the sciences. These changes have fundamentally altered what is possible in ecological analysis and are enabling a new synthesis and applications of data and method to recast ecological theory in a spatially and temporally explicit framework. In this book we argued that ecological theory and its dissemination is lagging significantly behind advances in data acquisition and analytical methodology. This is in part because of the inertia and resistance of established procedure and theory, and secondly because of the way the sciences are currently administered and trained.

We argued that ecological systems are fundamentally sensitive to details of spatial and temporal variability, and that ecological theory and analysis must greatly expand to address this robustly; new culture is needed. We argued that landscape ecology provides a natural context for this synthesis, but only when this field itself transcends the limitations of its often ideological history to embrace a generalized, flexible, multi-scale and multi-variate theoretical and analytical framework.

S.A. Cushman
US Forest Service, Rocky Mountain Research Station, 800 E Beckwith, Missoula, MT 59801, USA
e-mail: scushman@fs.fed.us

F. Huettmann (✉)
EWHALE lab- Biology and Wildlife Department, Institute of Arctic Biology, University of Alaska-Fairbanks, 419 IRVING I, Fairbanks, AK 99775-7000, USA
e-mail: fffh@uaf.edu

S.A. Cushman and F. Huettmann (eds.), *Spatial Complexity, Informatics, and Wildlife Conservation*
DOI 10.1007/978-4-431-87771-4_24, © Springer 2010

We then argued that informatics is the foundation for this framework, with special emphasis on the need for multi-scale, multi-variate, spatial databases to serve as a foundation for flexible analysis and powerful modeling. We reviewed a number of varieties of ecological data within the context of this integrated framework, noting the unique challenges and opportunities posed by GPS telemetry data on animal movement and molecular genetic data as examples. We discussed the powerful contributions to data collection and analysis enabled by modern remote sensing technology, and discussed the structure and management of ecological databases. Customized, open-source analytical software tools were identified as an area of rapid expansion and enormous opportunity. We presented a broad range of spatial and multi-variate statistical methods for analyzing animal ecology. We argued that landscape genetics, particularly when expanded into a gradient framework, provides an immense opportunity to explore the causes and consequences of population structure and for a better management maintaining genetics diversity world-wide. We presented several examples from different geographical regions and across the taxonomic spectrum of these different kinds of spatially sophisticated analysis and then discussed the prospects for integrating the best available science on spatial and temporal dynamics into natural resources management, and ended with a chapter on the integration of science and policy in animal management and conservation.

The reader must judge the degree to which we have maintained coherent connections and flow among the major examples and ideas in the book. Certainly, more thinking and assessment is necessary to fine-tune modules and visions presented here. That is what academics are good at and what universities should provide: science-based management of natural resources with an aim to achieve global sustainability. The main purpose of this chapter, however, is to look forward to where we think the changes we have argued are needed and that we have described in their initial stages will progress in the coming decades, and also how hopeful we can be about the actual prospects of effectively integrating advanced ecological monitoring and research into a truly sustainable global future.

24.2 Looking Forward

Looking forward we can anticipate several trends that we identified in this book that will continue. First, there will continue to be great advances in the collection of spatial data at fine grain over broad extents. Technological improvements have been very rapid in remote sensing and sensor networks, with new platforms being designed that will greatly expand the range of data products in terms of spectral, spatial and radiometric resolution, temporal frequency and spatial extent, and accuracy. LIDAR, RADAR and other aerial platforms, e.g. unmanned vehicles and sensors, are also improving and becoming more available for all users world-wide. This advancement will also include taxonomic species data and inventories.

The challenge for the future may be society's ability to continue the very large investments in earth observing satellite deployment in an era of increasingly limited discretionary funding and potentially declining tax revenue due to demographic changes in the populations of the United States and Europe. Some believe due to these economic and social forces that China (Russia, India, Brazil and Indonesia) may emerge as the leader of the next generation of deployment of remote sensing satellites. This remains to be seen, as China is now tied to the west, faces at least as severe demographic challenges to continued economic growth, various management and democracy challenges, plus a clear linkage with the export market economies of the United States and Europe. Despite recent failures of latest Landsat missions, it seems very likely to expect no major break in the continued deployment of more and more advanced earth observing remote sensing platforms, which will continue to provide transparent spatial and temporal data of ever higher quality and finer resolution. In turn this will continue to drive demand and progress in analytical methodology and ecological theory, as well as sustainable management dealing with the human footprint.

We also expect progress in computing power and the computationally-intensive analytical and simulation approaches that it enables to continue apace. We have higher confidence in this area, as it requires far less in terms of massive central-government investment and oversight than developing and deploying complex earth observing satellites. The advances in cloud computing and in immense remote networks of processors may provide super-computer-like computational power for anyone with a personal computer and an internet connection, allowing for in-time solutions to complex ecological optimization problems. It is important here though to push for equal access to such opportunities, and lessen the digital divide towards a truly global village view of applications, users and funding.

The major drivers of the changes to ecological theory we have discussed in this book are intimately linked to rises in computational technology. Much research that appears in 'traditional Ecology' is actually only possible due to EcoInformatics, and perhaps should be labeled as such. This has already set up a new culture, and our understanding of ecology, our universe, and ourselves within it.

We further argued that the traditional fixation on simple, equilibrium models of ecological systems is a historical legacy of the absence of powerful computing and simulation until the past several decades. This needs to be quickly reformed. The new context for modeling and analysis provided by modern computers is truly revolutionary and has led to immense changes in what questions are deemed tractable, and the sophistication and detail of ecological analysis and modeling. We argued in this book for an explicit focus on pattern–process relationships at multiple spatial scales across large extents in time and space. This was simply not even conceivable until powerful computation was coupled with extensive and fine scaled ecological data, and legally demanded, e.g. Biodiversity Convention. Together with online initiatives and opportunities, the combination of these has fundamentally altered the field of ecology, and with computational power it is expected to continue to grow at a rapid pace. Therefore, we can expect analytical and computational methodology to continue being a major driver of the advancement of ecological knowledge and the reframing of core ecological theory, all to be disseminated online for all citizens and decision-makers world-wide!

With the coupling of ever more powerful computing, with ever more extensive and detailed ecological datasets, we anticipate that many researchers will independently adopt the multi-scale spatial informatic data-cube model we presented in Chapter 6, and use it to implement the kinds of flexible, multi-scale, multi-variate space–time analyses and multi-species modeling efforts we described in Chapters 2, 4 and 6 and elsewhere in this volume. Technological advances being made as we write will enable the sequencing of entire genomes very rapidly and inexpensively. This will revolutionize landscape genetics, the study of geneflow, population fragmentation and adaptive evolution in complex and changing environments. Landscape and their resources can get optimized towards reaching best possible goals.

The generation gaps between current cutting-edge data, computing and analysis and established ecological theory might continue and perhaps expand. The rate of advance in data acquisition and computational power will likely increase, which will likely lead to further disequilibrium between established procedures and principles and the potential of currently emerging methods and datasets. On the other hand, one could argue that there will be a decrease in the dischord between orthodox ecology and the cutting edge of science. It may be argued that this dischord is at its maximum right now, as traditional modes of thought, governance and global structures, procedures of analysis and explanatory theories which were developed prior to the availability of GIS, remote sensing and computing are still dominant in the global culture of the science and its funding agencies. Over time, these advances in method and data are infiltrating orthodox ecological thought and in some cases, as described in Chapters 2 and 4, leading to major shifts in paradigms. We argue that this cultural shift from non-spatial, equilibrium paradigms of the twentieth century, to spatially and temporally explicit, nonequilibrium paradigms of the twenty-first century is the most important transition in the history of ecological thought. Once it has been accomplished then even very rapid advances in the quantity of data or improvements of the speed of computation will not lead to larger disequilibrium between established thought and emerging method. That is, once this *qualitative* change in paradigms is made, further quantitative change in data and method will not prove revolutionarily challenging and will not provoke further paradigm shifts for the foreseeable future. Therefore, we offer that the skeleton space–time–scale–pattern–process paradigm we outline in this book may emerge as the core of the ecological science of the next century. This is not a small feat, and we are very excited to see its development world-wide.

The arguments above are hopeful and optimistic. We unfortunately are less optimistic about our second theme of expected future changes. A second, although usually subsidiary, theme of this book is the potential of advances in ecological monitoring and research to guide management of the earth's natural resources and how it potentially could play an instrumental role in a global transition to a sustainable future. Earth is clearly facing the greatest mass extinction in 65 million years. Human population growth, one-sided resource exploitation, compounded with anthropogenic climate change are a potent combination driving this biodiversity crisis. It is not at all clear that the technical and theoretical advances we describe in this book will be decisively beneficial in mitigating this crisis. What is needed is massive public outcry demanding real change to environmental sustainability. There is a key role for science in this, as the public debate will not occur in earnest until the severity of the circumstances become more universally understood by the people, the media and governments. We feel that scientists have been far too timid and ineffectual in their efforts to communicate with the public and powers in charge. We feel strongly that scientists should not make policy. Politics is a process of valuation of alternative actions by the people; science ideally is knowledge of processes and conditions. While the political debate absolutely needs to be deeply informed by science, scientists should resist the desire to use science as a political tool. There is a very fine line here. Science should fully participate in all relevant key questions of this world, and always work at the core. That Earth is facing a dire biodiversity crisis is clear. That there needs to be bold and quick political action to address this is also clear. That science is the key source of knowledge to provide a basis for the political decisions that will address this crisis is also clear. The role of scientists is as communicators not as policy makers. The current view of most practicing scientists is that scientists create knowledge and knowledge will rise to the top because of its inherent value. As a result scientists generally allocate very little of their time and effort to communicating their findings to non-specialists. As a result, there is a massive disjunction in understanding of many major issues of global importance between the scientific community and the general population. This is where we feel scientists should focus their efforts with the aim of moving toward a more sustainable future. Scientists must inform the debate and provide reliable data and information. Simultaneously, there needs to be a much stronger response from the legal community and national and international governing bodies to ensure global sustainability, environmental justice and world peace.

Index

A

Active remote sensing systems, 212
Adaptive Management, 112
Adaptive spatial management, 435
Affirming the consequent, 5
Africa, 182, 349
African, 332
Aggregation
  error, 52
  statistics, 258
Aian Knowledge Network (AKN), 436
AIC. *See* Akaike's Information Criterion
Airborne Hyperspectral Imagery (HYMAP), 210
Airborne sensors, 210
Airborne Visible/Infrared Imaging Spectrometer (AVIRIS), 210
Ajubaj, 333
Akaike's Information Criterion (AIC), 280
Alaska, 238, 434
Alaska Ocean Observing System (AOOS), 238
Algorithm-based models, 292–299
Algorithm models, 274
American black bear *(Ursus americanus),* 100, 319
American National Standards Institute (ANSI), 224
Amphibians, 371–373
Anadyrsky Gulf, 421
Analytical computing, 246
Angola, 349, 361
Animal movement, 131–148
Antarctic, 236
Antarctic marine living resources, 235
Antarctic Master Directory (AMD), 239
Antarctic treaty, 236
Ararat, 333
Arc catalog, 229
Arctic, 234
Arctic Ocean Diversity Project (ArcOD), 235
Arctic seas, 409
Area Under ROC Curve (AUC), 300
Armenia, 331, 333
Artavan, 333
Australia, 179, 225, 239
Australia ANZLIC spatial land information council standard, 225
Autocorrelation, 131–148, 350
  analysis, 138
  of elephant movement, 143–146
  structure, range and variability, 139–143
AVHRR, 196
Avian flu, 266

B

Bagging, 294
Baltic sea, 410, 415
Barents sea, 234
Barrier, 349–367
Bat (*Pipistrellus* spp.), 290
Bayesian analysis, 287–288
Behavioural data, 176
Behaviour databases, 178
Belgium, 180
Beluga whale (*Delphinapterus leucas*), 410, 417, 423
Benthic biomass, 154
Benthic foragers, 151
Benthos, 151, 154
Bering and Chukchy Seas, 412
Bering sea, 154
Berkshire valley, 371
Biodiversity databases, 233
Biological Active Zones (BAZ), 415
Biological Data Profile (BDP), 224
Bird habitat preferences, 206
Black-footed ferrets *(Mustela nigripes),* 285

Bob Marshall Wilderness, 324
Boosted Regression Trees, 294
Boosting, 294
Boston harbor, 406
Botetti river, 361
Botswana, 139, 349
Brazil, 182
British Columbia, 283, 320
Brown bears *(Ursus arctos),* 186

C
Cabinet mountains, 325
Calanus finmarchicus, 400, 401
Canada, 180
Cape Cod Bay, 389, 390
Caprivi, 361
Carrots and sticks, 230
Caspian sea, 415
Categorical map analysis, 84
Caucasian leopard *(P. p. ciscaucasica),* 333
Caucasus, 342
Causal modeling, 100, 319, 324
Census of Antarctic Marine Life (CAML), 235
Census of Marine Life (CoML), 235
Cetacean–environment models, 393
Cetaceans, 387
Change-detection, 194
China, 182
Chukchi sea, 154, 236
Circuit theory, 319
Classification and Regression Trees (CART), 293–295, 388
Clementsian landscape ecology, 76–78
Climate change impacts, 430
Coarse filter, 116
Commission, 299
Compact Airborne Spectrographic Imager (CASI), 210
Competition-equilibrium theory, 25
Competitive exclusion, 22
Complete spatial randomness (CSR), 258
Computation, 36
Conditional logistic regression, 283
Confirmationist paradigm, 10
Connectedness, 74
Connectivity, 350, 370
Connectivity thresholds, 370, 379–380
Conservation, 429
Conservation implications of connectivity, 365
Content standard for digital geospatial metadata, 224
Contextual, 35
Continuous Phytoplankton Recorder (CPR), 235
Continuous surface, 86
Correlation length, 370, 375
Corridor, 69, 349–367
Cost distance, 322
Count-based models, 283–284
Counties, 433
Coyote *(Canis latrans),* 279
Criterion of falsification, 7

D
Darmanadzor, 333
Darwin Core schema, 225, 237, 239
Dastakert, 333
Data
  collection, 36, 113
  delivery, 440
  model, 274
  policies, 236
  quality act, 239
  sharing culture, 234
  storage and management, 246–248
Data base workflow, 238
Deductive inference, 7
Deep autocorrelation in animal movement, 147
Delmarva fox squirrel (DFS), 212
Denying the antecedent, 5
Detection probability, 338
Developing custom software, 251–253
DIF. *See* Directory Interchange Format
DiGIR. *See* Distributed Generic Information Retrieval
Directory Interchange Format (DIF), 224, 239
Disease spread, 255
Distributed Generic Information Retrieval (DiGIR), 237, 439
DNA sequences, 170
Documentation of geospatial datasets, 223–224
Down-scaling, 49
Dublin Core, 225
Dvinsky bay, 421
Dynamic landscape mosaic model, 71–72

E
Earth-Observing (EO) instruments, 195
Earth Observing System, 197
East Africa, 171
Ecological communities, 21–25
Ecological Metadata Language (EML), 224
Ecological Niche, 22, 274–275
Ecological Niche Factor Analysis (ENFA), 290
Ecological theory, 12

Ecosystem Approaches to Management (EAM), 389
Elephant, 139, 143–146, 349–367
Elk *(Cervus elaphus),* 278
Elpin, 333
Empirical samples, 113
Empirical studies of habitat fragmentation, 370
Engelmann spruce *(Picea engelmannii),* 323
Ensemble models, 294, 298–299
Entropy maximization, 295–296
Environmental data, 155
Epidemic waves, 268
Escherichia coli, 314
Euclidian distance, 262
Evolutionary significant units, 171–172
Executive Order 12906, 224
eXtensible markup language (XML), 224
Extent, 44

F
Federal Geographic Data Committee (FGDC), 224, 239
Field work, 194
Forest birds, 94
Forest Inventory and Analysis program (FIA), 433
Fractal analysis, 103–104
Fragmentation, 69, 369–383
Fragstats, 374–375
France, 180
Fugitive species, 26–27
Functional connectivity, 262
Future of analytical computing, 253

G
Gap analysis, 434
Gause, G.F., 21–25
GBIF. *See* Global Biodiversity Information Facility
Geary's *c,* 262
Genbank, 240
Gene flow, 100, 315
Generalized Additive Models (GAMs), 286–287
Generalized Dissimilarity Modelling (GDM), 291
Generalized Linear Model (GLM), 152, 279–286
Genes, 314
Genes under selection, 168
Genetic Algorithm for Rule-set Prediction (GARP), 297
Genetic algorithms, 296–297
Genetic data, 314, 316–318
Genetic distance, 317–318, 323
Geodatabase, 397
Geographical epidemiology, 266
Geographically weighted regression, 288
Geographic Information Systems (GIS), 14, 245, 431
GeophyloBuilder, 240
Georges Bank, 389
Geospatial One Stop (GOS), 225
Geostatistics, 249
Glacier National Park, 324
GLM. *See* Generalized Linear Model
Global Biodiversity Information Facility (GBIF), 177, 236–238
Global Change Master Directory (GCMD), 224
Global Forestry Information System (GFIS), 225
Global Ocean Observing System (GOOS), 234
Global Positioning Systems (GPS), 206, 245, 431
Global vegetation index, 208
Gradient, 54
  analysis of continuous field variables, 87–88
  attributes of categorical patterns, 86–87
  concept of landscape structure, 75–78
  modeling, 58, 124–127
  paradigm, 83–105
  paradigm of landscape structure, 88–92
  theory, 55–59
Grain, 44
Graph theory, 250
Gray whales *(Eschrichtius robustus),* 151
Greenland, 415
Greenness, 207
Grid-based sampling designs, 435, 437
Gulf of Maine (USA), 388, 389, 400, 403

H
Habitat, 92–96
  extent, 73
  isolation, 73–74
  loss, 69, 369
  subdivision, 73
  surrogacy, 118
Habitat capability (HC) models, 432
Habitat suitability index (HSI), 432
Harp seals (*Phoca groenlandica*)), 410
Helgoland, 235
Heligoland, 235
Heterogeneity, 83
Hierarchy, 44, 53
Hoosic valley, 371

Hot spot approach, 417
Howler monkeys *(Alouatta palliata)*, 186
H-resolution, 200
Huffaker, C., 29–30
Humpback whales *(Megaptera novaengliae)*, 185, 186, 389
Hyperion sensor, 210

I
Idaho, 320, 322
IKONOS, 196
Imputation, 125
India, 182
Individualistic concept, 76
Inductive inference, 7
Informatics, 13–17
Information-theoretic criteria, 280
Integrated multiple-scale monitoring, 124–127
Integrated Taxonomic Information System (ITIS), 229
Integrated satellite-airborne technology, 420
Integrating landscape and community ecology, 92
Interaction between habitat area and fragmentation, 381
Interaction between population size, dispersal ability and connectivity, 380–381
Intermediate disturbance, 27–28
International organization for standardization (ISO), 224
International Polar Year (IPY), 230, 235, 240
Invertebrate organisms, 151
IPY Data Information Service (IPY DIS), 240
Irani, 345
IRS (RESOURCESAT-1), 196
ISO 19115, 225
Isolation by landscape resistance gradients, 101
Isotropy, 256
ITIS (Integrated Taxonomic Information System), 229

J
Jeffreys Ledge, 389
Joint committee on antarctic data management (JCADM), 236

K
Kandalakshskiy bay, 421
Kapan, 333
Kappa, 299
Kara sea, 415
Kenai national wildlife refuge, 441
Kepler project, 439
k-fold cross-validation, 301
Khustup, 333
Konyushinsky shores, 423
Kriging, 264
Kwando river, 361

L
Lake Ladoga, 415
Land cover, 194, 202
Land managers, 429
Landsat-5 (TM), 196
Landsat-7 (ETM+), 196
Landsat imagery, 196, 201, 202, 205
Landsat Multi-Spectral Scanner (MSS), 202
Landsat Thematic Mapper (TM), 205
Landscape, 66
  ecology, 65–79, 83–105
  genetics, 100, 165, 313–326
  genomics, 165–172
  resistance, 349–367
  structure, 102
Laptop, 157
Leaf palatability, 210
Least cost
  dispersal, 372
  distance, 323
  path analysis, 354–355
Leopard *(Panthera pardus)*, 332
Lesser prairie chicken (*Tympanuchus pallidicinctus*), 202
LIDAR, 88–89, 212
Life-history space, 375
Limitations of categorical mapping, 84–86
Local Indicators of Spatial Association (LISA), 262
Localization, 127
Logical syllogisms, 5
Logistic regression, 273, 281–282
Long Term Ecological Eesearch program, 224
L-resolution, 200

M
Machine Learning, 153, 293, 388
Macro-characteristics, 116
Mad cow disease, 266
Manomet Bird Observatory (MBO) database, 390
Mantel
  correlogram, 138
  test, 320

Mapping, 248
Mapping gene flow, 319, 324
Mapping landscape resistance, 359
Mapping movement corridors, 363
MapServer, 440
Marine biodiversity data sets, 233
Marine Protected Areas (MPAs), 388
Markers under selection, 166
Markov Chain Monte Carlo, 288
MARS™ (Multivariate Adaptive Regression Splines) 399
Massachusetts, 370
Massachusetts bay, 389, 406
Matrix, 68
MAXENT (Maximum Entropy), 295
Medical geography, 266
Metadata
- clearinghouses, 177
- editors, 229
- editor tools, 229
- requirements, 230
- standards, 223–225

Mezen Bay, 420
Microsatellites, 169
Middle East, 345
Model evaluation, 299–302
Model selection, 280
Moderate Resolution Imaging Spectroradiometer (MODIS), 211
Molecular genetics, 165
Monitoring, 112, 116
Monitoring fixed grids, 123–124
Monmonier algorithm, 318
Montana, 320
Monte Carlo randomization, 358–359
Moose *(Alces alces),* 284
Moran's *I,* 262
Mosaic, 68
Movement path selection, 350
Movement path simulation, 134–137
Moving windows, 86
Multiple scale analysis, 43, 54
Multiple spatial scales, 404
Multi-Resolution Land Characteristics program (MRLC), 438
Multi-scale monitoring data-cube, 120–124
Multi-scale vegetation structure, 199–200
Multispectral sensors, 196
Multivariate Adaptive Regression Splines (MARS), 291–292, 388
Muskoxen *(Ovibos moschatus),* 186
Mutation, 314

N

Nakhichevan, 344
Namibia, 349, 350
National Biological Information Infrastructure (NBII), 224, 239
National Center for Ecological Analysis and Synthesis (NCEAS), 224
National Oceanic and Atmospheric Administration (NOAA), 224
National Resources Inventory (NRI), 433
National wildlife refuges, 436
Network analysis, 250
Neural networks, 292–293
Neutral genetic variation, 315
Neutral markers, 166
Nicaragua, 186, 187
Non-equilibrium, 117
- communities, 30–33
- dynamics, 276

Nonlinear relationships, 160
Nonstationarity, 20
Noravank, 333
Normalized Difference Vegetation Index (NDVI), 207
North Atlantic oscillation, 404
North Atlantic right whale *(Eubalaena glacialis),* 389
North Atlantic Right Whale Consortium (NARWC), 390
Northwest Atlantic, 387
Norwegian sea, 418
Nuvadi, 333, 337

O

Obis-Seamap, 237
Occupancy models, 285–286
Ocean Biogeographic Information System (OBIS), 236
Okavango Delta, 363, 365
Omission, 299
Onegskiy Bay, 421
Online databases, 177
Open access, 178
Open Geospatial Consortiums (OGC), 440
OpenModeler, 240
Optical sensors, 196
Orbview-3, 196
Oregon Coast Range, 28, 94
Organisation for the Economic and Commercial Development (OECD), 225
Overfishing, 233

P
Pair-wise genetic distance, 320
Palm springs ground squirrel *(Spermophilus tereticaudus chlorus),* 286
Paradox of plankton, 26–27
Partial Mantel test, 320
Particulate, 35
Passive remote sensing technology, 212
Patch, 68
  geometry, 73
  mosaic model, 68, 84
Pathogens, 171
Path randomization, 351–353
Pattern analysis, 249
Pattern-process relationships, 20
Persian leopard *(P. p. saxicolor),* 333
Pew oceans commission, 389
Poaching, 345
Point data, 260
Polar bears *(Ursus maritimus),* 186, 283
Polar regions, 234, 425
Polygon change analysis, 262
Population
  connectivity, 99–102
  genetics, 101, 313
  genomics, 166
PostGreSQL, 440
Prairie pothole region, 441
Predator mediated coexistence, 27–28
Primary production, 207
Priority Leopard Conservation Areas (PLECAs), 333, 344
Programme PRESENCE, 338
Pronghorn antelope *(Antilocapra americana),* 286
Proxy variable, 388

Q
Quantile regression, 286
Quantitative spatial analysis, 248–250
QuickBird-2, 196

R
RADARSAT-2, 214
Radiometric resolution, 197
Random effects models, 289–290
Random forest algorithm, 153
Random forests, 294–295
Randomization tests, 249–250
Range of scales, 387
Really simple syndication (RSS), 225
Receiver Operating Characteristic curve (ROC), 300
Refutationist paradigm, 8
Register of antarctic marine species (RAMS), 240
Regression Trees (CART™), 399
Relational Database Management (RDBM) system, 245, 246
Remeasurement, 114
Remote sensing, 194, 224, 438
Repositories, 223
Representative sampling, 113, 114
Resistant kernel model, 374
Resource selection function (RSF), 282, 432
Resource selection model, 276
Resource selection probability function (RSPF), 278, 282
Resource utilization functions, 284–285
Ringed seals (*Pusa hispida*), 410
Rio Convention, 177, 235, 236
Ripley's *K*-function, 258
Road ecology, 370
Road ecology, effects on fragmentation, 381–382
Rocky mountains, 321
Roe deer *(Capreolus capreolus),* 288
RSF. *See* Resource Selection Function
RSPF. *See* Resource Selection Probability Function
Russia, 236
Russian Arctic, 409, 418

S
Satellite telemetry, 132, 350
Scale, 13, 19, 43, 122, 275–276
Scaling ecological knowledge, 48–55
Scaling landscape resistance, 353
Science Environment for Ecological Knowledge (SEEK), 439
Scientific Committee on Antarctic Research (SCAR), 237
Scientific Committee on Antarctic Research and Marine Biodiversity Information Network (SCAR-MarBIN), 236
Scientific method, 3
Sea ducks (*Somateria* spp., *Melanitta* spp.), 151
Sea of Okhotsk, 415
SeaWIFS, 393
Seemingly Unrelated Regression (SUR) models, 267
Selection, 167
Selkirk Mountains, 325

Semivariance, 262
Sensitivity, 299
Sensitivity analysis, 301
Sheep species *(Ovis aries)*, 289
Shoreline data profile, 224
Side-Looking Radar (SLR), 413
Simulation, 36
Single nucleotide polymorphism, 169
Sioma National Park, 351, 361
Sisian, 333
Software of spatial ecological analysis, 246
Software transparency, documentation, and certification, 252
Solovetsky Archipelago, 423
Southern Ocean, 234
Space–Time Autoregressive Integrated Moving Average (STARIMA) models, 267
Spatial aggregation, 256
Spatial and temporal variability, 19
Spatial autocorrelation, 157, 262
Spatial complexity, 58, 113, 275–276
Spatial data management, 223–225
Spatial diffusion models, 267
Spatial epidemiology, 266
Spatial information system, 245
Spatial interpolation technique, 264
Spatially explicit environmental variables, 193
Spatially referenced ecological information, 115–116
Spatial monitoring programs, 431
Spatial non-stationarities, 266
Spatial pattern, 13
Spatial process, 131
Spatial variance, 262, 264
Species distribution, 193
Species distribution modeling, 273–302
Species richness, 205
Species richness gradients, 207
Species surrogacy, 117
Specificity, 299
Spectacled Eider (*Somateria fischeri*), 151
Spectral analysis, 104
Spectral resolution, 196
SPOT 2 (HRV), 196
SPOT 4 (HRVIR), 196
SPOT 4 (VGT), 196
SPOT 5 (HRG), 196
Spotted seals *(Phoca largha)*, 186
Static island biogeographic model, 68–71
Stationarity, 256
Statistical modelling, 273–302
Statistical power, 114
Steller's Eider (*Polysticta stelleri*), 151–152
Stellwagen Bank, 389, 390
Stellwagen Bank National Marine Sanctuary (SBNMS), 388
Stochastic simulation, 250
Structure, 318
Subalpine fir *(Abies lasiocarpa)*, 323
Supervised classification, 202, 205
Support Vector Machines (SVM), 297–298
Surface metrology, 103
Sustainable resource management, 403
Synthetic Aperture Radars (SAR), 413

T
Taconic mountains, 371
Temporal resolution, 196
Temporal variation, 47
Terra (ASTER), 196
Terra (MODIS), 196
Terrain ruggedness index, 339
Theoretical studies of habitat fragmentation, 369
Theory corroboration, 9
Toothfish (*Dissostichus eleginoides)*, 235
Townships, 433
TreeNet, 294

U
Underdetermination of theories by facts, 37
Unifying paradigm, 56
Unimodality of support, 364
Unimodal response, 92
United Kingdom (UK), 180
United States, 230, 430
Unsupervised classification, 206
Up-scaling, 48
Urts ridge, 333
USA, 180, 239, 434, 441
U.S. Commission on Ocean Policy, 389
U.S. Fish and Wildlife Service's habitat and population evaluation team, 441

V
Vayk, 333
Vegetation characteristics, 210
Vegetation structure, 213
Visual basic (VB), 393
Visualization, 248
Volterra, V., 21–25

W
Walrus *(Odobenus rosmarus)*, 151
Walruses, 410

Washington state, 324
Wavelet analysis, 104
Web Mapping Services (WMS), 440
Weighting, 353
Western New England, 371
West Nile virus, 266
White Sea, 410, 413, 414, 420
Wild fire index, 337
Wildlife journals, 183
Wildlife management, 151, 175
Wolverine *(Gulo gulo),* 319
Woodland caribou *(Rangifer tarandus caribou),* 277
World Register of Marine Species (WoRMS), 239
WorldView-1, 196
Wyoming, 119, 320

Y

Yellow bellied glider *(Petauroides australis),* 288
Yellowstone National Park, 323

Z

Zambia, 349, 351
Zangezur, 333
Zangezur ridge, 333
Zimny, 423

MIX
Papier aus verantwortungsvollen Quellen
Paper from responsible sources
FSC® C105338

If you have any concerns about our products,
you can contact us on
ProductSafety@springernature.com

In case Publisher is established outside the EU,
the EU authorized representative is:
Springer Nature Customer Service Center GmbH
Europaplatz 3, 69115 Heidelberg, Germany

Printed by Libri Plureos GmbH
in Hamburg, Germany